Environmental Systems Analysis with MATLAB®

Stefano Marsili-Libelli

Environmental Systems Analysis with MATLAB®

Stefano Marsili-Libelli

CRC Press
Taylor & Francis Group
Boca Raton London New York

CRC Press is an imprint of the
Taylor & Francis Group, an **informa** business

CRC Press
Taylor & Francis Group
6000 Broken Sound Parkway NW, Suite 300
Boca Raton, FL 33487-2742

First issued in paperback 2017

© 2016 by Taylor & Francis Group, LLC
CRC Press is an imprint of Taylor & Francis Group, an Informa business

No claim to original U.S. Government works

ISBN 13: 978-1-138-49078-9 (pbk)
ISBN 13: 978-1-4987-0635-3 (hbk)

Visit the Taylor & Francis Web site at
http://www.taylorandfrancis.com

and the CRC Press Web site at
http://www.crcpress.com

Dedication

To Patrizia,
who makes my day every day

Contents

Preface

God always forgives, man sometimes, nature never

Pope Francis

This book brings together systems theory and environmental science, by recasting environmental problems into a common system-theoretic methodology. In this way, simple models can be developed and new insights can be gained, which could not have been achieved otherwise. System theory is inherently a 'reductionist' discipline and its integration with the naturally 'holistic' environmental science at first was not easy. The first pioneering attempts at translating ecological principles into mathematical laws (Maynard Smith, 1974; Rinaldi et al., 1979) were met with a somewhat supercilious reaction by ecologists, who resented the intrusion of mathematicians (let alone engineers!) into their turf, and looked down on the attempt at translating ecological concepts into mathematical relations. Over the years, ego clashes have gradually receded and now mathematical modelling has become a well-established branch of ecology, in which this book is naturally set.

This book is about developing dynamical models of environmental processes, using a wide variety of mathematical methods. The first four chapters introduce the analytical tools for investigating environmental models and data. The emphasis is evenly balanced between the mechanistic and the data-driven approaches, assessing the merits and liabilities of both. Theory is introduced in its most accessible form and is applied to test cases based on first-hand data and experiences. While Chapters 1 and 2 illustrate with the basic concepts of modelling and identification, Chapter 3 deals with data processing, rarely considered in the environmental analysis, by introducing numerical techniques to improve the quality of the data and to extract the information they carry, in the context of model building and verification. Chapter 4 introduces the basic concepts of fuzzy logic and applies them to the modelling of environmental systems. Chapters 5 through 7 apply these methodologies to specific aspects of environmental modelling: population dynamics, flow systems, and environmental microbiology. The last chapter (Chapter 8) combines the notions of the previous chapters into the analysis of several aquatic ecosystems' case studies.

I am quite aware of the limitations of this book, because it reflects my own research experience, and therefore it is strongly oriented towards the aquatic environment, while it completely disregards other equally important subjects such as air pollution, groundwater, or solid waste, simply because I never worked in those areas.

Far from being a 'theory' book, its approach is eminently practical, in that every methodological concept is translated into a MATLAB® code. Nevertheless, this is not a recipe cookbook, but it takes the reader through a logical sequence from the basic steps of model building and data analysis to implementing these concepts into working computer codes, and then into assessing their results. Who should read this book? Certainly anyone who is looking for an introduction of the mathematical aspects of ecology, or anyone who wants to take a fresh look at known problems from a differing viewpoint. This book may represent a first encounter with mathematical modelling before going on to more advanced books in the specific field of interest.

I have set up a quiver full of arrows. Then it is up to the archer-reader to call the shots. As Figure 1 shows, this book spins a web of relations among system theory and environmental issues, moving from the predominantly methodological first four chapters to the more applicative subsequent part, where specific environmental problems are addressed. Clockwise from the top of Figure 1, Chapter 1 summarizes the basic concepts of system theory for both linear and nonlinear models, while Chapter 2 is devoted to model identification. Chapter 3 considers the data acquired from the field, how their quality can be improved and the embedded information extracted. Chapter 4 describes the fuzzy approach to modelling and data analysis, showing that fuzzy models in many cases have an advantage over their mechanistic counterparts, and how the fuzzy approach can be

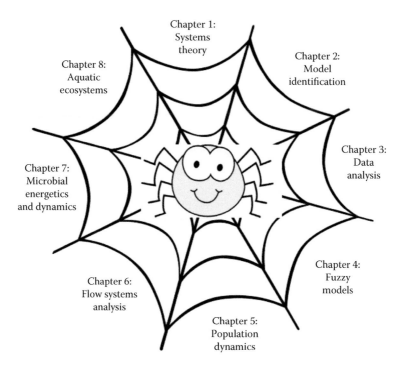

FIGURE 1 This book spins a web of relations (clockwise from the top) among system theory and environmental issues, moving from the predominantly methodological first four chapters to the more applicative subsequent chapters where specific environmental problems are addressed.

profitably used for data analysis and diagnosis. While the first four chapters illustrate the methodological aspects of environmental systems analysis, the subsequent chapters apply them to specific environmental problems. Chapters 5 deals with population dynamics, from single species to food webs, with an emphasis on investigating the conditions for species conservation, or on explaining unexpected behaviours like the catastrophic forest defoliation caused by the spruce budworm. Chapter 6 deals with flow modelling. Far from competing with hydraulic modelling, it considers some simple flow schemes, which provide the first level for aquatic ecosystem modelling. Chapter 7 introduces the basic concepts of environmental microbiology, both in the natural and man-made environment. I strived to show the similarities and the differences of microbial processes in either setting, without getting too much involved in the analysis of wastewater treatment systems, given the many excellent textbooks available in this area (Bastin and Dochain, 1990; Orhon and Artan, 1994; Olsson and Newell, 1999; Dochain and Vanrolleghem, 2001; Gujer, 2008). The final Chapter 8 is a wrap-up chapter, in which I discuss several case studies drawn from my own experience.

It has been said that any scientific book is partly autobiographical, and this one is no exception. In fact, being near the end of my academic career, I felt the need to gather the lecture material that I have amassed in over 30 years of environment analysis teaching. Educating has been an exciting experience all along, because I was fortunate enough to teach students with differing backgrounds and brilliant minds, ranging from automation to environmental engineering, which resulted in a fruitful cross-fertilization. Class projects frequently became master theses, and sometimes produced journal papers, with the students' involvement as co-authors. These mixed classes reflected my own mixed background, which blended my early electronic engineering training with the subsequent commitment to environmental issues. I felt that I owed this book to generations of students, who have now become accomplished professionals, and from whom I still receive a constant flow of affection and gratitude. Now, when we meet at professional gatherings or just over a pizza, they love to reminisce over their student days and how much we enjoyed working together and learning together. This book is my token of gratitude for the endless affection and inspiration I have been receiving from them all.

All the material in this book has been intensively student-tested and the companion software has been honed through many critical reviews by cohorts of students. I have found that MATLAB is an ideal teaching, and learning, tool in that it provides a platform where mathematical ideas can be quickly converted into ready-to-run codes. Unlike other more formal programming languages, such as Fortran or C/C++, which require setting up complex structures, in MATLAB the user can immediately translate the problem at hand into an executable code and get the results in no time. Many segments of the MATLAB code (referred to as *scripts*) are included in the boxes along the book to demonstrate the programming structure, while the complete software collection can be downloaded from the book web page in the CRC Press website and copied in the local hard disk. It is strongly suggested to maintain the original folder structure, with a master directory named `ESA _ Matlab`, in which each chapter folder contains the pertinent codes for that chapter, plus a 'tools' folder, which gathers all the utility functions common to all chapters. No installation is necessary, but once MATLAB is started, the master folder and its subfolders must be added on top of the MATLAB path. This software organization makes the path definition vital to retrieve the called functions. The exercises for each chapter have self-explanatory names and are referenced in the figure captions. Of course, this book should not be considered as a MATLAB primer, many of which are freely available in the web. So a certain basic MATLAB literacy is a prerequisite.

A brief clarification regarding an important mathematical notation. Throughout the book the imaginary unit is indicated by j, so that $j \times j = -1$. There was an old joke circulating in the past saying how you could tell a mathematician from an engineer. When asked about the imaginary unit, the mathematician would say $i \times i = -1$, while the engineer would claim that $j \times j = -1$, and this book goes for the latter, though MATLAB uses i. Additional material is available from the CRC Press Web site: http://www.crcpress.com/product/isbn/9781498706353.

Stefano Marsili-Libelli
University of Florence, Italy

MATLAB® is a registered trademark of The MathWorks, Inc. For product information, please contact:

The MathWorks, Inc.
3 Apple Hill Drive
Natick, MA 01760-2098 USA
Tel: +1 508 647 7000
Fax: +1 508 647 7001
E-mail: info@mathworks.com
Web: www.mathworks.com

Acknowledgements

Research is a constant swing from frustration to elation and back, and so is scientific book writing. Far from being a solitary activity, I am indebted to a vast number of people with whom I shared elation and to whom I am grateful for pulling me out of frustration. My first thanks goes to my wife Patrizia, a devoted and reliable life companion, for the loving empathy and constant encouragement, apart from discussing some mathematical aspects of the book, being a math teacher. My daughters Ilaria and Chiara provided an equally enthusiastic backing. Ilaria, an agricultural microbiologist, is also to be thanked for revising Chapter 7. Apart from the family circle, I am deeply grateful to several people for their vital contribution both as accomplished professional and as very close friends. Two very special persons stand out, without whom this endeavour could not have been accomplished: Irma S. Britton, CRC Press senior editor, and Professor Emeritus William D. Lakin, University of Vermont, Burlington, VT. Irma has been a dream editor, the one that any author would love to encounter. From her and her staff, I received endless support and encouragement. Her ability to answer my queries and smooth my workflow has been invaluable. Bill Lakin, a friend of long standing and an excellent mathematician, carefully reviewed the whole manuscript and provided valuable comments regarding both the English and the mathematical aspects. His support has been vital. I am also grateful to Dr. Alessandro Spagni, ENEA Bologna and University of Padua, co-author of many papers and careful reviewer of Chapter 7. A special thanks goes to Dr. Luca Palmeri, University of Padua, for being the first instigator of this book. Upon his suggestion, I was invited by CRC Press to write a brief endorsement of his book (Palmeri et al., 2014) and then I was asked whether I had a book proposal of my own. I said yes, CRC Press liked my proposal, and this is how it all started. Another precious instigator of this book has been Dr. Michela Mulas, Aalto University, Finland, who in 2012 invited me to teach a short course entitled 'Environmental Data Modeling and Analysis'. The material I prepared for that course later became the kernel of this book and her enthusiastic support for the initiative always spurred me throughout my toil. Much of the material presented in this book is the result of research carried out with my assistant Elisabetta Giusti, PhD. Her invaluable input, her curiosity, and her hard-nosed determination supported me in many ways during our decade-long fellowship rich in scientific advances, frank discussions, forthright personal interaction. I am also deeply indebted to Professor Antony Jakeman, Australian National University and Editor-in-Chief, *Environmental Modelling and Software* (EMS), for his loyal friendship, optimistic wisdom and constant encouragement. He asked me to join the EMS editorial board in 2004 and that job has been a constant source of reward, satisfaction and personal growth. I am also deeply grateful to Michelle Herbert, IWA Journals Editorial Co-ordinator, who sympathetically modulated my *Water Science & Technology* editorial work flow while I was busy with the book, and quickly provided the permission to reproduce several figures from my WST papers. I am also indebted to Design Science (www.dessci.com), particularly Bob Mathews and Derrick Fimon, for helping me getting the most out of MathType 6.9 for Windows, a superb math editor, which I used extensively to write the many equations appearing in the book. My PhD students Emanuele El Basri and Giacomo Barni are to be thanked for their helpful assistance and the revision of the MATLAB codes. I am also thankful to www.brainyquotes.com for providing me with a vast source of inspiring citations, some of which I used as chapter openers.

In further hindsight, I am grateful to my late parents, who taught me to love and respect Nature from my early days, and to the medical staff who helped me in my battle against cancer, particularly Professor Moretti, who took away the worst part of me, and Professor Amadori, who guided me through the painful path of chemotherapy with a strong hand, unabashed optimism, and a captivating smile. Teaching during these 'medical' months was not easy, and I am grateful to my family and to my students, who supported and encouraged me through that dire moments. I owe them all this new lease of life, which made the book possible. In that time I experienced first-hand Winston Churchill's quotation 'Success is not final, failure is not fatal: it is the courage to continue that counts'.

Scientific research is a never-ending struggle to further our knowledge, but there are times when we must pause and take stock of our progress. I hope that you will enjoy reading this book, as much as I did in writing it, and after turning the last page, lean back—as I have done after writing the last line—and take time to stand and stare.

> *What is this life if, full of care,*
> *We have no time to stand and stare.*
> *No time to stand beneath the boughs*
> *And stare as long as sheep or cows.*
> *No time to see, when woods we pass,*
> *Where squirrels hide their nuts in grass.*
> *No time to see, in broad daylight,*
> *Streams full of stars, like skies at night.*
> *No time to turn at Beauty's glance,*
> *And watch her feet, how they can dance.*
> *No time to wait till her mouth can*
> *Enrich that smile her eyes began.*
> *A poor life this if, full of care,*
> *We have no time to stand and stare.*

W. H. Davies
Leisure, *1911*

Florence,
July 30, 2015.

Author

Stefano Marsili-Libelli was born and educated in Florence, Italy. He received a *cum laude* MS degree in electronic engineering from the University of Pisa in 1973. Later in the same year, he joined the University of Florence on a post-graduate grant and has served in the Faculty of Engineering ever since, first as a technical assistant, then as an associate professor (since 1983), and finally as a full professor of environmental system modelling since 2000.

He has always been active in promoting the system theory approach to the study of environmental systems, both at a faculty level and in a broader context, teaching seminars and courses on environmental modelling and control at the University of Gent (Belgium), University of Leuven (Belgium), Institute of Hydroinformatics (IHE, Delft, the Netherlands), Institute of Environmental Biotechnology, TU Delft (the Netherlands), University of Glamorgan (Wales, UK), Aalto University (Finland) and University and Polytechnic of Valencia (Spain).

He spent study periods at the Institute of Hydrology, Wallingford (UK); International Institute of Applied Systems Analysis (IIASA), Laxenburg (Austria); University of Glamorgan (Wales, UK).

He established Erasmus bilateral links with the University of Gent (Belgium), University of Delft (the Netherlands) and University of Amsterdam (the Netherlands).

Apart from being a founding member of the environmental engineering curriculum at the University of Florence, and serving as the director of the Laboratory of Environmental Process Control, he joined several PhD faculties, among which the PhD curriculum in hydrodynamics and environmental modelling (Consortium among the Universities of Padua, Florence, Genoa and Trento) and the PhD curriculum of sanitary engineering (University of Rome 2). He is presently serving as a faculty member of the international PhD curriculum in civil and environmental engineering (University of Braunschweig, University of Florence, University of Pisa and University of Perugia).

Dr. Marsili-Libelli's teaching responsibilities have always been in the modelling and control of the environmental systems, particularly aquatic environment, both natural and man-made, developing models describing the ecology of lagoons and rivers, as well as new models of microbial kinetics to be applied in the control of wastewater treatment systems.

Other research interests include the modelling of the environment using the fuzzy sets approach and the identification of environmental models, for which he proposed new ad hoc optimization methods for parameter estimation and validation. Modelling applications range from the rivers and lagoon to subsurface constructed wetlands, anaerobic digesters, aerobic wastewater treatment processes and agricultural systems.

He is presently serving as an associate editor for the ISI international journals *Environmental Modelling & Software* (Elsevier) and *Water Science & Technology* (IWA Publishing).

He is member of the following scientific societies:

- International Environmental Modelling & Software society (iEMSs)
- International Society of Ecological Modelling (ISEM)
- International Water Association (IWA)
- Italian Society for Automation (ANIPLA)

Over the years, Dr. Marsili-Libelli has received the following awards:

- The paper Checchi N., Giusti E., Marsili-Libelli S. (2007). PEAS: A toolbox to assess the accuracy of estimated parameters in environmental models. *Environmental Modelling & Software* **22**: 899–913, was awarded the Best Paper Award 2007 by the *Int. J. Environmental Modelling & Software*.

- *Biennial Medal* awarded by the International Environmental Modelling & Software society (iEMSs) in 2008.
- Nomination to *iEMSs Fellow* in 2008.
- *Best paper award* granted by AssoAutomazione (Italian association for automation) in the area of Technological Innovation in the Water Sector, biennial Forum for remote control of public utilities, Rome, October 2009.
- Nominated *Reviewer of the year* 2010 by the Editorial Board of the *Int. J. Environmental Modelling & Software*.
- *Best paper award* granted by AssoAutomazione (Italian association for automation) in the area of Competition and Sustainability in the Public Utilities, biennial Forum for remote control of public utilities, Bologna, November 2013.

1 Introduction

I became more and more convinced that even Nature could be understood as a relatively simple mathematical structure, but how can it be that mathematics, being after all a product of human thought which is independent of experience, is so admirably appropriate to the objects of reality?

Albert Einstein

This book is about building mathematical models of environmental systems, and using these models to analyse their behaviours. Models are mind representations of reality. They are at the basis of modern science, pioneered by Galileo Galilei and Isaac Newton. Environmental models, no matter how elaborate, have the peculiarity of being simplified representations of Nature's complexity. Nevertheless, they helped scientists make inroads into understanding the functioning of the environment in which we are living. To help us in this endeavour, computers make our mind creations come alive, so that we can explore a whole paradigm of *what–if* scenarios, and check the correctness or fallacy of our assumptions.

1.1 ENVIRONMENTAL SYSTEMS MODELLING, THE BASIC CONCEPTS

Environmental models are abstract representations of reality that can be used to improve our understanding of the natural systems or to assist in taking management decisions. The task of the modeller is to translate non-numerical concepts and facts into mathematical equations that describe the system evolution in quantitative terms, as illustrated in Figure 1.1. The first pioneering attempts at turning ecological principles into mathematical laws focused on population dynamics (Gause, 1934; Morris and Miller, 1954; Leslie, 1957; May, 1974, 1976a; Maynard Smith, 1974). Since then mathematical modelling of ecosystems has won a growing acceptance in the field of ecology and is now an integral part of it (May, 1976b; Pielou, 1977; Casti, 1979; Begon and Mortimer, 1986; Hallam and Levin, 1986; Levin et al., 1989; Agren and Bosatta, 1996; Ricklefs and Miller, 1999; Clark and Mangel, 2000; Gotelli, 2001; Odum and Barrett, 2004; Rockwood, 2006; Pastor, 2008; Krebs, 2009; Legendre and Legendre, 2012; Mittelbach, 2012; Vandermeer and Goldberg, 2013).

Given the holistic nature of ecology, when we consider an ecosystem, we inevitably draw a line between the part of the ecosystem we want to study and what we want (or have) to leave out, thus making a somewhat arbitrary division between the *inside* and the *outside* of the subject of our study. No matter how we draw this boundary, however, the *outside* portion will continue to influence the evolution of the *inside* system, interacting with it through the boundary that we have drawn. We therefore create the somewhat artificial situation shown in Figure 1.2 and confine our study to the inner part of the ecosystem described by the variables $x(t)$. No matter where we set the border, the *outside* part will continue to influence its evolution by providing external inputs $u(t)$ crossing the ecosystem boundaries, whereas other dynamics $z(t)$, not directly related to the *inside* portion, will continue to evolve and influence the inner system behaviour.

Thus, when modelling an ecosystem, the following preliminary aspects must be addressed:

1. Decide where to put the boundary
2. Identify which inputs $u(t)$, either controllable or not, influence the *inner* ecosystem from the outside
3. Determine which external dynamics $z(t)$ are relevant for the inner system, either directly or indirectly

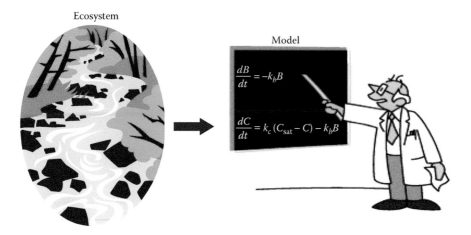

FIGURE 1.1   The modelling process: from natural facts to their mathematical representation.

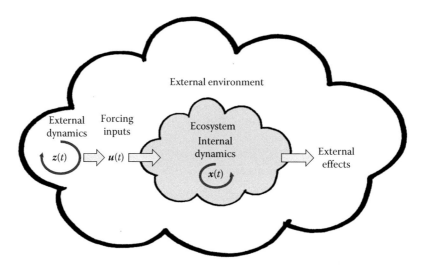

FIGURE 1.2   Ecosystems boundaries are stretchy and are usually set at our convenience.

Thus, three sets of equations (models) must be considered: the inner system dynamics, described by the state variables $x(t)$, the input functions $u(t)$ and the outer dynamics $z(t)$ for the portion related to the generation of the inputs $u(t)$. Consequently, the basic model of our ecosystem is composed of the three main blocks of the following equations:

$$\begin{cases} \text{Inner system dynamics,} & \dfrac{dx}{dt} = f(x, u) \\[2mm] \text{External inputs,} & u = f(z) \\[2mm] \text{External dynamics,} & \dfrac{dz}{dt} = h(z) \end{cases} \qquad (1.1)$$

Modelling can serve different purposes, and the end use of the model should be decided from the very start, because this decision will greatly affect the subsequent model development phase. Environmental modelling can be undertaken for the following purposes:

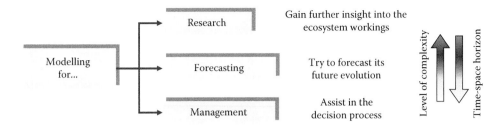

FIGURE 1.3 The level of modelling detail depends on the final use and involves differing levels of complexity and model horizon (the arrows point to the increasing direction).

- *Research*: To gain a deeper insight into the ecosystem workings.
- *Forecasting*: To try to glean the possible ecosystem evolution in the future.
- *Management*: To assist in the development of management policies, assess the costs and benefits of alternative solutions and so on.

In the scheme of Figure 1.3, forecasting and management are closely linked because management decisions are based on the generation of *what–if* scenarios, for which the forecasting capabilities of models are required.

1.2 ATTRIBUTES AND DICHOTOMIES OF ENVIRONMENTAL MODELS

Some basic features of environmental models will now be discussed to illustrate that apparent contradictions can, in fact, be reconciled by realizing that they represent complementary aspects of the same reality.

1.2.1 ENVIRONMENTAL MODELS AS BALANCES

First, it is important to understand that ecosystems are characterized by the flow of energy and matter (Pentz, 1972; Odum, 1983, 1988, 2000a; Odum and Odum, 2000; Agren and Bosatta, 1996; Ricklefs and Miller, 1999). In this sense, ecosystem models may be regarded as *double ledgers*, balancing inputs and outputs for both mass and energy. Figure 1.4 shows the main difference between energy and matter flows through the ecosystem. In particular, the energy flow is unidirectional and dissipative (lossy flow), whereas the materials flow is circular and conservative. Energy is carried through the ecosystem by matter, which is recycled between biotic and abiotic components.

The basic energy supply comes from the sun (solar energy), and in its unidirectional flow across the ecosystem, it is partially converted into matter, which may be regarded as an energy reservoir and transporter, never to come back again. Conversely, matter is neither created nor destroyed, but just transformed into matter of a different form. For example, inorganic carbon in the form of CO_2 is converted into organic carbon by autotrophic organisms through photosynthesis, and then back again into CO_2 by decomposers. Losses occur along the energy flow because the fraction that is not converted into matter is simply lost to the environment as heat or respiration. Thus, the net energy collected at the *high end* of the ecosystem (highly complex organisms) is just a small fraction of the incoming energy. Though matter can be regarded as an energy storage, and it can be expressed in the same units as energy (kcal or Joules), the *cost* of producing matter from energy greatly varies with the level of complexity of the organism. The concept of *transformity* (Odum, 1988; Bastianoni, 2008; Jorgensen, 2008) was introduced to quantify the energy required to produce a unit of mass with a certain level of organization, in this way denoting that the value of the produced matter depends on the energy required to obtain the transformation.

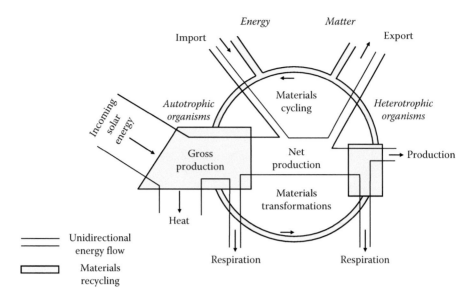

FIGURE 1.4 Materials and energy balances across an ecosystem.

Odum (1988) proposed this term by defining the *solar transformity* as the amount of solar energy (Joules) required to produce 1 Joule of matter of a certain complexity. So, while the incident solar radiation has obviously a transformity equal to 1, the more complex is an organism, the higher is its transformity. Thus, the *embodied energy* or *emergy* is a measure of the value of the matter of each organism in an ecosystem (Odum et al., 2000; Brown and Bardi, 2001; Brandt-Williams, 2002; Jorgensen et al., 2004). Palmeri et al. (2014) thoroughly discuss these concepts and report numerical values for the transformity of many common materials. For example, crude oil has a transformity of 5.4, whereas the transformity of *corn* (8.3), *caterpillar pupae* (200), or *aquaculture shrimp* (1300) denotes the increasing ecological values of these ecosystem components as they move up the evolutionary ladder.

An example will help to clarify the computation of energy flow in a simple ecosystem. Consider crop growth: besides water and nutrients, its subsistence is primarily based on the solar radiation, which provides the energy required to activate the photosynthetic process. Figure 1.5 shows some typical quantities to illustrate how this incoming energy is converted into organic matter through crop growth. It can be seen that only a very small fraction of the incoming solar energy is actually utilized by the plants (1.274%), which on the other hand are very efficient in converting that amount of energy into organic matter (82.5%). The resulting biomass (primary production) can be regarded as energy stored as matter, to be eventually utilized by other organisms (consumers), thus forming a food chain, as will be considered in Chapter 5.

The mass and energy balances of Figure 1.4 and the example of Figure 1.5 show the differing nature of the flow of energy and materials across the ecosystem boundaries. Figure 1.4 is a *qualitative* picture describing the fate of the ecosystem components, whereas Figure 1.5 represents a *quantitative* assessment of energy flow through a simple system composed of primary producers. Further, the numbers shown in Figure 1.5 are assumed to be constant in time (they could be regarded as long-term averages), whereas in fact the energy flow is always varying in time due to the daily and seasonal cycles. So static balances are always a simplification, which sometimes may suffice in modelling the ecosystem, but more often may prove inadequate. We may opt for a *static* or a *dynamical* model depending on the importance of the time variations. The following are some considerations that can help to decide the most appropriate approach:

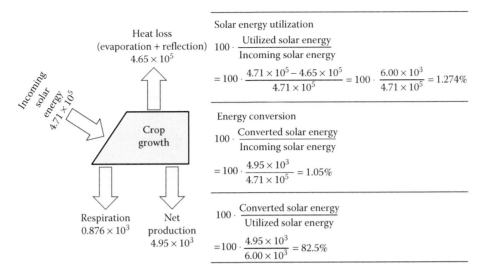

FIGURE 1.5 An example of energy flow across a simple ecosystem (crop growth) using typical energy values. It can be seen that just about 1% is eventually converted into organic matter but the conversion efficiency of primary producers is very high (82.5%). (Data from Pentz, M.J., *Science Foundation Courses: Unit 20 Species and Populations*, The Open University Press, Oxford, 86, 1972.)

- *Static balance*: Assumes that the system is at steady state, which may be reasonable on a long-term time horizon. Reconciles the *long-term* mass and energy flows.
- *Dynamical balance*: Keeps track of the short-term variations in the system. Reconciles the short-term *rates* of mass and energy conservation, taking into account the internal transformations.

Both balances should take into account not just the exchanges with the external world, but also the internal transformations, whereby one component is transformed into another through energy–matter conversion. It is up to the modeller to make the most appropriate choice for the problem at hand. Generally, a *static balance* is sufficient when the system is considered at steady state or when its dynamics is very slow. Usually, in this case, a long-term time horizon is considered and average quantities are used. In that balance, inputs and outputs should cancel out, considering the transformations between energy and mass, as in the example of Figure 1.5. On the other hand, on a short-term time horizon, the system total quantities may not be at equilibrium, with *storage* acting as a sort of flywheel to compensate for the imbalances. In this case, a *dynamical* balance is required, involving the *rates* rather than the *quantities*, as in Equation 1.2 where the time step Δt figures prominently.

$$\begin{bmatrix} \text{Rate} \\ \text{of} \\ \text{change} \end{bmatrix} = \underbrace{\frac{\Delta \, \text{storage}}{\Delta t} = \frac{\text{input}}{\Delta t} - \frac{\text{output}}{\Delta t}}_{\text{transport}} + \underbrace{\frac{\text{generation}}{\Delta t} - \frac{\text{utilization}}{\Delta t}}_{\text{internal transformations}} \qquad (1.2)$$

The basic features characterizing static and dynamical balances are summarized in Table 1.1. It can be seen that from the modelling viewpoint, the static balance is by far less demanding, because it just requires a balance of the average quantities, whereas the dynamical model requires further knowledge of the *rates* that govern the internal transformation. The pertinent mathematical tools are also different, because in the static case a set of simple algebraic equations will balance the

TABLE 1.1

Characteristics of Static and Dynamical Balances

Static Balance	Dynamical Balance
Assumes that all the inputs and process variables have a constant value	Takes into account time-varying inputs and the internal variability
Assumes static mass and energy balances	Uses dynamical mass and energy balances
Neglects internal storage and reaction processes	Includes internal storage and reaction processes
Uses algebraic equations	Uses dynamical equations

system, whereas in the dynamical case the modelling of time-varying rates requires the use of difference or differential equations.

As a concluding remark about the differing features of static and dynamical models, consider the two situations shown in Figure 1.6 where an ecosystem is driven by a time-varying input, and its output is to be monitored with reference to some threshold (perhaps representing some environmental limit, not to be exceeded). If the average input is considered and a static model is developed, based on static material and energy balances, then the upper time plot shows that the (average) model output appears to comply with the required limit. Instead, if the full input variability is considered (lower time plot) and a dynamical model is developed, the full output swings (and violations), previously lost in the static model, now become apparent. This new, more detailed analysis may reveal the inadequate environmental compliance and generally sheds more light into the system behaviour.

1.2.2 First-Principle versus Data-Driven Modelling

There are two basic approaches to modelling that may sometimes be combined, but otherwise rest on radically differing assumptions. Their main features are briefly summarized here, whereas both techniques will be treated in detail in subsequent chapters.

Mechanistic or *first-principle* modelling is the most conventional approach to model building. This approach constructs models on the basis of the available theory in the pertinent fields of knowledge (physics, chemistry, biology, etc.). The modelling task is facilitated by the prior availability of

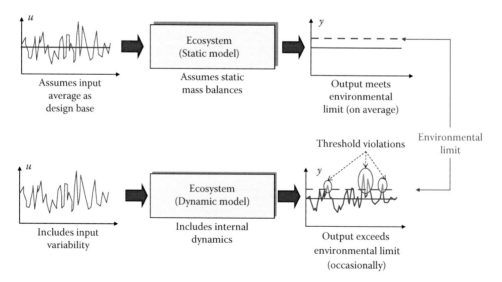

FIGURE 1.6   Differing insights provided by static and dynamical modelling.

the principles and equations required to set up the model, and the model-building process consists of selecting the right kind of equations and of adjusting the model parameters to obtain the best possible match between observations and model output. This latter aspect will be amply treated in Chapter 2. The mechanistic approach generally yields *parsimonious* (i.e. low complexity) models, whose inner workings are apparent and can be easily disseminated and understood by third parties. This approach is certainly appropriate when the aim of modelling is to gain a deeper phenomenological understanding of the system, and the reality being modelled is known to rest on solid physical principles.

Data-driven models, on the other hand, are based on the opposite assumption that nothing is known of the internal functioning of the system, and the aim is just to match the data and the model outputs. Of course, data-driven models have a structure of their own, but this is totally predetermined by the modeller in order to achieve the desired matching and has nothing to do with the actual inner structure of the system. This approach is preferable to the mechanistic method when the physical principles at the basis of the observed behaviour are not clearly understood, or when a mechanistic model would be too complex and/or too difficult to calibrate. These models are often called *black-box* models when the inner model structure is totally inaccessible, or *grey-box* models when a partial internal description is possible. There are several classes of data-driven models, depending on the mathematical approach: the most popular being the *principal components analysis, decision trees, neural networks (NNs)* and *fuzzy models (FMs)*, to name but a few. This list is surely neither exhaustive nor mutually exclusive, in the sense that two or more of these approaches can be combined to obtain a hybrid model that inherits the best properties of each. In all cases, the model structure depends on the method, and its dimension should be selected by the user seeking a compromise between model performance and computational complexity.

1.2.3 ENVIRONMENTAL MODELLING APPROACH VERSUS DATA AVAILABILITY

Because the purpose of environmental modelling is to reproduce the observed natural behaviour, any model requires data for its construction and verification. Of the techniques just mentioned, not all have similar requirements when it comes to data needs: as expected, mechanistic models will be less *data hungry* thanks to their internal predetermined structure, whereas data-driven models, relying solely on data for their functioning, will require considerably more data for their training and validation. A rough qualitative diagram relating the amount of data required to build a model versus the available knowledge is shown in Figure 1.7. Now the characteristics of some data-driven methods are briefly described.

Principal component analysis and *decision trees* are the simplest, often preliminary, analyses that can be applied to the data. They can hardly be considered as proper models, because they just perform data manipulation in order to extract information from the data. In this sense, they might be rather considered as *data-mining* (Witten and Frank, 2005) techniques aimed at putting order in the data, for example, determining a set of sub-regions in the data space by aggregating data according to some kind of similarity. In this way, these methods are often capable of revealing hidden structures in the data, and they may pave the way towards more sophisticated analyses. Apart from the *blind* approach, which disregards cause–effect relationships in the data, the other major limit of these methods is the lack of *time indexing*, so that the sequential nature of the data is lost and only ensemble analysis is performed. For this reason, with reference to Figure 1.7, they can be placed in an area where little knowledge and few data are available, in the sense that the limited results expected from this analysis can be achieved even with a limited amount of data. Of course, the insight gained into the problem is also equally limited, and at the end of this analysis, we will be left none the wiser about the causes behind the observed behaviour.

Coming to more organized approaches, *stochastic models (SMs)*, NNs and *fuzzy systems* all share the common feature of being truly data-driven, in the sense that their inner structure (or complexity) is decided by the user, possibly with an eye to the amount of available data and the purpose of modelling, but totally disconnected from the nature of the environmental process.

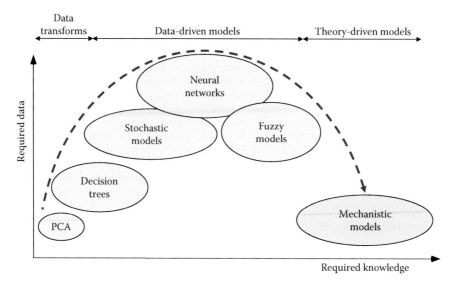

FIGURE 1.7   Possible modelling paradigms in the knowledge versus data domain.

The purpose of SMs (Astrom, 1970; Ljung, 1999; Box et al., 2008) is to reproduce the observed time-varying input–output behaviour without any attempt to explain the inner system structure from a mechanistic viewpoint. Their use should be preceded by a careful structural analysis to establish the cause–effect relationship in data, that is, to define the system inputs (causes) and outputs (effects) in order to understand which is which. SMs generally have a linear structure consisting of three parts, which represent the contribution of the past outputs (auto-regressive, AR, part), of the deterministic inputs (exogenous part), and the influence of unmeasurable white noise on the system. The possible combination of past noise samples (moving average, MA, part) makes this term *auto-correlated* in the sense that it is partially predictable on the basis of past observations. This is not true for *white* noise, which is by definition unpredictable, because each sample is uncorrelated with all the previous samples. The concept of autocorrelation will be considered in detail in Chapters 2 and 3. Partially correlated noise is also referred to as *coloured* noise, as opposed to the wholly unpredictable *white* noise. A simple ARMA exogenous (ARMAX) model has the form

$$y_t = \underbrace{a_1 y_{t-1} + \cdots + a_n y_{t-n}}_{\text{AR}} + \underbrace{b_1 u_{t-1} + \cdots + b_m u_{t-m}}_{\text{deterministic exogenous input}} + \underbrace{\varepsilon_t + c_1 \varepsilon_{t-1} + \cdots + c_n \varepsilon_{t-n}}_{\substack{\text{MA part} \\ \text{(coloured noise)}}} \tag{1.3}$$

where the model parameters $(a_1,\ldots,a_n,b_1,\ldots,b_m,c_1,\ldots,c_n)$ must be estimated in order to achieve the best possible agreement between the observed and the model outputs. The degrees of freedom of the model (1.3) consist of deciding how many past samples (n and m) should be considered in each part, considering that the MA part is generally—but not necessarily—of the same order as the AR part. Early modellers put a premium on keeping the model complexity low, and this resulted in very simple models, spawning the enormously successful field of adaptive control (Astrom, 1970), where the controller is based on a model that could be recalibrated in real time thanks to its simple structure.

SMs became popular in the 1970s when control engineers were confronted with difficult processes, such as a gas furnace, paper mill and arc welding, which defied a first-principle approach, given the complexity and variability of the involved dynamics. The calibration of SMs generally requires a large amount of data, usually partitioned in calibration and validation data sets. The calibration of the MA part posed special problems because the past noise samples $(\varepsilon_t, \varepsilon_{t-1}, \ldots, \varepsilon_{t-n})$

were by definition unavailable. This difficulty was solved with the introduction of auxiliary variables called *instrumental variables* (Young, 1970). This theory was later structured and refined to become a MATLAB® toolbox (Young, 2006; Tych and Young, 2012).

NNs came later and share with SMs the narrow goal of reproducing an observed behaviour. Exploiting the parallel progress in computing power, NNs have a more complex structure and require more sophisticated calibration techniques than SMs. They also introduce inner nonlinearities to make their behaviour more flexible. The workings of NNs were basically inspired by the basic element of the nervous system, the neuron, which reacts by activating its output (perception) when the input stimulus exceeds a given threshold. There are several neurological factors that can shape the neural response, and to account for this behaviour, a nonlinear activation function was introduced, an example of which is shown in Figure 1.8a. The input information is propagated to the output through an array of neurons, generally organized in an input layer, a hidden layer and an output layer, as shown in Figure 1.8b. The presence of the intermediate (hidden) layer greatly amplifies the flexibility of the NNs. The key to the success of the NNs is undoubtedly the *backpropagation* (BP) algorithm (Rumelhart and McClelland, 1986; Goldberg, 1989; Press et al., 1986; Dahlquist and Bjorck, 2003) used to tune the weights of each neuron (often called *synapsis* to follow the nervous system analogy) in approximating the training data. The BP algorithm is an extension of a gradient descent in the synaptic space, sometimes ending in a local minimum, as its detractors pointed out. This criticism notwithstanding, the BP algorithm, and its many subsequent modifications, made the fortune of the NNs. For a thorough introduction to NNs, the reader is referred to the excellent survey by Lippmann (1987). NNs are extensively used in environmental modelling where very complex phenomena like algal blooms (Wilson and Recknagel, 2001) or the dynamics of aquatic insect populations (Obach et al., 2001) can be conveniently modelled with relatively simple NN structures.

FMs, which will be thoroughly described in Chapter 4, introduce the concept of *approximate reasoning* into the model, which is composed of a collection of logical rules. The model output is obtained as a weighted sum of the individual response of an array of classifiers, similar to the neural response, which in this case is *softened* by the concept of *degree of membership* obtained by comparing the data to a predefined set of prototypical behaviours. FMs possess a higher degree of structure than the previous models, and this can be partially traded for a lesser data availability. For this reason, FMs have been put on the descending arc in the modelling paradigm of Figure 1.7. FMs are often combined with NNs to form neuro FMs (Kosko, 1992; Jang, 1993; Brown and Harris, 1994; Jang et al., 1996; Tsoukalas et al., 1997a, b; Fuller, 1999). In a nutshell, fuzzy reasoning consists of a collection of *modus ponens* implications whereby the *degree of truth* of the *antecedents* implies the degree of truth of the *consequent*. Contrary to conventional wisdom, for which a concept is either *true* or *false*, in fuzzy reasoning, the *degree of truth* runs smoothly from 0 (false) to true (1). Having defined a set of reference concepts (*prototypes*), the input variables are *fuzzified* by comparing them with the prototypes to obtain their *degree of truth*. Through an *implication* operator, these *antecedent degrees of truth* are projected onto the *consequent*. All of the degrees of truth resulting from each rule are then combined through *defuzzification* to yield the reasoning outcome, which in our case is

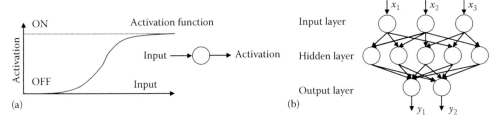

FIGURE 1.8 Typical neuron activation response (a) and a generic NN structure (b).

the model output. All these operations will be described in detail in Chapter 4. Formally, the generic *i*th fuzzy rule can be written as

$$R_i: \text{if} \underbrace{\left(x_1 \text{ is } A_1\right) \text{and} \left(x_2 \text{ is } A_2\right) \text{and},\ldots,\text{and} \left(x_n \text{ is } A_n\right)}_{\text{antecedents}} \underbrace{\text{then}}_{\text{implication}} \underbrace{\left(y_i \text{ is } B_i\right)}_{\text{consequent}} \qquad (1.4)$$

where the degrees of truth of the antecedents $\left(x_1,\ldots,x_n\right)$ with respect to their prototypes $(A_1,\ldots,A_n)$ are combined (*and*-ed) prior to the implication (*then*) onto the consequent. Graphically, the fuzzy reasoning scheme is shown in Figure 1.9, in which the triangles represent the prototype membership functions.

The last paradigm in Figure 1.7 is the mechanistic (or *first-principle*) approach, which sets up the model by sifting into the appropriate knowledge domain. Given the vast knowledge embedded in the principles, much less is to be *learned* from the data, which are nevertheless necessary to produce a model that agrees with the observations, as explained in Chapter 2. A concluding comment on Figure 1.7 regards data versus information richness; in other words plentiful data are not necessarily good data. There are situations in which data are abundant, but carry little information, so that the model will not *learn* much from those data. When dealing with experimental time series, the premium will be put on the richness of information carried by the data, rather than in their abundance, as will be explained in Chapter 3. Several caveats should be considered when performing a preliminary data screening. In deciding whether a data-driven approach should be preferred to a mechanistic model, consider that the former is basically a *data description*, whereas the latter is a *system description*. In the first case, the series of experimental data should be closely examined to check their informative content and to make sure that they cover all the possible situations that we want to model, considering that data-driven models have less generalization capabilities than their mechanistic counterpart. Conversely, if a *system description* is pursued (mechanistic model), the generalization property is intrinsic to its structure, so that less data are required for its calibration, though they will still play a crucial role in testing and validating the model.

Remember that a good model can trade a large amount of *data information* for a small amount of *system information* and that a complex model is not necessarily more accurate and reliable than a simple one! In fact, very large models calibrated with very large data sets risk data *over-fitting*, meaning that the model response is systematically too close to the data, creating the illusion of a

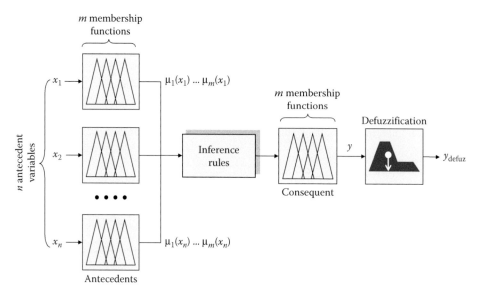

FIGURE 1.9   Functional diagram of a fuzzy reasoning scheme.

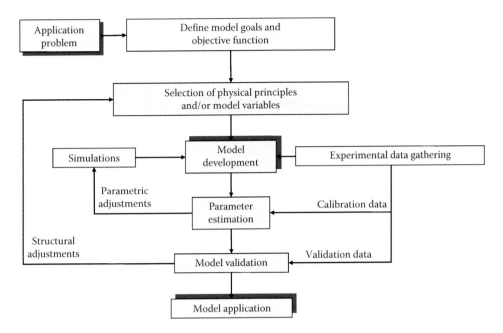

FIGURE 1.10   The model-building process as an iterative procedure.

perfect model, but in fact just producing a replication of the data, and the noise that comes with them, and consequently it lacks generality.

With these caveats and limitations in mind, the model-building process can be summarized in Figure 1.10. It refers to mechanistic model building, but with small changes it can also be adapted to data-driven models. Starting from the upper left corner, there must be a motivation for building a model, be it a lofty scientific endeavour or a more mundane economic need. In all cases, the modelling effort should be appropriate for the model purpose, avoiding any attempt to *overkill* the problem by setting up a model that is more complex (and costly) than strictly necessary. Quoting Albert Einstein: 'Everything should be made as simple as possible, but not simpler.'

Having defined the purpose (and budget) of the application, an objective function should be defined by which the model performance will be assessed, declaring beforehand which margin of error is acceptable. This is a critical step for the modeller, because it will set the yardstick by which the model will be eventually accepted or rejected. The modeller should be very honest in setting this acceptability mark, because too low a threshold will have little credibility, whereas a too high mark will be hardly ever met, possibly leading to the rejection of an otherwise useful model. So, be careful!

Considering the first iterative loop of Figure 1.10, the mathematical equations required to set up the model should now be selected. In the case of data-driven models, this step can be replaced by a selection of the most suitable approach of Figure 1.7. The *model development* core box involves the selection of the pertinent physical principles at the basis of the model and its adaptation to the case at hand. This is achieved through the inner loop of Figure 1.10, where calibration data are used to adjust the model parameter so that its response, obtained by simulation, is a good approximation of the observed behaviour according to a figure of merit that will be specified later. After the best possible agreement between model response and calibration data has been achieved, the procedure moves to the outer loop, in which another independent data set (validation data) is used to cross-check the model validity. Unlike the outer loop, in this case no further parameter adjustments are made, and the agreement between validation data and model response is again tested. This second check is aimed at assessing the model generality, represented by the ability to explain data with which it was not previously trained. Normally, this second match will be slightly worse than the calibration match, but some statistical tests should be passed before the model is validated. If validation fails, first the validity of this second data set should be scrutinized: are they equally representative of the typical model operational behaviour, albeit different from the calibration data? Are they general enough? Is the signal/noise ratio comparable with the calibration data? If these questions are positively resolved, then the model structure should be reconsidered and its components changed accordingly until the new structure successfully passes both the calibration and validation tests, after which it can be released.

1.3 PRACTICAL MECHANISTIC MODELLING

The notions of the previous sections are enough to introduce some general guidelines for the model-building process. In this section *mechanistic* (or *first-principle*) models will be considered as they are naturally related to the familiar principles of physics, chemistry and other branches of science. In mechanistic models, the first and most relevant aspect is the selection of the pertinent physical laws describing the system behaviour, so attention will now be focussed on specifying the *structure* of these models, leaving the parameter calibration task (a formidable one indeed!) for Chapter 2. Going back to the model-building diagram of Figure 1.10, the central block can now be expanded to reveal the finer inner structure of Figure 1.11.

As already discussed in Section 1.2.1, environmental models involve mass and energy dynamical equilibria, as described by Equation 1.2, to balance input–output flows and internal transformations (generation/utilization) within the control volume delimiting the system boundaries, as shown in Figure 1.12.

The dynamical balance Equation 1.2 is repeated here for convenience.

$$
\begin{bmatrix} \text{Rate} \\ \text{of} \\ \text{change} \end{bmatrix} = \frac{\Delta \text{storage}}{\Delta t} = \underbrace{\frac{\text{input}}{\Delta t} - \frac{\text{output}}{\Delta t}}_{\text{transport}} + \underbrace{\frac{\text{generation}}{\Delta t} - \frac{\text{utilization}}{\Delta t}}_{\text{internal transformations}}
\tag{1.5}
$$

In mathematical terms, this implies that differential (or difference) equations must be written for the state $\left(x \in \mathbb{R}^n \right)$ and output $\left(y \in \mathbb{R}^q \right)$ model variables in the form of Equation 1.6. In the continuous-time case, the explicit reference to time t will be dropped for simplicity in the following.

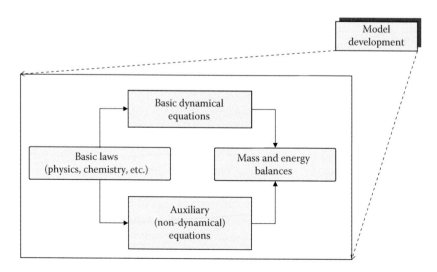

FIGURE 1.11   The finer inner structure of the mechanistic model building: selecting the appropriate equation set.

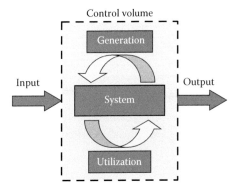

FIGURE 1.12   Rate balances are at the foundation of any dynamical model. The dashed box defines the control volume delimited by the system boundaries.

$$\text{Continuous-time} \begin{cases} \dfrac{dx}{dt} = f(x,u,z) \\ y = g(x,u,z) \end{cases} \quad \text{discrete-time} \begin{cases} x_{t+1} = f(x_t,u_t,z_t) \\ y_t = g(x_t,u_t,z_t) \end{cases} \quad (1.6)$$

In Equation 1.6, the vector-valued functions $f(\bullet): \mathbb{R}^n \to \mathbb{R}^n$ and $g(\bullet): \mathbb{R}^n \to \mathbb{R}^q$ may involve not only the state variables $(x \in \mathbb{R}^n)$ and the inputs $(u \in \mathbb{R}^m)$, but also other variables $(z \in \mathbb{R}^g)$ that are required by the rate equations, but do not appear on the left-hand-side derivative (or difference) of Equation 1.6. Unless these variables are expressed in terms of (x), the model equations are not *matched* in the sense that they cannot be solved because the state variables (x), whose time derivatives form the left side of the state equations, may not be fully matched by the expressions on the right side. In this case, the basic equations in Equation 1.6 must be complemented with *auxiliary* algebraic equations $x = h(x,u,z)$ in order to obtain a matching set of dynamical equations in which all the state variables (x) appear on both sides of the equations, that is,

$$\text{Continuous-time} \begin{cases} \dfrac{dx}{dt} = f_1(x,u) \\ y = g_1(x,u) \end{cases} \quad \text{discrete-time} \begin{cases} x_{t+1} = f_1(x_t,u_t) \\ y_t = g_1(x_t,u_t) \end{cases} \quad (1.7)$$

1.3.1   An Example of a Simple Mechanistic Model

Let us consider a cylindrical tank with constant cross section A, whose input flow (F_i) can be manipulated, whereas the output flow is controlled by gravity, the water height (h) and the friction coefficient of the output pipe k_v, as shown in Figure 1.13.

Assuming a constant specific weight of the liquid γ, the *mass* balance around the control volume (i.e. the volume of the liquid in the tank) reduces to a *volume balance*

$$A\frac{dh}{dt} = F_i - F_o \quad (1.8)$$

If we limit ourselves to this basic Equation 1.8, little progress can be achieved because the water height (h) is not directly related to the output flow (F_o), but if an auxiliary equation relating it to the water height is introduced, for example,

$$F_o = k_v \sqrt{h} \quad (1.9)$$

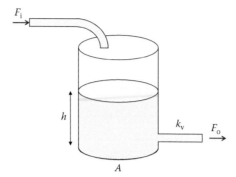

F_i

h

k_v F_o

A

FIGURE 1.13   A simple example illustrating the need for auxiliary equations.

then a *matched* model is obtained, in the sense that the state variable (h) appears on both sides of the state equation

$$A\frac{dh}{dt} = F_i - k_v\sqrt{h} \tag{1.10}$$

which can now be solved for the height h based on the knowledge of the input F_i.

1.3.2 Mass Balances

After this very simple preliminary example, we consider the general task of writing material balances, which are at the basis of reactor modelling (Stephanopoulos, 1984; Seborg et al., 1989). Not only the rate of the *total mass* in the system must balance out, but also the rate of each component. Going back to the example tank of Figure 1.13, suppose that the input stream F_i carries a dissolved substance 'a' with concentration C_a^i and that this substance is subject to transformations, for example, it may be involved in a chemical reaction inside the tank that results in a concentration change. Then, in addition to the general mass rate balance Equation 1.8, the rate of change for the substance 'a' (M_a) must also balance. This leads to the additional equation

$$\frac{dM_a}{dt} = \frac{d(VC_a)}{dt} = C_a\frac{dV}{dt} + V\frac{dC_a}{dt} = F_iC_a^i - F_oC_a \pm Vr_a \tag{1.11}$$

where:
  r_a is the reaction rate of the substance 'a'
  the mass M_a has been expressed as the product of concentration (C_a) and the control volume (V)

It will also be assumed that the tank is perfectly mixed, so that the concentration (C_a) is the same inside and at the output of the tank. Substituting the volume derivative of Equation 1.8 into Equation 1.11 yields

$$V\frac{dC_a}{dt} = -C_a(F_i - F_o) + F_iC_a^i - F_oC_a \pm Vr_a \tag{1.12}$$

Rearranging the concentration terms (the terms F_oC_a and $-F_oC_a$ cancel out), and dividing by the control volume V, the concentration rate balance is obtained as

$$\frac{dC_a}{dt} = \frac{F_i}{V}(C_a^i - C_a) \pm r_a \tag{1.13}$$

Equation 1.13 presents the same *matching* problem previously encountered in Equation 1.8, because the reaction rate r_a does not contain any explicit reference to the concentration C_a. Though a full discussion on chemical kinetics will be given in Chapter 7, for the time being let us introduce an *auxiliary equation* assuming that the reaction rate r_a follows a *first-order kinetics*, meaning that the reaction velocity is proportional to the concentration C_a, that is, $r_a = -k_a C_a$. The minus sign indicates that the substance 'a' *disappears* from the system as a consequence of the chemical reaction and some other product is formed instead, for which another equation should be written for mass continuity. However, as far as Equation 1.13 is concerned, substituting the explicit expression for the rate r_a yields

$$\frac{dC_a}{dt} = \frac{F_i}{V}\left(C_a^i - C_a\right) - k_a C_a$$

$$= \frac{F_i}{V} C_a^i - C_a \left(\frac{F_i}{V} + k_a\right) \tag{1.14}$$

In Equation 1.14, the first term represents the input, while the dynamics of the state variable $\left(C_a\right)$ is controlled by two parameters with a very clear physical meaning: the first term F_i/V, often referred to as the *dilution rate*, represents the effect of the hydraulics, whereas the second term (k_a) accounts for the effect of the chemical reaction. As a further comment, it should be noted that the output flow F_o does not explicitly appear in Equation 1.14, but it is still governed by the previous Equation 1.8, linking the height h (and ultimately the volume V) to the output flow F_o. Both equations should be treated in parallel, so that the complete model in the two state variables (h and C_a) has the following final form:

$$\begin{cases} \dfrac{dh}{dt} = \dfrac{1}{A}\left(F_i - k_v\sqrt{h}\right) \\[3mm] \dfrac{dC_a}{dt} = \dfrac{F_i}{Ah} C_a^i - C_a\left(\dfrac{F_i}{Ah} + k_a\right) \end{cases} \tag{1.15}$$

In general, if there are n mass components to be modelled in the system, $(n + 1)$ mass balances will be required—a global mass balance and one specific rate balance for each component, as given in the following equation:

$$\begin{cases} \dfrac{dV}{dt} = F_i - F_o \\[3mm] \dfrac{dC_1}{dt} = \dfrac{F_i}{V} C_1^i - \dfrac{F_i}{V} C_1 \pm r_1\left(C_1,\ldots,C_n\right) \\[3mm] \dfrac{dC_2}{dt} = \dfrac{F_i}{V} C_2^i - \dfrac{F_i}{V} C_2 \pm r_2\left(C_1,\ldots,C_n\right) \\[3mm] \vdots \\[3mm] \dfrac{dC_n}{dt} = \dfrac{F_i}{V} C_n^i - \dfrac{F_i}{V} C_n \pm r_n\left(C_1,\ldots,C_n\right) \end{cases} \tag{1.16}$$

A final remark should be made concerning the dimensions of the quantities involved in the mass balance rate Equation 1.16. Since they refer to mass variations over time $\left[M \cdot T^{-1}\right]$, each term in the balance must be consistent with these dimensions. Equation 1.15 seems to violate this principle, but a closer inspection proves the contrary. In the following equivalences, the brackets [.] denote *dimensions of* the quantities indicated therein. The basic mass balance Equation 1.8 was derived from

$$\frac{dm}{dt} = \dot{m}_i - \dot{m}_o \rightarrow \frac{d\gamma V}{dt} = \gamma_i F_i - \gamma F_o \tag{1.17}$$

where the mass flow $\dot{m}$ (the dot indicating time derivative) is equal to the volumetric flow $F\left[L^3 \cdot T^{-1}\right]$ multiplied by the fluid-specific weight (or mass density) γ as $\dot{m} = \gamma F$. A dimensional check can now be done

$$\left[\dot{m}\right] = \left[MT^{-1}\right]$$

$$\left[\gamma F\right] = \left[ML^{-3}L^3T^{-1}\right] = \left[MT^{-1}\right] \tag{1.18}$$

where M, L and T denote the dimensions of mass, length and time, respectively. Thus, we can conclude that Equation 1.17 is *dimensionally* correct. Of course, this is just a *necessary* condition, because a balance may be dimensionally correct but *conceptually* wrong for other reasons (wrong choice of equations, wrong balance, etc.).

1.3.3 Energy Balances

We now turn our attention to the *energy* balance, considering that the energy rate of change must equal the algebraic sum of the input–output energy flows and the energy exchange across the control volume of Figure 1.12

$$\underset{\substack{\text{input} \\ \text{energy} \\ \text{flow}}}{\frac{dE}{dt}} = \underset{\substack{\text{input} \\ \text{energy} \\ \text{flow}}}{\dot{E}_i} - \underset{\substack{\text{energy} \\ \text{rate} \\ \text{in the} \\ \text{system}}}{\dot{E}} \pm \underset{\substack{\text{heat} \\ \text{exchange} \\ \text{rate}}}{\dot{Q}} \pm \underset{\substack{\text{mechanical} \\ \text{work} \\ \text{exchange} \\ \text{rate}}}{\dot{W}} \tag{1.19}$$

where the dot superscript stands for time rate. If the energy exchanges involve only changes in the fluid temperature, then the situation can be described by the simple reactor of Figure 1.14.

In this case, Equation 1.19 simplifies into

$$\frac{dE}{dt} = \gamma V c_p \times \frac{dT}{dt} = \gamma F c_p T_i - \gamma F c_p T \pm \dot{Q} \tag{1.20}$$

where c_p is the specific heat capacity $\left[c_p\right] = \left[L^2T^{-2}\Theta^{-1}\right]$ of the fluid, and its preferred SI units are [kcal Kelvin^{-1}]. Dividing by $\gamma V c_p$ shows that, in this simplified case, the energy balance reduces to a temperature balance

$$\frac{dT}{dt} = \frac{F}{V}T_i - \frac{F}{V}T \pm \frac{\dot{Q}}{\gamma V c_p} \tag{1.21}$$

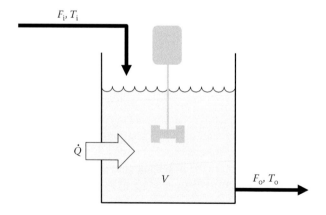

FIGURE 1.14 A simple continuous flow tank reactor with heat exchange.

A dimensional check of Equation 1.21 can be carried out in the same way as with Equation 1.17, namely

$$\left[\frac{dT}{dt}\right]=\left[\Theta T^{-1}\right] \qquad\qquad [\Theta]=\text{kcal}$$

$$\left[\frac{F}{V}T\right]=\left[L^3T^{-1}L^{-3}\Theta\right]=\left[\Theta T^{-1}\right] \qquad\qquad (1.22)$$

$$\left[\frac{\dot{Q}}{\gamma Vc_p}\right]=\left[\underbrace{ML^2T^{-3}}_{\dot{Q}}\underbrace{M^{-1}L^3}_{\gamma}\underbrace{L^{-3}}_{V}\underbrace{L^{-2}T^2\Theta}_{c_p}\right]=\left[\Theta T^{-1}\right]$$

which shows that all the terms in Equation 1.21 have consistent dimensions.

It is worthwhile to point out the formal similarity between the mass balance Equation 1.13 and the heat balance Equation 1.21. Although the former establishes a mass balance involving kinetic rates, the latter balances heat transfer rates. In the first case, the state variable is the concentration C_a, whereas in the second it is the temperature T. Figure 1.15 shows the role played by each term in either equation: the input–output terms have the same *dilution rate* coefficient, whereas the inner dynamics is represented by the kinetic rate in the first case and the heat transfer rate in the second.

As to the heat exchange mechanisms in the reactor tank, Figure 1.16 illustrates two possibilities, although more on this can be found in Stephanopoulos (1984). In the first case (Figure 1.16a), heat is directly supplied by an electrical resistance. In the case of a two-way heat exchange, the resistance can be replaced by a Peltier cell in which the direction of the heat flux can be reversed by inverting the direction of the current. In this case, the electric power (P) is directly proportional to the heat transfer rate, that is, $\dot{Q}=\eta P$, and Equation 1.21 can be rewritten as

$$\frac{dT}{dt}=\frac{F}{V}T_i-\frac{F}{V}T\pm\eta\frac{P}{\gamma c_p} \qquad\qquad (1.23)$$

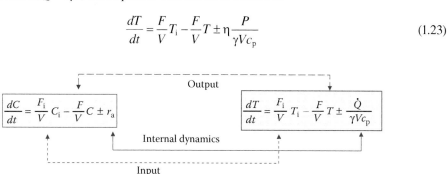

FIGURE 1.15   Formal similarity between the mass balance and the energy balance equations.

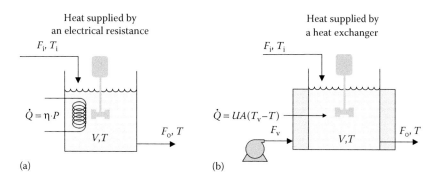

FIGURE 1.16   Differing possibilities of external heat supply: (a) through an electrical resistance and (b) through a jacket heat exchanger.

where η is the efficiency of the heat generating/absorbing device, with the $\pm$ accounting for a possible bidirectional heat exchange.

Figure 1.16b shows another possibility, much used in process engineering, in which the reactor is surrounded by a thermal jacket carrying a heating or coolant fluid with temperature T_v and flow rate F_v. In this case, the heat flux is determined by the temperature difference of the two fluids $(T_v - T)$ multiplied by the heat transfer coefficient U of the jacket wall of surface A. In this case, Equation 1.21 becomes

$$\frac{dT}{dt} = \frac{F}{V}T_i - \frac{F}{V}T \pm \frac{UA(T_v - T)}{\gamma V c_p} \tag{1.24}$$

where the heat exchange coefficient U can be related to the flow in the exchanger F_v by an empirical relation such as $U = k_h F_v^b$ with $0.8 \le b < 1$.

1.4   ENVIRONMENTAL MODELS AS DYNAMICAL SYSTEMS

The basic purpose of modelling is the understanding of the system working and the prediction of its future evolution. Before a reliable forecast can be produced, though, the model should be calibrated by comparing the actual and simulated outputs when the model is driven by the same experimental inputs that caused the observed output. Once a reasonable agreement between the actual and simulated model responses has been obtained, as will be explained in Chapter 2, forecasting is possible, provided an estimate of the future inputs is available. For this, the synthesis of inputs will be considered in Chapter 3. Supposing that t_o represents the current time, up to which the actual output is available, any attempt to estimate the system output for $t > t_o$ can be regarded as *prediction* (or *forecasting*). Hence, it is important to provide *credible* system inputs on which the prediction will be based. This consideration takes us back to Figure 1.2, the selection of boundaries and the outer activities producing the external inputs $\boldsymbol{u}(t)$. The synthesis of such *credible* system inputs will be considered in Chapter 3 (Figure 1.17).

Dynamical systems have the fundamental property of evolving in time and space, but their evolution is not random. Their structure, together with their past conditions, determines their actual state and influences their future. For this reason, the variables describing the evolution of a dynamical system are necessarily *time-indexed*. An elegant—and efficient—way of modelling these characteristics in mathematical terms is the use of difference or differential equations, which keep track of the system evolution through their *state*, a set of system variables that globally represent the system

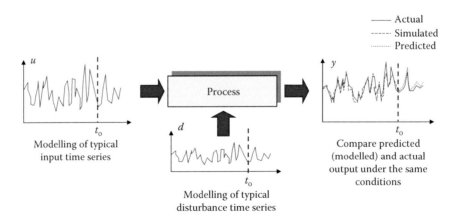

FIGURE 1.17   Modelling an environmental system implies the availability of typical input time series, both controllable and uncontrollable.

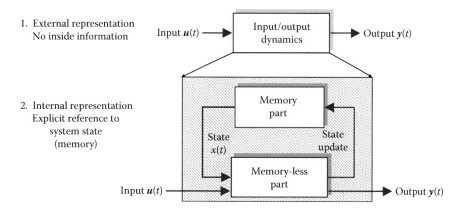

1. External representation
   No inside information

Input $\boldsymbol{u}(t)$ → Input/output dynamics → Output $\boldsymbol{y}(t)$

2. Internal representation
   Explicit reference to
   system state
   (memory)

Memory part

State $x(t)$

State update

Memory-less part

Input $\boldsymbol{u}(t)$ → → Output $\boldsymbol{y}(t)$

FIGURE 1.18 An internal system representation, separating the memory part from the memory-less input–output link.

history (Athans et al., 1974). Figure 1.18 shows the two possible representations of dynamical systems, external and internal, given as follows:

1. *The external representation* simply relates the system inputs and outputs, without any reference to the internal organization of the information. Given the memory characteristics of the dynamical system, this representation is not very economical since it should keep track of the entire history of the system evolution.
2. *The internal representation* is much more efficient, because it condenses the past system history in the *system state* $\boldsymbol{x}(t) \in \mathbb{R}^n$, a set of internal variables that enables the full reconstruction of the previous system evolution. The external input–output behaviour of point 1 can be obtained by combining the system state with a memory-less part, which produces the system output $\boldsymbol{y}(t) \in \mathbb{R}^q$, based in the state $\boldsymbol{x}(t) \in \mathbb{R}^n$ and the input $\boldsymbol{u}(t) \in \mathbb{R}^m$.

System evolution can be described in either continuous time or discrete time. In the former case, the time variable is considered as a continuous variable, that is, $t \in \mathbb{R}^+$, whereas in the latter time varies in discrete lumps that can be related to the ordinal sequence of integer numbers, that is, $t \in \mathbb{Z}^+$. In the first instance, the system will be modelled with *differential equations* (Quinney, 1987; Dormand, 1996; Butcher, 2003), whereas in the second *difference equations* (Elaydi, 2005; Goldberg, 2010) are used. In either case, Figure 1.18b shows the inner mechanism through which the memory and memory-less parts communicate. The form of the system update mechanism depends on the nature of the time-indexing. If the model is in discrete time, then the updating mechanism consists of projecting the current state $\boldsymbol{x}(t)$ onto the next state $\boldsymbol{x}(t + \Delta t)$, where Δt is the time step, whereas in the continuous-time case the state projection requires the evaluation of the state time derivative. In either case, the following notations can be introduced:

Discrete-time system $\Omega(x)$: $\boldsymbol{x}_t \rightarrow \boldsymbol{x}_{t+\Delta t}$

State update equation $\begin{cases} \boldsymbol{x}_{t+\Delta t} = \boldsymbol{f}(\boldsymbol{x}_t, \boldsymbol{u}_t) \\ \\ \boldsymbol{y}_t = \boldsymbol{g}(\boldsymbol{x}_t, \boldsymbol{u}_t) \end{cases}$ $\quad t \in Z^+, \quad \boldsymbol{x} \in \mathbb{R}^n, \quad \boldsymbol{u} \in \mathbb{R}^m, \quad \boldsymbol{y} \in \mathbb{R}^q$ (1.25)

Output equation

A discrete-time model, defined by (1.25), has the following basic features:

- The time variable is constrained to assume only a set of discrete values at intervals Δt.
- The system update occurs only at those discrete times $\boldsymbol{x}_t \rightarrow \boldsymbol{x}_{t+\Delta t}$.

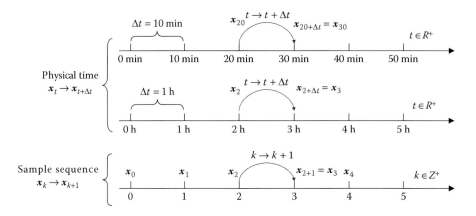

FIGURE 1.19 Relationship between physical time and sample sequence.

- The system updating function $f(x,u)$ yields the next value of the state $x_{t+\Delta t}$ given the current state x_t.
- No state transition is defined for any other time $t \notin \mathbb{Z}^+$.

To better understand the difference between the model time step Δt and the physical time t, Figure 1.19 shows the correspondence between the physical time and the sequence of samples $t \in \mathbb{Z}^+$. In the latter case, the notion of physical time is lost and is substituted by the sequence index, that is,

$$
\begin{array}{ccccc}
x_t & x_{t+\Delta t} & x_{t+2\Delta t} & x_{t+3\Delta t} & \cdots \\
\updownarrow & \updownarrow & \updownarrow & \updownarrow & \updownarrow \\
x_0 & x_1 & x_2 & x_3 & \cdots
\end{array}
\tag{1.26}
$$

In the following, when dealing with discrete-time sequence, once the sampling interval Δt has been defined, the sequence notation will be used, so that the succession of states will be denoted by their sample number, as in the bottom row of Equation 1.26. Conversely, a continuous-time system is described by a set of differential equations as Equation 1.27. In the following, we shall introduce the notation of the upper dot to indicate time derivative, that is, $\dot{x} \triangleq \frac{dx}{dt}$

Continuous-time systems $\Omega(x): x(t) \to \dot{x}(t)$

$$
\begin{aligned}
\text{State update equation} \quad & \begin{cases} \dot{x} = f(x,u) \\ \\ y = g(x,u) \end{cases} \quad \forall t \in \mathbb{R}^+, \quad x \in \mathbb{R}^n, \quad u \in \mathbb{R}^m, \quad y \in \mathbb{R}^q \\
\text{Output equation} \quad &
\end{aligned}
\tag{1.27}
$$

- The time variable may assume any value in $t \in \mathbb{R}^+$.
- The system state $x(t)$ is continuously updated and is defined for any $t \in \mathbb{R}^+$.
- The updating mechanism requires the use of the time derivative of the system state $\dot{x}$.

1.5 LINEAR TIME-INVARIANT DYNAMICAL SYSTEMS

A very popular and much studied class of dynamical systems is the one in which the state update and output functions appearing in Equations 1.25 and 1.27 are linear and the system parameters are constant in time. A very lucid account of linear systems in the wider context of mathematical modelling and optimization is given in specialized textbooks (Athans et al., 1974; Banks, 1986). In Nature almost each and every phenomenon is governed by nonlinear laws, nevertheless, the huge wealth of knowledge available on linear systems makes these system very suitable to approximate complex, nonlinear behaviours. Therefore, we are now including a brief summary of the main results in linear time-invariant (LTI) theory that will help us in analysing natural system, though we must always be aware that LTI represents an approximation and that often more sophisticated analytical tools are required. Now, we start by considering discrete-time LTI systems that can be represented by the following mathematical objects:

$$\text{State update equation} \begin{cases} \boldsymbol{x}_{t+1} = \boldsymbol{A} \cdot \boldsymbol{x}_t + \boldsymbol{B} \cdot \boldsymbol{u}_t \\ \\ \boldsymbol{y}_t = \boldsymbol{C} \cdot \boldsymbol{x}_t + \boldsymbol{D} \cdot \boldsymbol{u}_t \end{cases} \quad t \in \mathbb{Z}^+, \quad \boldsymbol{x} \in \mathbb{R}^n, \quad \boldsymbol{u} \in \mathbb{R}^m, \quad \boldsymbol{y} \in \mathbb{R}^q \quad (1.28)$$

$$\text{Output equation}$$

where the subscript t indicates time and the matrices $\boldsymbol{A} \in \mathbb{R}^{n \times n}$, $\boldsymbol{B} \in \mathbb{R}^{n \times m}$, $\boldsymbol{C} \in \mathbb{R}^{q \times n}$, $\boldsymbol{D} \in \mathbb{R}^{q \times m}$ are the linear operators that provide the required state update and input–output mappings. Their dimensions are compatible with the dimension of the state ($\boldsymbol{x}$), input ($\boldsymbol{u}$) and output ($\boldsymbol{y}$) vectors. The inner system structure of Figure 1.18 can then be specified as in Figure 1.20. Here, the memory element is represented by the unit delay operator $\left(z^{-1} \right)$ that shifts back the state by one time step

$$z^{-1} \cdot \boldsymbol{x}_{t+1} = \boldsymbol{x}_t$$

$$z^{-1} \cdot \boldsymbol{x}_t = \boldsymbol{x}_{t-1}$$

$$z^{-1} \cdot \boldsymbol{x}_{t-1} = \boldsymbol{x}_{t-2} \quad (1.29)$$

$$\vdots$$

The state update mechanism is represented by the matrix $\boldsymbol{A}$ that projects the current state onto the next time step, that is, $\boldsymbol{x}_{t+1} = \boldsymbol{A} \cdot \boldsymbol{x}_t$. Matrices $\boldsymbol{B}$, $\boldsymbol{C}$ and $\boldsymbol{D}$ simply add the appropriate input–output contributions to obtain the full system representation Equation 1.28.

Likewise, continuous-time linear dynamical systems can be described by the following differential–algebraic equations:

$$\text{State update equation} \begin{cases} \dot{\boldsymbol{x}} = \boldsymbol{A} \cdot \boldsymbol{x} + \boldsymbol{B} \cdot \boldsymbol{u} \\ \\ \boldsymbol{y} = \boldsymbol{C} \cdot \boldsymbol{x} + \boldsymbol{D} \cdot \boldsymbol{u} \end{cases} \quad \forall t \in \mathbb{R}^+ \quad (1.30)$$

$$\text{Output equation}$$

where the explicit dependence of $\boldsymbol{x}$ and $\boldsymbol{u}$ on t has been omitted for clarity. The generic inner structure of Figure 1.18 now turns into the one shown in Figure 1.21. In this case, the memory element is represented by the integral of the state derivative $\boldsymbol{x}(t) = \int \dot{\boldsymbol{x}} \cdot dt$, and the state update is performed by the system matrix $\boldsymbol{A}$ as in the discrete-time case, that is, $\dot{\boldsymbol{x}} = \boldsymbol{A} \cdot \boldsymbol{x}$, whereas the connecting matrices $\boldsymbol{B}$, $\boldsymbol{C}$ and $\boldsymbol{D}$ play the same role as in the discrete-time case.

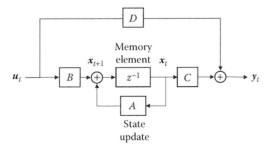

FIGURE 1.20 Inner structure of a discrete-time linear dynamical system.

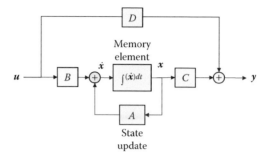

FIGURE 1.21 Inner structure of a continuous-time LTI dynamical system. The explicit time dependence has been omitted for clarity.

Equations 1.28 and 1.30 are *local representation* because they describe the *local* behaviour of the system. In the discrete-time case, the next state x_{t+1} can be computed via Equation 1.28 from x_t and the input u_t, but how could this be extended to compute, say, x_{t+10}? Likewise, given the continuous-time system description of Equation 1.30, how can we evaluate the system response from a given time t_0 to any other time t_s in the future? In other words, we need tools to analyse the system and predict its behaviour *at large*, and not only *locally*. The next few paragraphs will briefly summarize the main results of the dynamical LTI system theory. The fundamental assumption is that the system properties can be gleaned from its structure, without the need to actually compute the system evolution, either via calculus or by numerical means. This is the great strength of the analysis that follows. Unfortunately, the differing nature of discrete-time and continuous-time systems prevents a common analysis, so that these analysis must follow separate paths, as described next.

1.6   DISCRETE-TIME LTI DYNAMICAL SYSTEM

Rewriting the first of Equation 1.28 without any external input $u(t) = 0$, that is, considering an *autonomous (or unforced) system,* yields

$$x_{t+1} = A \cdot x_t \tag{1.31}$$

Now, iterating Equation 1.31 from the initial $t = 0$, the following sequence is obtained:

$$x_1 = Ax_0$$

$$x_2 = Ax_1 = A \times A \times x_0$$

$$x_3 = Ax_2 = A \times A \times A \times x_0 \tag{1.32}$$

$$\vdots$$

which can be generalized into the *global system representation*

$$x_t = \underbrace{A \times \cdots \times A \times A}_{t} \times x_0 = A^t \times x_0 \tag{1.33}$$

because it determines the system state at any future time $x(t)$ from the initial state $x_0 = x(t_0)$. The matrix A^t is referred to as the *transition matrix* because it relates states at differing times $t_2 > t_1$, for example,

$$x_{t_2} = A^{(t_2 - t_1)} x_{t_1} \tag{1.34}$$

TABLE 1.2

Properties of the Transition Matrix for Discrete-Time Systems

Property	Definition	Proof
Identity	$\Phi(t,t) = I$	$A^{(t-t)} = A^0 = I$
Composition	$\Phi(t_1,t_2) \cdot \Phi(t_2,t_3) = \Phi(t_1,t_3)$ $t_1 > t_2 > t_3$	$A^{(t_2-t_1)} \cdot A^{(t_3-t_2)} = A^{t_2-t_1+t_3-t_2} = A^{t_3-t_1}$
Reversibility (provided that A^{-1} exists)	$\Phi^{-1}(t_1,t_2) = \Phi(t_2,t_1)$	$\left(A^{(t_2-t_1)}\right)^{-1} = A^{(t_1-t_2)}$
Commutability	$A \cdot \Phi(t_1,t_2) = \Phi(t_1,t_2) \cdot A$	$A \cdot A^{(t_2-t_1)} = A^{(t_2-t_1+1)} = A^{(t_2-t_1)} \cdot A$

The transition matrix $\Phi(t_1,t_2) = A^{(t_2-t_1)}$ has the properties shown in Table 1.2, which can be easily demonstrated with elementary matrix algebra.

Of all these properties, *reversibility* is particularly interesting because it shows that the flow of time can be reversed, so that both forward and backward simulation of the system can be equally performed.

1.7   STABILITY OF DISCRETE-TIME DYNAMICAL LTI SYSTEMS

A fundamental property of dynamical systems is *stability of the equilibrium*, defined as a persistently quiescent state of the system. *Stability* is a property of the equilibrium and implies that once it is reached, it is maintained indefinitely, or, if some external perturbation occurs, the system returns to it after a time-limited transient. The notion of equilibrium is obviously common to linear and nonlinear systems, and for discrete-time system, it can be defined as

$$\overline{x}: \ \overline{x} = f(\overline{x}) \tag{1.35}$$

The time index in Equation 1.35 can be dropped because equilibrium is by definition not time-dependent, that is, $x_t = x_{t+1} = \cdots = x_{t+n} = \overline{x}$. The equilibrium definition (1.35) will be considered again when dealing with nonlinear systems, which may have multiple equilibria, but for now let us consider the implications for LTI discrete-time unforced systems, for which Equation 1.35 becomes

$$\overline{x} = A \cdot \overline{x} \tag{1.36}$$

In this case, $\overline{x} = 0$ is an equilibrium, and if $I - A$ is non-singular, this is the *unique equilibrium*. The question is if and how this equilibrium point will be reached and maintained. To answer this question, we concentrate on the properties of the system matrix A. As an example, let us introduce the following two two-dimensional discrete-time systems:

$$x_{t+1} = A \cdot x_t$$

$$\Sigma_1: A_1 = \begin{bmatrix} -0.14 & -0.55 \\ -0.6 & 0.4 \end{bmatrix} \quad x_0 = \begin{bmatrix} 1 \\ 2 \end{bmatrix} \tag{1.37}$$

$$\Sigma_2: A_2 = \begin{bmatrix} -1.14 & -0.55 \\ -0.6 & 0.4 \end{bmatrix} \quad x_0 = \begin{bmatrix} 1 \\ 2 \end{bmatrix}$$

The two system matrices differ by the values of the upper left element (a_{11}). Their behaviour starting with the same initial condition $x_0 = \begin{bmatrix} 1 & 2 \end{bmatrix}^T$ is strikingly different: Σ_1 produces a stable trajectory with both state components quickly reaching zero, whereas Σ_2 is clearly unstable, with both state

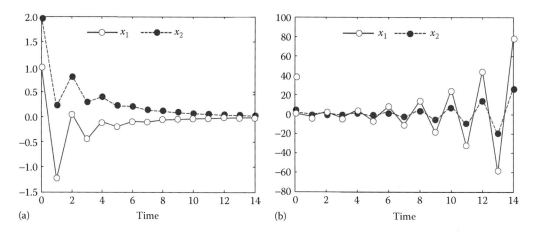

FIGURE 1.22   Trajectories of a stable (a) and an unstable (b) two-dimensional system.

variables becoming unbounded, as shown in Figure 1.22. Could this behaviour be predicted without resorting to simulation? The answer is obviously *YES* and lies in the matrix *A* itself.

1.7.1 Who's Who in the System

The behaviours of Figure 1.22 could have been predicted by analysing the system matrix *A*. Precisely, we know from linear algebra that a square matrix $A \in \mathbb{R}^{n \times n}$ may be viewed as a linear operator that shifts and resizes vectors $x \in \mathbb{R}^n$. But there are privileged vectors (*eigenvectors*) that resist the influence of *A* and maintain their direction in $\mathbb{R}^n$, being subject only to a resizing according to the corresponding *eigenvalues*. The precise definition of these quantities arises from the eigenvector equation

$$A v_i = \lambda_i v_i \tag{1.38}$$

which states that the application of the matrix *A* to the *eigenvector* v_i is equivalent to multiplying it by the corresponding *eigenvalue* λ_i. The eigenvectors then define invariant directions and can be assumed as a basis in the $\mathbb{R}^n$ Euclidean space, so that all other vectors can be expressed as their linear combinations. In particular, let the initial state vector x_0 be expressed as a function of the eigenvectors $(v_1, v_2, ..., v_n)$ as shown in Figure 1.23.

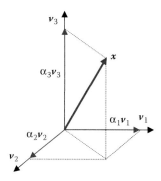

FIGURE 1.23   Any vector $x \in \mathbb{R}^n$ can be represented as a linear combination of the eigenvectors $(v_1, v_2, ..., v_n)$.

$$x_0 = \alpha_1 v_1 + \alpha_2 v_2 + \cdots + \alpha_n v_n = \sum_{i=1}^{n} \alpha_i v_i \tag{1.39}$$

Now, if we apply Equation 1.39 to the definition of autonomous system Equation 1.31, we can see that the evolution of the system state $x(t)$ can be expressed as a linear combination of eigenvalues and eigenvectors of the matrix A, where the constants $(\alpha_1, \alpha_2, \ldots, \alpha_n)$ depend on the particular choice of x_0.

$$x_t = A^t x_0 = A^t \sum_{i=1}^{n} \alpha_i v_i = A^{t-1} \sum_{i=1}^{n} \alpha_i A v_i = A^{t-1} \sum_{i=1}^{n} \alpha_i \lambda_i v_i$$

$$= A^{t-2} \sum_{i=1}^{n} \alpha_i \lambda_i^2 v_i = \cdots = \sum_{i=1}^{n} \alpha_i \lambda_i^t v_i \tag{1.40}$$

Equation 1.40 clearly shows that the evolution of the state x_t of an autonomous system is entirely determined by the eigenvalues and eigenvectors of the system matrix A. This result is particularly relevant for determining the stability of a system because in the last summation of Equation 1.40 any eigenvalue with magnitude greater than 1 will produce a growing term λ_i^t as time increases, so that the corresponding eigenvector v_i will be progressively amplified, eventually leading to instability, whereas any eigenvalue with magnitude smaller than 1 will contribute with a term that vanishes with time. It should be stressed that one unstable eigenvalue is enough to generate system instability, because it will provide a growing contribution that will eventually dwarf all the other stable modes that converge to zero. If in Equation 1.40, the initial condition x_0 coincides with one eigenvalue, then the system trajectory is bound to follow that eigenvalue and the contribution of all the others will be zero. This can be easily demonstrated by setting to zero all the α coefficients but one. Repeated application of Equation 1.39 with $x_0 = v_i$ yields

$$x_0 = v_i$$

$$x_1 = Ax_0 = Av_i = \lambda_i v_i$$

$$x_2 = Ax_1 = A \times \lambda_i v_i = \lambda_i^2 v_i$$

$$x_3 = Ax_2 = A \times \lambda_i^2 v_i = \lambda_i^3 v_i \tag{1.41}$$

$$\vdots$$

$$x_t = \lambda_i^t v_i$$

showing that the system trajectory will never abandon the direction defined by v_i, its magnitude being only altered by the eigenvalue factor λ_i^t as time progresses. To better understand the influence of eigenvalues on the system evolution, let us consider again the unstable system Σ_2 of Equation 1.37 and compute its eigenvectors and eigenvalues

$$A = \begin{bmatrix} -1.14 & -0.55 \\ -0.6 & 0.4 \end{bmatrix} \Rightarrow \begin{cases} v_1 = \begin{bmatrix} -0.9448 \\ -0.3276 \end{bmatrix} & \lambda_1 = -1.3307 \\ v_2 = \begin{bmatrix} 0.3029 \\ -0.9530 \end{bmatrix} & \lambda_2 = 0.5907 \end{cases} \tag{1.42}$$

Clearly, v_1 is the unstable eigenvector, because the corresponding eigenvalue λ_1 has magnitude greater than 1, whereas v_2 is the stable one. Now, instead of selecting an arbitrary initial condition x_0, let us

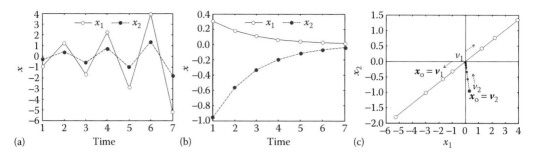

(a) (b) (c)

FIGURE 1.24 System trajectories initiated along an eigenvector are bound to remain on it. In (a) the initial condition of the system (1.42) was selected parallel to the unstable eigenvector λ_2, whereas in (b) the initial condition was parallel to the stable eigenvector λ_1. In (c), the white dots represent the system evolution along the unstable v_1, whereas the blue dots show the evolution along the stable eigenvector v_2.

initialize the system (1.42) with one of its eigenvectors. In Figure 1.24a, the system was initialized with the unstable eigenvector v_1 and the resulting system evolution is clearly unstable. In Figure 1.24b, the initial state was instead set equal to the stable eigenvector v_2. In this case, the trajectory cannot be influenced by v_1 and the state tends to zero along v_2. Figure 1.24c shows both simulations in the state space: when $x_0 = v_1$, the trajectory remains on the line defined by v_1 and becomes unstable, whereas when $x_0 = v_2$ the trajectory ends in the origin regardless of the unstable eigenvalue v_1.

To compute the eigenvalues of the matrix A, let us elaborate on the defining Equation 1.38

$$A \cdot v = \lambda \cdot v \quad \rightarrow \quad (A - \lambda I) \cdot v = 0 \tag{1.43}$$

Because we are looking for non-zero eigenvectors v, solving Equation 1.43 implies that the matrix $A - \lambda I$ must be singular and therefore its determinant must be zero:

$$\det(A - \lambda I) = 0 \tag{1.44}$$

The eigenvalues are the roots of the polynomial Equation 1.44, which has in general n roots (eigenvalues), not necessarily distinct, either real or complex conjugate because A is a real matrix. Having established that the evolution of an autonomous (unforced) system is entirely determined by its eigenvalues, a correspondence can be established between the location of the eigenvalues in the complex plane and the system stability, considering that an eigenvalue with magnitude less than 1 is stable, one with magnitude greater than 1 is unstable, and those with magnitude equal to 1 are marginally stable (they neither converge to zero nor diverge to infinity). Figure 1.25 shows the correspondence between the eigenvalues location in the complex plane $\lambda = \sigma + j\omega$, where j is the imaginary unit defined as $j^2 = -1$, and the unforced system response. The portion of the complex plane *inside* the unit circle $(|\lambda| < 1)$ represents the stability region, in the sense that any eigenvalue inside this area produces a stable response. In general, the eigenvalue location determines the system behaviour as sketched in Figure 1.25. With reference to the letters in the figure, the cases of Table 1.3 may occur.

It is interesting to note that the same damped oscillatory behaviour can be obtained with three different eigenvalue locations (a, b, h), with the damping decreasing as the eigenvalue approaches the unit circle boundary. In general, the eigenvalues inside the unit circle are *well behaved* in the sense that they produce a stable response, whereas those outside the unit circle will cause the system to diverge because they produce an unbounded response. Obviously, just a single unstable eigenvalue suffices in destabilizing the whole system, because its growing response will soon overwhelm the decreasing contributions of the stable eigenvalues. As a concluding remark,

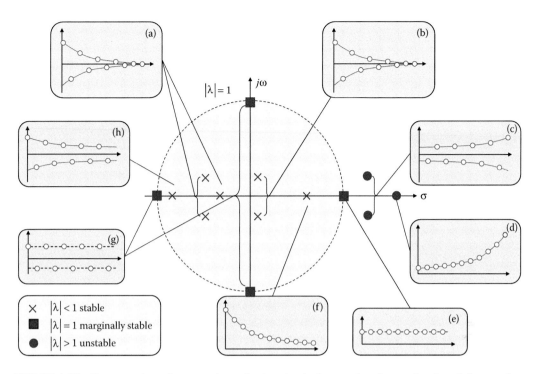

FIGURE 1.25 Correspondence between eigenvalue location in the complex plane and unforced discrete-time system response.

TABLE 1.3

Correspondence between Eigenvalue Position and System Response for a Discrete-Time System as Shown in Figure 1.25

Case	Eigenvalue	$(\lambda = \sigma \pm j\omega)$	System Response
a	Stable, negative real	$(\lambda = \sigma)$ $(-1 < \sigma < 0)$	Damped oscillations
b	Stable, complex conjugate	$(\lambda_{1,2} = \sigma \pm j\omega)$ $(-1 < \sigma < 1)$ $(-j < \omega < j)$	Damped oscillations
c	Unstable, complex conjugate	$(\lambda_{1,2} = \sigma \pm j\omega)$ $(|\sigma| > 1)$	Unbounded oscillations
d	Unstable, positive real	$(\lambda = \sigma)$ $(\sigma > 1)$	Unbounded monotonic response
e	Simply stable, positive real	$(\lambda = \sigma)$ $(\sigma = 1)$	Constant response
f	Stable, positive real	$(\lambda = \sigma)$ $(0 < \sigma < 1)$	Damped monotonic response

(Continued)

TABLE 1.3 (*Continued*)

**Correspondence between Eigenvalue Position and System Response
for a Discrete-Time System as Shown in Figure 1.25**

Case	Eigenvalue	$(\lambda = \sigma \pm j\omega)$	System Response
g	Simply stable	$(\lambda = \sigma)$	Sustained bounded oscillations
		$(\sigma = -1)$	
		$(\lambda = \pm j)$	
		$(\sigma = 0)$	
h	Stable, negative real	$(\lambda = \sigma)$	Damped oscillations
		$(-1 < \sigma < 0)$	

stability is a much coveted property in engineering because an unstable technological system will almost surely wreak havoc in an orderly organization. In nature, instead, instability is often associated with growth and may be a desirable characteristic, at least to some extent. So, we must resist the temptation to label stability (or any other system property for that matter) with *human* good/bad connotations.

1.7.2 Phase-Plane Portrait of Two-Dimensional Discrete-Time LTI Systems

An easy graphical way to represent the evolution of two-dimensional system is the so-called phase-plane portrait (PPP), which consists of plotting the two state components $x_{1,t}$ and $x_{2,t}$, one versus the other, instead of plotting each of them versus time. Figure 1.24c has anticipated this representation, in which the characteristic traits of the system behaviour emerge more clearly, and the PPP provides a powerful and intuitive analysis tool that will be much used in the following, especially when dealing with two-species ecosystems (Chapter 5). Figure 1.26a shows how the system time evolution can be translated into its phase portrait just by plotting one state variable versus the other. Of course, this representation is limited to two-dimensional systems, but in this context it represents an effective graphical tool for analysing the system behaviour.

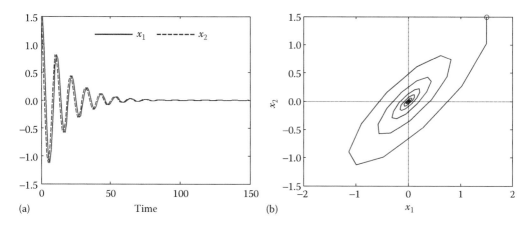

FIGURE 1.26 Translating the system time behaviour (a) into its PPP (b) by plotting x_1 versus x_2.

1.7.3 FORCED LTI DISCRETE-TIME SYSTEM

Having clarified that *stability* is an intrinsic property of an autonomous system, we now reintroduce the forcing term and consider the full system Equation 1.28 repeated here for convenience

$$x_{t+1} = A \cdot x_t + B \cdot u_t \qquad (1.45)$$

In what way will the input affect the system behaviour? Certainly, the input u_t will not affect stability, though a stable system may become unstable if it is subject to an unbounded input, but it will then surely influence system evolution and its steady state. To obtain the *global representation with input*, let us repeat the iteration scheme of Equation 1.33, but this time using the full system of Equation 1.45

$$x_1 = A \cdot x_0 + B \cdot u_0$$

$$x_2 = Ax_1 + Bu_1 = A\left[Ax_0 + Bu_0\right] + Bu_1$$

$$= A^2 x_0 + ABu_0 + Bu_1$$

$$x_3 = Ax_2 + Bu_2 = A\left[A^2 x_0 + ABu_0 + Bu_1\right] + Bu_2 \qquad (1.46)$$

$$= A^3 x_0 + A^2 Bu_0 + ABu_1 + Bu_2$$

$$\vdots$$

By inspection, it can be seen that the first term in each summation is the transition matrix $A^t \cdot x_0$ already encountered as Equation 1.33, whereas the other terms form a summation of the past inputs up to time t weighted by decreasing powers of A. Equation 1.46 can be separated into two terms

$$x_t = \underbrace{A^t \cdot x_0}_{\text{unforced response}} + \underbrace{\sum_{i=0}^{t-1} A^{t-(i+1)} Bu_i}_{\text{input forcing}} \qquad (1.47)$$

with the first term representing the contribution of the initial condition x_0 and the second the influence of all the past inputs. This second term is referred to as a *convolution summation*. Now, if a constant input $\bar{u}$ is applied, the system steady state $\bar{x}$ must satisfy the condition

$$\bar{x} = A \cdot \bar{x} + B \cdot \bar{u} \qquad (1.48)$$

Solving for $\bar{x}$ yields

$$\bar{x} = \left(I - A\right)^{-1} \cdot B \cdot \bar{u} \qquad (1.49)$$

So, the steady state of a stable forced discrete-time LTI will linearly depend on the steady-state input value $\bar{u}$.

1.7.4 STEADY STATE OF LTI DISCRETE-TIME SYSTEMS WITH CONSTANT INPUT

As an example consider the following two-dimensional system with one input

$$\begin{bmatrix} x_{1,t+1} \\ x_{2,t+1} \end{bmatrix} = \begin{bmatrix} 0 & 1 \\ -0.7 & 1 \end{bmatrix} \times \begin{bmatrix} x_{1,t} \\ x_{2,t} \end{bmatrix} + \begin{bmatrix} 1 \\ 2 \end{bmatrix} \times u_t \qquad (1.50)$$

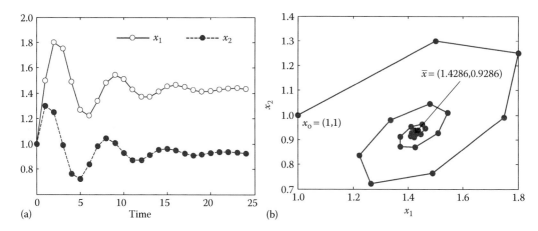

FIGURE 1.27   Time plot (a) and PPP (b) of the example model Equation 1.51. The steady state $\bar{x}$ is approached by damped oscillations because the eigenvalues are complex conjugate with magnitude less than 1.

Assuming a constant value for the input $\bar{u} = 0.5$, the steady state can be computed according to Equation 1.49 as

$$
\begin{aligned}
\bar{x} &= \left(\begin{bmatrix} 1 & 0 \\ 0 & 1 \end{bmatrix} - \begin{bmatrix} 0 & 1 \\ -0.7 & 1 \end{bmatrix} \right)^{-1} \times \begin{bmatrix} 1 \\ 2 \end{bmatrix} \times 0.5 \\
&= \begin{bmatrix} 0 & 1.4286 \\ -1.0000 & 1.4286 \end{bmatrix} \times \begin{bmatrix} 1 \\ 2 \end{bmatrix} \times 0.5 = \begin{bmatrix} 1.4286 \\ 0.9286 \end{bmatrix}
\end{aligned}
\tag{1.51}
$$

An inspection of the system eigenvalues $\lambda_{1,2} = 0.5 \pm j0.6708$ ensures its stability because $|\lambda_{1,2}| = 0.8367 < 1$, so that the steady state $\bar{x}$ is reached from any initial condition with a damped oscillatory response (Table 1.3, case b) because the eigenvalues are complex conjugate. The time plot and the PPP of this example are shown in Figure 1.27.

Equation 1.47 represents the *global response* of the system and in principle yields the system evolution for any time horizon given the initial condition and the input u. However, this equation is hardly ever used in practice, both because of the computational difficulties involved and for its limited use in providing information about the system response, which can be obtained by eigenvalues analysis, as we have seen. As to the time evolution, numerical methods are preferred, and this will be all the more true for continuous-time system for which the equivalent of Equation 1.47 requires even more complex calculus.

1.8   CONTINUOUS-TIME LINEAR DYNAMICAL SYSTEM

We now go on to consider dynamical systems for which time is now a continuous variable, and the system evolution is described by the vector differential Equation 1.30 with constant matrices (A, B, C, D), repeated here for convenience.

$$
\begin{cases}
\text{State update equation} & \dot{x} = A \cdot x + B \cdot u \\
& \qquad\qquad\qquad\qquad \forall t \in \mathbb{R}^+ \\
\text{Output equation} & y = C \cdot x + D \cdot u
\end{cases}
\tag{1.52}
$$

Again, let us consider the special case of an autonomous system

$$\dot{x} = A \cdot x \tag{1.53}$$

Equation 1.53 can be solved by separation of variables in a formally similar way to scalar differential homogeneous equations

$$\frac{dx}{x} = A \cdot dt \quad \rightarrow \quad x = e^{At} \cdot x_0 \tag{1.54}$$

where x_o is the initial condition. Equation 1.54 is the *global system representation* for LTI continuous-time system, equivalent to Equation 1.33 in the discrete-time case. The matrix exponential in Equation 1.54 can be approximated by a Taylor series expansion

$$e^{At} = I + A \cdot t + \frac{1}{2!}A^2 \cdot t^2 + \frac{1}{3!}A^3 \cdot t^3 + \cdots = \sum_{k=0}^{\infty} \frac{1}{k!}A^k \cdot t^k \tag{1.55}$$

In Equation 1.54, e^{At} represents the system *transition matrix* in much the same way as that defined by Equation 1.34 for discrete-time LTI systems and for which we can define similar properties, illustrated in Table 1.4.

1.8.1 STABILITY OF CONTINUOUS-TIME LTI SYSTEMS

We now investigate the stability of the equilibrium for unforced continuous-time systems, as we have done in Section 1.7 for discrete-time systems. Since the system is now described by a vector differential equation, a necessary condition for equilibrium is that the time derivative vanishes as $t \to \infty$, that is,

$$\lim_{t \to \infty} \dot{x} = 0 \implies A \cdot x = 0 \implies x = 0 \tag{1.56}$$

As with discrete-time systems, we can now prove that the stability of the equilibrium $x = 0$ depends on the eigenvalues of the matrix A. Following the same approach already used for discrete-time systems, the unforced system evolution can be expressed as a function of its eigenvalues. To begin with, consider the case in which the initial condition coincides with the ith eigenvector $x_0 = v_i$. Expanding

---

TABLE 1.4

Properties of the Transition Matrix $\Phi(t_1, t_2) = e^{A(t_2 - t_1)}$ for Continuous-Time Systems

Property	Definition	Proof
Identity	$\Phi(t, t) = I$	$e^{A(t-t)} = e^0 = I$
Composition	$\Phi(t_1, t_2) \cdot \Phi(t_2, t_3) = \Phi(t_1, t_3) \quad t_1 > t_2 > t_3$	$e^{A(t_2 - t_1)} \cdot e^{A(t_3 - t_2)} = e^{A(t_2 - t_1 + t_3 - t_2)} = e^{A(t_3 - t_1)}$
Reversibility	$\Phi^{-1}(t_1, t_2) = \Phi(t_2, t_1)$	$\left[e^{A(t_2 - t_1)} \right]^{-1} = e^{A(t_1 - t_2)} = e^{-A(t_2 - t_1)}$
Commutability	$A \cdot \Phi(t) = \Phi(t) \cdot A$	$A \cdot e^{At} = A \cdot \left(I + A \cdot t + \frac{1}{2!}A^2 \cdot t^2 + \ldots \right)$
		$= A + A^2 \cdot t + \frac{1}{2!}A^3 \cdot t^2 + \ldots$
		$= \left(I + A \cdot t + \frac{1}{2!}A^2 \cdot t^2 + \ldots \right) \cdot A = e^{At} \cdot A$

the transition matrix as in Equation 1.55, and remembering that if v_i is an eigenvector it satisfies the equation $Av_i = \lambda_i v_i$, then $A^k v_i = \lambda^k v_i$ for any integer $k > 0$ and

$$x_i = e^{At} \cdot v_i$$

$$= x_i = e^{At} v_i = v_i + A v_i t + \frac{1}{2!} A^2 v_i t^2 + \frac{1}{3!} A^3 v_i t^3 + \cdots \tag{1.57}$$

$$= v_i + \lambda_i v_i t + \frac{1}{2!} \lambda_i^2 v_i t^2 + \frac{1}{3!} \lambda_i^3 v_i t^3 + \cdots = \sum_{k=0}^{\infty} \frac{\lambda_i^k t^k}{k!} v_i = e^{\lambda_i t} v_i$$

Equation 1.57 shows that if the initial state coincides with an eigenvector, its evolution will depend on that eigenvector v_i alone, with the corresponding eigenvalue λ_i acting as a modulating coefficient. This result is the continuous-time equivalent of Equation 1.41 in the discrete-time case. Now, since the eigenvectors form a basis in $\mathbb{R}^n$, the state x can be expressed as a linear combination of the eigenvectors, that is,

$$x = \beta_1 v_1 + \beta_2 v_2 + \cdots + \beta_n v_n = \sum_{i=1}^{n} \beta_i v_i \tag{1.58}$$

Substituting Equation 1.57 yields

$$x = \sum_{i=1}^{n} \beta_i x_i = \beta_1 e^{\lambda_1 t} v_1 + \beta_2 e^{\lambda_2 t} v_2 + \cdots + \beta_n e^{\lambda_n t} v_n = \sum_{i=1}^{n} \beta_i e^{\lambda_i t} v_i \tag{1.59}$$

where the coefficients β_i depend on the initial conditions. Equation 1.59 is the continuous-time equivalent of Equation 1.40 and shows that the behaviour of the autonomous system is totally determined by the eigenvalues of the system matrix A, which can be computed again with Equation 1.44

$$\det(\lambda I - A) = 0 \tag{1.60}$$

giving rise to relations between eigenvalues and system responses similar to those established for discrete-time systems. However, given the exponential form of the transition matrix e^{At}, the conclusions will be radically different. With reference to the complex plane $\lambda = \sigma \pm j\omega$, the stability boundary is given in this case by the imaginary axis $(\pm j\omega)$. In fact, any eigenvalue with negative real part will give an exponentially damped contribution in Equation 1.59

$$e^{-(\sigma \pm j\omega)t} \tag{1.61}$$

while an eigenvalue with positive real part will produce an unbounded term

$$e^{(\sigma \pm j\omega)t} \tag{1.62}$$

By similar considerations that have led to Table 1.3 and Figure 1.25, the following correspondences between eigenvalues position in the complex plane and system response can be determined for a continuous-time system, as shown in Figure 1.28 and Table 1.5. In general, the following conclusions can be drawn:

- An eigenvalue with positive real part $(\sigma > 0)$ produces an unstable behaviour $(e^{\sigma t})$.
- An eigenvalue with negative real part $(\sigma < 0)$ produces a stable behaviour $(e^{-\sigma t})$.

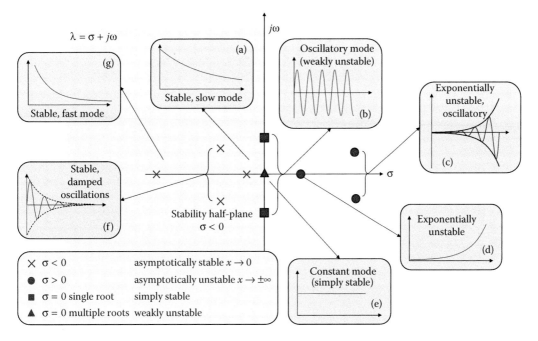

FIGURE 1.28 Correspondence between eigenvalue location in the complex plane and autonomous continuous-time system response. The stability region is the left-hand half complex plane (light blue area).

TABLE 1.5

Correspondence between Eigenvalue Position and System Response for a Continuous-Time System as Shown in Figure 1.28

Case	Eigenvalue	$(\lambda = \sigma \pm j\omega)$	System Response
a	Negative real small	$\lambda = -\sigma$	Monotonic, slowly decreasing
b	Imaginary conjugate	$\lambda = \pm j\omega$	Sustained bounded oscillations (weakly unstable)
c	Complex conjugate with positive real part	$\lambda_{1,2} = \sigma \pm j\omega$	Unbounded oscillations (strongly unstable)
d	Positive real	$\lambda = \sigma$	Unbounded monotonic growth
e	Zero	$\lambda = 0$	Constant response (weakly unstable)
f	Complex conjugate with negative real part	$\lambda_{1,2} = -\sigma \pm j\omega$	Damped oscillations
g	Negative real large	$\lambda = -\sigma$	Monotonic, fast decreasing

---

- A real eigenvalue with multiplicity k greater than 2 produces a contribution $t^{k-1}e^{\lambda t}$.
- A single couple of purely imaginary eigenvalues $(\pm j\omega)$ produces a sustained bounded oscillation.
- Multiple purely imaginary eigenvalues $(\pm j\omega)$ produce an oscillatory unstable behaviour.

The case of multiple roots is further clarified in Figure 1.29 showing three different cases of multiple eigenvalues, depending on the sign of the real part. If the eigenvalue is stable $(\sigma < 0)$, then as its multiplicity k increases, so does the *hump* of the response. This kind of behaviour will often be encountered when modelling cascaded reactors, which are often convenient approximations of hydraulic systems, an example of which is treated in Section 1.8.3.1. The other two cases are *pathological* situations of unstable systems and are not normally encountered in models of environmental systems.

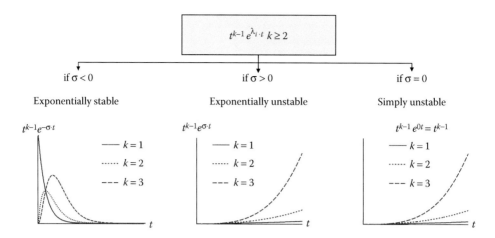

FIGURE 1.29 Response produced by an eigenvalue of multiplicity k, depending on its real part.

1.8.2 FORCED LTI CONTINUOUS-TIME SYSTEM

Having studied the intrinsic system property of stability, we now reintroduce the input term in the first equation (1.52) and seek its formal solution, which is similar to Equation 1.47 with the difference that in place of the summation, we now have a *convolution integral* representing the input contribution

$$x(t) = \underbrace{e^{At} x_0}_{\text{unforced response}} + \underbrace{\int_0^t e^{A(t-\sigma)} bu(\sigma)\, d\sigma}_{\text{input forcing}} \tag{1.63}$$

As with discrete-time LTI system, linearity allows for the superposition of the effects, so that the first contribution, controlled by the initial conditions x_0 and the matrix A, is summed to the input contribution, represented by the convolution integral. Equation 1.63, like its discrete-time version Equation 1.47, is conceptually important to investigate the system properties, but is hardly ever used to determine the system response in practice because the input u may not have a simple mathematical form, and therefore, the integral may not have an analytical solution. As will be fully treated in the following, in most cases numerical methods will be preferred to simulate the system behaviour. However, in very simple cases (e.g. a constant input), Equation 1.63 can actually be solved. Let us consider a single-input system with a constant value u_0. In this case, the convolution integral in Equation 1.63 can be handled via Taylor series expansion of the transition matrix, whereby each term can be separately integrated

$$\int_0^t e^{A(t-\sigma)} bu_0\, d\sigma = \int_0^t e^{At} e^{-A\sigma} bu_0\, d\sigma = e^{At} \int_0^t e^{-A\sigma} bu_0\, d\sigma$$

$$= e^{At} \int_0^t \left(I - A\sigma + \frac{A^2\sigma^2}{2!} - \frac{A^3\sigma^3}{3!} + \cdots \right) bu_0\, d\sigma \tag{1.64}$$

$$= e^{At} \left(\int_0^t I\, d\sigma - \int_0^t A\sigma\, d\sigma + \int_0^t \frac{A^2\sigma^2}{2!}\, d\sigma - \int_0^t \frac{A^3\sigma^3}{3!}\, d\sigma + \cdots \right) bu_0$$

1.8.3 SOME SIMPLE RESPONSE CALCULATIONS: STEP RESPONSE AND IMPULSE RESPONSE

Now, consider the even simpler case in which the system matrix A reduces to a scalar, that is, $A = -a$. Thus, the system dynamics reduces to the scalar differential equation

$$\dot{x} = -a \cdot x + b \cdot u_0 \tag{1.65}$$

Assuming that the initial condition is zero, that is, $x_0 = 0$, the system (1.65) can be thought to be driven abruptly from a zero steady state $(x = 0)$ to another steady state determined by the constant input u_0 applied at $t = 0+$. This situation is often referred to as *step response*, meaning that the input abruptly changes from 0 to u_0 in a step-wise mode. The first term in Equation 1.63 vanishes, and the convolution integral becomes

$$x(t) = \int_0^t e^{-a(t-\sigma)} b u_0 \, d\sigma = e^{-at} b u_0 \int_0^t e^{a\sigma} \, d\sigma = e^{-at} b u_0 \left. \frac{e^{a\sigma}}{a} \right|_0^t$$

$$= e^{-at} b u_0 \left(\frac{e^{at} - 1}{a} \right) = \frac{b u_0}{a} \left(1 - e^{-at} \right) \tag{1.66}$$

The response of this system is graphically presented in Figure 1.30, showing that the steady-state value depends not only on the input value u_0 but also on the system parameters a and b, whereas the speed with which the asymptotic value is reached depends only on a, the system time constant, as shown by Equation 1.66.

Now consider again the system (1.65), but this time the input is represented by a Dirac $\delta(t)$ impulse, which can be defined as follows:

$$\text{(i)} \quad \delta(t) = 0 \quad \forall t \neq 0$$

$$\text{(ii)} \quad \delta(0) = \infty \tag{1.67}$$

$$\text{(iii)} \quad \int_{-\infty}^{\infty} \delta(t) \, dt = \int_0^{0+} \delta(t) \, dt = 1$$

As shown in Figure 1.31, the Dirac impulse can be viewed as the limit of a *distribution of rectangular pulses* of duration a and amplitude $1/a$. When a tends to zero, the amplitude tends to infinity but the strength of the impulse (height times duration) is still unit.

Basically, the Dirac impulse is non-zero only at $t = 0$ and has unit area. Now suppose again that the system starts from zero initial condition, so that its evolution is entirely determined by the convolution integral, which can be solved by using the above properties (1.67) of the Dirac pulse

$$x(t) = \int_0^t e^{A(t-\sigma)} b \delta(\sigma) d\sigma = e^{At} b \int_0^{0+} e^{-A\sigma} \delta(\sigma) d\sigma$$

$$= e^{At} b \int_0^{0+} \delta(\sigma) d\sigma = e^{At} b \tag{1.68}$$

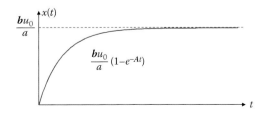

FIGURE 1.30 Response of the system (1.65) to a constant input u_0 starting with zero initial condition.

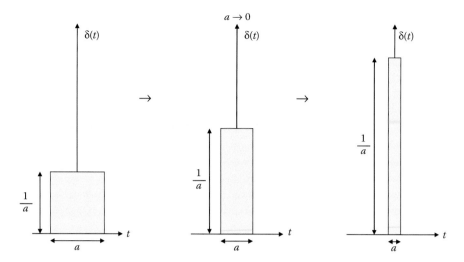

FIGURE 1.31   The Dirac impulse can be viewed as a *distribution of pulses* of constant unit area, but decreasing duration.

Therefore, the state and output of the system are

$$x(t) = e^{At} b$$

$$y(t) = c\, e^{At}\, b \tag{1.69}$$

By inspection of Equation 1.69, it appears that the application of a Dirac impulse is equivalent to altering the initial system condition from **0** to **b**, the input matrix. From then on, the subsequent evolution is that of an autonomous system starting from the new initial condition **b**.

1.8.3.1   An Application of an Impulse-Drive Linear Model: The Nash Reservoir

A simple yet effective linear model is due to the famous mathematician John Nash, who inspired the successful movie *A Beautiful Mind* starring Russell Crowe in the title role and who was killed in a car crash in May 2015. Nash developed a simple model of a reservoir by cascading a certain number of elementary reservoirs (Rinaldi et al., 1979), as illustrated in Figure 1.32. The system states are the volumes of each tank and the outflow is proportional to the last volume, that is, $q = k \cdot x_n$. The dynamics of each reservoir is given by the difference between the incoming and outgoing flows, similar to the example of Section 1.3.1

$$\dot{x}_i = k \cdot x_{i-1} - k \cdot x_i \tag{1.70}$$

The system representation of the ensemble of n tanks in series is therefore

$$\begin{bmatrix} \dot{x}_1 \\ \dot{x}_2 \\ \dot{x}_3 \\ \dots \\ \dot{x}_n \end{bmatrix} = \begin{bmatrix} -k & 0 & 0 & \dots & 0 \\ k & -k & 0 & \dots & 0 \\ 0 & k & -k & \dots & 0 \\ \dots & \dots & \dots & \dots & 0 \\ 0 & 0 & 0 & \dots & -k \end{bmatrix} \cdot \begin{bmatrix} x_1 \\ x_2 \\ x_3 \\ \dots \\ x_n \end{bmatrix} + \begin{bmatrix} 1 \\ 0 \\ 0 \\ \dots \\ 0 \end{bmatrix} \cdot u \tag{1.71}$$

$$q = \begin{bmatrix} 0 & 0 & 0 & \dots & k \end{bmatrix} \cdot x$$

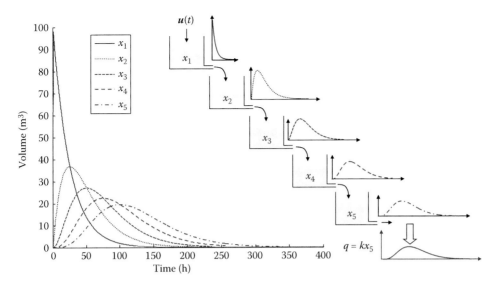

FIGURE 1.32 Schematic representation of the Nash reservoir model, consisting of a number of cascaded tanks. The volume variations in response to a flow pulse are shown in the left-hand diagram.

To obtain the impulse response of the system (1.71), we can use the result of Equation 1.69 and initialize the system with the initial condition $x(0) = b$ appearing in Equation 1.71. Considering as an example, the cascade of five reservoirs, the state impulse response is shown in Figure 1.32, whereas the output variations for several values of the reservoir constant k are shown in Figure 1.33. The analysis of the system can be carried a step further. Consider again the output impulse response in the form of the second Equation 1.69 and notice that the matrix A of the system (1.71) is lower triangular; therefore, it has n eigenvalues, all equal to $-k$. Thus, we can express the impulse response as

$$y(t) = c\, e^{At}\, b = \frac{1}{T(n-1)!} \left(\frac{t}{T}\right)^{n-1} e^{-\frac{t}{T}} \quad \text{where } T = \frac{1}{k} \tag{1.72}$$

The time of the maximum flow can be computed by setting the output derivative to zero and solving for t to yield

$$t_{\max} = (n-1)T \tag{1.73}$$

This result will be considered again in Chapter 6, when dealing with the impulse response of real reactors. As a concluding remark of this section, analysing the system response to apparently

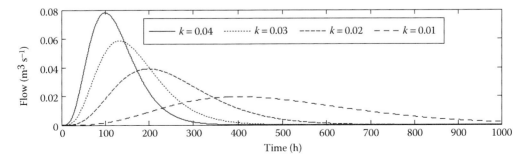

FIGURE 1.33 Impulse response of the Nash reservoir model for several values of the flow constant k.

unrealistic inputs might appear to be an arid academic exercise. The contrary is in fact true because step and impulse inputs (or their real-world approximations) will be often encountered in practice, and the notions learned here will undoubtedly become useful in due course.

1.8.4 STEADY STATE OF LTI CONTINUOUS-TIME SYSTEMS WITH CONSTANT INPUT

Let us now consider again the basic system Equation 1.52 and determine its steady state in the case of constant input. A necessary condition for the equilibrium is that the time derivative vanishes, that is, $\dot{x} = 0$, so that the system Equation 1.52 becomes

$$0 = A \cdot \bar{x} + B \cdot \bar{u} \quad \Rightarrow \quad \bar{x} = -A^{-1} \cdot B \cdot \bar{u} \tag{1.74}$$

As an example, consider the following system:

$$\begin{bmatrix} \dot{x}_1 \\ \dot{x}_2 \end{bmatrix} = \begin{bmatrix} 0 & 1 \\ -0.3 & -1.1 \end{bmatrix} \times \begin{bmatrix} x_1 \\ x_2 \end{bmatrix} + \begin{bmatrix} 1 \\ 2 \end{bmatrix} \times u \tag{1.75}$$

The system is stable because the eigenvalues of the system matrix A are negative real ($\lambda_1 = -0.5$; $\lambda_2 = -0.6$). Selecting $\bar{u} = 0.5$, the steady state can be computed according to Equation 1.74 as

$$\bar{x} = -A^{-1}bu = -\underbrace{\begin{bmatrix} -3.6667 & -3.3333 \\ 1 & 0 \end{bmatrix}}_{A^{-1}} \times \underbrace{\begin{bmatrix} 1 \\ 2 \end{bmatrix}}_{b} \times \underbrace{(0.5)}_{\bar{u}} = \begin{bmatrix} 5.1667 \\ -0.5 \end{bmatrix} \tag{1.76}$$

Figure 1.34 shows the time behaviour (a) and the corresponding PPP (b) of the system (1.75) in reaching the steady state computed via Equation 1.76.

1.8.5 PHASE-PLANE BEHAVIOUR OF TWO-DIMENSIONAL CONTINUOUS-TIME LTI SYSTEMS

This analysis is the continuous-time counterpart of that presented in Section 1.7.2 for two-dimensional systems. Plotting the two components of the state one versus the other is an effective graphical method to analyse the system behaviour. As a motivation for considering this tool, let us anticipate that the stability of many two-dimensional ecosystems, like population dynamics, can be

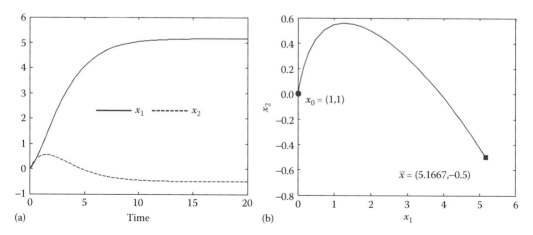

FIGURE 1.34 Time plot (a) and PPP (b) of the system (1.75) to reach the steady state $\bar{x}$ of Equation 1.76 from the initial condition x_0 when driven by a constant input $\bar{u} = 0.5$.

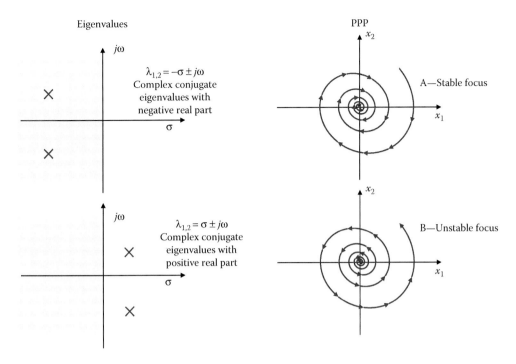

FIGURE 1.35   PPP in the case of complex conjugate eigenvalues with either negative or positive real part.

handled very efficiently with this method. To get a *feeling* of the effect of each system eigenvalue on its evolution, let us reconsider the time portraits of Figure 1.28. So, let us now consider again a two-dimensional unforced system

$$\begin{bmatrix} \dot{x}_1 \\ \dot{x}_2 \end{bmatrix} = \begin{bmatrix} a_{11} & a_{12} \\ a_{21} & a_{22} \end{bmatrix} \times \begin{bmatrix} x_1 \\ x_2 \end{bmatrix} \tag{1.77}$$

The following figures show the correspondence between the location of the eigenvalues in the complex plane and the system evolution in the phase plane. In the case of complex conjugate eigenvalues (Figure 1.35), using complex trigonometry, the unforced response can be shown to be an exponentially damped sinusoid

$$x(t) = e^{\lambda_1 t} + e^{\lambda_2 t} = e^{-\sigma t + j\omega t} + e^{-\sigma t - j\omega t} = e^{-\sigma t}\left(e^{j\omega t} + e^{-j\omega t}\right) = e^{-\sigma t}\, 2\cos(\omega t) \tag{1.78}$$

where the damping rate is controlled by the real part $(e^{-\sigma t})$ and the oscillation has a period equal to $T = \frac{2\pi}{\omega}$. So the system trajectory, called a *focus,* is a spiral that converges to the origin if $\sigma < 0$ or diverges if $\sigma > 0$.

Figure 1.36 shows the behaviour produced by two real eigenvalues: if the real part is negative, then the trajectories will tend to the origin *without encircling it* because the oscillatory term is missing $(\omega = 0)$.

The cases presented in Figure 1.37 are surely the most interesting: in the case of real eigenvalues with opposite signs, the unstable one will eventually prevail making the whole system unstable. In fact, the corresponding unstable eigenvector (v_2) will *attract* the trajectories, whereas the contribution of the stable one (v_1) will tend to zero. We shall encounter a *saddle* case when dealing with the steady-state regime of diffusive reactors, in Chapter 6.

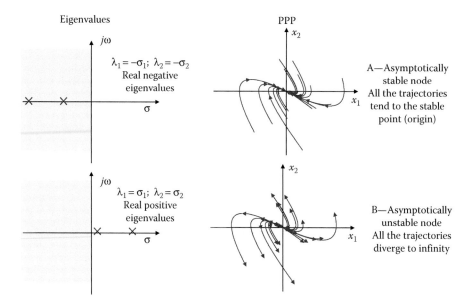

FIGURE 1.36   PPP in the case of real eigenvalues either both negative or positive.

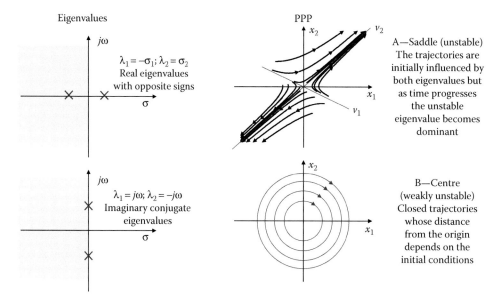

FIGURE 1.37   PPP in the case of real eigenvalues with opposite signs or in the case of purely imaginary conjugate eigenvalues.

1.9   NONLINEAR SYSTEMS

Though linear systems can explain many natural phenomena, the majority of natural systems are governed by nonlinear laws, for which we need other analysis tools. Consequently, it is necessary to include here a brief survey of nonlinear systems, recalling the notions that will be used in the following. For a full treatment of nonlinear systems, the reader is referred to specialized textbooks (Cook, 1986; Drazin, 1992; Kolk and Lerman, 1992; Strogatz, 1994; Kaplan and Glass, 1995). It appears that there is no better way of defining a nonlinear system than just *a system that is not linear*. This negative definition does not shed much light onto the subject, and the first attempt at studying a nonlinear system is usually to try to approximate it with a linear system. This approach has the merit of enabling the use of many of the tools that were described in Section 1.5, though we shall see that this approach has its limits. Thus, the first technique that will be introduced here is the *linearization* method.

1.9.1 LINEARIZATION OF A NONLINEAR SYSTEM

Given a generic nonlinear dynamics, either discrete-time or continuous-time,

$$\text{Discrete-time} \begin{cases} x_{t+1} = f(x_t, u_t) \\ y_t = g(x_t, u_t) \end{cases} \quad \text{continuous-time} \begin{cases} \dot{x} = f(x, u) \\ y = g(x, u) \end{cases} \quad \begin{matrix} x \in \mathbb{R}^n \\ u \in \mathbb{R}^m \\ y \in \mathbb{R}^q \end{matrix} \tag{1.79}$$

linearization means finding an approximating LTI system in the neighbourhood of an equilibrium (or fixed) point $(\bar{x}, \bar{u})$, where Equation 1.79 becomes

$$\text{d.t.} \begin{cases} \bar{x} = f(\bar{x}, \bar{u}) \\ \bar{y} = g(\bar{x}, \bar{u}) \end{cases} \quad \text{c.t.} \begin{cases} 0 = f(\bar{x}, \bar{u}) \\ \bar{y} = g(\bar{x}, \bar{u}) \end{cases} \tag{1.80}$$

Let us consider the discrete-time case first. Decomposing each quantity into its equilibrium and incremental part, that is, $x_t = \bar{x} + \tilde{x}_t$ and $u_t = \bar{u} + \tilde{u}_t$, a Taylor series expansion of the first of Equation 1.79 around the fixed point $(\bar{x}, \bar{u})$ yields

$$x_{t+1} = \bar{x} + \tilde{x}_{t+1} \cong f(\bar{x}, \bar{u}) + \left.\frac{\partial f}{\partial x}\right|_{(\bar{x}, \bar{u})} \tilde{x}_t + \left.\frac{\partial f}{\partial u}\right|_{(\bar{x}, \bar{u})} \tilde{u}_t + O_2(\tilde{x}_t) \tag{1.81}$$

which, neglecting the second-order terms $O_2(\tilde{x}_t)$ and noticing that the two terms $\bar{x}$ and $f(\bar{x}, \bar{u})$ cancel out at equilibrium, simplifies into

$$\tilde{x}_{t+1} = \left.\frac{\partial f}{\partial x}\right|_{(\bar{x}, \bar{u})} \tilde{x}_t + \left.\frac{\partial f}{\partial u}\right|_{(\bar{x}, \bar{u})} \tilde{u}_t \tag{1.82}$$

In Equation 1.82, the two matrices

$$J = \left.\frac{\partial f}{\partial x}\right|_{(\bar{x}, \bar{u})} = \begin{bmatrix} \dfrac{\partial f_1}{\partial x_1} & \dfrac{\partial f_1}{\partial x_2} & \cdots & \dfrac{\partial f_1}{\partial x_n} \\ \dfrac{\partial f_2}{\partial x_1} & \dfrac{\partial f_2}{\partial x_2} & \cdots & \dfrac{\partial f_2}{\partial x_n} \\ \cdots & \cdots & \cdots & \cdots \\ \dfrac{\partial f_n}{\partial x_1} & \dfrac{\partial f_n}{\partial x_2} & \cdots & \dfrac{\partial f_n}{\partial x_n} \end{bmatrix}_{(\bar{x}, \bar{u})} \in \mathbb{R}^{n \times n} \quad \left.\frac{\partial f}{\partial u}\right|_{(\bar{x}, \bar{u})} = \begin{bmatrix} \dfrac{\partial f_1}{\partial u_1} & \dfrac{\partial f_1}{\partial u_2} & \cdots & \dfrac{\partial f_1}{\partial u_m} \\ \dfrac{\partial f_2}{\partial u_1} & \dfrac{\partial f_2}{\partial u_2} & \cdots & \dfrac{\partial f_2}{\partial u_m} \\ \cdots & \cdots & \cdots & \cdots \\ \dfrac{\partial f_n}{\partial u_1} & \dfrac{\partial f_n}{\partial u_2} & \cdots & \dfrac{\partial f_n}{\partial u_m} \end{bmatrix}_{(\bar{x}, \bar{u})} \in \mathbb{R}^{n \times m} \tag{1.83}$$

determine the dynamics of the approximating linear system in the neighbourhood of the equilibrium $(\bar{x}, \bar{u})$. The output equation can be linearized in a similar manner to yield

$$\tilde{y}_t = \left.\frac{\partial g}{\partial x}\right|_{(\bar{x}, \bar{u})} \tilde{x}_t + \left.\frac{\partial g}{\partial u}\right|_{(\bar{x}, \bar{u})} \tilde{u}_t \tag{1.84}$$

In the continuous-time case,

$$\frac{d(\bar{x} + \tilde{x})}{dt} \cong f(\bar{x}, \bar{u}) + \left.\frac{\partial f(x, u)}{\partial x}\right|_{(\bar{x}, \bar{u})} \tilde{x} + \left.\frac{\partial f(x, u)}{\partial u}\right|_{(\bar{x}, \bar{u})} \tilde{u} + O_2(\tilde{x}) \tag{1.85}$$

As with Equation 1.81, being $\frac{d\bar{x}}{dt} = f(\bar{x},\bar{u}) = 0$ according to Equation 1.80 and neglecting the higher order terms of the expansion O_2, Equation 1.85 simplifies into

$$\dot{\tilde{x}} = \left.\frac{\partial f(x,u)}{\partial x}\right|_{(\bar{x},\bar{u})} \tilde{x} + \left.\frac{\partial f(x,u)}{\partial u}\right|_{(\bar{x},\bar{u})} \tilde{u} \tag{1.86}$$

which is the linear approximation of the continuous-time nonlinear system (1.79) in the neighbourhood of the fixed point $(\bar{x},\bar{u})$, formally equivalent to the discrete-time linearization of Equation 1.82. The output equation can be similarly linearized by a Taylor series expansion

$$\tilde{y} = \left.\frac{\partial g(x,u)}{\partial x}\right|_{(\bar{x},\bar{u})} \tilde{x} + \left.\frac{\partial g(x,u)}{\partial u}\right|_{(\bar{x},\bar{u})} \tilde{u} \tag{1.87}$$

Obviously, the validity of the two linearized system Equations 1.82 and 1.87 is limited to the small perturbations $(\tilde{x},\tilde{u})$ around the equilibrium $(\bar{x},\bar{u})$, which determine the system matrices, in particular of the system matrix J, often referred to as the *Jacobian* of the system, and formally defined by Equation 1.83.

1.9.2 STABILITY OF NONLINEAR SYSTEMS

The notion of stability has a much wider meaning for nonlinear systems than for linear systems and depends on the equilibria $(\bar{x},\bar{u})$, which may be more than one. Considering an autonomous system, a general definition of stability is given as follows.

An equilibrium $\bar{x}$ is (locally) asymptotically stable if given any initial condition x_0 for which $|x_0 - \bar{x}| < \varepsilon$, there exists a constant $\delta > 0$ for which $|x - \bar{x}| < \delta$, asymptotically meaning that $\lim_{t\to\infty} x(t) = \bar{x}$.

$$|x_0 - \bar{x}| < \varepsilon \quad \exists \delta : |x - \bar{x}| < \delta \quad \text{with } \varepsilon, \delta > 0 \tag{1.88}$$

As Figure 1.38a shows, this sufficient condition implies that any trajectory originating in the neighbourhood of the equilibrium remains confined in its vicinity, in spite of small perturbations that tend to displace x from the equilibrium $\bar{x}$. Figure 1.38b portrays the *domain of attraction* $B(\bar{x})$ of the equilibrium $\bar{x}$, defined as the region including all the initial conditions x_0 that produce trajectories converging into $\bar{x}$.

$$B(\bar{x}) \triangleq \left\{ \forall x_0 : \lim_{t\to\infty} x(t) \to \bar{x} \right\} \tag{1.89}$$

In general, an autonomous nonlinear system may have more than one equilibrium, each with its own domain of attraction.

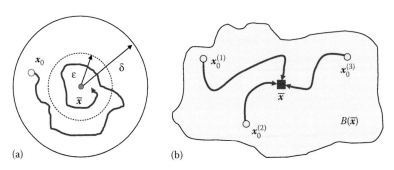

(a)   (b)

FIGURE 1.38 Graphical representation of a stable equilibrium (a) and of its domain of attraction (b).

1.9.3   Local Stability Analysis of Autonomous Nonlinear Systems

The linearization just introduced can be used to determine the local system stability by using the tools already developed for linear systems in Sections 1.7 and 1.8.1. If the Jacobian matrix is non-singular, some conclusions can be drawn by examining its eigenvalues. The non-singularity condition is necessary for the linear term to be dominant and its behaviour to be determined by its eigenvalues; otherwise, the system behaviour might be controlled by the higher order terms of the Taylor expansion and the linear approximation would be meaningless.

If J is non-singular, then some conclusions about stability can be drawn by eigenvalues inspection, as summarized in Table 1.6. It can be seen that the linear analysis is far from exhaustive, because many important cases may remain unanswered when the eigenvalues are placed on the boundary of the (linear) stability region. Moreover, the possibility of more than one equilibrium, each with its own domain of attraction, gives the analysis only a local validity. For the cases in which the linear analysis is conclusive and for two-dimensional systems, the characterization of the equilibrium in terms of the eigenvalues position follows the classification of Figures 1.35 through 1.37.

1.9.4   Limit Cycles

Unlike their linear counterpart, autonomous nonlinear systems may have other equilibria in the form of closed curves, which correspond to a periodic solution that retraces the same trajectory indefinitely. The limit cycle, occurring for $(\sigma_i = 0)$, is the typical equilibrium for which the linear analysis fails, but it should not be confused with the sustained bounded oscillations examined in Table 1.3 case g, for discrete-time LTI systems, and in Table 1.5 case b, for continuous-time LTI systems, shown in Figure 1.37b. In fact, the two behaviours are radically different: in the linear case, the amplitude of the cycle depends on the initial conditions, whereas the limit cycle is a *structural property*, which has its own shape irrespective of the initial conditions. Moreover, a limit cycle is an *isolated* closed trajectory, in the sense that neighbouring trajectories will either converge to it (*stable limit cycle*) or diverge from it (*unstable limit cycle*).

In Figure 1.39a, the stable limit cycle (left) attracts all of the trajectories, both from the inside and the outside. Conversely, an unstable limit cycle Figure 1.39b will repel any trajectory, so that all outside trajectories will tend to infinity, but what about the inner trajectories? They must leave the unstable limit cycle but cannot terminate in $\bar{x}$ because this is an unstable fixed point. So inside an unstable limit cycle, there must exist another stable limit cycle collecting the trajectories escaping from the outer unstable limit cycle. Conversely, a stable limit cycle may contain an unstable one (right), in which case the inner trajectories repelled by the inner unstable limit cycle will

---

TABLE 1.6

Classification of Stability by System Linearization

Eigenvalues of J		$\dot{\bar{x}} = J(\bar{x}) \cdot \tilde{x}$	$\bar{x}$
Discrete-Time	**Continuous-Time**		
$\|\lambda_i\| < 1$	$\sigma_i < 0$	Asymptotically stable	Asymptotically stable
$\|\lambda_i\| < 1$	$\sigma_i > 0$	Asymptotically unstable	Asymptotically unstable
$\|\lambda_i\| = 1$	$\sigma_i = 0$	Simply stable	No conclusion
Simple eigenvalue	Simple eigenvalue		
$\|\lambda_i\| = 1$	$\sigma_i = 0$	Weakly unstable	No conclusion
At least one multiple eigenvalue	At least one multiple eigenvalue		

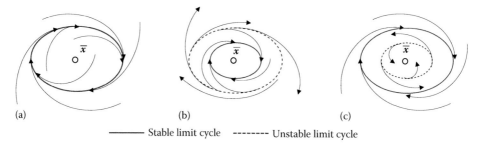

(a)   (b)   (c)

———— Stable limit cycle   - - - - - - Unstable limit cycle

FIGURE 1.39 Limit cycles as periodic solutions of a nonlinear autonomous system associated with the equilibrium $\bar{x}$. The limit cycles may be stable and isolated (a), but if there is an unstable limit cycle, then there must be an inner stable limit cycle to collect the inbound escaping trajectories (b), or an outer stable limit cycle (c). The thin lines are sample trajectories either converging to a limit stable cycle or leaving it, if unstable.

terminate in $\bar{x}$ itself, which is now a stable fixed point, as shown in Figure 1.39c. Limit cycles are very important in Nature because they describe systems that produce self-sustained oscillations, even in the absence of external periodic inputs, such as the cardiac rhythm, or the fluctuations often observed in population dynamics.

1.9.5 ISOCLINES

A powerful graphical tool for stability analysis, most often used with two-dimensional continuous-time autonomous systems, is the method of *isoclines*. As the word implies, *isoclines* are *constant slope* loci in the phase plane, which are intersected by the system trajectories with a prescribed slope. Taking the ratio between the two state equations of a two-dimensional system yields the definition of slope in the phase plane:

$$
\begin{aligned}
\frac{dx_1}{dt} &= f_1(x_1, x_2) \\
\frac{dx_2}{dt} &= f_2(x_1, x_2)
\end{aligned}
\Rightarrow \quad \text{slope}: \frac{dx_2}{dx_1} = \frac{\dfrac{dx_2}{dt}}{\dfrac{dx_1}{dt}} \Rightarrow
\begin{cases}
\text{horizontal isocline} & \dfrac{dx_2}{dt} = 0 \\[2mm]
\text{vertical isocline} & \dfrac{dx_1}{dt} = 0
\end{cases}
\tag{1.90}
$$

Of particular importance are the isoclines on which the derivative of either state variable vanishes (dashed lines in plots b and c in Figure 1.40). In general, there is no reason why the isoclines should be straight lines. Their intersection, if it exists, determines the system equilibrium because at that point both derivatives are zero. In Chapter 5, extensive use of the isoclines will be made, showing that there may be more than one intersection point, corresponding to multiple equilibria.

1.9.6 GRAPHICAL STABILITY ANALYSIS OF ONE-DIMENSIONAL AUTONOMOUS SYSTEMS

Another graphical method for analysing the stability of one-dimensional autonomous systems is now considered. The rationale for this approach is that an equilibrium must satisfy both the system dynamics and the quiescent condition.

1.9.6.1 Graphical Stability of One-Dimensional Discrete-Time Systems

Let us begin by considering a discrete-time system. Equilibrium means that the following two equations must be simultaneously satisfied:

$$
\begin{cases}
x_{t+1} = f(x_t) & \text{system dynamics} \\
x_{t+1} = x_t & \text{equilibrium condition}
\end{cases}
\tag{1.91}
$$

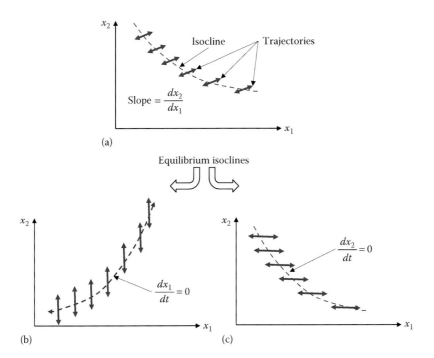

FIGURE 1.40 Definition of isoclines (dashed curves) as constant slope loci; (a) shows an example of a generic isocline, whereas (b) and (c) show equilibria isoclines where the derivative of either x_1 or x_2 vanishes.

So, we might just as well drop the t index, because at the equilibrium both of Equation 1.91 become independent of time. Figure 1.41 is a graphical representation of Equation 1.91, and the equilibrium point is represented by a dot at the intersection between the function $f(x_t)$ representing the system dynamics and the static assumption $x_{t+1} = x_t$. Generally, dealing with ecological quantities, we are interested only in positive values of the variable, and therefore, we seek equilibria in the first quadrant $(x > 0)$. The following method, introduced by May (1976a, b, 2001) in connection with population dynamics, can be used to assess the stability of a scalar nonlinear discrete-time system. Figure 1.41a shows the graphical location of the equilibrium at the intersection of the two equalities in Equation 1.91, whereas Figure 1.41b shows how an iteration can be started from the initial

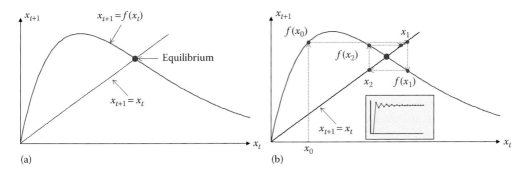

FIGURE 1.41 Graphical representation of equilibrium for a one-dimensional discrete-time system (a) and its stability analysis by successive iterations (b). The inset in (b) shows the time response of the system.

condition x_0 to evaluate the corresponding $f(x_0)$, which in turn yields $x_1 = f(x_0)$ and so on. The following sequence is thus generated:

$$x_0 \rightarrow f(x_0) \rightarrow x_1 \rightarrow f(x_1) \rightarrow x_2 \rightarrow f(x_2) \cdots \tag{1.92}$$

which converges to the equilibrium point only if the following inequality, the so-called Lipschitz condition (Athans et al., 1974), is satisfied

$$\left| \frac{f(x_{t+1}) - f(x_t)}{x_{t+1} - x_t} \right| = \left| \frac{\Delta f}{\Delta x} \right| < 1 \tag{1.93}$$

The inset in Figure 1.41b shows the time response of the system, whose damped oscillations terminate at the equilibrium.

Thus, stability can be assessed by testing the slope of the function $f(x)$ at the equilibrium point, as shown in Figure 1.42. Taking the limit of Equation 1.93 for $\Delta x \rightarrow 0$, the following stability conditions are obtained:

$$
\begin{array}{ll}
\text{Unstable equilibrium} & \lim_{\Delta x \to 0} \left| \frac{\Delta f}{\Delta x} \right| = \left| \frac{df}{dx} \right| > 1 \\[2mm]
\text{Limit cycle} & \lim_{\Delta x \to 0} \left| \frac{\Delta f}{\Delta x} \right| = \left| \frac{df}{dx} \right| = 1 \\[2mm]
\text{Stable equilibrium} & \lim_{\Delta x \to 0} \left| \frac{\Delta f}{\Delta x} \right| = \left| \frac{df}{dx} \right| < 1
\end{array}
\tag{1.94}
$$

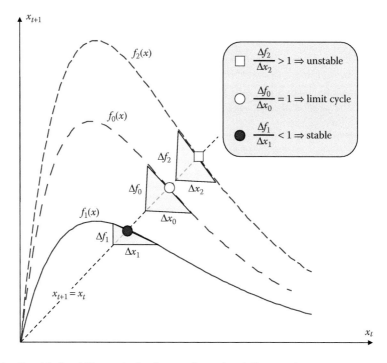

FIGURE 1.42 Graphical stability analysis of a one-dimensional discrete-time system.

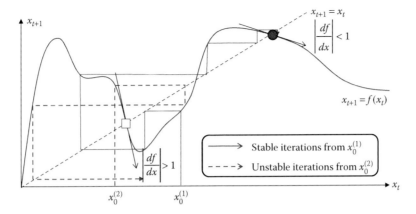

FIGURE 1.43 Graphical analysis of a stable (blue dot) equilibrium and an unstable one (white square). The solid lines show the iterations starting from $x_0^{(1)}$ converging to the stable equilibrium, whereas the dashed lines starting from $x_0^{(2)}$ diverge from the unstable equilibrium.

Because in principle there is no constraint on the shape of $f(x)$, there may be more than one intersection with the quiescent line $x_{t+1} = x_t$. Figure 1.43 shows a rather fanciful $f(x)$ with two equilibria, whose stability can be graphically analysed using Equation 1.94. It can be seen that the iterations relative to the unstable equilibrium (white square) diverge, whereas those around the stable point (blue dot) converge.

An interesting question regards the choice of the initial condition x_0 leading to either equilibrium. This brings us back again to the concept of the *domain of attraction* of the equilibrium, sketched in Figure 1.39. There is no general rule to determine its boundary, which must be obtained by numerical simulations.

It is interesting that limit cycles can be detected with this method because they correspond to iterations that are exactly superimposed. Numerically, this implies that $\left|\frac{df}{dx}\right| = 1$, as shown in Figure 1.44, where successive iteration retraces the same path indefinitely.

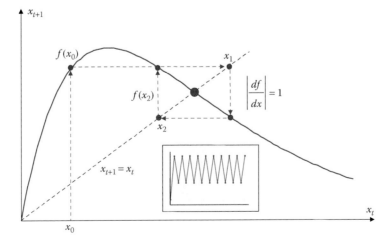

FIGURE 1.44 Graphical representation of the limit cycle condition. The time response is shown in the inset.

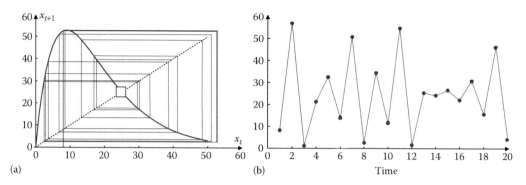

FIGURE 1.45 An example of a discrete-time system with a *chaotic* behaviour. The iteration portrait (a) shows that each step reaches a new point on $f(x)$ around the equilibrium point (white square) that is never reached. The corresponding time plot (b) is *apparently* random.

There is just one additional case to be examined and that is when the derivative of $f(x)$ is much larger than one. In this case, the iterations may span the whole quadrant in a chaotic manner, without ever passing by the same point twice, as shown in Figure 1.45. A chaotic trajectory may be mistaken for a random behaviour, but it is not so. In fact, whereas infinite realizations of a random process never replicate themselves exactly, the chaotic trajectory is *deterministic* because if initialized with the same starting condition, it will exactly replicate itself.

One-dimensional discrete-time systems have the peculiarity of giving rise to chaotic behaviour, whereas chaos can only develop in continuous-time system of at least order 3 (Strogatz, 1994). A remarkable example of a single discrete-time equation with chaotic behaviour will be studied next in the context of population dynamics (May, 1976a).

1.9.6.2 Graphical Stability of One-Dimensional Continuous-Time Systems

Now we turn to the analysis of *continuous-time* one-dimensional autonomous system, for which the equilibrium conditions, similar to Equation 1.91, are

$$0 = f(\bar{x}) \tag{1.95}$$

Thus, the equilibrium points are determined by the intersections of the system dynamics $f(x)$ with the real positive axis, instead of the 1:1 line as in the discrete-time case. Linearizing the system dynamics $\dot{x} = f(x)$ according to Equation 1.86 around an equilibrium point $\bar{x}$ defined by Equation 1.95 yields

$$\dot{\tilde{x}} = \left. \frac{\partial f}{\partial x} \right|_{\bar{x}} \tilde{x} \tag{1.96}$$

where $\left. \frac{\partial f}{\partial x} \right|_{\bar{x}}$ represents the one-dimensional Jacobian of the system, and therefore, the criteria of Table 1.6 apply, of course with the exception of complex conjugate solutions because the system is one dimensional and has therefore only one solution. So, we can conclude that the local stability of the equilibrium point can be decided by the following rules, similar to Equation 1.94:

$$
\begin{aligned}
\text{Stable equilibrium} \qquad & \frac{df}{dx} < 0 \\
&\\
\text{Unstable equilibrium} \qquad & \frac{df}{dx} > 0
\end{aligned}
\tag{1.97}
$$

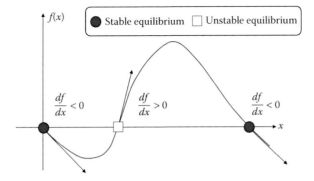

FIGURE 1.46   Graphical stability analysis of a one-dimensional autonomous continuous-time system.

while the linearization does not yield any definitive conclusion for $\frac{df}{dx} = 0$. This condition is graphically depicted in Figure 1.46 where $f(x)$ intersects the x-axis three times, two of which with negative derivative (blue dots), corresponding to a stable equilibrium, and one with positive derivative (white square), denoting instability. These concepts will be at the basis of the analysis of population dynamics considered in Chapter 5.

1.9.7   GLOBAL STABILITY ANALYSIS OF AUTONOMOUS NONLINEAR SYSTEMS

We have seen in Section 1.9.3 the limits of the local stability analysis based on linearization, also referred to as *Lyapunov first method*. Now, we define an equilibrium $\bar{x} \in \mathbb{R}^n$ to be *globally asymptotically stable* if all the trajectories originating in its domain of attraction, defined in Figure 1.38b, converge to $\bar{x}$. Although this notion has already been introduced, an operational definition based on the Lyapunov function will now be considered (Athans et al., 1974; Cook, 1986; Strogatz, 1994). Given a continuous-time system with an equilibrium at the origin (the equilibrium can be moved to the origin by a suitable change of coordinates, if needed)

$$\begin{cases} \dot{x} = f(x) \\ 0 = f(0) \end{cases} \tag{1.98}$$

the second, or direct, Lyapunov method is based on the analysis of a continuous scalar function $V(x)$ such that its time derivative along a trajectory is given by

$$\dot{V}(x) = \sum_{i=1}^{n} \frac{\partial V}{\partial x_i} \dot{x}_i = \sum_{i=1}^{n} \frac{\partial V}{\partial x_i} f_i(x) \tag{1.99}$$

Sufficient conditions for the origin $x = 0$ to be *globally asymptotically stable* are

$$\begin{cases} V(0) = 0 \\ V(x) > 0 \quad \forall x \neq 0 \\ \dot{V}(x) < 0 \quad \text{along a trajectory} \end{cases} \tag{1.100}$$

The Lyapunov function $V(x)$ can be regarded as an *energy function* whose minimum lies at the equilibrium. The main difficulty in applying the direct Lyapunov method is finding a suitable function $V(x)$, which is generally defined as a quadratic form of the state x. For linear systems, a commonly used function is $x^T \cdot P \cdot x$ with $P > 0$ a positive definite matrix. Applying this function to the continuous-time system $\dot{x} = A \cdot x$ yields

$$\frac{dV}{dt} = \dot{x}^T P x + x^T P \dot{x} = (Ax)^T Px + x^T PAx$$

$$= x^T A^T Px + x^T PAx = x^T (A^T P + PA)x \tag{1.101}$$

so that

$$\frac{dV}{dt} = -\dot{x}^T Qx \text{ with } Q > 0$$

so the system is asymptotically stable if a positive definite matrix Q can be found such that

$$(A^T P + PA) = -Q \quad \text{with} \quad Q > 0 \tag{1.102}$$

For a nonlinear system, finding a Lyapunov function is not so straightforward. Krasovskii's method (Krasovskii, 1956; Gu et al., 2003; Kharitonov and Zhabko, 2003; Stojanovic et al., 2007) suggests using the quadratic norm of the state derivative

$$V(x) = \left\| \frac{dx}{dt} \right\|_2 = f^T(x) \cdot f(x) \tag{1.103}$$

If its time derivative can be computed as

$$\dot{f}(x) = \frac{\partial f}{\partial x}\frac{\partial x}{\partial t} = F(x) \cdot \dot{x} \quad \text{with} \quad F(x) = \text{Jacobian matrix}$$

then

$$\dot{V}(x) = \dot{f}^T \cdot f + f^T \cdot \dot{f} = f^T \cdot F^T \cdot f + f^T \cdot F \cdot f = f^T (F^T + F)f \tag{1.104}$$

therefore

$$\dot{V} < 0 \Leftrightarrow F^T + F < 0$$

Further, if

$$\lim_{|x| \to \infty} f^T(x) \cdot f(x) \to \infty \tag{1.105}$$

then the origin is globally asymptotically stable. The power of the Krasovskii method is that the *global* stability of the nonlinear system $\dot{x} = f(x)$ can be analysed by just considering its Jacobian, as in the local analysis. In particular, it suffices that the matrix $F^T + F$ has eigenvalues with a negative real part. As an example, consider the following nonlinear autonomous system

$$\begin{cases} \dfrac{dx_1}{dt} = -x_1 \\ \dfrac{dx_2}{dt} = x_1 - x_2 - x_2^3 \end{cases} \tag{1.106}$$

for which the origin is obviously an equilibrium. Applying the Krasovskii theorem, we get

$$f(x) = \begin{bmatrix} -x_1 \\ x_1 - x_2 - x_2^3 \end{bmatrix} \Rightarrow F(x) = \begin{bmatrix} -1 & 0 \\ 1 & -1 - 3x_2^2 \end{bmatrix}$$

$$F^T(x) + F(x) = \begin{bmatrix} -2 & 1 \\ 1 & -2 - 6x_2^2 \end{bmatrix}$$

(1.107)

The eigenvalues of $F^T + F$ have negative real part for any value of x_2. Moreover, because

$$f^T(x) \cdot f(x) = x_1^2 + \left(x_1 - x_2 - x_2^3\right)^2 \Rightarrow \lim_{|x| \to \infty} f^T(x) \cdot f(x) \to \infty$$

(1.108)

the origin is globally asymptotically stable.

1.10 NUMERICAL ASPECTS AND THE ROLE OF MATLAB

In the previous sections, we have examined the structural properties of systems, stressing that most of them can be gleaned from the analysis of the local (or *closed-form*) representation, Equation 1.25 for discrete-time case systems and Equation 1.27 in the continuous-time case, without the need of actually solving the model equations. In fact, the knowledge of the analytical solution is neither practical nor desirable because the main model characteristics, such as stability and type of response, can be inferred from the model structure itself. Further, the system inputs hardly ever have an analytical form, which makes the application of calculus impossible. This is all the more true for nonlinear systems, for which ad hoc methods must be used, given the limited applicability of the linear approximation.

Given these premises, an efficient numerical tool is required to carry out the modelling exercise. Its purpose is the production of environmental scenarios and providing answers to the *what-if* questions so often encountered in environmental decision-making. The availability of a computational platform is also essential for other purposes connected to the modelling practice, such as parameter calibration, system optimization and signal analysis. This platform should provide adequate functionalities to deal with these problems and yet present a user-friendly interface for easy programming and ample graphical capabilities. The MATLAB platform (The Mathworks Inc. www.mathworks.com) possesses all these features and is widely accessible in the academic world, either in teaching laboratories, central campus servers or as a low-cost student version. Public domain MATLAB-like platforms are also available, such as Octave, O-Matrix, or Mlab. A thorough review of these and other similar software can be found at http://www.scientific-computing.com/review1.html.

It is beyond the scope of this book to provide a basic MATLAB tutorial, for which introductory books exist (see, e.g. Gilat, 2005; Pratap, 2006) and manuals are freely available over the Internet, either directly from the MATLAB Documentation centre in The Mathworks website, Wikipedia or even Youtube. MATLAB codes are referred to as *scripts* because, like a script, the MATLAB interpreter scans and executes each line in the sequence they are written. The freedom of specifying and handling objects such as matrices, vectors, strings and Booleans without any prior formal definition, like in C, makes programming easy and the resulting code immediately intelligible. The efficiency of MATLAB in handling vector and matrices lies in the powerful primitive functions that process each of these objects in a compact way using array operators. As an example, Box 1.1 compares the time required to compute the function $\sin(x)/x$ in 10^5 points equally spaced between −5 and 5. First, the

---

BOX 1.1 A SIMPLE COMPUTATIONAL EXAMPLE SHOWING THE TIME SAVED BY USING ARRAY OPERATIONS INSTEAD OF INDEXED LOOPS

```
clear
% generate 100,000 equally spaced points
x=linspace(-10,10,100000);
% Compute the sin(x)/x function
% using array operations
tic
y=sin(x)./x;
toc
% Print the elapsed time
% Compute the sin(x)/x function
% indexed variables in a loop
tic
for i=1:length(x)
    yy(i)=sin(x(i))/x(i);
end
toc
% Print the elapsed time
%---------------------------

Elapsed time is 0.001727 seconds.
Elapsed time is 10.970103 seconds.

Elapsed time ratio = 6.3780e+003
```

---

function evaluation is done using array operators that are characterized by a period (.) preceding the arithmetic operators. In this way, array multiplication A.*B produces a new array in which each element is the product of the corresponding elements in A and B. The same applies for division (./) and power (.^), whereas it coincides with the arithmetic operations for addition and subtraction. The results shown at the bottom of Box 1.1 are quite striking, because the first method is 6378.0 times faster than the *conventional* loop operation, in which each component of the array is individually addressed. This is why MATLAB users are strongly encouraged to use vector/matrix operations and avoid loops. In fact, MATLAB is fairly weak in handling iterative loops, where each instruction will be interpreted every time the loop is iterated. However, in the case of discrete-time models, loops are inevitable because at each step a new input sample is injected into the system and an update of the state vector is performed.

1.10.1 MATLAB HANDLING OF DISCRETE-TIME MODELS

Discrete-time systems, such as Equation 1.28, involve the iteration of the state vector x_t to obtain the state at the next time step x_{t+1}. Simulation of a discrete-time model then reduces to functional evaluations and vector update, where the newly obtained vector x_{t+1} is substituted to the *old* before the next iteration is started. Of course, an initial value for x_t must be provided, as well as a termination criterion in terms of the maximum number of iterations. As an example, consider the model (1.50) and suppose that it is driven by a random input. The following Box 1.2 shows the MATLAB script that simulates the system and plots the results, which are presented in Figure 1.47.

BOX 1.2   MATLAB SCRIPT TO COMPUTE EQUATION 1.50 ITERATIVELY

```
% Discrete-time model eq. (4.24)
%
clear
% Definition of system matrices
A=[0 1;-0.24 1];
b=[1 2]';
Tmax=50;
% Initial condition
x=[5 5]';
% Prepare a matrix of results
X=x;
for t=1:Tmax
    % Model iteration
    x_new=A*x+b*rand(1);
    % Store the result for plotting
    X=[X x_new];
    % replace the new state for the old
    x=x_new;
end

figure(1)
plot(X(1,:),'-o','MarkerFaceColor','w','MarkerSize',8)
set(gca,'FontName','Arial','FontSize',14)
hold on
plot(X(2,:),':ob','MarkerFaceColor','b','MarkerSize',8)
legend('x_1','x_2','location','north','orientation','horizontal')
xlabel('time')
ylabel('x_1,x_2')
```

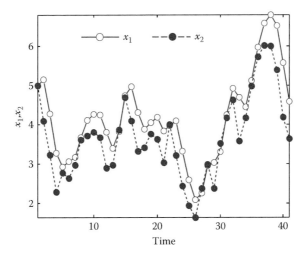

FIGURE 1.47   System (1.50) response to a random input starting from the initial condition $x = \begin{bmatrix} 5 & 5 \end{bmatrix}^T$.

BOX 1.3   MATLAB SCRIPT TO SIMULATE THE RICKER MODEL

```
IT_max=20; % Maximum number of steps
K=20; % Carrying capacity
Rmax=6.8; % Growth rate
x=0.5*K; % Initial population
Xs=zeros(1,IT_max); % Prepare empty vector of results
Xs(1)=x; % Set the initial population
        % as the first vector element
% Start simulation
for t=1:IT_max
   x1=x*Rmax^(1-x/K);
   Xs(t)=x; % Store partial result
   x=x1; % Exchange the new state for the old
end

Figure(1)
plot(Xs,'-o')
xlabel('time')
ylabel('Population')
```

Notice that, thanks to MATLAB capability for handling vectors and matrices, the model iteration `x_new=A*x+b*rand(1)` is performed *vector-wise* with the asterisk '*' implementing the matrix multiplication, without having to worry about individual components.

Another example of a discrete-time equation that can be simulated by direct iteration is the Ricker single-species growth equation, which will be considered later in Chapter 5

$$x_{t+1} = x_t \times R_{max}^{\left(1 - \frac{x_t}{K}\right)}$$

(1.109)

This equation describes the evolution of a population x_t as a function of its carrying capacity K and its growth rate R_{max}. In spite of its innocent look, Equation 1.109 may exhibit surprisingly differing behaviours depending on the value of R_{max}, as already pointed out in Section 1.9.6.1. The pertinent MATLAB script is listed in Box 1.3, and the model responses for two differing values of R_{max} are shown in Figure 1.48. In Figure 1.48a, the growth rate R is such that it drives the population to a steady state through damped oscillations, whereas in b the large value of R produces a pseudo-random chaotic behaviour.

1.10.2 MATLAB Handling of Continuous-Time Models

The simulation of continuous-time model is considerably more complex because it involves the numerical solution of differential equations. Basically, there are two kinds of differential equations: those that start from an *initial condition* and move forward, and those for which the *initial and final points* of the solution are given. The first kind gives rise to the so-called initial-value problem, whereas the second defines a *boundary-value* problem. Although it is fairly straightforward to perform *forward* integration from a given initial condition, composed of the initial values of the function and of its derivatives, it is more difficult to satisfy the boundary condition, in

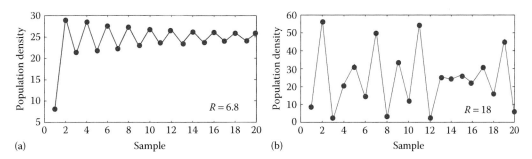

FIGURE 1.48 Differing behaviours of the Ricker equation (1.109) as a function of the growth rate R. In (a), a moderate value of R produces a damped oscillatory stable response, whereas in (b) a large R value results in a chaotic behaviour.

which the conditions are specified at both ends of the integration domain. In this case, successive approximation of the initial derivatives is required, in order to satisfy the conditions at the other extreme of the integration interval. The successive approximation of the initial derivative originates the so-called shooting method, reminiscing of the improving accuracy in hitting a target represented by the final (boundary) condition, as will be considered in Chapter 6. The two methods are sketched in Figure 1.49 for a first-order differential equation.

Since we are interested in the solution starting from an initial condition x_0 onwards, we shall consider only initial-value problems for the time being, although boundary-value problems may be encountered in river quality modelling and will be treated in Chapter 6. In practice, given a differential equation $\dot{x} = f(x)$, we want to extend the solution from $x(t)$ to $x(t+h)$. This requires an approximation of the derivative to project the solution from the current time t to the next integration time $t + h$, with h being the integration step. The crudest (and inaccurate) numerical approximation of the derivative is the Euler method, which projects the current value of the solution $x(t)$ along the tangent at $f(x)$ so that $x(t+h)$ is approximated as shown in Figure 1.50, where the error in estimating $x(t+h)$ is quite evident.

Instead of moving from t to $t + h$ in a single step, more accurate methods use intermediate steps to improve the derivative projection. The fourth-order Runge–Kutta (FORK) method is a very popular (and accurate) algorithm that approximates $x(t+h)$ based on intermediate function evaluations in

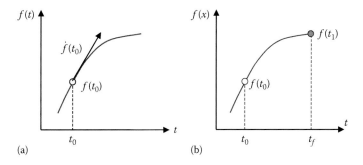

FIGURE 1.49 Possible integration conditions of a first-order differential equation. In the initial-value problem (a), the solution is specified on the basis of the initial value of the function and its derivative, whereas in the boundary-value problem (b), the value of the function is specified at the boundaries of the solution domain.

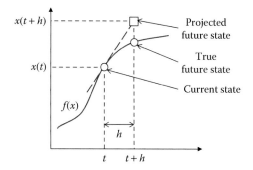

FIGURE 1.50 Illustration of the Euler numerical integration method.

the middle of the integration interval $(t + h/2)$, as shown in Figure 1.51a. The intermediate weights $(k_1, \ldots, k_4)$ and the next solution point $x(t + h)$ are computed by Equation 1.110

$$x(t+h) = x(t) + \frac{h}{6}(k_1 + 2k_2 + 2k_3 + k_4)$$

$$\begin{cases} k_1 = f(t, x(t)) \\ k_2 = f\left(t + \frac{h}{2}, x + \frac{h}{2}k_1\right) \\ k_3 = f\left(t + \frac{h}{2}, x + \frac{h}{2}k_2\right) \\ k_4 = f(t + h, x + h \cdot k_3) \end{cases} \tag{1.110}$$

There is a vast body of literature concerning numerical integration methods based on the FORK algorithm, one of its most important variations being the adaptation of the integration step size h, as shown in Figure 1.51b, depending of the shape of the function $f(x)$ to maintain a constant integration error, or the ability to deal with *stiff* systems of differential equations (Quinney, 1987), involving simultaneous equations with widely differing time scales. In this latter case, the difficulty is represented by the choice of the integration step h, because *fast* equations would require a short step h, but this would unnecessarily hamper the solution of *slow* equations for which a longer h would

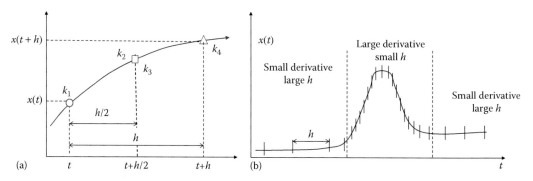

FIGURE 1.51 Graphical illustration of the FORK integration method (a) and the need to adapt the integration step to the function shape (b).

suffice. For a deeper understanding of the fascinating world of numerical methods, the reader is referred to the abundant available literature (see, e.g. Athans et al., 1974; Chapra and Canale, 1991; Press et al., 1986; Dormand, 1996; Dahlquist and Bjorck, 2003). In particular, differential equations are accurately treated by Butcher (2003), whereas boundary-value problems are considered by Quinney (1987).

Coming to the use of MATLAB for solving systems of ordinary differential equations, there are a number of different solvers and it is up to the user to select the most appropriate method for the problem at hand. These methods fall in two main categories: fixed step and variable step, based on either the Euler or Runge–Kutta algorithm just explained, with the latter method offering several versions, depending on the order of the approximation. There are also specific methods for stiff systems of equations.

1.10.3 INTRODUCING THE SIMULINK TOOLBOX

There are two ways for handling initial-value problems in MATLAB: one is to use the appropriate primitive, with the following syntax [t, x] = ode45('model_name',[time_span], x0) where ode45, among others, is the differential equation solver using a FORK method, and the function 'model_name' contains the model definition. The other, immensely more powerful, approach is to build the model using the Simulink® toolbox and running it from a calling MATLAB script. The great advantage of building the model with Simulink is the friendly graphical editor, the huge paradigm of available functions and the wide choice of integration methods. Further, a Simulink-implemented model controlled by a MATLAB script can be linked to all the other functionalities of the platform, such as optimization, signal analysis and graphics.

There are of course other simulation platforms to build and simulate nonlinear dynamical models. One that gained some favour in the ecological community is STELLA (http://www.iseesystems.com), which, however, does not compare to MATLAB in terms of functional capabilities and ease of use. Also, STELLA uses a graphical language that follows the notation introduced by Odum (1983), which won little acceptance outside the ecological circles, because the resulting diagrams are not as intuitive as the Simulink models.

The combined MATLAB/Simulink approach will be used throughout this book, but first let us try the purely MATLAB-based approach, if anything, to discover its limitations. As an example, consider the well-known logistic equation describing the growth of a single-species population (Chapter 5 will be devoted to population dynamics). This continuous-time differential equation describes the growth rate of a single species as

$$\frac{dx}{dt} = r \cdot x \cdot \left(1 - \frac{x}{K}\right) \tag{1.111}$$

where:
  r is the growth rate
  K is the population carrying capacity, that is, its limit population density

These parameters, though conceptually equivalent to those defined in Equation 1.109, have a significantly different meaning and role, as will be explained in Chapter 5. Solving this equation with MATLAB requires the prior definition of the right-hand side of Equation 1.111 as a MATLAB function. Then, a MATLAB script is defined in which the ode45 solver calls this function to compute the solution. The two pieces of code (function and main script) are listed in Box 1.4. They must be saved as separate file, with the logist function being invoked by the calling MATLAB script. Let us examine the pros and cons of this modelling approach: packing all the model equation in a single function is certainly handy, but lacks flexibility. However, the main limitation of this method is the inability to interact with the model during runtime, for

BOX 1.4   MATLAB SCRIPT FOR THE DIRECT SOLVING OF THE LOGISTIC EQUATION

```
% Example of Logistic equation
% direct solution with the
% Matlab solver ODE45
global r, K % make r and K visible to Logist
t0=0;tfin=40; % Simulation interval
K=10;          % Carrying capacity
x0 = 0.5;      % Initial population
% Call the ODE solver in the loop
for r=0.2:0.1:0.4
[t,x] = ode45('logist',[t0 tfin], x0);
% Plot of the solution

Figure(1)
plot(t,x)
hold on
end
% Plot the results
xlabel('time')
ylabel('population')
title('Logistic equation with ODE45 solver')

function xp = logist(t,x)
global r, K% Repeat the global statement in the
         % receiving function
% Logistic equation
xp = r*x*(1-x/K);
```

example, changing the inputs and/or the parameters during the simulation. The latter is a fundamental aspect of the model identification techniques treated in Chapter 2. Another difficulty is parameters passing the definition of the `logist` function requires the input arguments to be the time and the integration variable, that is, (t, x). If we want to experiment with the effect of varying the growth rate r or the carrying capacity K, these parameters should be passed to the function by declaring them `global`, enabling the function `logist` to access their values, set by the MATLAB script, in the common memory area reserved to MATLAB-named `Workspace`. This method of parameter passing will be instrumental when using the combined MATLAB/ Simulink approach for parameter estimation and optimization. As a bottom line, the simulation of continuous time models using the direct MATLAB integration is rather clumsy and is definitely unsuitable for the kind of problems that we shall be treating in this book. Conversely, the Simulink graphic environment represents a very attractive alternative for handling all kinds of models, either continuous-time, discrete-time or a hybrid combination of the two. This toolbox has a very intuitive user interface consisting of a window into which graphic symbols representing computational functions can be dragged from the library palette and connected (*wired*) to form a diagram that represents the model. Behind the graphical interface, Simulink generates an m-file, invisible to the user, that is run by the MATLAB interpreter like any other MATLAB script when the pertinent command (`sim`) is invoked. The computation of a Simulink diagram is based on the *data-flow* approach, whereby the output of each block is produced as soon as all the required data are available at its inputs. Simulink has a powerful sorting algorithm, which schedules the block evaluation order to comply with the data flow of the diagram. The logical data availability implicit in the sorting requires that no block in a loop has contemporary data at its input and outputs, in which case an *algebraic loop* error is generated. To avoid this, each loop should contain at least one integrator, or another memory element like a delay block, to avoid the simultaneity issue.

Though the generated Simulink file (with extension *.mdl) can be run on its own, it is very convenient to control its execution from a calling MATLAB script as previously described, so that the MATLAB/Simulink model combination has a nested structure as illustrated in Figure 1.52. The Simulink model is invoked with the command sim, with the following minimal syntax:

$$[t, x, y] = sim(\text{'Model\_Name'},t\_final); \qquad (1.112)$$

where the left-hand side contains three vectors, respectively, containing the simulation times (t) and the corresponding integrator outputs (x) and model outputs (y). The model name is a string referring to the model that we want to simulate (without the.mdl extension), from 0 until the final time t _ final. Other options controlling the simulation may be added by the expert user.

The example of Figure 1.52 refers again to the logistic model, so that this approach can be compared with the previous one, based on the ode45 MATLAB solver. In this simple example, the two methods produce identical results, as Figure 1.53 shows, but with the model complexity growing, the combined MATLAB/Simulink approach provides superior flexibility and is definitely the preferred approach to ecosystem modelling.

In fact, the MATLAB environment provides the amplest freedom in specifying the simulation conditions and in plotting the results, whereas the Simulink toolbox, with its a graphical representation, provides the highest flexibility in model building. Further, running the model from a calling MATLAB script enables full control of the simulation in the most diverse conditions. One last advantage of the Simulink approach is the ability to feed the model with numerically defined input functions, a feature that is of paramount importance when calibrating models subject to experimental inputs.

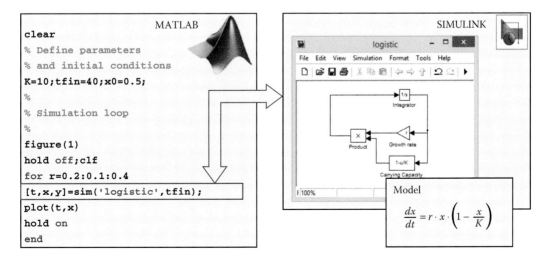

FIGURE 1.52 Nested structure of Simulink-implemented model controlled by a MATLAB script. The arrow shows the connection between the two environment, with Simulink launched by the MATLAB command sim.

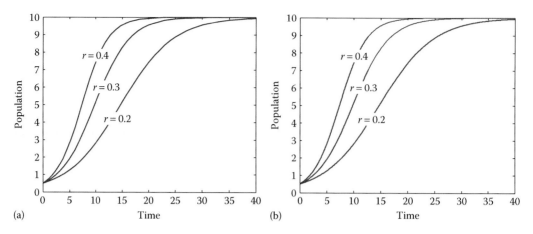

FIGURE 1.53 Simulation of the logistic growth equation with the direct `ode45` solver (a) or via Simulink (b), with the `ode45` (Dormand-Prince) integration method. The results obviously coincide.

1.10.4 Two Simple Benchmark Models

Chapter 2 will deal with model identification, and of course some example models will be required to demonstrate the various properties and procedures. Therefore, we now introduce two very simple, but much studied, models that will be used as benchmarks in Chapter 2.

1.10.4.1 The Monod Kinetics

This model, which will be treated in great detail in Chapter 7, is amply described in the biochemical and environmental literature (Maynard Smith, 1974; Holmberg, 1982; Campbell, 1983; Roels, 1983; Bailey and Ollis, 1986; Battley, 1987; Orhon and Artan, 1994; Cloete and Muyima, 1997; Jorgensen and Bendoricchio, 2001), and it is at the basis of all the models describing the interaction between a biodegradable substrate and a microbial mass that feeds upon it. The full derivation of the model will be described in Chapter 7, but for now it suffices to introduce its basic equations and show its MATLAB implementation. The model equations describing the substrate consumption and the biomass growth in a batch, that is, a closed vessel, are

$$\begin{cases} \text{Substrate} & \dfrac{dS}{dt} = -\dfrac{1}{Y}\dfrac{\mu_{max}S}{K_s + S}X \\[3mm] \text{Biomass} & \dfrac{dX}{dt} = \dfrac{\mu_{max}S}{K_s + S}X - b_h X \end{cases} \tag{1.113}$$

Though the derivation of the model and the biochemical meaning of its parameters will be explained in due course, some basic considerations can be anticipated here:

- The substrate is supposed to be totally biodegradable, and therefore it decays to zero, being consumed by the biomass. Its rate of consumption follows the same dynamics of the biomass growth and the proportion (yield factor) Y denotes the fact that for each unit of consumed substrate, Y units of biomass are produced, that is, $Y = \frac{\Delta X}{\Delta S}$.
- A fixed fraction (b_h) of the biomass decays.
- The biomass growth rate cannot exceed its maximum value (μ_{max}).
- The relative speed of the reaction is represented by the so-called half-saturation constant (K_s), where a small value implies that the biomass grows at a high rate (near μ_{max}) even with a limited substrate availability.

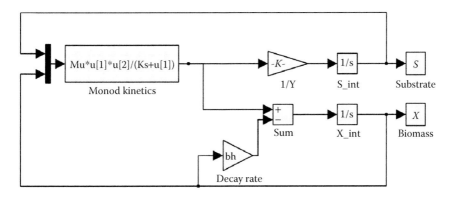

FIGURE 1.54 Simulink model implementing the basic Monod Equation 1.113 as `Monod_model.mdl` in the `ESA_Matlab\Exercises\Chapter_1` folder.

BOX 1.5   MATLAB SCRIPT LAUNCHING THE BATCH MONOD MODEL SIMULATION

```
% Nominal parameter values
Ks=20;Y=0.5;Mu=0.5;bh=0.03;
% Batch simulation horizon
tfin=150;
% Initial conditions
So=10;Xo=0.1;
% Launch the simulation
[t,x,y]=sim('Monod_model',[0 tfin]);
figure(1)
plot(t,x(:,1),'b',t,x(:,2),':b')
xlabel('time')
ylabel('concentration (mg/l)')
legend('Substrate','Biomass')
```

As in the previous examples, Equation 1.113 will be implemented with a Simulink model, shown in Figure 1.54, controlled by a MATLAB script (Box 1.5).

The resulting batch simulation is shown in Figure 1.55. Here, the substrate is gradually consumed, providing nourishment for the biomass, which starts growing exponentially after an initial *incubation* period. When the substrate is not sufficient to sustain the rising population, the biomass stops growing and begins to decline until all the available substrate in the batch, including the decay material produced by the dead biomass, is exhausted.

This model is at the basis of many biotechnical and environmental microbial models and will be used in Chapter 2 as an important benchmark for model identification.

1.10.4.2   The Streeter & Phelps Water Quality Model

Another fundamental, and tutorially relevant, model is the river water quality dynamics due to Streeter and Phelps (S&P) to explain the dissolved oxygen (DO) sagging curve in a river, downstream of a pollution point source (Rinaldi et al., 1979; Tchobanoglous and Schroeder, 1985; Chapra, 1997). This oversimplified model describes the interaction between a biodegradable

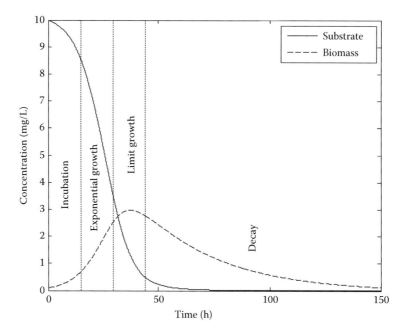

FIGURE 1.55 Simulation results for the Monod batch model Equation 1.113 obtained with the `Go_Monod.m` launch MATLAB script in the `ESA_Matlab\Exercises\Chapter_1` folder. After an initial incubation period, the biomass grows exponentially until it reaches its maximum development. After that it begins to decay for the lack of substrate.

substance (similar to the substrate in the Monod kinetics) expressed as the biochemical oxygen demand (BOD) and the DO in the river, which is used to oxidize this pollutant. Again, postponing the in-depth description and extensions of the model to Chapters 7 and 8, its basic equations are

$$\begin{cases} \text{BOD} & \dfrac{dB}{dt} = -k_b \cdot B \\[2mm] \text{DO} & \dfrac{dC}{dt} = k_c \left(C_{\text{sat}} - C \right) - k_b \cdot B \end{cases} \tag{1.114}$$

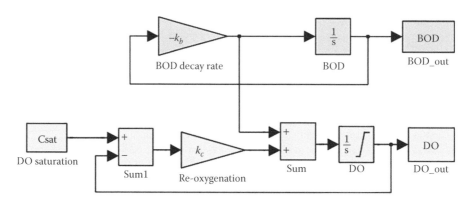

FIGURE 1.56 Simulink model implementing the S&P Equation 1.114. This Simulink model (`SP_reach.mdl`) can be retrieved from the `ESA_Matlab\Exercises\Chapter_1\Streeter&Phelps` folder.

BOX 1.6   MATLAB SCRIPT LAUNCHING THE S&P MODEL

```
Csat=8; % DO saturation
Qm=10; % Uprstream flow
Bm=7; % Upstream BOD
Cm=Csat % Saturated upstream DO
Qs=0.5; % Sewer flow
Bs=40; % Sewer BOD
Kb=0.15; % BOD decay rate
Kc=0.20; % river re-oxygenation rate
ts=50; % Maximum flow time
% Compute upstream conditions
Bo=(Bm*Qm+Qs*Bs)/(Qm+Qs)
Co=Cm*Qm/(Qm+Qs)
%
% Reach simulation
%
[t,x,y]=sim('SP_reach',[0 ts]);
```

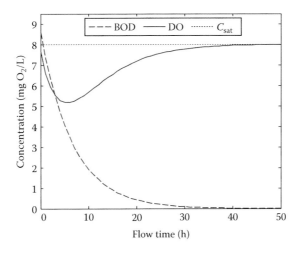

FIGURE 1.57   Evolution of the S&P model for a single pollution point source.

In Equation 1.114, the time is to be considered as *flow time*, that is, the time taken by the water to travel downstream and C_{sat} is the DO saturation concentration. Again, a Simulink model can be set up to implement the model Equation 1.114, as shown in Figure 1.56. The MATLAB launch script is listed in Box 1.6, whereas Figure 1.57 shows the system evolution. The DO sag downstream of the discharge is eventually recovered through the natural re-oxygenation, as shown in the simulation of Figure 1.57.

1.11   WORKED EXAMPLES

This section presents some worked examples relative to the theory presented in this chapter. The exercises include both the theory and the MATLAB programming. The code developed in each exercise can be retrieved from the related software collection. To avoid mistaking litres for number 1, the former will be indicated with a capital L whenever ambiguity may arise.

1.11.1 SIMULATING THE TIME-VARYING FLOW IN A CYLINDRICAL TANK

With reference to the system illustrated in Figure 1.13, consider a cylindrical tank with a cross section of 20 cm², subject to square-wave input, which varies between 1 and 3 L/min with a frequency of 10^{-3} L/min. The outlet friction coefficient is $k_v = 1.2$. Build the Simulink model and simulate the system around its steady-state value. What is the effect of varying any of the system parameters?

Solution
The model equation to be used is the global mass balance Equation 1.10, where h represents the height of the liquid

$$A\frac{dh}{dt} = F_i - k_v\sqrt{h} \tag{1.115}$$

The steady-state value can be obtained by setting the derivative of h to zero in that equation so that the equilibrium point can be determined as

$$h_0 = \left(\frac{F_b}{k_v}\right)^2 \tag{1.116}$$

where F_b is the average flow, that is, $F_b = \frac{1+3}{2} = 2\,(\text{L/min})$. Substituting the numerical values yields $h_0 = 2.7778$ cm. The Simulink model is shown in Figure 1.58, and the corresponding MATLAB launch script is shown in Box 1.7, whereas its response to a square wave input is shown in Figure 1.59.

It is left to the reader to experiment with differing values of the parameters. In particular, what is the effect of doubling the frequency of the square wave?

The software for this exercise can be found in `Exercises\Chapter _ 1\Ex.1.11.1 _ Tank` and consists of the launch script `Go _ tank.m` and of the Simulink model `tank.mdl`.

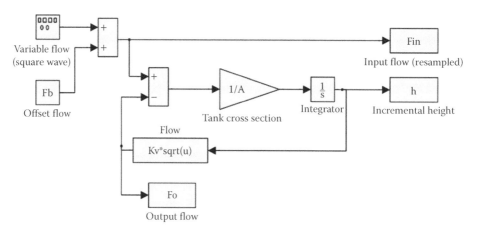

FIGURE 1.58 Simulink model of Exercise 1.11.1 `tank.mdl` in the `ESA_Matlab\Exercises\Chapter_1\Ex.1.11.1_Tank` folder.

BOX 1.7   MATLAB LAUNCH SCRIPT FOR EXERCISE 1.11.1

```
A=20;      % Tank cross-section (cm^2)
Fi=1;      % Input variable flow (L/min)
Kv=1.2;    % Outlet friction coefficient
Fb=2;      % Offset flow (L/min)
w=0.001;   % Square wave frequency (1/min)
tfin=2000;% Final simulation time (min)
ho=(Fb/Kv)^2; % Initial condition (steady-state)
% Launch the simulation
[t,x,y]=sim('tank',tfin);
figure(1)
set(1,'Position',[78    140    866    459])
subplot(2,1,1)
plot(t,h)
hold on
plot([0 tfin],[ho ho],':k')
ylabel('height (cm)')
subplot(2,1,2)
plot(t,Fin,':b',t,Fo)
hold on
plot([0 tfin],[Fb Fb],':k')
ylabel('Flow (L/min)')
legend('Fi','Fo','location','northoutside',
...'orientation','horizontal')
xlabel('time (min)')
```

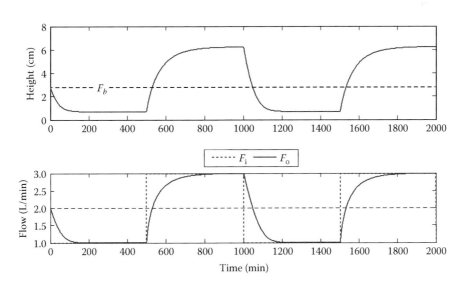

FIGURE 1.59   Model response of Exercise 1.11.1. This graph was produced by the MATLAB script Go _ tank.m in the ESA _ Matlab\Exercises\Chapter _ 1\Ex.1.11.1 _ Tank folder.

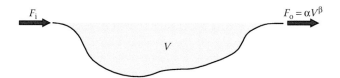

FIGURE 1.60   The reservoir considered in Exercise 1.11.2.

1.11.2   SIMULATING THE FLOW DYNAMICS IN A RESERVOIR

Consider the reservoir of Figure 1.60 where, unlike in the previous exercise, the input and output flows are at the same level. The dynamical equation describing the volume variation can be written as

$$\frac{dV}{dt} = F_i - F_o = F_i - \alpha V^\beta \tag{1.117}$$

where the exponential law in the second term αV^β is the equivalent of the square root term of Equation 1.115 and represents the outlet rating curve, defined by the parameters α and β, which are supposed to be known $\left(\alpha = 2.4 \times 10^{-2}; \beta = 0.875\right)$. Build the Simulink model and simulate the system around its steady-state value.

Solution

The steady-state volume can be obtained by setting the derivative of Equation 1.117 to zero and solving for V_0

$$V_0 = e^{\left\{\left[\ln(F_b)-\ln(\alpha)\right]/\beta\right\}} \tag{1.118}$$

where, as in the previous exercise, F_b represents the average input flow. The Simulink model implementing Equation 1.117, with the initial condition provided by Equation 1.118, is shown in Figure 1.61, while the MATLAB launch script is listed in Box 1.8. The model response is shown in Figure 1.62. Regarding the launch script of Box 1.8, the input data vector can be replicated by concatenating the original data set as many times as required, as shown in lines 27–32 where Cr is the repetition parameter.

The software for this exercise can be found in Exercises\Chapter _ 1\Ex.1.11.2 _ Lake and consists of the launch script Go _ lake.m and of the Simulink model Lake _ volume.mdl. Several data files are provided for the simulation.

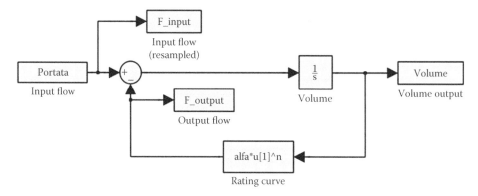

FIGURE 1.61   Simulink model for Exercise 1.11.2. The resampled input flow has as many samples as integration steps, regardless of the number of samples in Portata. The Simulink model Lake _ volume.mdl can be retrieved from the ESA _ Matlab\Exercises\Chapter _ 1\Ex.1.11.2 _ Lake folder.

BOX 1.8 MATLAB LAUNCH SCRIPT FOR EXERCISE 1.11.2

```
g=uigetfile('Flow*.mat','Load calibration data');
eval(['load ' g]);
Cr=input('Number of cyclic repetitions ');
% Rating curve parameters
alfa=2.4e-2;    % Flow coefficient
beta=0.875;     % Flow exponent
ts=Portata(:,1);
q=Portata(:,2);

Ts=ts;Q=q;
if Cr>1,
    for j=1:Cr
       Ts=[Ts;ts+Ts(end)];
       Q=[Q;q];
    end
end
Fb=mean(Q); % Mean input flow
Vo=exp((log(Fb)-log(alfa))/beta); % Steady state volume
tfin=Ts(end);
Portata=[Ts Q];
% Launch the simulation
[t,x,y]=sim('Lake_volume',tfin);
figure(1)
set(1,'Position',[252 61 844 579])
set(gca,'FontName','Arial','FontSize',14)
subplot(211)
plot(t,F_input,':b',t,F_output,'b','LineWidth',2)
set(gca,'FontName','Arial','FontSize',14)
legend('Input flow','Output flow',…
'location','northoutside','orientation','horizontal')
ylabel('Flow (m^3/d)')
subplot(212)
plot(t,Volume,'LineWidth',2)
set(gca,'FontName','Arial','FontSize',14)
ylabel('Volume (m^3)')
xlabel('time (d)')
```

1.11.3 ANALYSIS OF A DISCRETE-TIME SYSTEM

Given the discrete-time system described by the following dynamical model

$$\begin{bmatrix} x_1(t+1) \\ x_2(t+1) \end{bmatrix} = \begin{bmatrix} 0 & 1 \\ 0.25 & 0 \end{bmatrix} \times \begin{bmatrix} x_1(t) \\ x_2(t) \end{bmatrix} + \begin{bmatrix} 0 \\ 1 \end{bmatrix} \times u(t) \tag{1.119}$$

determine the eigenvalues of the system (1.119) and discuss its stability; indicate the kind of response of the autonomous system $(u = 0)$; determine its equilibrium when the system is subject to a constant input $\bar{u} = 1$.

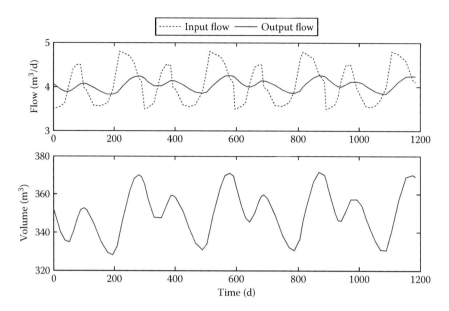

FIGURE 1.62 Reservoir model response of Exercise 1.11.2. This graph was produced by the MATLAB script Go _ Lake.m in the ESA _ Matlab\Exercises\Chapter _ 1\Ex.1.11.2 _ Lake folder.

Solution

The system is represented in canonical controllable form (Athans et al., 1974), hence the system matrix can be written explicitly as a function of the eigenvalues (λ_1, λ_2) as

$$A = \begin{bmatrix} 0 & 1 \\ -\lambda_1\lambda_2 & \lambda_1 + \lambda_2 \end{bmatrix} \tag{1.120}$$

and the characteristic polynomial can be written by inspection as

$$s^2 - (\lambda_1 + \lambda_2)s + \lambda_1 \cdot \lambda_2 = 0 \tag{1.121}$$

because $a_{21} = -\lambda_1\lambda_2$ and $a_{22} = \lambda_1 + \lambda_2$. Hence, the eigenvalues are the roots of the equation

$$s^2 - 0.25 = 0 \rightarrow \lambda_1 = 0.5 \ \lambda_2 = -0.5 \tag{1.122}$$

Because both eigenvalues are inside the unit circle, the system is asymptotically stable. However, since $\lambda_2 = -0.5$, there will be a damped oscillating response. It is left to the reader as an exercise to write the pertinent MATLAB script and run the simulation.

1.11.4 MATLAB SIMULATION OF A DISCRETE-TIME SYSTEM

Given the discrete-time system described by the following system equation

$$\begin{bmatrix} x_1(t+1) \\ x_2(t+1) \end{bmatrix} = \begin{bmatrix} 0.2 & 0.3 \\ -5 & 0.1 \end{bmatrix} \times \begin{bmatrix} x_1(t) \\ x_2(t) \end{bmatrix} + \begin{bmatrix} 0 \\ 1 \end{bmatrix} \times u \tag{1.123}$$

implement a MATLAB script to simulate the system evolution with a constant input $u = 6.15$ for $0 \leq t \leq 20$ and initial conditions $x(0) = \begin{bmatrix} 1 & -0.5 \end{bmatrix}^T$. Compute also the equilibrium state $\bar{x}$.

Solution

The MATLAB script, which simulates the system for the given time length, is listed in Box 1.9. In the first lines, the eigenvalues and eigenvectors of the system matrix are computed to assess the system stability. Then, a simulation loop is set up for the required ten time steps. At each iteration, the new state x _ t1 replaces the old state x. Finally, the equilibrium state is computed with Equation 1.49. The reader is encouraged to develop the relevant MATLAB script.

1.11.5 A Two-Reach S&P River Quality Model with Point Discharges

This exercise is aimed at extending the S&P single-reach model to a two-reach system, each with an upstream point discharge.

Solution

The S&P model equation to be used for each reach has already been defined as Equation 1.114, and therefore, the same Simulink model of Figure 1.56 can be used without changes. The new aspect of the exercise is the combination of the two reaches, as shown in Figure 1.63, which will require a change in the initial conditions at the beginning of each reach where the new point source of pollutant enters the river.

BOX 1.9   MATLAB SCRIPT FOR EXERCISE 1.11.4

```
A=[0.2,0.3;-5,0.1]; % System matrix
disp('Eigenvalues and Eigenvectors')
[v,e]=eig(A)
% System definition
u=6.15;% Constant input
b=[0;1];
t=0;
x=[1 -0.5]'; % Initial conditions
xo=x;
X=[];T=[];
while t<20        %Simulation loop
    x_t1=A*x+b*u;
    X=[X; x'];
    x=x_t1;
    T=[T t];
    t=t+1;
end
figure(1)
plot(X(:,1),X(:,2),'-ob')
hold on
plot(xo(1),xo(2),'ob')
% Computation of the final equilibrium
I2=eye(2);
A1=inv(I2-A);
x_eq=A1*b*u
plot(x_eq(1),x_eq(2),'sb',…
'MarkerSize',8,'MarkerFaceColor','b')
```

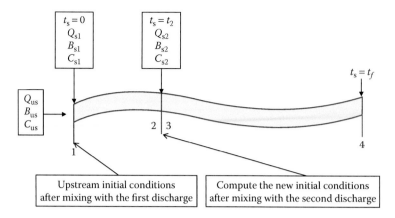

FIGURE 1.63 Schematization of the two-reach river system with point discharges, considered in Exercise 1.11.5 and implemented in the MATLAB script `Go_SP_2reach.m` in the `ESA_Matlab\Exercises\Chapter_1\Streeter&Phelps` folder.

The extension of the basic S&P equations to this combined system requires that the computation is broken down in the following steps:

1. Compute the upstream initial conditions at section 1 by mixing the upstream quality with that of the first discharge, that is,

$$B_1 = \frac{B_{us}Q_{us} + Q_{s1}B_{s1}}{Q_{us} + Q_{s1}} \quad C_1 = \frac{C_{us}Q_{us} + Q_{s1}C_{s1}}{Q_{us} + Q_{s1}} \tag{1.124}$$

2. Integrate the S&P equations up to point 2, that is, from 0 to t_2, immediately upstream of the second discharge
3. Stop the integration there and compute the initial conditions for the second reach by mixing the final quality of the first reach, at point 2, with the second discharge, that is,

$$B_3 = \frac{B_2(Q_{us} + Q_{s1}) + Q_{s2}B_{s2}}{Q_{us} + Q_{s1} + Q_{s2}} \quad C_3 = \frac{C_2(Q_{us} + Q_{s1}) + C_{s2}Q_{s2}}{Q_{us} + Q_{s1} + Q_{s2}} \tag{1.125}$$

4. Use these new initial conditions to integrate the S&P equations along the second reach, until the final point is reached. Notice that the integration interval for this second simulation will be from 0 to $t_f - t_2$ and not t_f

There is no need to redefine a new Simulink model with respect to the single-reach model already shown in Figure 1.56, which can be used to simulate each reach, provided the appropriate initial conditions are recomputed before simulating the next reach, as outlined in step 3. The launch MATLAB script is shown in Box 1.10, where the plotting details have been omitted for clarity. The Simulation results are shown in Figure 1.64, where the discontinuities caused by the changes in the initial conditions are clearly visible.

The software for this exercise can be retrieved from the `ESA_Matlab\Exercises\Chapter_1\Streeter&Phelps` folder and consists of the launch script `Go_SP_2reach.m` and of the Simulink model `SP_reach.mdl`.

BOX 1.10 SIMULATION OF THE TWO-REACH S&P RIVER QUALITY MODEL, EXERCISE 1.11.5

```
% Two-reach Streeter & Phelps water quality modelling
clear
close all
clc
prompt={'Upstream flow [m^3/s]',...
        'Upstream BOD [mg/l]',...
        'Upstream DO [mg/l]',...
        'Upstream discharge flow [m^3/s]',...
        'Upstream discharge BOD [mg/l]',...
        'Upstream discharge DO [mg/L]',...
        'DO Saturation'};
def={'5','3','8','0.5','30','0','8'};
dati=inputdlg(prompt,'Upstream loading conditions',1,def,'on');
dati=char(dati);
Qm=str2num(dati(1,:));
Bm=str2num(dati(2,:));
Cm=str2num(dati(3,:));
Qs=str2num(dati(4,:));
Bs=str2num(dati(5,:));
Cs=str2num(dati(6,:));
Csat=str2num(dati(7,:));
% Upstream initial conditions
Q1=Qm+Qs
Bo=(Bm*Qm+Qs*Bs)/Q1
Co=(Cm*Qm+Cs*Qs)/Q1
prompt={'Kb [1/h]',...
        'Kc [1/h]',...
        'Reach total length [h]',...
        'Second discharge location [h]',...
        'Second discharge flow [m^3/s]',...
        'Second discharge BOD [mg/l]',...
        'Second discharge DO [mg/L]'};
def={'0.12','0.15','40','15','0.35','20','2'};
dati=inputdlg(prompt,'River system parameters',1,def,'on');
dati=char(dati);
Kb=str2num(dati(1,:));
Kc=str2num(dati(2,:));
L=str2num(dati(3,:));
ts2=str2num(dati(4,:));
Qs2=str2num(dati(5,:));
Bs2=str2num(dati(6,:));
Cs2=str2num(dati(7,:));
% Simulation of the first reach
Boo=Bo;Coo=Co; % Save upstream conditions for later
[t,x,y]=sim('SP_reach',[0 ts2]);
last=length(tsim);
BOD1=BOD;DO1=DO;t1=tsim;
B1=BOD(last)
C1=DO(last)
```

Box 1.10 continued

Box 1.10 continued

```
% Second discharge mixing
Qm1=Qm+Qs
Bo=(B1*Qm1+Qs2*Bs2)/(Qm1+Qs2)
Co=(C1*Qm1+Cs2*Qs2)/(Qm1+Qs2)
% Simulation of the second reach
[t,x,y]=sim('SP_reach',[0 L-ts2]);
last=length(tsim);
B_finale=BOD(last)
C_finale=DO(last)
% Combine the two reaches
BOD=[BOD1' BOD'];
DO=[DO1' DO'];
tsim=[t1' (tsim+ts2)'];
figure(1)
hold off;clf
plot(tsim,BOD,':b',tsim,DO,'b','LineWidth',2)
a=axis;
axis([a(1) a(2) 0 ceil(max(Boo,Coo))])
xlabel('Reach length (h)')
ylabel('Concentration (mg/l)')
```

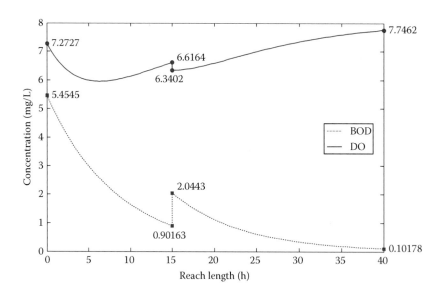

FIGURE 1.64 Simulation results for the two-reach S&P model considered in Exercise 1.11.5. This graph was produced by the MATLAB script Go_SP_2reach.m in the ESA_Matlab\Exercises\Chapter_1\Streeter&Phelps folder.

1.11.6 A S&P RIVER QUALITY MODEL WITH NON-POINT POLLUTION

In this exercise, we consider a single-reach river system with an upstream point discharge and where a fraction of the downstream reach is subject to a diffuse pollution. How do we include this distributed input into the model?

Solution

Contrary to the previous exercise, the non-point pollution acts as an input to the BOD equation, rather than changing its initial conditions. Further, the units for this input are no longer in mg/L but in mg/L per unit of flow time, that is, mg/L h. The modified Simulink model is shown in Figure 1.65, where the constant distributed BOD input has been implemented as the difference of two step functions, the first switching on at the beginning of the distributed load and the other at the end. For the rest, the model is the same as that shown in Figure 1.56 and the launch MATLAB script is shown in Box 1.11.

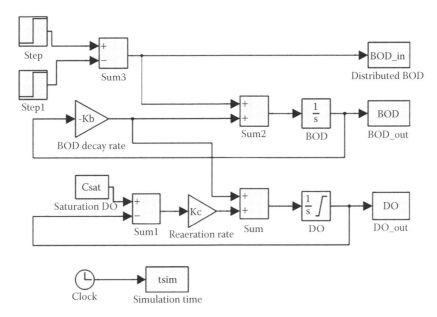

FIGURE 1.65 A modified S&P Simulink model accommodating a non-point BOD input, as considered in Exercise 1.11.6. This Simulink model (SP _ distr _ const.mdl) can be retrieved from the ESA _ Matlab\Exercises\Chapter _ 1\Streeter&Phelps folder.

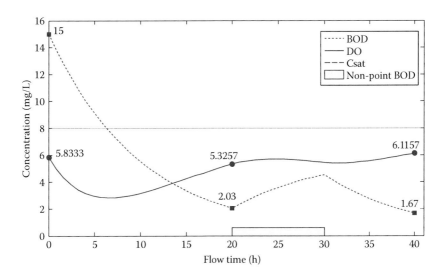

FIGURE 1.66 Evolution of BOD and DO in a reach with a partial non-point BOD input, considered in Exercise 1.11.6. This graph was produced by the MATLAB script Go _ SP _ distr _ const.m in the ESA _ Matlab\Exercises\Chapter _ 1\Streeter&Phelps folder.

BOX 1.11   SIMULATION OF A SINGLE-REACH S&P RIVER QUALITY MODEL WITH A PARTIAL NON-POINT BOD INPUT, EXERCISE 1.11.6

```
prompt={'Upstream BOD [mg/l]',...
       'Upstream DO [mg/l]',...
       'DO Saturation [mg/l]',...
       'Upstream flow [m^3/s]',...
       'Upstream discharge flow [m^3/s]',...
       'Upstream discharge BOD [mg/l]',...
       'Upstream discharge DO [mg/l]',...
       'Kb BOD decay rate [1/h]',...
       'Kc Reaeration rate [1/h]',...
       'Final flow time [h]',...
       'Initial non-point BOD flow time [h]',...
       'Final non-point BOD flow time [h]',...
       'Non-point BOD value [mg/L.h]'};
def={'10.0','7.0','8','5.0','1.0','40','0','0.1','0.15','40','20',
'30','0.6'};
dati=inputdlg(prompt,'River system definition',1,def,'on');
dati=char(dati);
Bm=str2num(dati(1,:));
Cm=str2num(dati(2,:));
Csat=str2num(dati(3,:));
Qm=str2num(dati(4,:));
Qs=str2num(dati(5,:));
Bs=str2num(dati(6,:));
Cs=str2num(dati(7,:));
% Initial BOD
Bo=(Qm*Bm+Qs*Bs)/(Qm+Qs);
% Initial DO
Co=(Cm*Qm+Cs*Qs)/(Qm+Qs);
Kb=str2num(dati(8,:));
Kc=str2num(dati(9,:));
tfin=str2num(dati(10,:));
T1=str2num(dati(11,:));
Tf=str2num(dati(12,:));
Bu=str2num(dati(13,:));
% Simulation
[t,x,y]=sim('SP_distr_const',tfin);
figure(1)
hold off;clf
plot(tsim,BOD,':b',tsim,DO,'b')
a=axis;
axis([0 tfin+1 0 max(Csat+1,Bo+1)])
xlabel('Flow time (h)')
ylabel('Concentration (mg/l)')
hold on
plot([0 tfin],[Csat Csat],'--b')
fill(tsim,BODin,[223/255 233/255 0.999])
legend('BOD','DO','Csat','non-point BOD')
title('Non-point BOD Streeter & Phelps model')
```

Box 1.11 continued

Box 1.11 continued

```
BOD1=interp1(tsim,BOD,T1)
DO1=interp1(tsim,DO,T1)
plot(0,Bo,'sb','MarkerSize',8,'MarkerFaceColor','b')
text(0,Bo,['  ',num2str(Bo)],'vert','bottom')
plot(T1,BOD1,'sb','MarkerSize',8,'MarkerFaceColor','b')
text(T1,BOD1,[num2str(BOD1),'  '],'vert','bottom','hor','right')
plot(T1,DO1,'ob','MarkerSize',8,'MarkerFaceColor','b')
text(T1,DO1,[num2str(DO1),'  '],'vert','bottom','hor','right')
plot(0,Co,'ob','MarkerSize',8,'MarkerFaceColor','b')
text(0,Co,['  ',num2str(Co)],'vert','bottom')
Bf=BOD(length(BOD));
plot(tfin,Bf,'sb','MarkerSize',8,'MarkerFaceColor','b')
text(tfin,Bf,[num2str(Bf),'  '],'vert','bottom','hor','right')
Cf=DO(length(DO));
plot(tfin,Cf,'ob','MarkerSize',8,'MarkerFaceColor','b')
text(tfin,Cf,[num2str(Cf),'  '],'vert','bottom','hor','right')
```

The system evolution is depicted in Figure 1.66, showing that the effect of the distributed BOD input is that of increasing the BOD and decreasing the DO. In this sense, a reach with a non-point BOD input can be detected by a reversal of the BOD/DO trend normally observed downstream of a point BOD source pollution, where the BOD decreases exponentially and the DO produced the well-known sag downstream of the discharge point.

The software for this exercise can be found in 'Exercises\Chapter_1\Streeter&Phelps' and consists of the launch script Go _ SP _ distr _ const.m and of the Simulink model SP _ distr _ const.mdl.

2 Identification of Environmental Models

Identification is the process of constructing a mathematical model of a dynamical system from observation and prior knowledge.

John Norton (1986)

2.1 INTRODUCTION

The last part of the quotation (the *prior knowledge*) of this all-encompassing definition has already been treated in Chapter 1, where we considered the various model building alternatives. Now, we concentrate on the other aspect of the above definition, which consists of a reconciliation between observations and model response.

As we have seen in Chapter 1, environmental models consist of a set of dynamical equations in which some numerical parameters appear. These constants are instrumental in shaping the model response, thus making its behaviour as close as possible to the observed data that the model is intended to mimic. Identification is the procedure through which the model parameters are assigned the numerical values that make the model response 'as close as possible' to the observed behaviour. The qualitative statement (*as close as possible*) is illustrated in Figure 2.1 and will soon be defined in precise mathematical terms, but for now let us just consider it as the fundamental endeavour of the identification problem.

2.1.1 FORMAL STATEMENT OF THE IDENTIFICATION PROBLEM

The parameter calibration problem can be formally stated in the following terms: given a model (acting as a constraint to the optimization algorithm) and a set of experimental observations, find the set of parameter values that minimize the sum of squared errors between the observed data and the model response, as illustrated in Figure 2.1. Thus, the three basic ingredients in our problem are as follows:

$$\text{The model:} \begin{cases} \dfrac{dx}{dt} = f(x, u_{\exp}, P) & \quad x \in \mathbb{R}^n \\[2mm] & \quad y \in \mathbb{R}^q \\[2mm] y = g(x, u_{\exp}, P) & \quad P \in \mathbb{R}^{n_p} \end{cases} \tag{2.1}$$

$$\text{The data:} \quad Y^{\exp} = \begin{bmatrix} y_{t_1}^{\exp} \\ y_{t_2}^{\exp} \\ \vdots \\ y_{t_k}^{\exp} \\ y_{t_N}^{\exp} \end{bmatrix} \in \mathbb{R}^q \tag{2.2}$$

where $(t_1, t_2, \ldots, t_k, \ldots, t_N)$ are the N sampling instants.

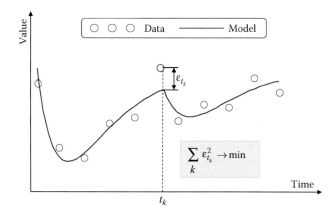

FIGURE 2.1 The fundamental goal of model identification: minimize the sum of squared errors between the data and the model response.

and

The error functional: $E(\boldsymbol{P}) = \sum_{k=1}^{N} \left[\boldsymbol{y}_{t_k}(\boldsymbol{P}) - \boldsymbol{y}_{t_k}^{\exp} \right]^T \boldsymbol{Q}_k \left[\boldsymbol{y}_{t_k}(\boldsymbol{P}) - \boldsymbol{y}_{t_k}^{\exp} \right] \quad \boldsymbol{Q}_k \in \mathbb{R}^{q \times q}$ \hfill (2.3)

The data of Equation 2.2 refer to the q system outputs $\boldsymbol{y} \in \mathbb{R}^q$. In some lucky cases, the whole system state may be directly observable, that is, $\boldsymbol{y} = \boldsymbol{x} \in \mathbb{R}^n$, in which case the error functional (2.3) becomes

$$E(\boldsymbol{P}) = \sum_{k=1}^{N} \left[\boldsymbol{x}_{t_k}(\boldsymbol{P}) - \boldsymbol{x}_{t_k}^{\exp} \right]^T \boldsymbol{Q}_k \left[\boldsymbol{x}_{t_k}(\boldsymbol{P}) - \boldsymbol{x}_{t_k}^{\exp} \right] \quad \boldsymbol{Q}_k \in \mathbb{R}^{n \times n} \qquad (2.4)$$

Thus, the optimization problem can be stated as the search for the optimal parameters $\hat{\boldsymbol{P}}$ that minimizes the error functional (2.3) or (2.4), having the model (2.1) as a constraint, and using the experimental data (2.2), namely

$$\hat{\boldsymbol{P}} = \arg \min_{\boldsymbol{P}} E(\boldsymbol{P}) \text{ subject to model } (2.1) \qquad (2.5)$$

However, before taking up the task of finding the 'best' parameter values, there are some preliminary steps that should be performed. These operations include:

- Data gathering, validation and accuracy characterization
- Error functional definition
- Sensitivity analysis, to assess the parameter influence on the output
- Numerical optimization, using a suitable search method
- Validation of the estimates, to assess their consistency

2.1.2 The Quality of the Data

Choosing the right data is crucial for a successful identification. The data of Equation 2.2 should be fully representative of the actual system behaviour and reasonably noise-free, or being affected by noise with more or less known characteristics. The matrix $\boldsymbol{Q}_k$ in the error functional (2.3) or (2.4) has the role of weighing the data according to their accuracy and is generally defined as the reciprocal of the measurement errors. Suppose that at each sampling instant t_k, each measurement is replicated m times, so that we have a data set arranged as in Equation 2.6:

$$\leftarrow N \text{ samples} \rightarrow$$

$$\mathbf{Y}^{\text{exp}} = \begin{bmatrix} \mathbf{y}_{t_1}^{(1)} & \mathbf{y}_{t_2}^{(1)} & \cdots & \mathbf{y}_{t_N}^{(1)} \\ \mathbf{y}_{t_1}^{(2)} & \mathbf{y}_{t_2}^{(2)} & \cdots & \mathbf{y}_{t_N}^{(2)} \\ \vdots & \vdots & \vdots & \vdots \\ \mathbf{y}_{t_1}^{(m)} & \mathbf{y}_{t_2}^{(m)} & \cdots & \mathbf{y}_{t_N}^{(m)} \end{bmatrix} \quad m \text{ replicates} \tag{2.6}$$

Averaging the m replicates taken at each measuring time t_k yields

$$\bar{y}_{t_k} = \frac{1}{m} \sum_{i=1}^{m} y_{t_k}^{(i)}$$

$$\sigma_{t_k} = \left[\frac{1}{m-1} \sum_{i=1}^{m} \left(y_{t_k}^{(i)} - \bar{y}_{t_k} \right)^2 \right]^{\frac{1}{2}} \tag{2.7}$$

which is repeated for each measured quantity, so that the data matrix becomes

$$\mathbf{Y}^{\text{exp}} = \begin{bmatrix} \bar{\mathbf{y}}_{t_1} \pm \sigma_{t_1} \\ \bar{\mathbf{y}}_{t_2} \pm \sigma_{t_2} \\ \vdots \\ \bar{\mathbf{y}}_{t_N} \pm \sigma_{t_N} \end{bmatrix} \tag{2.8}$$

An example is shown in Figure 2.2. In general, the quantities in Equation 2.8 can be computed, even when the number of replicates at each t_k varies from sample to sample.

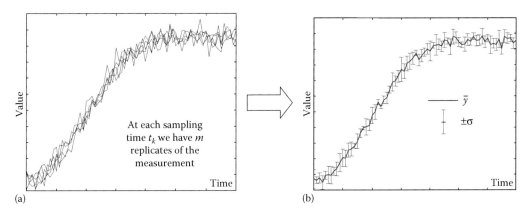

(a)   (b)

FIGURE 2.2 If each measurement is replicated m times (a), we can condense this information by taking the average and standard deviation of the set of samples at each time t_k. In the (b) plot, the same information is conveyed by the set average ($\bar{y}$) and standard deviation ($\pm\sigma$).

If we do not have replicates at each sampling instant, and therefore we cannot compute Equation 2.7 directly, we can estimate the standard deviation representing the measurement accuracy with the technical specifications supplied by the instrument manufacturer. In any case, the weighting matrix Q_k is defined as a diagonal matrix of the reciprocal variances, that is,

$$Q_k = \begin{bmatrix} \dfrac{1}{\sigma_{t_k}^2(1)} & 0 & \cdots & 0 \\ 0 & \dfrac{1}{\sigma_{t_k}^2(2)} & \cdots & 0 \\ 0 & 0 & \cdots & 0 \\ 0 & 0 & \cdots & \dfrac{1}{\sigma_{t_k}^2(w)} \end{bmatrix} \tag{2.9}$$

where the matrix dimension w is equal to either q or n, depending on whether output or state measurements are considered. It is important to remember that for each sampling instant t_k, a specific Q_k matrix is defined on the basis of the local error statistics. However, if the matrix reflects just the instrument accuracy, then it is constant over all the sampling instants, and the subscript k may be dropped. Figure 2.3 shows two short data series, one with a variable accuracy for each measurement and the other with constant accuracy. In either case, the matrix Q reflects the measurement accuracy, in the sense that accurate measurements produce a matrix with large diagonal elements, and vice versa. We shall get back to the quality of the observations, when a link will be established between data noise and parameter accuracy.

2.1.3 Input Information Content

The model data may be generated in response to a particular time evolution of the input. The ability of an input time series of activating all the system modes for as long as required to achieve a good parameter estimation is called *persistent excitation* (Bellman and Åström, 1970; Mareels et al., 1987; Shimkin and Feuer, 1987; Ljung and Glad, 1994; Ding and Chen, 2005; Willems et al., 2005). Several input classes were defined as 'persistently exciting' if their power spectrum

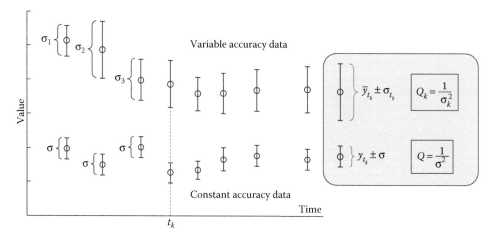

FIGURE 2.3 An example of a data set with variable accuracy (above) and with constant accuracy (below): the shaded area on the right shows the average and standard deviation characterizing the matrix Q. In the case of constant variance, the index k is dropped.

is never less than a prescribed quantity for any given time window. The step input is the simplest persistently exciting input, but more elaborate (and efficient) choices range from white noise to pseudo-random binary sequences that mimic the behaviour of white noise, though assuming only two levels.

2.1.4 THE ERROR FUNCTIONAL

The definition of the error functional (2.3) or (2.4) deserves some comment. First of all, we refer to each of them as a *functional*, rather than a *function*, because it encompasses the whole time evolution over the sampling instants $(t_1,...,t_N)$. The optimization procedure that will be described in the next section is intended to reach the minimum of that functional, and thus find the 'optimal' value of the parameters $\hat{P}$, which represent the best approximation to the otherwise unknown 'true' parameters. If there are no systematic modelling errors, the difference between the data and the model response is an approximation of the data variance

$$\sigma^2 \sim s^2 = \frac{E(\hat{P})}{N - n_p} = \frac{1}{N - n_p} \sum_{k=1}^{N} \left[y_{t_k}(P) - y_{t_k}^{\exp} \right]^T Q_k \left[y_{t_k}(P) - y_{t_k}^{\exp} \right] \tag{2.10}$$

and it would be nice to recast this object into a framework that would enable some statistical consideration. If the number of data N is large enough, the errors $\left[y_{t_k}(P) - y_{t_k}^{\exp} \right]$ can be considered as normally distributed, and their sum of squares is χ^2-distributed with $N - n_p$ degrees of freedom (Press et al., 1986; Bates and Watts, 1988). Thus, an error functional such as in Equation 2.10 can be viewed as a χ^2 *goodness-of-fit* statistics to assess the significance of the calibration. The limit χ^2 value can be computed in MATLAB® with the primitive function X = chi2inv(alpha, DF), where alpha is the confidence level and DF is the number of degrees of freedom.

In the previous section, the weighting matrix Q_k was assumed to represent the measurement accuracy (or uncertainty), but other kinds of weighting factors can be adopted when a special emphasis is required on a portion of the data. Figure 2.4 shows four possible weighting methods. Reconsidering the weighting matrix Q_k no longer associated with the measurement noise, but as a weighting factor emphasizing some feature of the data, the following are but a few options in its new role. It is assumed, as in Equation 2.9, that the matrix Q_k is diagonal $Q_k = \text{diag}(w_k)$.

1. *Proportional weighting*: The importance of the data grows with their magnitude, attributing more importance to the largest data with a proportional factor α,

$$w_k = \alpha \left| y_{t_k}^{\exp} \right| \tag{2.11}$$

2. *Inverse weighting*: The importance of the data is inversely proportional to their value, with a factor β,

$$w_k = \beta \frac{1}{\left| y_{t_k}^{\exp} \right|} \tag{2.12}$$

3. *Derivative weighting*: The importance of the data is proportional to the magnitude of their derivative, giving more importance to the fast changing data, with a factor γ,

$$w_k = \gamma \left| \frac{y_{t_{k+1}}^{\exp} - y_{t_k}^{\exp}}{t_{k+1} - t_k} \right| \tag{2.13}$$

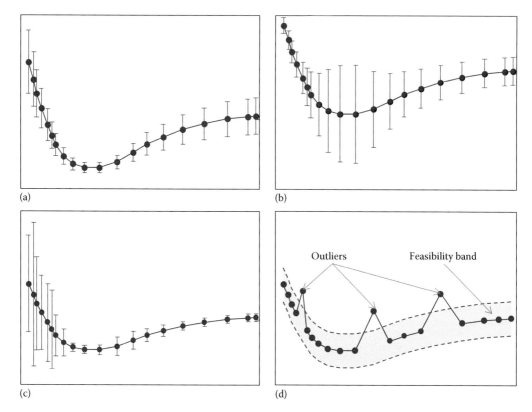

FIGURE 2.4   Four possible definitions of error weighting factors w_k: (a) proportional weighting; (b) inverse weighting; (c) derivative weighting; and (d) robust weighting.

4. *Robust weighting*: The data outside a feasibility band are considered as *outliers* (considered in detail in Chapter 3) and severely scaled down, being considered as unreliable data or an abnormally large deviation,

$$
w_k = \begin{cases} \left(\dfrac{1}{1+\left(y_{t_k}^{\exp}\right)^2} \right)^2 & \text{if } y_{t_k}^{\exp} \text{ is outside the feasibility band} \\[6mm] 1 & \text{if } y_{t_k}^{\exp} \text{ is inside the feasibility band} \end{cases} \tag{2.14}
$$

2.2   SENSITIVITY ANALYSIS AS AN IDENTIFIABILITY TEST

Before actually embarking on the—sometimes troublesome—task of finding the 'best' parameter values, in the sense outlined in the previous section, it is strongly advisable to have a closer look at the system structure and determine what is the role played by each parameter. This preliminary analysis is often referred to as *sensitivity analysis* and is very useful in simplifying the subsequent task of the actual parameter estimation.

The central role of the sensitivity analysis is to make sure that the system is identifiable, that is, its parameters can be estimated on the basis of the available data. *Structural identifiability* was initially aimed at making sure that the model structure was such as to allow its parameters to be estimated. Lack of identifiability may occur when a parameter variation has no effect on the observed system output (*real* structural identifiability), but more often it occurs because the wrong data are used for the estimation. In the past, research was limited to the former motive, and analytical methods (Pohjanpalo, 1978) were proposed to make certain that the model structure was such as to guarantee model identifiability. The limit of those methods, apart from the extensive calculus involved, was a too clear-cut yes/no answer, whereas, in practice, it is more reasonable to consider identifiability as a gradual model property, jointly considering the model structure and the available data (Van Tongeren, 1995; Kong et al., 1996; Dochain and Vanrolleghem, 2001; Marsili Libelli et al., 2001; Petersen et al., 2001; Reichert and Vanrolleghem, 2001; De Pauw, 2005; Sin et al., 2005; Lindenschmidt, 2006; Spindler, 2014). Consequently, present-day identifiability tests involve *both* the structure *and* the data. In this section, we shall concentrate on the former, having already considered the data in Section 2.1.2.

Sensitivity is defined as the variation in the observed variable (either output or state) relative to the parameter variation, that is,

$$\begin{cases} \text{State sensitivity } S^x_{p_i} & \lim_{\Delta p_i \to 0} = \dfrac{\Delta x}{\Delta p_i} = \dfrac{\partial x}{\partial p_i} \\[3mm] \text{Output sensitivity } S^y_{p_i} & \lim_{\Delta p_i \to 0} = \dfrac{\Delta y}{\Delta p_i} = \dfrac{\partial y}{\partial p_i} \end{cases} \qquad (2.15)$$

Depending on the context in which the parameter variation Δp_i is considered, differing aspects of its influence on the model are brought to the fore. There are two complementary approaches to sensitivity: static and dynamical, and both will be examined.

2.2.1 STATIC SENSITIVITY

If we want to know the effect of a parameter variation on the whole simulation horizon of our model, the simplest way to compute this effect is to perform a number of simulations with perturbed parameter values. Because it is difficult to discern the effect of simultaneous parameter variations, let alone present them graphically, it is advisable to vary only one or two parameters at a time and assess the results, which will take the form of curves or surface relating a model figure of merit to the parametric perturbation. This figure of merit can be defined by borrowing the basic error functional definitions (2.3) or (2.4). However, in this case, no data are involved, so we should drop the accuracy matrix Q_k, because we shall *not* be comparing data versus model output, but rather *nominal* versus *perturbed* model trajectories, defined, respectively, as simulations performed by keeping the selected parameter at its nominal value, or by perturbing it. Thus, the previous functional definitions (2.3) or (2.4) reduce to the following comparison of model trajectories obtained with perturbed model parameters:

$$\Delta E = \begin{cases} \sum_k \left(x^{\text{pert}}_k - x^{\text{nom}}_k \right)^T \left(x^{\text{pert}}_k - x^{\text{nom}}_k \right) & \text{State sensitivity functional} \\[3mm] \sum_k \left(y^{\text{pert}}_k - y^{\text{nom}}_k \right)^T \left(y^{\text{pert}}_k - y^{\text{nom}}_k \right) & \text{Output sensitivity functional} \end{cases} \qquad (2.16)$$

In Equation 2.16, the summations are extended to a set of k samples freely selected along the trajectories, as shown in Figure 2.5, and not necessarily coinciding with the experimental samples. In the left plot (a), the nominal and perturbed trajectories are plotted, starting with the same initial conditions. An arbitrary set of points is selected to compute the difference between each perturbed trajectory and the nominal one. In the right plot (b), the squared sums resulting from Equation 2.16 are plotted as a function of the perturbation increment Δp_i. There is no compelling reason to select equally spaced perturbations, but it helps in the computations.

The shape of the sensitivity curves thus obtained may have the forms (a) or (b) in Figure 2.6. A fairly flat curve (a) obviously implies a low sensitivity because large parametric variations result

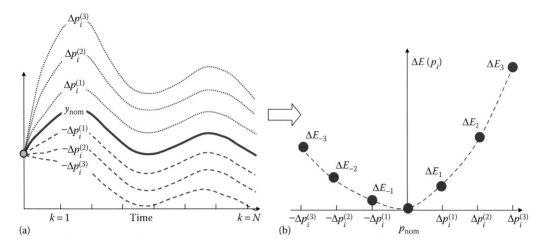

(a)

(b)

FIGURE 2.5 An example of producing perturbed trajectories (all starting with the same initial conditions) and computing the sensitivity functional ΔE (a). In (b), the values of ΔE are plotted as a function of the varying magnitude of the perturbation Δp_i.

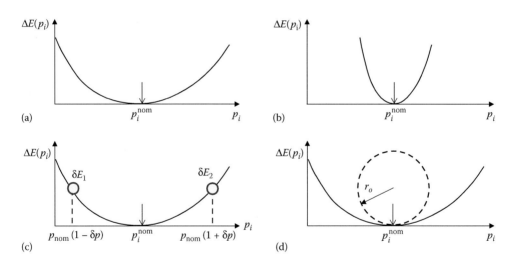

FIGURE 2.6 Possible look of the single-parameter sensitivity outlooks for a generic parameter p_i. In (a), a fairly flat curve indicates a low sensitivity, whereas the steeper curve in (b) denotes a high sensitivity. Two possible numerical formalizations are shown in (c) and (d). Parameter sensitivities can be ranked by considering the average increment of ΔE corresponding to symmetrical parameter perturbations $\pm\delta p$ (c), or by computing the curvature $(1/r_o)$ of the sensitivity curve in the neighbourhood of p_{nom}, as in (d).

in a moderate increase of the sensitivity. Conversely, curve (b) denotes a high sensitivity. The schemes in (c) and (d) indicate two possible quantifications of the sensitivity.

2.2.1.1 Static Sensitivity by Error Increment

Figure 2.6c takes the ratio of the error functional increment $\Delta E(p_i)$ to the parameter perturbation δp_i as a measure of the sensitivity to p_i. To remove the dependence on the relative magnitude of the parameter, select a common percentage increment as $p_{nom}(1 \pm \delta)$, and compute the corresponding increments δE_1 and δE_2 of the functional. The sensitivity can then be defined as the average

$$S_{p_i} = \frac{\delta E_1 + \delta E_2}{2} \tag{2.17}$$

In this way, sensitivities can be directly compared, with larger values corresponding to higher sensitivities.

As an example, Figure 2.7 shows the static sensitivities of the Monod kinetic model (Section 1.10.4.1)

$$\begin{cases} \text{Substrate} & \dfrac{dS}{dt} = -\dfrac{1}{Y}\dfrac{\mu_{max}S}{K_s + S}X \\[2ex] \text{Biomass} & \dfrac{dX}{dt} = \dfrac{\mu_{max}S}{K_s + S}X - b_h X \end{cases} \tag{2.18}$$

caused by the perturbation of one parameter at a time around the nominal values $\left(\mu_{max} = 0.5,\, K_s = 20,\, Y = 0.5,\, b_h = 0.03\right)$. In this case, both state variables (S and X) are observed, so the sensitivity functional (2.16) is computed as

$$E_{SX} = \sum_k \left(S_k^{\text{pert}} - S_k^{\text{nom}}\right)^2 + \sum_k \left(X_k^{\text{pert}} - X_k^{\text{nom}}\right)^2 \tag{2.19}$$

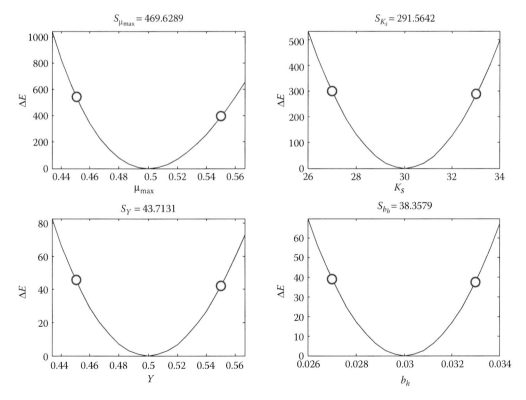

FIGURE 2.7   Static sensitivity of the Monod kinetics model perturbing only one parameter at a time by the same relative amount $\pm \delta p / p_{\text{nom}} = \pm 10\%$. The dots represent the corresponding variations δE, while the sensitivity computed with Equation 2.17 is shown on top of each plot.

Obviously, each sensitivity curve is zero when the parameter is at its nominal value. Notice the magnitude spread of the four vertical scales, emphasizing the differing influence of each parameter on the model response.

2.2.1.2 Static Sensitivity by Curvature

The method outlined in Figure 2.6d considers instead the curvature of the functional $\Delta E(p_i)$ in the neighbourhood of the nominal value p_{nom}. This quantity is defined as the reciprocal of the radius, which is computed via the second derivative:

$$\Phi = \frac{1}{r_o} = \left(\frac{\partial^2 \Delta E}{\partial p_i^2} \bigg|_{p_{\text{nom}}} \right)^{-1} \tag{2.20}$$

Numerically, the second derivative can be approximated as follows (Abramowitz and Stegun, 1970):

$$r_o = \left(\frac{\partial^2 \Delta E(\delta)}{\partial \delta^2} \right) \cong \frac{-\Delta E(-2\delta) + 16\Delta E(-\delta) - 30\Delta E(0) + 16\Delta E(\delta) - \Delta E(2\delta)}{12\delta^2} \tag{2.21}$$

where δ represents a relatively small increment of p_i around its nominal value p_{nom}.

2.2.1.3 Monte Carlo Sensitivity

Instead of performing simulation with a parameter perturbed by fixed increments, the Monte Carlo approach can be followed, which consists of producing a large number of simulations with the perturbed parameter drawn from a statistically defined population. The distribution of the target quantity, for example, the value of a state variable at a certain time along the simulation, is statistically analysed in comparison with the statistics of the perturbed parameter.

Suppose that the perturbed parameter is drawn from a normal distribution with mean equal to the nominal value and a variance comparable with the expected range of variation, that is, $p_i \sim N(p_{i_{\text{nom}}}, \sigma_{p_i})$. Then, the Monte Carlo simulations will produce a distribution of the target variable. By comparing the variances of the two distributions, the influence of the perturbed parameter can be assessed.

As an example, consider again the Monod kinetics (2.18), and let us perform a set of Monte Carlo simulations by extracting the parameter K_s from a normal distribution of known characteristics, that is, $K_s \sim N(K_{s,\text{nom}}, \sigma_{K_s})$. Setting the target variable as the biomass concentration at a given time during the simulation, the result of Figure 2.8 is obtained. The influence of K_s on that particular aspect of the model can be assessed by comparing the relative magnitude of the variances. In general, the behaviour of Figure 2.9 can be analysed, with a high-sensitivity target quantity exhibiting an amplification of the parameter variance, whereas a low-sensitivity target will compress the parameter variance.

2.2.1.4 Joint Sensitivity of Parameter Couples

Extending the perturbation method to a parameter couple may reveal interesting aspects of the error surface, and possible inter-parametric interactions. Consider again the Monod model (2.19), but this time two parameters, μ_{max} and K_s, are jointly perturbed. Figure 2.10 presents two views of the static sensitivity portrait.

In the 3D plot (a), the minimum is placed in a flat valley, and this should be taken as a warning of possible numerical problems for the search algorithm. The contour portrait (b) shows several contours of the previous surface, again emphasizing the narrow valley where the minimum is located. This kind of visualization can reveal interesting aspects in the error function surface, such as parameter correlation, that are likely to cause numerical difficulties for the search algorithm.

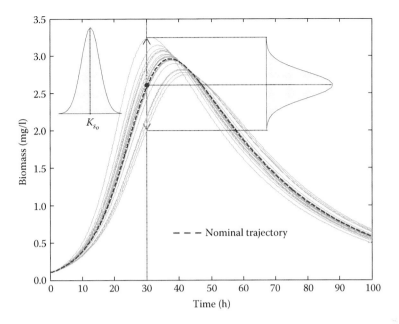

FIGURE 2.8 An example of Monte Carlo sensitivity assessment of the biomass fluctuation at 30 h into the batch, caused by a random K_s drawn from a normal distribution of known characteristics.

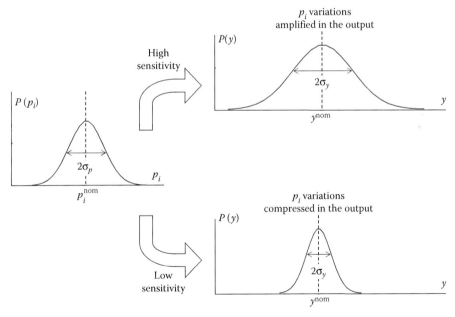

FIGURE 2.9 The sensitivity of a target quantity (in the case the model output y) can be assessed by the amplification or compression of the variance of the parameter distribution.

Knowing the shape of the error surface may help select the favourable zones where the search can be started, or avoid unfavourable regions, which might cause the search to go astray.

The shape of the error functional obviously depends on the observed variables. Figure 2.11 compares the Monod error functional for different choices of the output variables. Here, in addition to the full error functional E_{SX} of Equation 2.19, we consider two other cases in which only one state variable is observed:

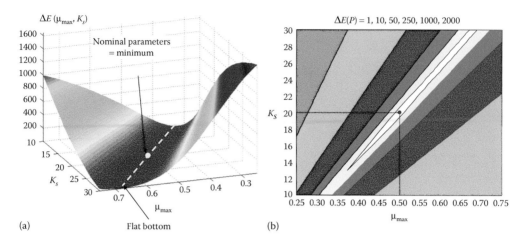

FIGURE 2.10 Static sensitivity portraits of the Monod model with respect to the (μ_{max}, K_s) parameter couple. The 3D plot (a) shows that the minimum is placed in a flat trough, which may cause numerical problems to the search algorithm seeking the optimal parameters. The contour portrait (b) shows some levels of the same error function, emphasizing the narrow valley where the minimum is located.

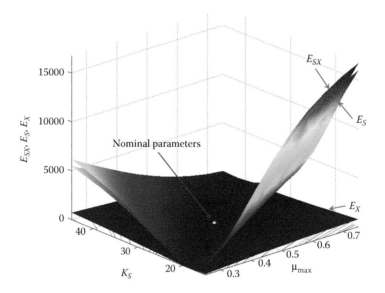

FIGURE 2.11 Differing shapes of the error functional, depending on which output variables are considered.

$$E_S = \sum_k \left(S_k^{pert} - S_k^{nom} \right)^2$$

$$E_X = \sum_k \left(X_k^{pert} - X_k^{nom} \right)^2$$

(2.22)

The three possible cases of Equations 2.19 and 2.22 are plotted in the subspace of the two critical parameters μ_{max} and K_s. In the first case, E_{SX} of Equation 2.19, both state variables are observed, whereas in the other two cases (E_S or E_X) of Equation 2.22 only one of them is available. Figure 2.11 indicates that the substrate (S) gives the largest contribution to the overall sensitivity, whereas the relative weight of the biomass (X) is comparatively minor. Thus, substrate is the most important variable in the estimation. In some sense, Figure 2.11 reverses the sensitivity problem that was initially posed in terms of the parameters, whereas it is now considered from the observed variables viewpoint.

2.2.2 Dynamic (Trajectory) Sensitivity

A different approach is now considered to show the influence of each parameter along the system evolution. Deriving the definition of sensitivity (2.15) with respect to time and exchanging the order of the derivatives yields

$$\frac{d}{dt}\left(\frac{\partial x}{\partial p_i}\right) = \frac{\partial}{\partial p_i}\left(\frac{dx}{dt}\right) = \frac{\partial f(x, p_i)}{\partial p_i} = \frac{\partial f(x, p_i)}{\partial x}\bigg|_{x_{\text{nom}}} \cdot \frac{\partial x}{\partial p_i} + \frac{\partial f}{\partial p_i} \tag{2.23}$$

producing the dynamic sensitivity system

$$\dot{S}^x_{p_i} = J_n \cdot S^x_{p_i} + \frac{\partial f}{\partial p_i} \quad \text{with } J_n = \frac{\partial f(x, p_i)}{\partial x}\bigg|_{x_{\text{nom}}} \tag{2.24}$$

The sensitivity dynamical equation (2.24) is to be solved in conjunction with the system equations, because the matrix J_n is similar to the Jacobian, but it is not computed at the equilibrium because it follows the nominal trajectory x_{nom} obtained with nominal parameter values. The importance of this new sensitivity definition lies in its relation to system evolution, because it reveals the influence of each parameter along the system trajectory, hence its name of *trajectory sensitivity*. In other words, it provides information on the parts of the trajectory in which the parameter influence is greatest. The data collected during this time are therefore the most effective in parameter estimation. Thus, trajectory sensitivity can be used to plan a data acquisition campaign, concentrating the sampling effort at the times when the data are most effective for estimating a given parameter.

This concept will be considered again in due course, when discussing the validity of the estimates and how their accuracy can be maximized. For now, let us elaborate on Equation 2.24. To obtain the sensitivity system, calculus must be applied to obtain the Jacobian matrix J_n and the other term $\partial f/\partial p_i$, which can be regarded as an input. Of course, $\partial f/\partial p_i$ is obtained by evaluating the derivative of the system function with respect to the parameter of interest p_i. Figure 2.12 shows a

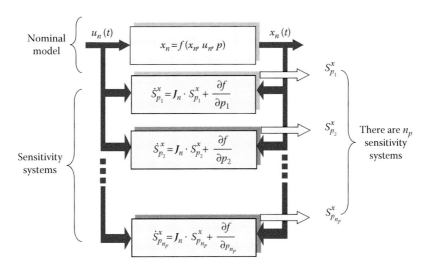

FIGURE 2.12 Trajectory sensitivity system, including the model evolving along its nominal trajectory and n_p sensitivity systems fed with the nominal state trajectory, incorporated in J_n.

general trajectory sensitivity system composed of the system model, evolving with nominal parameters, and a bank of n_p sensitivity systems, one for each parameter of interest. Notice that the latter systems are not autonomous, because the pseudo-Jacobian matrix $\boldsymbol{J}_n$ must be fed with the nominal state trajectory $\boldsymbol{x}_{nom}$ along the system evolution. The last term in each sensitivity system, $\partial f / \partial p_i$, is the derivative of the system equations with respect to each parameter and only the column of the relevant parameter p_i should be used.

As an example, consider again the Monod model (2.18), and suppose that we want to set up the sensitivity system for its four parameters. Using the notation of Equation 2.24, the sensitivity terms take the form

$$
\boldsymbol{J}_n = \left(\frac{\partial f}{\partial x} \right)_{x_{nom}} = \begin{bmatrix} \dfrac{\partial f_1}{\partial S} & \dfrac{\partial f_1}{\partial X} \\[2ex] \dfrac{\partial f_2}{\partial S} & \dfrac{\partial f_2}{\partial X} \end{bmatrix} = \begin{bmatrix} -\dfrac{1}{Y} \cdot \dfrac{\mu_{max} \cdot K_s}{\left(K_s + S \right)^2} \cdot X & -\dfrac{1}{Y} \cdot \dfrac{\mu_{max} \cdot S}{K_s + S} \\[3ex] \dfrac{\mu_{max} \cdot K_s}{\left(K_s + S \right)^2} \cdot X & \dfrac{\mu_{max} \cdot S}{K_s + S} - b_h \end{bmatrix} \tag{2.25}
$$

$$
\frac{\partial f}{\partial \boldsymbol{P}}\bigg|_{x_{nom}} = \begin{bmatrix} \dfrac{\partial f_1}{\partial \mu_{max}} & \dfrac{\partial f_1}{\partial K_s} & \dfrac{\partial f_1}{\partial Y} & \dfrac{\partial f_1}{\partial b_h} \\[2ex] \dfrac{\partial f_2}{\partial \mu_{max}} & \dfrac{\partial f_2}{\partial K_s} & \dfrac{\partial f_2}{\partial Y} & \dfrac{\partial f_2}{\partial b_h} \end{bmatrix} = \begin{bmatrix} -\dfrac{1}{Y} \cdot \dfrac{X \cdot S}{K_s + S} & \dfrac{1}{Y} \cdot \dfrac{\mu_{max} \cdot X \cdot S}{\left(K_s + S \right)^2} & \dfrac{1}{Y^2} \cdot \dfrac{\mu_{max} \cdot X \cdot S}{K_s + S} & 0 \\[3ex] \dfrac{X \cdot S}{K_s + S} & \dfrac{-\mu_{max} \cdot X \cdot S}{\left(K_s + S \right)^2} & 0 & -X \end{bmatrix} \tag{2.26}
$$

These equations generate four sensitivity systems, one for each parameter, which can be composed as follows:

Sensitivity to μ_{max}:

$$
\begin{bmatrix} \dot{S}^S_{\mu_{max}} \\[2ex] \dot{S}^X_{\mu_{max}} \end{bmatrix} = \begin{bmatrix} -\dfrac{1}{Y} \cdot \dfrac{\mu_{max} K_s}{\left(K_s + S \right)^2} \cdot X & -\dfrac{1}{Y} \cdot \dfrac{\mu_{max} \cdot S}{K_s + S} \\[3ex] \dfrac{\mu_{max} K_s}{\left(K_s + S \right)^2} \cdot X & \dfrac{\mu_{max} \cdot S}{K_s + S} - b_h \end{bmatrix} \times \begin{bmatrix} S^S_{\mu_{max}} \\[2ex] S^X_{\mu_{max}} \end{bmatrix} + \begin{bmatrix} -\dfrac{1}{Y} \cdot \dfrac{X \cdot S}{K_s + S} \\[3ex] \dfrac{X \cdot S}{K_s + S} \end{bmatrix} \tag{2.27}
$$

Sensitivity to K_s:

$$
\begin{bmatrix} \dot{S}^S_{K_s} \\[2ex] \dot{S}^X_{K_s} \end{bmatrix} = \begin{bmatrix} -\dfrac{1}{Y} \cdot \dfrac{\mu_{max} K_s}{\left(K_s + S \right)^2} \cdot X & -\dfrac{1}{Y} \cdot \dfrac{\mu_{max} \cdot S}{K_s + S} \\[3ex] \dfrac{\mu_{max} K_s}{\left(K_s + S \right)^2} \cdot X & \dfrac{\mu_{max} \cdot S}{K_s + S} - b_h \end{bmatrix} \times \begin{bmatrix} S^S_{K_s} \\[2ex] S^X_{K_s} \end{bmatrix} + \begin{bmatrix} \dfrac{1}{Y} \cdot \dfrac{\mu_{max} \cdot X \cdot S}{\left(K_s + S \right)^2} \\[3ex] \dfrac{-\mu_{max} \cdot X \cdot S}{\left(K_s + S \right)^2} \end{bmatrix} \tag{2.28}
$$

Sensitivity to Y:

$$
\begin{bmatrix} \dot{S}^S_Y \\[2ex] \dot{S}^X_Y \end{bmatrix} = \begin{bmatrix} -\dfrac{1}{Y} \cdot \dfrac{\mu_{max} K_s}{\left(K_s + S \right)^2} \cdot X & -\dfrac{1}{Y} \cdot \dfrac{\mu_{max} \cdot S}{K_s + S} \\[3ex] \dfrac{\mu_{max} K_s}{\left(K_s + S \right)^2} \cdot X & \dfrac{\mu_{max} \cdot S}{K_s + S} - b_h \end{bmatrix} \times \begin{bmatrix} S^S_Y \\[2ex] S^X_Y \end{bmatrix} + \begin{bmatrix} \dfrac{1}{Y^2} \cdot \dfrac{\mu_{max} \cdot X \cdot S}{K_s + S} \\[3ex] 0 \end{bmatrix} \tag{2.29}
$$

Sensitivity to b_h:

$$\begin{bmatrix} \dot{S}_{b_h}^S \\ \dot{S}_{b_h}^X \end{bmatrix} = \begin{bmatrix} -\dfrac{1}{Y} \cdot \dfrac{\mu_{max} K_s}{(K_s + S)^2} \cdot X & -\dfrac{1}{Y} \cdot \dfrac{\mu_{max} \cdot S}{K_s + S} \\ \dfrac{\mu_{max} K_s}{(K_s + S)^2} \cdot X & \dfrac{\mu_{max} \cdot S}{K_s + S} - b_h \end{bmatrix} \begin{bmatrix} S_{K_d}^S \\ S_{K_d}^X \end{bmatrix} + \begin{bmatrix} 0 \\ -X \end{bmatrix} \qquad (2.30)$$

Apart from the considerable calculus involved, the exact computation of sensitivities with the above method requires the joint simulation of the original model, to generate the nominal trajectory, and of the sensitivity systems for the parameters of interest, all arranged as in Figure 2.12. If the model consists of a considerable number of equations and/or parameters, this may represent a formidable challenge. Fortunately there is a way to obtain equally accurate sensitivities in a much simpler way.

2.2.2.1 Approximate Trajectory Sensitivity

If we relax the basic hypothesis under which the sensitivity system (2.24) was derived, that is, $S_p^y = \lim_{\Delta p \to 0} (\Delta y / \Delta p) = (\partial y / \partial p)$, and instead assume that each parameter is perturbed by a finite increment Δp, then a handy way of computing an approximate trajectory sensitivity is available (Petersen, 2000; Petersen et al., 2003; De Pauw, 2005). Consider perturbing one parameter p by a very small (but finite) amount $\Delta p = \pm \delta$, and perform the simulations with perturbed parameters, in addition to the nominal simulation with reference parameter value p_{nom}. Then, the sensitivity S_p^y is computed as the average of the two side sensitivities S^+ and S^-:

$$\left. \begin{array}{l} p_{nom}(1+\delta) \to y_{pert}^+ \\ p_{nom} \to y_{nom} \\ p_{nom}(1-\delta) \to y_{pert}^- \end{array} \right\} \to \left. \begin{array}{l} S^+ = \dfrac{y_{pert}^+ - y_{nom}}{\delta \cdot p_{nom}} \\[2mm] S^- = \dfrac{y_{nom} - y_{pert}^-}{\delta \cdot p_{nom}} \end{array} \right\} \to S_p^y = \dfrac{S^+ + S^-}{2} \qquad (2.31)$$

If the incremental perturbation $\Delta p = \pm \delta$ is small enough, then Equation 2.31 is an accurate approximation of the trajectory sensitivity system (2.24) without having to go through all the involved calculus. A word of caution is due here regarding very complex models, for which the choice of δ may become critical, and its optimal value should be determined (De Pauw, 2005; Iacopozzi et al., 2007). As an example, Figure 2.13 compares the exact (2.27) and approximate (2.31) sensitivities of the Monod kinetics to the maximum growth rate μ_{max}. Though for $\delta = 0.05$ there is still some discrepancy between the two sensitivities, for a smaller value $(\delta = 0.0001)$ the two trajectories coincide for all practical purposes. Figure 2.14 shows the trajectory sensitivities for the complete set of parameters in the Monod kinetics.

The trajectory sensitivity approach can be used for planning the data collection with a view towards parameter estimation. In fact, the instants in which the magnitude of the sensitivity to a parameter is large are those in which that parameter has the highest influence on the system response, and hence the data collected in those time lapses will carry the highest amount of information for its estimation. Figure 2.15 shows the most effective time brackets for collecting substrate and biomass data for μ_{max} estimation.

In a similar way, the sensitivities of the Streeter & Phelps (S&P) model

$$\begin{cases} \text{BOD} & \dfrac{dB}{dt} = -K_b \cdot B \\[3mm] \text{DO} & \dfrac{dC}{dt} = K_c (C_{sat} - C) - K_b \cdot B \end{cases} \qquad (2.32)$$

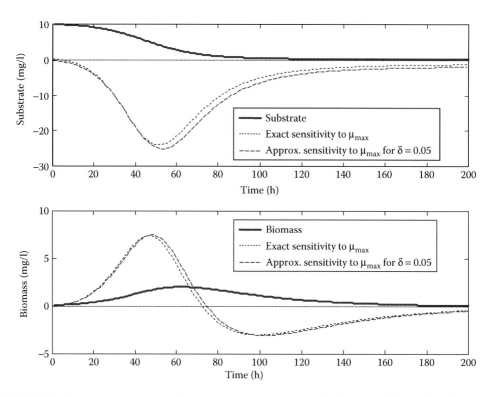

FIGURE 2.13 Comparison of exact and approximate trajectory sensitivities of the Monod kinetics to μ_{max}. The approximate sensitivities for $\delta = 10^{-4}$ are superimposed to the exact ones, while some deviation is discernible for $\delta = 0.05$.

with respect to its parameters K_b and K_c, can be obtained, as shown in Figure 2.16, for the single reach case with upstream point source. The shaded areas along the BOD and DO curves indicate the most favourable portion of the system evolution for gathering data used to estimate the two model parameters K_b and K_c. The sensitivity of BOD to K_c was not traced, being identically zero, because this parameter does not appear in the BOD model equation.

2.2.2.2 Sensitivity Ranking

The graphical representation of trajectory sensitivities yields a visual appraisal of the role played by each parameter along a certain system evolution. This information can be used as a basis for planning a data collection campaign, but when it comes to ranking the parameters according to their sensitivity, a numerical figure of merit is certainly preferable. This numerical information can be extracted from the sensitivity trajectories by sampling them at regular intervals and computing their root-mean-square (RMS) value, that is,

$$\xi_i = \sqrt{\frac{1}{N} \sum_{j=1}^{q} \sum_{k=1}^{N} S_{p_i}^{y_j}(k)^2} \quad i = 1, \ldots, n_p \tag{2.33}$$

Extending the computation of Equation 2.33 to all of the n_p parameters in the model, they can be ranked according to their scores ξ, so that we can decide which are the most sensitive and limit the estimation to that critical subset. The N sampling times in Equation 2.33 can be freely selected and are not required to coincide with the actual sampling times.

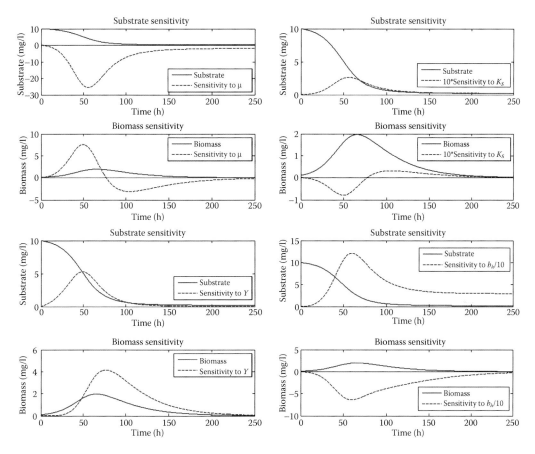

FIGURE 2.14 Trajectory sensitivities of the Monod kinetic to the four parameters computed with the approximate method and $\delta = 10^{-4}$. Note the largely differing magnitudes, which are unimportant in the sensitivity assessment.

2.3   GRADIENT-FREE OPTIMIZATION METHODS (DIRECT SEARCH)

Having assessed the shape of the error functional and ranked the parameter sensitivities, we are now in a position to move onto the actual parameter estimation, limiting the procedure to the subset of parameters with the highest sensitivity. In fact, the number of parameters that can successfully be estimated depends on their identifiability and the amount of available data. A very rough rule of thumb suggests that for each parameter, at least a dozen data should be available, so the amount of parameters that can actually be estimated is determined not only by their intrinsic identifiability, but also by the data availability. A more compelling reason for using as many data as possible is based on the fact that the statistics used to judge the quality of the estimation χ^2 and F hold if the number of data is sufficiently large.

Having resolved this preliminary selection problem, we now turn our attention to practical methods for searching the minimum of the error functional (2.4) using a gradient-free method for two good reasons. First, the analytical solution to the model (2.1) is almost always unavailable, so the gradient could not be obtained via calculus. Second, in narrow valley situations, such as encountered with the Monod kinetics and shown in Figure 2.10, the gradient method, projecting the search in the direction of maximum slope, may bounce back and forth between the almost parallel walls of the valley. Figure 2.17 compares the performance of gradient and search methods in coping with the narrow valley problem. Clearly, the search method has an edge, and is to be preferred in this context.

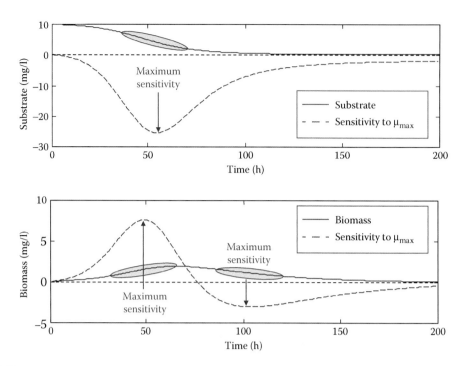

FIGURE 2.15   Use of the trajectory sensitivity to select the information-rich data for the estimation of μ_{max}. In this case, the favourable time brackets of substrate and biomass almost coincide, with an additional biomass window in the endogenous phase.

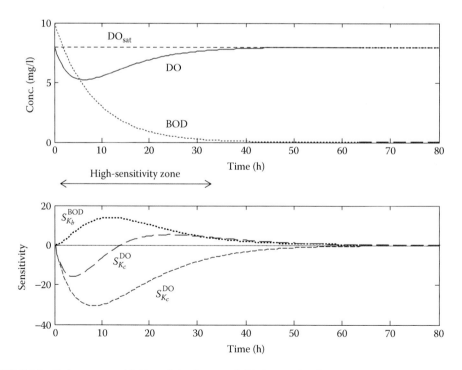

FIGURE 2.16   Trajectory sensitivities of the Streeter & Phelps model. The high sensitivity zone is located around the DO_sag.

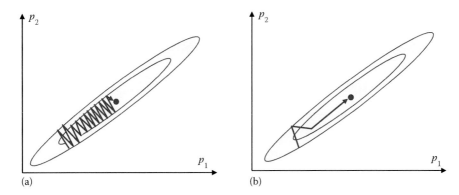

FIGURE 2.17 Problems created by a 'narrow valley' error functional. The gradient-based method (a) makes little progress because its projection is orthogonal to the contour lines. A search method (b) reaches the minimum quickly once the valley direction has been detected.

The optimization of Equation 2.4 is usually solved through a numerical search combined with a simulation engine to obtain the model output for each new trial of the parameter vector P. The simplex algorithm, which will soon be described in detail, is an improved version of the classical flexible polyhedron (Nelder and Mead, 1965; Himmelblau, 1972), with an optimized expansion (Marsili-Libelli and Castelli, 1987; Marsili-Libelli, 1992). Other efficient search methods are the quasi-Newton (QN) method (Himmelblau, 1972; Press et al., 1986). The algorithm implemented in the MATLAB Optimization Toolbox uses the Broyden, Fletcher, Goldfarb, and Shanno (BFGS) version (Shanno, 1970). It should be recalled that BFGS is just one implementation of the QN optimization method. Alternative methods exist, such as the Davidson, Fletcher, and Powell (DFP) algorithm (Himmelblau, 1972), although the BFGS method can be regarded as the best performing one. The QN algorithm is well suited for highly correlated problems because it approximates the error functional with the Hessian matrix (second order), and not with the Jacobian (first order), as does the Gauss–Newton method. The Levenberg–Marquardt algorithm should also be mentioned as one of the most popular nonlinear optimization methods, although it has been criticized as inadequate for a wide range of parameter estimation problems (De Pauw, 2005).

Another important aspect to consider is that all of the above optimization methods are inherently *unconstrained*, although parameter estimation sometimes requires constraints on parameter values, for example positivity. There are several reasons that prevent the introduction of constraints in the simplex search (Marsili-Libelli et al., 2003; Checchi and Marsili-Libelli, 2005; Marsili-Libelli and Checchi, 2005). First of all, constraints may disrupt the convergence process, introduce a wide discretion margin and produce a less realistic covariance matrix. Further, if the search is left unconstrained, an incorrect problem formulation can be easily detected, for example, if negative or 'wild' parameter values are found. The problems arising from imposing constraints are well treated by Ruano et al. (2007).

2.3.1 Optimized Flexible Polyhedron

All the direct search methods originated from the basic flexible polyhedron algorithm (Nelder and Mead, 1965; Himmelblau, 1972; Kuester and Mize, 1974). The basic idea is to evaluate the error functional for several combinations of the parameter values, discard the worst and substitute it with a better one, whose selection is the core of the search algorithm. Given n_p parameters, the 'flexible polyhedron' has $n_p + 1$ vertices $\left(P_1, P_2, \ldots, P_{n_p}, P_{n_p+1} \right)$, so that it covers a convex nondegenerate region of the parameter space (the 'hull'). After testing the error functional with each of the $n_p + 1$ combinations, a ranking is made, and the worst one is replaced with a better one.

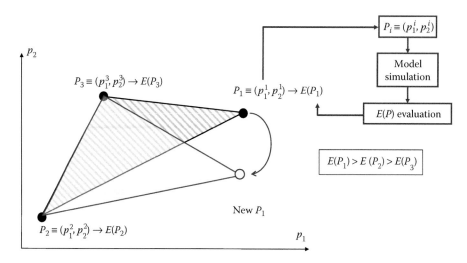

FIGURE 2.18   Substitution of the worst point (vertex) of the old polyhedron (hatched triangle). In this case, the ranking resulted in P_1 being the worst point, and it is replaced by the new P_1. Convexity requires the polyhedron to have $n_p + 1$ vertices.

The problem is to make an educated guess to select the new parameter vector, which will be referred to simply as 'point', in the parameter space. Figure 2.18 shows the worst point replacement procedure in a two-parameter space: after evaluating $E(P)$ for all the vertices by model simulation, the worst is discarded and replaced by a new one. The question now is how to select a new point that will improve the search.

2.3.1.1   Polyhedron Reshaping

The simplex polyhedron is called 'flexible', and not without reason. In fact, it is conceived to change its shape and adapt to the error surface in its search for the minimum. Once all the vertices have been tested and ranked as $(P_{max}, \ldots, P_{min})$, the worst point P_{max} is excluded from the set, and the *centroid* of the remaining n_p points is computed as their average

$$P_{cen} = \frac{1}{n_p} \sum_{\substack{i=1 \\ i \neq max}}^{n_p} P_i \tag{2.34}$$

One of the following operations is then performed, depending on the relative merit of each point, along the search direction defined by the line joining P_{max} to P_{cen}:

Reflection: This preliminary operation defines a new point P_r along the direction $(P_{max} \rightarrow P_{cen})$ as

$$P_r = P_{cen} + \alpha (P_{cen} - P_{max}) \quad \alpha > 0 \tag{2.35}$$

$E(P_r)$ is then compared with the previous values. Based on its relative merit, one of the three operations of Figure 2.19 is made:

Expansion: If $E(P_r) < E(P_{min})$, the reflected point is better than the previous best (P_{min}), so the search is continued along the same direction, and a new *expanded* point P_{exp} is determined as

$$P_{exp} = P_{cen} + \gamma (P_{cen} - P_{max}) \quad \gamma > 1 \tag{2.36}$$

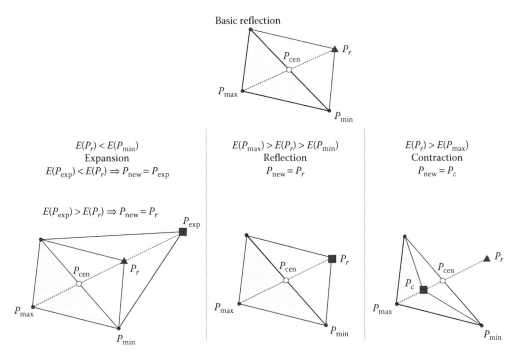

FIGURE 2.19 The three basic operations through which the simplex changes its shape starting with the reflected point (filled triangle). The final new point is indicated by the filled square, while a hollow circle indicates the centroid.

Now, the new point P_{new} is determined, depending on the comparison between P_r and P_{exp}, as follows:

$$P_{new} = \begin{cases} P_{exp} & \text{if } E(P_{exp}) < E(P_r) \\ P_r & \text{if } E(P_{exp}) > E(P_r) \end{cases} \qquad (2.37)$$

Reflection: If the merit of the reflected point is intermediate between the maximum and minimum points, then the reflected point already computed with Equation 2.35 is retained as the new point, that is,

$$P_{new} = P_r \text{ if } E(P_{max}) > E(P_r) > E(P_{min}) \qquad (2.38)$$

Contraction: If the expansion does not bring any improvement, that is, if $E(P_r) > E(P_{max})$, then the new point is selected inside the old polyhedron according to

$$P_{new} = P_c = P_{cen} + \beta(P_{cen} - P_{max}) \quad 0 < \beta < 1 \qquad (2.39)$$

Finally, If $E(P_c) > E(P_{max})$, then the simplex must be shrunk to sink deeper into the bottom of the error surface. This is accomplished through a

Reduction: All the vertices are brought closer to the best point (P_{min}) by reducing all of its sides by a fixed amount

$$P_i^{(new)} = P_i^{(old)} + \delta(P_i^{(old)} - P_{min}) \quad 0 < \delta < 1 \quad i = 1, \ldots, n_p + 1 \text{ and } i \neq \min \qquad (2.40)$$

2.3.2 TERMINATION CRITERION

To terminate the search, a test for flatness is made on the polyhedron. In fact, when all the vertices yield nearly the same functional level within a specified tolerance ε, it is assumed that the search has reached the bottom of the surface, that is, it is close to the minimum. Therefore, the terminating criterion can be stated as

$$\varepsilon \geq \frac{1}{n_p} \sum_{i=1}^{n_p+1} \left[E\left(\boldsymbol{P}_i\right) - E\left(\boldsymbol{P}_{\text{cen}}\right) \right]^2 \tag{2.41}$$

It might be suggested that the search can be stopped when the error has reached a certain value. Optimistically (or naively), one might suggest to stop the search when $E\left(\boldsymbol{P}_{\min}\right) = 0$, but this value would never be reached in practice, because there will always be a residual error due to the noise affecting either the data and/or model approximations. Setting a nonzero value for $E\left(\boldsymbol{P}_{\min}\right)$ would not work either, because there is no clue on how to set an appropriate value. Hence, the constraint (2.41) emerges as the only sensible stopping criterion. To elaborate further on the behaviour of the search near the minimum, consider that the data actually used for the estimation are the sum of the 'true' model error $y(t)$ and the measurement noise $v(t)$, as shown in Figure 2.20 for the case of a single output and a single parameter. Thus, the minimum of the error functional is the combination of two error terms, due to the model mismatch and/or to the presence of measurement noise, that is,

$$E\left(p_{\min}\right) = \begin{cases} \text{perfect model \& no noise} & E_{\text{mod}} = 0 \;\&\; \sigma = 0 & E\left(p_{\min}\right) = 0 \\ \text{perfect model \& noise} & E_{\text{mod}} = 0 \;\&\; \sigma \neq 0 & E\left(p_{\min}\right) = \left(N - n_p\right)s^2 \\ \text{model error \& noise} & E_{\text{mod}} \neq 0 \;\&\; \sigma \neq 0 & E\left(p_{\min}\right) = E_{\text{mod}} + \left(N - n_p\right)s^2 \end{cases} \tag{2.42}$$

Figure 2.21 shows the effects of modelling and measurement errors following the scheme of Figure 2.20.

The bottom curve represents the ideal case of perfect model and noiseless measurements. In this case, the estimated value of the parameter $p\left(p_{\min}\right)$ converges to its true value p_{true} and $E_1\left(p_{\text{true}}\right) = 0$. The second curve represents the case of perfect model and noisy measurements. In this case, the residual error is just the noise variance, that is, $E_2\left(p_{\min}\right) = \left(N - 1\right)s^2$. In the third curve, there are both modelling $\left(E_{\text{mod}}\right)$ and measurement errors $\left(\sigma^2 \sim s^2\right)$, so the minimum error will be the sum of both factors, that is, $E_3\left(p_{\min}\right) = E_{\text{mod}} + \left(N - 1\right)s^2$. If both errors are moderate, then the algorithm will still converge to the true parameter value $p_{\min} = p_{\text{true}}$, but if modelling and/or measurement

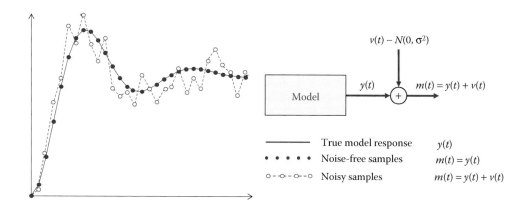

FIGURE 2.20 Modelling the measurement error as an additive noise affecting the output.

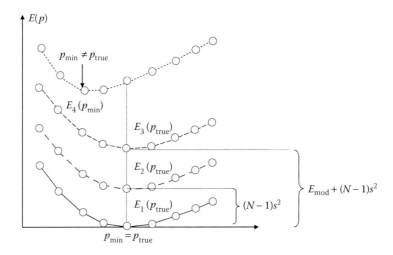

FIGURE 2.21 Effect of model and measurement error on the estimation accuracy, in the single parameter case.

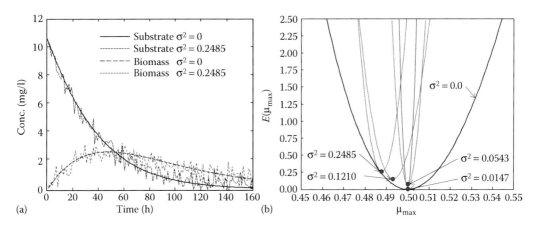

FIGURE 2.22 An example of how noisy measurement may shift the minimum of the error functional. The noiseless and noisy measurement are shown on the left (a), while (b) depicts the changes in the error functional due to the noisy measurements. If the noise is moderate, the minimum still coincides with the true parameter value, but it is shifted away as the noise increases.

errors are large, the minimum of the error functional may be shifted, so that the estimator will no longer converge to the true value (upper curve). An example of the effect of noise on the estimation is provided in Figure 2.22, where the maximum growth rate (μ_{max}) of the Monod kinetics is estimated with 'clean' and noisy data. The static sensitivity curve is affected by noise that also shifts the minimum variance that only coincides with the true value for very low noise.

2.3.2.1 Optimized Simplex Search

One of the weaknesses of the basic simplex search is that the polyhedron reshaping parameters $(\alpha, \beta, \gamma, \delta)$ are constant throughout the search. On the other hand, the search could be improved by adjusting the local expansion to reach the local minimum in the current search direction, as shown in Figure 2.23. This improvement was introduced by setting up a unidirectional search (Marsili-Libelli and Castelli, 1987; Marsili-Libelli, 1992) based on the Fibonacci interval elimination (Himmelblau, 1972). This approach results in fewer, but more effective expansions, and has proved particularly efficient in coping with the 'narrow valley' problem.

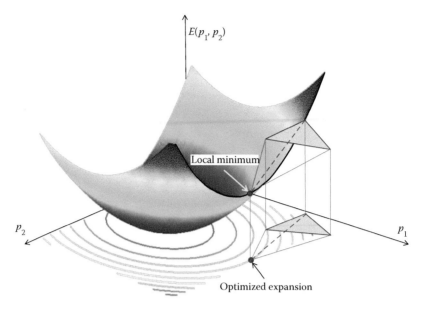

FIGURE 2.23   Optimized search along the reflection line. The black line indicates the section of the $E(\boldsymbol{P})$ surface that includes the search direction, along which the local minimum is reached.

2.3.3   Genetic Algorithm

The other major weakness of the search algorithm is that it explores a limited range of the domain at any one time, so there is the risk that the polyhedron becomes trapped in a local minimum, and terminates the search there if the stopping criterion (2.41) based on its flatness is met. For this reason, it is advisable to run a global optimization algorithm prior to the simplex search. This algorithm, though not as accurate as the simplex, will not be sidetracked by local minima or other undesirable error function features, and it will quickly find the approximate location of the global minimum. This information can then be used to start the flexible polyhedron algorithm.

Genetic algorithms (GAs) represent a highly efficient search procedure for the determination of global extremal points of multivariable functions, imitating the patterns of genetic reproduction in living organisms. GAs work by successive modifications of a collection (or *population* in genetics parlance) of parameter combinations in the search space, after appropriate coding and bundling them into numerical strings termed *chromosomes*. The initial, randomly chosen, set of chromosomes evolves through a number of operations (*selection, crossover* and *mutation*), so that subsequent generations produce more and more chromosomes in the neighbourhood of the global optimum, according to a previously defined *fitness* functional. Thus, GAs are the mathematical version of the 'survival of the fittest' strategy: they evolve, improving the qualities of the population towards a given fitness objective, and for this reason no terminating criterion such as Equation 2.41 can be defined. The population simply evolves until the programmed number of generations is reached, and the best chromosome in the last generation is assumed to be the overall best.

GAs were introduced by Goldberg (1989) and Holland (1992). A more recent interesting publication (Poli et al., 2008) can be found on the internet (http://www.gp-field-guide.org.uk) to get an overview of the method. GAs differ from conventional search algorithms in the following aspects:

- GAs work with a *coding* of the parameters, not the parameters themselves.
- GAs' search is highly parallel, using a collection (*population*) of points (parameters combinations), not a single one, thus thoroughly covering the search region and avoiding being trapped into local extrema.

- GAs use probabilistic, not deterministic, transition rules to modify the initial population over successive generations.
- GAs have a high implicit parallelism, making them numerically very efficient.

Thanks to their derivative-free, highly parallel search method, GAs are highly insensitive to discontinuities and irregularities in the fitness function.

2.3.3.1 Parameter Coding

This very important preliminary operation involves bundling all the parameters into a single string (*chromosome*). The most frequently used conversion transforms the real numbers into decimal integers corresponding to a binary coding. Having defined the range of the parameter values as $R(r_{min}, r_{max})$ and the number of bits of the base-2 integer coding (β), a given decimal number p is converted into a base-10 integer as

$$g = \text{INT}\left[\frac{p - r_{min}}{r_{max} - r_{min}}(2^\beta - 1)\right] \tag{2.43}$$

As an example, suppose that the selected range is $R = (-1, 1)$, and that we are going to use $\beta = 12$ bits for the coding. Then, the number $p = 0.2345$ would be converted as

$$g = \text{INT}\left[\frac{0.2345 - (-1)}{1 - (-1)}(2^{12} - 1)\right] = \text{INT}\left[\frac{1.2345}{2}(4096 - 1)\right] = 2528 \tag{2.44}$$

As a counterproof, let us transform the result of Equation 2.44 back into the original number. Inverting Equation 2.43 yields

$$p = r_{min} + (r_{max} - r_{min})\frac{g}{(2^\beta - 1)} \tag{2.45}$$

Substituting the values of Equation 2.44

$$p = -1 + [1 - (-1)]\frac{2528}{(4096 - 1)} = -1 + 2\frac{2528}{4095} = 0.2347 \tag{2.46}$$

we recover the original p, within the precision provided by the number of bits (12 in this case). In fact it can be seen that the last digit in Equation 2.46 is not correctly recovered. To conserve the n-decimal precision of the original values $v_{10} = 10^{-n}$, the equivalent binary precision $v_2 = 2^{-\beta}$ should use a number of digits β equal to

$$\beta = \text{int sup}\left(\frac{n}{\log_{10} 2}\right) = \left\lceil\frac{n}{0.69314718}\right\rceil \tag{2.47}$$

In MATLAB, the *int sup* is implemented by the `ceil` (ceiling) upper round-off operator.

Once the parameters have been converted into integers, they are organized in *chromosomes* by arranging them into a single string. In this operation, the boundaries among individual parameters are lost, and the string is considered (and processed) as a whole entity. Suppose that we want to code a vector of five parameters specified with a 4 decimal places precision in the range $(-1, 1)$. Then, Equation 2.47 suggests that to maintain the given precision, 14 bits are required, in fact, `ceil(4/log10(2))` `=` `14`. Thus, the conversion that conserves the original precision yields

$$P = [0.2345 \ -0.2363 \ 0.7867 \ -0.4593 \ 0.3745]$$

$$\Downarrow \tag{2.48}$$

$$G = [40451|25231|58523|17716|45039]$$

where the bars delimit the field of each parameter, but will be ignored by the GA, as G will be regarded as a single string.

2.3.3.2 Basic Evolutionary Mechanisms

The GA has three basic mechanisms to 'improve' the quality (or the 'fitness' in GA parlance) of the *population*, as the collection of the various chromosomes is called, all borrowed from the world of genetics. The optimization procedure starts from a population of randomly chosen chromosomes and generates successive populations, applying the operators of *reproduction* and *mutation*. Reproduction is composed of two operations: *selection* and *crossover*.

Selection stimulates the generation of offspring from the best-fitting chromosomes. Each chromosome is evaluated in terms of its relative fitness, in much the same way as each point of the simplex is evaluated with respect of the error functional $E(P)$. Then, each chromosome receives a rating based on the ratio of its fitness to the average fitness, which determines how many offspring it is allowed to generate. In this way, the chromosomes with the highest fitness produce more offsprings, whereas those with below average fitness are destined for extinction.

Crossover consists of splitting a chromosome pair at a random location with reciprocal exchange of one half, thus causing a mutual flow of information between pairs. In this way, in the next generation there will be two new chromosomes with new gene combinations. Both these operations are illustrated in Figure 2.24.

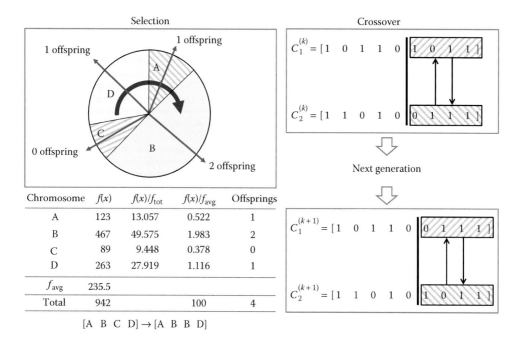

Chromosome	$f(x)$	$f(x)/f_{tot}$	$f(x)/f_{avg}$	Offsprings
A	123	13.057	0.522	1
B	467	49.575	1.983	2
C	89	9.448	0.378	0
D	263	27.919	1.116	1
f_{avg}	235.5			
Total	942		100	4

$$[A \ B \ C \ D] \rightarrow [A \ B \ B \ D]$$

FIGURE 2.24 Fundamental operations on chromosomes: in the selection, each individual chromosome is allowed to reproduce in proportion to its fitness, whereas in the reproduction, chromosome couples exchange genetic material, giving birth to new chromosomes. The figures in the table are just for demonstrating the selection method and do not refer to any example in particular.

Mutation introduces a random gene change in a chromosome. Usually, the mutation likelihood is kept constant at a low value for all genes, but fitness-dependent mutation has also been proposed (Marsili-Libelli and Alba, 2000).

2.3.3.3 Similarity Templates

As the population evolves following the above operations, it may happen that some gene patterns are more successful than other, and should therefore be preserved, or even privileged, in their propagation. These patterns are called *similarity templates* or *schema,* and are defined, for example, as

$$H = [1 \quad 0 \quad 1 \quad * \quad 1 \quad * \quad 1 \quad *] \tag{2.49}$$

where the symbol '*' indicates a 'don't care' gene, meaning that it may take any value, whereas the other genes are to be preserved, and therefore should not be involved in crossover and mutation. The length of a schema is computed as the difference between the positions of the last and the first fixed gene. Hence, in the example of Equation 2.49, the length is $\delta(H) = 7 - 1 = 6$. Longer schemes are more difficult to preserve because they are more likely to be disrupted by crossover. On the other hand, the length is a measure of the scheme importance.

Goldberg (1989) and Holland (1992) have demonstrated how schema evolve through successive generations. Let $m(H, k)$ be the number of chromosomes with a given template H at the kth generation. Then, its number at the next generation will be

$$m(H,k+1) = m(H,k)\frac{f(H)}{\bar{f}} \quad \text{where} \quad \bar{f} = \frac{1}{N}\sum_{i=1}^{N} f_i \tag{2.50}$$

and $\bar{f}$ is the mean fitness of the population at the kth generation. Equation 2.50 shows that schema with fitness above the average will expand in the next generation, whereas schema with fitness below the average will produce fewer and fewer offspring in successive generations. The propagation Equation 2.50 can be rewritten, considering a schema H that is above the average by an amount $c \cdot \bar{f}$ as

$$m(H,k+1) = m(H,k)\frac{\bar{f} + c \cdot \bar{f}}{\bar{f}} = (1+c) \cdot m(H,k) \tag{2.51}$$

Considering the evolution of the population from the start $(k = 0)$ yields

$$m(H,t) = m(H,0) \cdot (1+c)^k \tag{2.52}$$

which means that above-average schema increase exponentially over the generations, whereas underperformers disappear at the same rate. The problem with long schema is their possible breakdown due to crossover, thus losing the distinctive features embedded in their pattern. For this reason, attention is focused on short templates, or building blocks, which are less likely to be disrupted by crossover. The previous fundamental theorem can be viewed in terms of building blocks, meaning that a schema is a collection of many short similarity templates that propagate independently and exponentially over generations, depending on their fitness.

A schema may be disrupted by crossover, and its robustness depends on its length, with long schema being more likely to succumb to crossover. The probability that a schema of length δ in a chromosome of ℓ genes is cut across is given by

$$p_\delta = \frac{\delta(H)}{\ell - 1} \tag{2.53}$$

so its survival will be

$$p_s = 1 - p_d = 1 - \frac{\delta(H)}{\ell - 1} \qquad (2.54)$$

If a random crossover occurs inside the schema with probability p_c, then the survival probability becomes

$$p_s \geq 1 - p_c \cdot \frac{\delta(H)}{\ell - 1} \qquad (2.55)$$

Assuming that selection and crossover are independent random processes, then the probability that a schema survives in the next generation is given by the combination of the two terms (2.50) and (2.55). If the effect of mutation is also added, assuming that it can change only one gene at any one time with probability $p_m \ll 1$, the global dynamics of the schema is given by

$$m(H, k+1) \geq m(H, k) \underbrace{\frac{f(H)}{\bar{f}}}_{\substack{\text{reproduction} \\ \text{survival}}} \left(1 - \underbrace{p_c \cdot \frac{\delta(H)}{\ell - 1}}_{\substack{\text{crossover} \\ \text{survival}}} - \underbrace{p_m \cdot \delta(H)}_{\substack{\text{mutation} \\ \text{survival}}} \right) \qquad (2.56)$$

2.3.3.4   GA in MATLAB

The MATLAB Optimization Toolbox provides an implementation of the GA with the solver named ga. Its syntax is similar to the syntax of the other optimization functions implemented in the toolbox, and its complete form is

```
[P,fval,exitflag]

= ga(errorfcn,np,A,b,Aeq,beq,LB,UB,@nonlcon,options)
```

$$(2.57)$$

The solver ga returns the n_p-dimensional solution $\hat{P}$, together with the value fval of the error functional $E(\hat{P})$, at the solution P, and the flag exitflag indicates the exit conditions. As to the input parameters, the first one calls the error functional, defined as either a MATLAB function file or an anonymous function, using the function handler '@'. In the latter case, please refer to the MATLAB help for its definition. The second input n_p is the dimension of the parameter vector P. If an unconstrained optimization is performed, all the other input parameters should be set as empty quantities '[]'. In the case of constrained optimization, the parameters A and b define the linear inequalities A × P ≤ b, whereas Aeq and beq defined the linear equalities Aeq x P = beq, with the vectors LB and UB representing respectively the lower and upper bounds for the solution LB≤P≤UB.

The last input, options, is a structure that includes all the solver settings, such as the maximum number of generations ('generations') or the maximum duration of the optimization process ('TimeLimit'). The creation of the initial population can also be defined here (CreationFcn), together with the number and the type of the populations (PopulationSize and PopulationType), the selection of the fittest individuals (SelectionFcn) and the method used for the mutations (MutationFcn).

The CreationFcn option generates the initial population. Setting it to 'uniform' creates an initial random population with uniform distribution, whereas 'feasible population' defines an initial well-distributed population that satisfies all the constraints. Choosing 'custom', a customized generation method can be defined with the data type indicated in PopulationType. This option is used to set the data type of P, which can be either 'doubleVector' for real variables `bitString` if they are bit strings or `custom` if they are user defined, in which case the user must specify the creation, mutation and crossover functions. If the `PopulationType` option is set to `bitString` or `custom`, linear and nonlinear constraints cannot be specified. The `PopulationSize` option defines the number of chromosomes of which the population is composed. A large population means a more accurate search of the solution with a more accurate exploration of the domain, but, of course, slows down the computation. For each option, the toolbox offers a set of pre-defined methods, but it is also possible to include user-defined methods. All these specifications can be created using the `gaoptimset` prior to the actual optimization. More details of the `ga` solver can be found in the MATLAB Help.

2.3.4 PERFORMANCE ASSESSMENT OF THE COMBINED ESTIMATION METHODS APPLIED TO THE MONOD KINETICS

The combined performance of the above search algorithms is now tested on the Monod kinetics, assumed as our benchmark. Though we have already studied the shape of the error functional and ruled out the existence of local minima, at least in the region of interest, nevertheless, starting the simplex search from a good point would surely expedite the procedure. A good test for any estimation algorithm is to generate data by simulating the model with known parameters and use them as experimental data. If the algorithm is correct, the estimated parameters should coincide, within inevitable round-off errors, with the original parameters used to generate the data.

We start the estimation by generating a population of parameters, and use a GA to explore the domain. After a certain number of generations, the GA will indicate the location of the minimum with a precision depending on the size of the population (N) and the number of generation. At this stage, we do not need to set very stringent requirements, as the main objective is to avoid pitfalls into which the simplex may fall, such as local minima. Consequently, when the GA obtains a rough estimate of the minimum, we can use this to initialize the simplex, which will provide more precision.

Figure 2.25 shows the two phases of the estimation of the two main (and most troublesome) parameters in the Monod kinetics. In the first phase, illustrated in Figure 2.25a, a GA with a population of 30 chromosomes selects the region around the minimum, and its best chromosome after 20 iterations is $\left(\mu_{\max} = 0.4587, K_s = 26.8417\right)$. In Figure 2.25b, the simplex algorithm, initialized with the GA result, reaches the exact parameter values. Notice that the search is predominantly concentrated along the valley bottom.

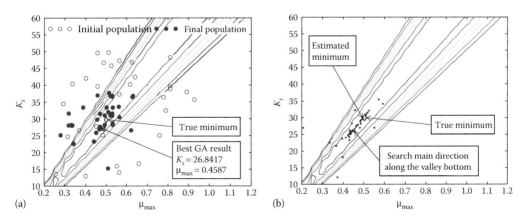

FIGURE 2.25 Estimation of the parameter couple ($\mu_{\max}, K_s$). In (a), the GA explores the domain with a population of 30 chromosomes. This preliminary phase ends after the prescribed number of generations. At that time most of the points (solid dots) are closer to the minimum than the initial population (hollow circles). In (b), the best chromosome is selected as the starter for the simplex, which reaches the true parameters by exploring predominantly along the valley bottom.

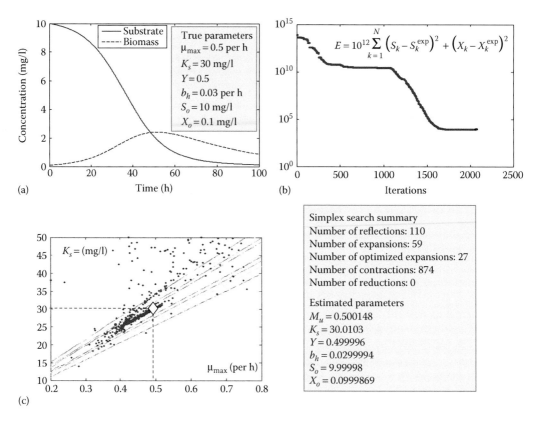

FIGURE 2.26 Full estimation of the Monod kinetics, including the initial conditions, with noise-free data (a). The decrease of the error functional (b) is uneven and very slow in the final part. In (c), the search progress is shown in the (μ_{max}, K_s) subspace, terminating almost at the true values (hollow diamond). Notice the privileged search direction along the valley bottom.

A full estimation (parameters plus initial conditions), shown in Figure 2.26, requires considerably more iterations, and the decrease of the error functional (b) is marked by several plateaus, where the algorithm seems to make little progress. These are the phases when a new search direction is tested by the optimized expansion sub-algorithm of Section 2.3.2.1.

In the case of noisy data, shown in Figure 2.27a, the error functional is raised by an amount equal to the total measurement variance, provided that there are no modelling errors, as shown in Figure 2.21. As Figure 2.27b shows, the minimum of the error functional (E_{min}) equals the total noise variance on the data, while it retains its smoothness thanks to its averaging properties. Thus,

$$E\left(\boldsymbol{P}\right) = \frac{1}{N-1}\left[\sum_{k=1}^{N}\left(S_k^{exp} - S_k\right)^2 + \sum_{k=1}^{N}\left(X_k^{exp} - X_k\right)^2\right] \rightarrow \min_{\boldsymbol{P}}\left(E\right) \approx \sigma_S^2 + \sigma_X^2 \qquad (2.58)$$

However, numerical inaccuracies may creep into the computation of the error functional, whose surface may become 'ripply', as shown in Figure 2.28. The simplex may be trapped inside its many tiny local minima, if its size is too small as a result of many consecutive contractions and/or reductions.

Figure 2.29b shows the more elaborate search pattern of the simplex in finding the minimum, ending the exploration near the true parameters in spite of the noise affecting the data, Figure 2.29c, which is more evident in the inset (d).

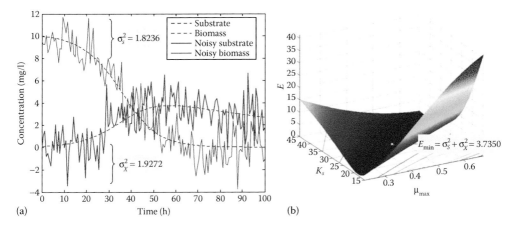

FIGURE 2.27   When noisy measurements are used, as in (a), the minimum (E_{min}) of the error functional E equals the total measurement variance, as shown in (b), provided that there is no inherent model error. Due to the averaging property of E, the functional retains its smoothness, regardless of the noisy measurements. This result is in agreement with Figures 2.12 and 2.22b.

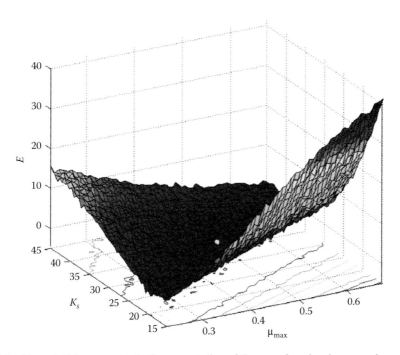

FIGURE 2.28   Numerical inaccuracies in the computation of the error functional may produce a ripply surface with many small local minima, in which the search algorithm can be trapped.

2.3.5   MATLAB Software Organization

We now turn our attention to the software implementation of the estimation algorithm. Figure 2.30 shows how the various modules are organized. In addition to the main script, there are two functions (simplex and error functions) and the Simulink® model. The main script specifies the nature of the problem (experimental data, model definition, simplex options, etc.). It also defines the *global* variables that will be made visible to the error function, and then passed on to the Simulink model. In fact, global variables, defined in a script, are normally available to the called Simulink model

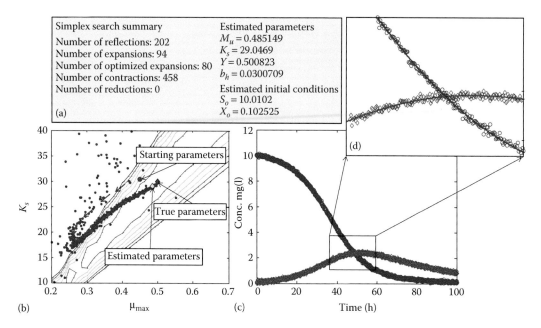

FIGURE 2.29   A full Monod kinetics calibration, including all the four parameters and the initial conditions, using noisy data. The calibration results are summarized in (a), whereas in (b) the arrows show the progress of the search path. The noisy data are shown in (c) with the inset (d) providing a magnified view of the noise near the maximum growth time.

through the common memory area called Workspace, but in this case the model is called by a function, which does not possess the prerogatives to publish its internal variables to the Workspace. Hence, the variables created in the error function would not be visible by the Simulink, and thus the need to define them as *global*. All the involved MATLAB modules (main and functions) should have the same *global* declaration, with the same variable names, before any executable statement. As Figure 2.30 shows, the main script loads the experimental data, defines the model and the simplex options, and then invokes the simplex function that supervises the optimization and releases control back to the main script after the search is terminated, because either the minimum has been reached (within the prescribed precision) or the maximum number of iterations has been exceeded. There may be a downstream processing section, where results are displayed and the estimates are validated, organized as shown in Box 2.1.

2.3.5.1   Organization of the Error Function

This function plays a crucial role in the estimation and should be specified by the user because it unbundles and passes the parameter to the Simulink model while supervising the computation of the actual estimation error. The basic structure of the pertinent MATLAB function is listed in Box 2.2.

Its salient features are the following: unbundling of the parameter vector to pass the real parameter names to the Simulink model, launching the simulation, synchronizing the simulation outputs with the experimental sampling times and computing the error functional.

Two aspects deserve further comment: the presence of the global statement and the synchronization. As to the first, a MATLAB function does not possess the prerogative to pass variables to another function (in this case a Simulink model), but the global declaration solves this parameter passing problem. The other aspect is synchronization, which requires the model outputs to be available at exactly the same instants as the experimental samples. This aspect is further clarified in Figure 2.31, showing how the model outputs are interpolated to be evaluated at the same sampling times as the experimental data.

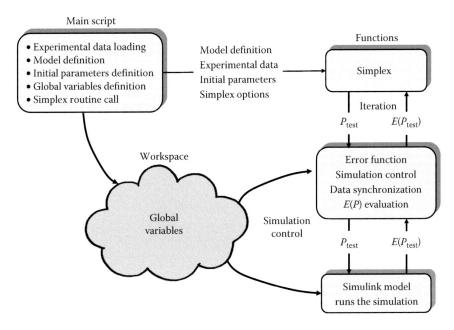

FIGURE 2.30 Software organization for the simplex search. The variables declared as 'global' are made available to the Simulink model through the calling error function.

BOX 2.1 BASIC STRUCTURE OF THE MAIN SCRIPT FOR THE ESTIMATION OF THE MONOD KINETIC PARAMETERS

There are three main segments to be considered: loading the experimental data, performing the optimization and the optional downstream processing of the results.

```
% Global parameters definition
global Mu Ks Kd Y So Xo
% Load experimental data
%
g=uigetfile('Mono*.mat','Load calibration data');
eval(['load' g]);
%
% Setting optimization options
options(1)=1;     % Print intermediate results
options(2)=0.01;  % Error tolerance on parameters
options(3)=0.01;  % Simplex flatness tolerance
options(14)=5000; % Maximum allowable iterations
% Optimization call
[Par_Best,options]=simplex('Ferr_Monod',Par,options,[],Model_
Name,tfin,Data_exp);
ParBest=ParCal;
% Downstream processing
% . . . . . . . . . . . . . . . . . . . .
```

**BOX 2.2   BASIC STRUCTURE OF THE MATLAB FUNCTION
COMPUTING THE ERROR FUNCTIONAL**

Possible plotting instructions have been omitted for clarity.

```
function E = Ferr_Monod(Par,Model_Name,tfin,Data_exp)
global Mu Ks Y bh So Xo
% Unbundle parameters
Ks=Par(1); Mu=Par(2); Y=Par(3); Kd=Par(4);
% Launch simulation
[t,x,y]=sim(Model_Name,tfin);
% Synchronize data and simulations
% Mono are the simulation outputs
% Mono(:,1) is the Substrate
% Mono(:,2) is the Biomass
% Data_exp(:,1) contains the experimental sampling times
% Data_exp(:,2) contains the experimental samples
% The simulation output is interpolated
% at the experimental data sampling times Data_exp(:,1)
ys1=interp1(Mono(:,1),Mono(:,2),Data_exp(:,1));
ys2=interp1(Mono(:,1),Mono(:,3),Data_exp(:,1));
% Compute the error functional
s1=sum((Data_sperim(:,2)-ys1).^2); % Substrate error
s2=sum((Data_sperim(:,3)-ys2).^2); % Biomass error
E=s1+s2;
```

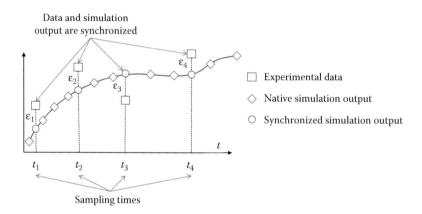

FIGURE 2.31   Synchronizing the simulation output to the experimental data is crucial for the correct evaluation of the error functional. Normally, simulation steps are decided by Simulink to comply with the integration error specifications, but for a correct computation of the errors (ε) between data and model outputs the latter must be evaluated at the sampling instants by interpolation.

2.4   VALIDATION OF THE ESTIMATES

Having gone through the previous computational tasks, we are only halfway to completing a serious parameter estimation job. We now must assess the results in terms of their credibility and reliability. To this end, two complementary approaches are possible, and *both* should be pursued, in the sense that each provides one kind of assessment, which the other cannot yield. Figure 2.32 provides an overview of the two approaches, with various procedures to validate the identification.

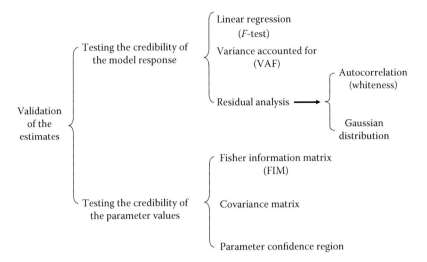

FIGURE 2.32   Procedures for the validation of the identified model.

These tests can be divided into two main categories: non-parametric and parametric. In the first category, the focus is on the agreement between the data and the model response, whereas in the second, the parameter values are assessed *per se*, no matter how good the model fit is. We begin with considering the first category of tests, which could be regarded as preliminary in the sense that if the model fails these tests, then it is unlikely to pass the second category, and its structure should be reconsidered. Before actually embarking on the analysis, however, it should be considered that validation is performed on a statistical basis, and therefore we should view the results of our analysis from a statistical standpoint. This requires the formulation of a hypothesis (the null hypothesis H_0), and its acceptance or rejection will be based on a statistical rationale. In this case, our null hypothesis can be formulated as

$$H_0 : \text{we cannot rule out data/model agreement} \tag{2.59}$$

which is to say, in the typically oblique statistical jargon, that we consider our model correct if we cannot prove it wrong. The null hypothesis H_0 is subject to two types of errors:

- Type I error: failing to accept a correct H_0
- Type II error: failing to reject a wrong H_0

Of course, we should also specify the significance level α of taking a wrong decision, that is, we would observe values greater than the one produced by the test only α of the time by chance. Normally, we set this margin of error at $\alpha = 5\%$. The two errors are not symmetrical, and they depend on the method used to test H_0. Generally, there is an ample tolerance in accepting H_0.

2.4.1   NON-PARAMETRIC TESTS: TESTING THE CREDIBILITY OF THE MODEL RESPONSE

The purpose of these tests is to check the agreement between data and model response without considering the correctness of the parameter values, which will be treated in the next set of validation tools. A wide review of non-parametric methods for testing the performance of environmental models can be found in Bennett et al. (2013).

2.4.1.1   Variance Accounted For

This test is well known under several alternative names, being also referred to as *Coefficient of Determination* or *Nash–Sutcliffe Model Efficiency*. It is not a very strict test that compares the inner data variability to the model residuals, namely

$$\text{VAF} = 100\left[1 - \frac{\sum_k \left(y_k^{\text{exp}} - y_k^m\right)^2}{\sum_k \left(y_k^{\text{exp}} - \overline{y}\right)^2}\right] \quad \text{where} \quad \begin{matrix} y^m = \text{model output} \\ y^{\text{exp}} = \text{experimental data} \\ \overline{y} = \text{mean}\left(y_k^{\text{exp}}\right) \end{matrix} \quad (2.60)$$

According to Equation 2.60, variance accounted for (VAF) yields the percentage of data variability explained by the model, so that the remaining part can be attributed to either noise or unmodelled dynamics. The numerator is the sum of squared modelling errors (residuals), whereas the denominator represents the fluctuations of the experimental data around their mean value. In other words, it tests the ability of the model to beat the squared data variability. Not a big deal indeed, but nevertheless it may yield a rough indication of model performance, which can be considered acceptable if VAF is above the 80%–90% range. Very bad models may even yield a negative VAF when the fraction in Equation 2.60 is greater than 1, in which case the differences between data and model are greater than the inner data variability, denoting a dramatic modelling failure. In any case, this test should not be considered exhaustive, and more accurate validation tools are required, as described in the following section.

2.4.1.2   Regression Line *F*-Test

A simple and yet effective way to check for model/data agreement is to plot the model outputs versus the data and see how they distribute with respect to the 1:1 line that would indicate perfect agreement. It should be considered that this procedure removes the time indexing from the analysis, and we just test the direct correspondence of each data/model output pair, and time synchronization is assumed as a prerequisite. A perfect matching would result in all of the points lying exactly on the 1:1 line, that is, the line with unit slope passing through the origin. We then reformulate our null hypothesis as a specific case of Equation 2.59:

$$H_0: \text{the regression line does not significantly differ from the } 1:1 \text{ line} \quad (2.61)$$

In MATLAB, the regression line can be computed with the following procedure. First, a polynomial of first degree (the regression line) is fitted to the data with the `polyfit` command with the following syntax:

$$\text{ab = polyfit(x,y,1);} \quad (2.62)$$

where x and y are the model and data vectors and 1 is the degree of the fitted line of the form $\hat{y} = b \cdot x + a$. Command (2.62) produces the regression line coefficients arranged in descending order, so that b is the slope and a the intercept. To trace the regression line on the graph, the command `polyval` is used

$$\text{yr = polyval(ab,x);} \quad (2.63)$$

which computes the regressed values of the data (`yr`) based on the coefficients `ab` and the model outputs (`x`). Figure 2.33 shows the scatter diagram produced by pairing data and model outputs. A first rough check can be performed by computing the correlation coefficient r between the x and y data series:

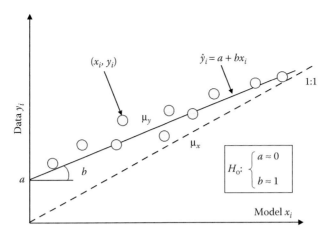

FIGURE 2.33 Producing a scatter diagram from pairs of data/model responses. The departure from the perfect agreement (1:1 line) is then statistically tested by examining the regression line coefficients (a, b).

$$r = \frac{\sum_{k=1}^{N}(x_k - \bar{x})(y_k - \bar{y})}{\sqrt{\sum_{k=1}^{N}(x_k - \bar{x})^2 \sum_{k=1}^{N}(y_k - \bar{y})^2}} \qquad \begin{cases} x_k = \text{model output} & \bar{x} = \text{mean}(x_k) \\ y_k = \text{data} & \bar{y} = \text{mean}(y_k) \end{cases} \qquad (2.64)$$

but it will soon become apparent that this indicator is of little help, because there are no statistical limit values against which it can be tested. A better test (Haefner, 2005) is to draw a regression line on the data and test its equivalence to the 1:1 line by examining the fitted line coefficients (a, b). Thus, our null hypothesis (H_o) is that the slope of the regression line is (statistically) equal to 1 *and* its intercept is (statistically) zero *simultaneously*. To test this H_o, a quantity following the F-statistics was proposed (Haefner, 2005):

$$F_{2,N-2}^{\alpha} = \frac{Na^2 + 2a(b-1)\sum_{k=1}^{N}x_k + (b-1)^2\sum_{k=1}^{N}x_k^2}{2S_{yx}^2}$$

where   (2.65)

$$S_{yx}^2 = \frac{\sum_{k=1}^{N}(y_k - \hat{y}_k)^2}{N-2} \quad \text{with} \quad \hat{y}_k = b \cdot x_k + a$$

The denominator term S_{yx}^2 is the residual mean squared error, equal to the difference between the data (y_k) and their equivalent on the regression line $(\hat{y})$.

The null hypothesis H_o cannot be rejected if Equation 2.65 yields a value smaller than the threshold value of the F-statistics for that confidence level and degrees of freedom. The latter can be directly obtained with MATLAB as `finv(0.95,2,N-2)`, meaning that we would observed by chance a larger value (i.e., we would reject the model) only $1 - 0.95 = 5\%$ of the times.

Equation 2.65 is not without pitfalls, which in many cases may lead to the rejection of a good model (type I error). For example, if the model is very good, the denominator term S_{yx}^2 is very small, producing a large F that would lead to (an incorrect) model rejection. The test also tends to fail when the number of data N is large, resulting in a very large first term in the numerator. A last warning regards the slope b, which may cause the second numerator term to change sign, depending on b being smaller or larger than one. In the latter case, the factor $(b-1)$ may lower the value of F, increasing the chance of accepting models with slope smaller than 1.

2.4.1.3 Further Regression Line Statistics

Several model quality indicators, in addition to the F-test, are reviewed by Haefner (2005) to validate the model and to detect the components mainly responsible for validation failure. Among them, the ones based on the *mean square error prediction* (MSEP) can be considered

$$\text{MSEP} = \frac{1}{N}\sum_k (x_k - y_k)^2 = (\overline{x} - \overline{y})^2 + (S_x - rS_y)^2 + (1-r^2)S_y^2 \tag{2.66}$$

where:

$\overline{x}$ and $\overline{y}$ are the mean values of the model outputs and of the data
r is the correlation coefficient
S_x and S_y are the standard deviations

Normalizing Equation 2.66 to MSEP yields

$$1 = \frac{(\overline{x} - \overline{y})^2}{\text{MSEP}} + \frac{(S_x - rS_y)^2}{\text{MSEP}} + \frac{(1-r^2)S_y^2}{\text{MSEP}} = \text{ME} + \text{SE} + \text{NC} \tag{2.67}$$

The first term (ME) accounts for the differences between means, the second (SE) regards slope differences, as it compares the two standard deviations through the correlation coefficient, while the third (NC) accounts for the effect of noise, that is, the random component of the error. Figure 2.34 reviews some cases of success and failure in applying the F-test to the regression line, with the aid of the additional indicators in Equation 2.67. In case (a), a good model, coinciding with the 1:1 line, is correctly accepted, in spite of the masking noise with standard deviation $\sigma = 0.4$. Of the three indicators, NC is the largest, indicating the presence of noise as the disturbing element. In case (b), a bad model (both slope and intercept are wrong) is rightly rejected, in spite of the same noise as before affecting the data. The ME and SE indicators show that both the intercept and the slope are responsible for the rejection. Cases (c) and (d) present the typical statistical errors: in (c), a type I error is made by rejecting a good model. This is caused by the very small term in the denominator of the F-statistics (Equation 2.65), where the very small residuals contribute to a large F. Conversely, case (d) shows the failure to reject a wrong model because of the large noise, as confirmed by the dominant value of the NC factor. In all cases, the correlation coefficient r of Equation 2.64 would yield practically no clue.

2.4.2 ANALYSIS OF THE RESIDUALS

The essence of model identification is to define the best possible model structure (or should we settle for 'plausible'?) and leave out the unwanted disturbance (generally referred to as 'noise'). In this quest, we strive to include all of the possible dynamics that we believe the system to possess, and to exclude the random fluctuations due to measurement errors or changes in the environmental conditions. Because they are—by definition—*random,* we hope that if our model is structurally correct and includes all the systems dynamics, what is left out (the *residuals*) is just noise. Further, we make the assumption that the noise samples are uncorrelated and come from a normal distribution, although the latter assumption is fairly difficult to meet in environmental models. Thus, the task of separating the structural part of the system from the noise is illustrated in Figure 2.35, where the residuals are analysed for autocorrelation (or the lack of it), suspecting that autocorrelated residuals may be due to unmodelled dynamics, in which case the model structure should be reconsidered.

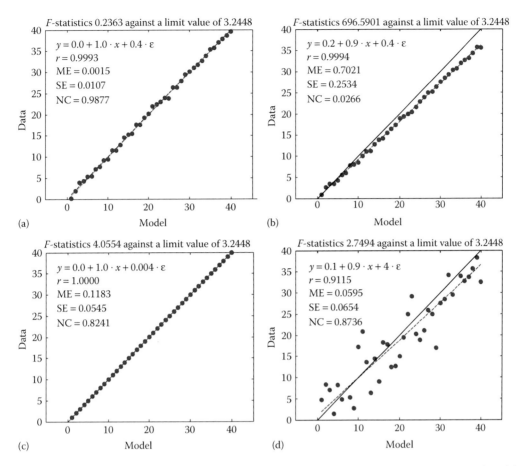

FIGURE 2.34 Merits and liabilities of applying the F-statistics to the regression line. In (a), a good model is accepted in spite of the masking noise (ε). In (b), a bad model (both the intercept and slope are wrong) is rejected. In (c), an error is made by rejecting a good model (too good in fact), whereas in (d) a bad model is accepted because of the large masking noise.

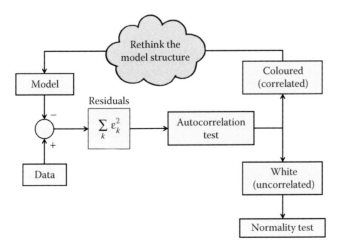

FIGURE 2.35 Road map in the analysis of the residuals: autocorrelated residuals may suggest an inadequate model structure and should lead to a reassessment of its structure.

2.4.2.1 Autocorrelation Test

This subject is amply treated in the major stochastic systems identification literature (Ljung, 1999; Box et al., 2008), so it will be only briefly recalled here. The autocorrelogram of a time series is obtained by multiplying its elements by their replicas shifted by k places (lags), as illustrated in Figure 2.36, and normalizing by their variance:

$$\rho_k = \frac{\sum_{i=1}^{N-k} \varepsilon_i \cdot \varepsilon_{i+k}}{\sum_{i=1}^{N} \varepsilon_i^2} \tag{2.68}$$

The statistically zero confidence band can be computed as $\pm \frac{1.96}{\sqrt{N-k}}$, where N is the number of samples in the time series. As Figure 2.36 shows, the confidence band gradually widens with k, because as the lag increases, the number of samples contributing to the autocorrelation decreases, and hence the tolerance increases. If all the elements of the autocorrelogram computed by Equation 2.68 are inside the zero confidence band, then the residuals are uncorrelated and we can consider them as purely random (or *white*) noise. Conversely, autocorrelated residuals may arise from unmodelled dynamics, which should lead to a revision of the model structure.

2.4.2.2 Normality Test

The second test that we may want to perform on the residuals is their normality, which reinforces the hypothesis that they are truly random errors coming from a normal distribution, unless we have reasons to believe that the residuals may have been generated by some other distribution (e.g., lognormal, Weibull, etc.). Generally, this check is made by comparing the cumulative distribution obtained from the histogram of the residuals with a known distribution. The one-sample Kolmogorov–Smirnov (KS) test (Massey, 1951; Marsaglia et al., 2003) compares the samples of a data vector x with a standard normal distribution y, and the null hypothesis H_0 is that they originate

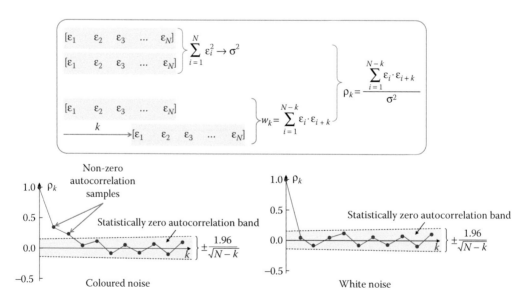

FIGURE 2.36 Definition of autocorrelation of a time series. Each element of the autocorrelogram is obtained by multiplying the samples of the series with their replica shifted by k places and normalized by the samples variance ($k = 0$). Below, a sample autocorrelogram is shown indicating some correlation (the one-lag sample is outside the statistically zero band).

from the same standard distribution. After computing the cumulative distributions $W_1(x)$ and $W_2(y)$, the test statistics consider the maximum difference $\max\left(\left|W_1(x)-W_2(y)\right|\right)$ and compares this with the maximum allowable difference. The MATLAB function for this test is

$$h \; = \; \texttt{kstest(x,y,alpha)} \qquad\qquad (2.69)$$

where the output h is 1 if the test rejects the null hypothesis at the alpha significance level, or 0 otherwise. The interesting aspect of the KS test is that the comparative distribution can be specified by the user, often using the normal distribution fitted to the data. As an alternative, the Lilliefors test (Lilliefors, 1967) can be used. In this test, the null hypothesis is that the data vector x comes from a normal distribution. The MATLAB syntax is

$$h \; = \; \texttt{lillietest(x,alpha)} \qquad\qquad (2.70)$$

and, as with the KS test, the answer is 1 if the null hypothesis is rejected at the alpha significance level. Unlike the KS test, in which the null distribution must be completely specified, this test can be used when the null distribution must be estimated. The Lilliefors test applies the same KS statistics to $\max\left(\left|W(x)-\hat{W}(x)\right|\right)$, where $W(x)$ is the empirical cumulative distribution obtained from the histogram of the x data vector and $\hat{W}(x)$ is the normal distribution fitted to the same data. Figure 2.37 shows the steps for applying the Lilliefors test: first, the histogram is constructed by counting the number of occurrences in a particular range of values. These frequencies form the histogram, which is then fitted with a normal distribution. The Lilliefors test is then computed by applying the KS statistics to the maximum difference between the cumulative distributions.

The example shown in Box 2.3 performs a Lilliefors test on two sets of numbers, the first coming from a normal distribution, produced by the internal MATLAB random number generator (randn), and the second generated by a uniform random number generator (rand). In both cases, the test correctly detects the agreement or disagreement with the normal distribution.

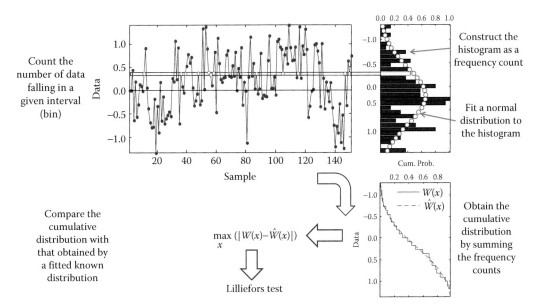

FIGURE 2.37 Histogram construction and comparison of the empirical and fitted cumulative distributions for the Lilliefors test.

BOX 2.3   AN EXAMPLE OF LILLIEFORS TEST FOR A DATA VECTOR OF 10,000 NUMBERS COMING FROM A NORMAL OR UNIFORM DISTRIBUTION

Both distributions are correctly detected by the test.

```
% Generate random
% numbers with a
% Normal distribution
X=randn(10000,1);
%Construct the histogram
hist(X)
%Apply the Lilliefors test
lillietest(X)

ans = 0 (Normality accepted)
```

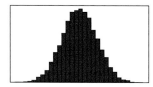

```
% Generate random
% numbers with a
% uniform distribution
X=rand(10000,1);
%Construct the histogram
hist(X)
%Apply the Lilliefors test
lillietest(X)
ans = 1 (Normality rejected)
```

2.4.3 PARAMETRIC TESTS: ASSESSING THE CREDIBILITY OF THE PARAMETER VALUES

Now, we turn our attention to the second series of tests indicated in Figure 2.32 and challenge the validity of the estimated parameters. Unlike the examples presented so far, where the data were generated by a model with known parameters, and the game was to guess their values having the 'true' values as a yardstick, in real life there is no such thing as 'real' parameters, so we have no way of comparing our results with a set of reference values. All we can hope for is one or more indicators that tell us that the estimated values are 'plausible'. This leads to the definition of a 'confidence region', within which we cannot reject the null hypothesis H_0 stating that the estimated parameters coincide with the 'real' ones.

2.4.3.1 Exact and Approximate Confidence Regions

Confidence regions are of primary importance because they provide a way to assess the accuracy of the parameter estimates $\hat{P}$. They represent the set of parameter values producing a model response within prescribed statistical boundaries. Because the error functional described by Equation 2.3 or 2.4 represents the 'closeness' of the experimental data to the fitted model, it is justifiable to base the confidence region on the contours of $E(P)$, so that any level $E(P) = c \cdot E(\hat{P})$ corresponds to a differing degree of confidence, depending on the increment c, for which it is difficult to specify a statistically significant value unless N is large, in which case $\Delta E = E(P) - E(\hat{P})$ has the necessary χ^2 asymptotic properties to apply the F-statistics (Seber and Wild, 1989). In that case, the confidence region of the estimated parameters $\hat{P}$ can be related to the F-statistics by the equality

$$\Delta E = E(P) - E(\hat{P}) = E(\hat{P})\, \frac{n_p}{N - n_p}\, F^{\alpha}_{n_p, N - n_p} \tag{2.71}$$

For parameters satisfying Equation 2.71, the null hypothesis cannot be rejected at the α confidence level. Figure 2.38 illustrates how the level increase ΔE projects the confidence region onto the parameter space.

The numerical difficulty in estimating the exact confidence regions has been considered by Vanrolleghem and Keesman (1996) and Dochain and Vanrolleghem (2001) who, on the basis of a previous work by Lobry and Flandrois (1991), proposed a successive contraction method to find the value of $E(P)$ corresponding to the prescribed value of the F-statistics. In the following, an approximate method for computing the confidence region will be discussed.

The other important aspect of the confidence region is its relationship with the parameter covariance matrix. In the linear model case,

$$y = X \cdot P + \varepsilon \quad y \in \mathbb{R}^{n \times 1};\ X \in \mathbb{R}^{n \times n_p};\ P \in \mathbb{R}^{n_p \times 1};\ \varepsilon \in \mathbb{R}^{n \times 1} \tag{2.72}$$

where $\varepsilon \sim N(0, I\sigma^2)$ is a random noise with normal distribution, the parameters $\hat{P}$ and the covariance matrix C can be directly estimated by the normal equations (Ljung, 1999)

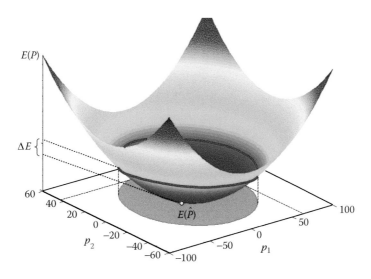

FIGURE 2.38 A 3D view of the exact parameter confidence region, represented by the shaded area in the (p_1, p_2) plane, defined by a statistically significant increase ΔE of the error functional defined by Equation 2.71.

$$\begin{cases} \hat{\boldsymbol{P}} = \left(\boldsymbol{X}^T \cdot \boldsymbol{X} \right)^{-1} \cdot \boldsymbol{X}^T \cdot \boldsymbol{y} \\ \boldsymbol{C} = \sigma^2 \left(\boldsymbol{X}^T \cdot \boldsymbol{X} \right)^{-1} \end{cases} \tag{2.73}$$

In addition to providing the link between measurement uncertainty and estimated parameters accuracy, the covariance matrix $\boldsymbol{C}$ defines the *parameter confidence region*. This region, often referred to as *confidence ellipsoid*, can be defined as the region in the n_p parameter space that would contain the 'real' parameters for a given error margin, for example, $100(1 - \alpha)\%$. For parameter values inside this region we cannot reject the null hypothesis that the estimated values are not significantly different from the 'real' ones. For this reason, the confidence ellipsoid is defined in terms of the *F*-statistics (Bates and Watts, 1988), with the covariance matrix $\boldsymbol{C}$ provided by the second Equation 2.73,

$$\left\{ \boldsymbol{P} : \left(\boldsymbol{P} - \hat{\boldsymbol{P}} \right)^T \boldsymbol{C}^{-1} \left(\boldsymbol{P} - \hat{\boldsymbol{P}} \right) \leq n_p F^{\alpha}_{n_p, N-n_p} \right\} \tag{2.74}$$

The *F*-statistics has a confidence level of $100(1 - \alpha)\%$ with n_p parameters and $N - n_p$ degrees of freedom. Unfortunately, for nonlinear models, there is no such direct method as Equation 2.73 to compute the parameter covariance matrix $\boldsymbol{C}$, but there are two differing ways to approximate the matrix $\boldsymbol{C}$, by extending the results of the linear theory, or through a second-order expansion of the error functional $E(\boldsymbol{P})$. However, first we must introduce a significant quantity that plays a central role in the parameter accuracy assessment: the FIM.

2.4.3.2   The FIM

This matrix, which plays a fundamental role in model identification and represents the link with the parameter sensitivity of Section 2.2, can be derived by perturbing the error functional in the neighbourhood of the optimal parameters $\hat{\boldsymbol{P}}$ as follows:

$$E\left(\hat{\boldsymbol{P}} + \delta\boldsymbol{P} \right) = \sum_{k=1}^{N} \left[\boldsymbol{y}_{t_k} \left(\hat{\boldsymbol{P}} + \delta\boldsymbol{P} \right) - \boldsymbol{y}_{t_k}^{\exp} \right]^T \boldsymbol{Q}_k \left[\boldsymbol{y}_{t_k} \left(\hat{\boldsymbol{P}} + \delta\boldsymbol{P} \right) - \boldsymbol{y}_{t_k}^{\exp} \right] \tag{2.75}$$

The output variation produced by the parameter perturbation $\hat{\boldsymbol{P}} \rightarrow \hat{\boldsymbol{P}} + \delta\boldsymbol{P}$ can be approximated by a first-order expansion

$$\boldsymbol{y}_{t_k} \left(\hat{\boldsymbol{P}} + \delta\boldsymbol{P} \right) \approx \boldsymbol{y}_{t_k} \left(\hat{\boldsymbol{P}} \right) + \left[\frac{\partial \boldsymbol{y}_k(\boldsymbol{P})}{\partial \boldsymbol{P}} \right] \times \delta\boldsymbol{P} \tag{2.76}$$

Substituting Equation 2.76 into Equation 2.75 yields

$$E\left(\hat{\boldsymbol{P}} + \delta\boldsymbol{P} \right) \cong E\left(\hat{\boldsymbol{P}} \right) + \delta\boldsymbol{P}^T \left[\sum_{k=1}^{N} \left(\frac{\partial \boldsymbol{y}_{t_k}}{\partial \boldsymbol{P}} \right)^T \boldsymbol{Q}_k \left(\frac{\partial \boldsymbol{y}_{t_k}}{\partial \boldsymbol{P}} \right) \right] \delta\boldsymbol{P} \tag{2.77}$$

If we want to increase the accuracy of the error functional, we should maximize the quantity in brackets in Equation 2.77, which represents the sensitivity of the error functional in the neighbourhood of the minimum $\hat{\boldsymbol{P}}$. This quantity is referred to as the FIM,

$$\mathbf{FIM} = \sum_{k=1}^{N} \left(\frac{\partial \boldsymbol{y}_{t_k}}{\partial \boldsymbol{P}} \right)^T \boldsymbol{Q}_k \left(\frac{\partial \boldsymbol{y}_{t_k}}{\partial \boldsymbol{P}} \right) \tag{2.78}$$

which is a function of *both* the model sensitivity $\left(\partial \mathbf{y}_{t_k}/\partial \mathbf{P}\right)$ *and* the measurement noise through the matrix $\mathbf{Q}_k$. It will be shown in the next section that this matrix can be used to approximate the parameter covariance matrix $\mathbf{C}$ and to plan the most favourable conditions for data sampling in such a way to maximize the estimation accuracy. It is clear from Equation 2.77 that the sensitivity of the error functional can be increased by making the FIM term as large as possible. While the $\mathbf{Q}_k$ matrix is determined by the measurement conditions and equipment, the sensitivity term can be manipulated to some extent by choosing the sampling times t_k to render the FIM matrix as large as possible. In the following, several criteria for maximizing the FIM will be illustrated, constituting a whole theory of experiment planning known as optimal experiment design (Fedorov, 1972; Dochain and Vanrolleghem, 2001).

A simple example will now show how to compute the FIM in practice. Consider a model with one output, two parameters, and three sampling times $\left(t_1, t_2, t_3\right)$. Then, assuming a constant measurement variance s^2, the FIM definition of Equation 2.78 yields

$$\mathbf{FIM} = \frac{1}{s^2}\left(\begin{bmatrix}\dfrac{\partial y_{t_1}}{\partial p_1}\\[2mm]\dfrac{\partial y_{t_1}}{\partial p_2}\end{bmatrix}\times\begin{bmatrix}\dfrac{\partial y_{t_1}}{\partial p_1}&\dfrac{\partial y_{t_1}}{\partial p_2}\end{bmatrix}+\begin{bmatrix}\dfrac{\partial y_{t_2}}{\partial p_1}\\[2mm]\dfrac{\partial y_{t_2}}{\partial p_2}\end{bmatrix}\times\begin{bmatrix}\dfrac{\partial y_{t_2}}{\partial p_1}&\dfrac{\partial y_{t_2}}{\partial p_2}\end{bmatrix}+\begin{bmatrix}\dfrac{\partial y_{t_3}}{\partial p_1}\\[2mm]\dfrac{\partial y_{t_3}}{\partial p_2}\end{bmatrix}\times\begin{bmatrix}\dfrac{\partial y_{t_3}}{\partial p_1}&\dfrac{\partial y_{t_3}}{\partial p_2}\end{bmatrix}\right)$$

$$= \frac{1}{s^2}\left(\begin{bmatrix}\left(\dfrac{\partial y_{t_1}}{\partial p_1}\right)^2&\dfrac{\partial y_{t_1}}{\partial p_1}\dfrac{\partial y_{t_1}}{\partial p_2}\\[3mm]\dfrac{\partial y_{t_1}}{\partial p_2}\dfrac{\partial y_{t_1}}{\partial p_1}&\left(\dfrac{\partial y_{t_1}}{\partial p_2}\right)^2\end{bmatrix}+\begin{bmatrix}\left(\dfrac{\partial y_{t_2}}{\partial p_1}\right)^2&\dfrac{\partial y_{t_2}}{\partial p_1}\dfrac{\partial y_{t_2}}{\partial p_2}\\[3mm]\dfrac{\partial y_{t_2}}{\partial p_2}\dfrac{\partial y_{t_2}}{\partial p_1}&\left(\dfrac{\partial y_{t_2}}{\partial p_2}\right)^2\end{bmatrix}+\begin{bmatrix}\left(\dfrac{\partial y_{t_3}}{\partial p_1}\right)^2&\dfrac{\partial y_{t_3}}{\partial p_1}\dfrac{\partial y_{t_3}}{\partial p_2}\\[3mm]\dfrac{\partial y_{t_3}}{\partial p_2}\dfrac{\partial y_{t_3}}{\partial p_1}&\left(\dfrac{\partial y_{t_3}}{\partial p_2}\right)^2\end{bmatrix}\right)$$

$$(2.79)$$

which can be generalized to the case of k measurements,

$$\mathbf{FIM} = \frac{1}{s^2}\begin{bmatrix}\sum_k\left(\dfrac{\partial y_{t_k}}{\partial p_1}\right)^2&\sum_k\dfrac{\partial y_{t_k}}{\partial p_1}\dfrac{\partial y_{t_k}}{\partial p_2}\\[3mm]\sum_k\dfrac{\partial y_{t_k}}{\partial p_2}\dfrac{\partial y_{t_k}}{\partial p_1}&\sum_k\left(\dfrac{\partial y_{t_k}}{\partial p_2}\right)^2\end{bmatrix}=\frac{1}{s^2}\begin{bmatrix}\sum_k\left(S_k^{p_1}\right)^2&\sum_k S_k^{p_1}S_k^{p_2}\\[3mm]\sum_k S_k^{p_1}S_k^{p_2}&\sum_k\left(S_k^{p_2}\right)^2\end{bmatrix} \qquad (2.80)$$

Figure 2.39 shows the synchronization between output sampling points and sensitivity computation to implement the numerical evaluation of the FIM according to Equation 2.80.

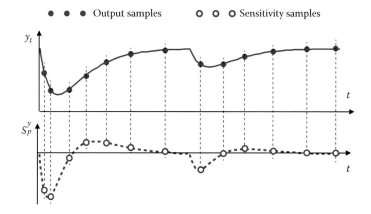

FIGURE 2.39 Output and sensitivity sampling points for the practical FIM computation.

2.4.3.3 Approximations of the Covariance Matrix

For nonlinear models, no equivalent exists for the normal Equation 2.73, and therefore there is no exact method for computing the covariance matrix, but there are two possible approximations. The first one (Ljung, 1999; Petersen, 2000; Dochain and Vanrolleghem, 2001) approximates C with the inverse of the FIM:

$$C_{\text{FIM}} = \text{FIM}^{-1} = s^2 \left[\sum_{k=1}^{N} \left(\frac{\partial \mathbf{y}_{t_k}}{\partial \mathbf{P}} \right)^T \left(\frac{\partial \mathbf{y}_{t_k}}{\partial \mathbf{P}} \right) \right]^{-1} \tag{2.81}$$

As an alternative, C can be approximated by a second-order expansion of the objective error function in the neighbourhood of the minimum $E(\hat{\mathbf{P}})$ involving the Hessian matrix (Seber and Wild, 1989; Press et al., 1986; Marsili-Libelli, 1992; Marsili-Libelli et al., 2003; Checchi and Marsili-Libelli, 2005)

$$C_H = 2s^2 \mathbf{H}^{-1} \quad \text{where} \quad \mathbf{H} = \frac{\partial^2 E}{\partial \mathbf{P} \partial \mathbf{P}^T} \bigg|_{\hat{P}} \tag{2.82}$$

Specific numerical algorithms for computing these approximations exist (Press et al., 1986; Marsili-Libelli et al., 2003; De Pauw, 2005). Now, we can go back to the approximate definitions of the parameter confidence regions, which can be defined as follows:

$$\text{Jacobian Conf. Reg.} \quad \left\{ \mathbf{P} : \left(\mathbf{P} - \hat{\mathbf{P}} \right)^T C_{\text{FIM}}^{-1} \left(\mathbf{P} - \hat{\mathbf{P}} \right) \leq n_p F_{n_p, N_{\text{exp}} - n_p}^{\alpha} \right\}$$

$$\text{Hessian Conf. Reg.} \quad \left\{ \mathbf{P} : \left(\mathbf{P} - \hat{\mathbf{P}} \right)^T C_H^{-1} \left(\mathbf{P} - \hat{\mathbf{P}} \right) \leq n_p F_{n_p, N_{\text{exp}} - n_p}^{\alpha} \right\} \tag{2.83}$$

with the approximate covariance matrices C_{FIM} and C_H computed with Equations 2.81 and 2.82, respectively. Each Equation 2.83 is a quadratic function representing the approximate confidence ellipsoid in the n_p parameter space. This can be shown by diagonalizing the equivalent covariance matrix, that is, $C^{-1} = U \Lambda U^T$, where U is the matrix of eigenvectors of C^{-1}, and $\Lambda = \text{diag}(\lambda_1, \lambda_2, \ldots, \lambda_{n_p})$ is the diagonal matrix of its eigenvalues. Introducing the change of coordinates $z = U^T (\mathbf{P} - \hat{\mathbf{P}})$ and shifting the origin in $\hat{\mathbf{P}}$, Equation 2.83 can be written as an n_p-dimensional ellipsoid:

$$\sum_{k=1}^{n_p} \lambda_k z_k^2 = n_p F_{n_p, N_{\text{exp}} - n_p}^{\alpha} \tag{2.84}$$

with axis lengths inversely proportional to $\sqrt{\lambda_k}$. Therefore, the smaller the eigenvalue, the larger the uncertainty of the corresponding parameter.

For direct visual inspection, Equation 2.84 can be projected onto a two-dimensional subspace. Let (p_i, p_j) be the parameter couple of interest. Then, Equation 2.83 becomes

$$\begin{bmatrix} p_i - \hat{p}_i & p_j - \hat{p}_j \end{bmatrix} \begin{bmatrix} C(i,i) & C(i,j) \\ C(j,i) & C(j,j) \end{bmatrix}^{-1} \begin{bmatrix} p_i - \hat{p}_i \\ p_j - \hat{p}_j \end{bmatrix} = 2F_{2, N_{\text{exp}} - n_p}^{\alpha} \tag{2.85}$$

Rewriting Equation 2.84 with $n_p = 2$ yields

$$\lambda_1 z_1^2 + \lambda_2 z_2^2 = 2F_{2, N_{\text{exp}} - 2}^{\alpha} \tag{2.86}$$

Using the eigenvalues u and v obtained from the similarity transformation

$$\begin{bmatrix} C(i,i) & C(i,j) \\ C(j,i) & C(j,j) \end{bmatrix}^{-1} = \begin{bmatrix} u & v \end{bmatrix} \mathrm{diag}(\lambda_1 \ \lambda_2) \begin{bmatrix} u \\ v \end{bmatrix} \tag{2.87}$$

The equations to trace the confidence ellipse centred in $(\hat{p}_i, \hat{p}_j)$ can be defined in polar coordinates ξ_i and ξ_j as

$$\xi_i = \sqrt{\frac{2F_{2,N_{\exp}-2}^{\alpha}}{\lambda_1}} u_1 \cos(\phi) - \sqrt{\frac{2F_{2,N_{\exp}-2}^{\alpha}}{\lambda_2}} u_2 \sin(\phi) + \hat{p}_i$$

$$\text{for } 0 \le \phi \le 2\pi \tag{2.88}$$

$$\xi_j = \sqrt{\frac{2F_{2,N_{\exp}-2}^{\alpha}}{\lambda_1}} v_1 \cos(\phi) - \sqrt{\frac{2F_{2,N_{\exp}-2}^{\alpha}}{\lambda_2}} v_2 \sin(\phi) + \hat{p}_j$$

In either case, the confidence interval δp_i of each individual parameter p_i can be computed as

$$\delta p_i = \pm t_{N-n_p}^{\alpha} \cdot \sqrt{C(i,i)} \tag{2.89}$$

where:
$t_{N-n_p}^{\alpha}$ is Student's t distribution for the given confidence level $100(1-\alpha)\%$
$N - n_p$ are the degrees of freedom

This statistics is consistent with the multivariate F distribution for $n_p = 1$, because $t_{N-n_p}^{\alpha} = \sqrt{F_{1,N-n_p}^{\alpha}}$. Substituting either C_{FIM} or C_H in place of C in Equation 2.89 yields the approximate confidence bounds of the estimated parameters $\hat{p}_i$. It is important to notice that though δp_i refers to a single parameter, it takes into account the full n_p-dimensional confidence region through the matrix C.

2.4.3.4   A Parameter Estimation Reliability Test

In addition to providing different confidence regions, the two approximations C_{FIM} and C_H provide a reliability test for the estimated parameters based on their inherent conceptual difference. This test is fully presented and discussed in specific papers (Marsili-Libelli et al., 2003; Checchi and Marsili-Libelli, 2005; Marsili-Libelli and Checchi, 2005; Checchi et al., 2007). The FIM approximation C_{FIM} is based on the sensitivities, whereas the Hessian approximation C_H depends on the shape of the error surface. For nonlinear systems, C_H is the most comprehensive approximation because it includes the effect of the curvature, reflecting the degree of nonlinearity induced by the model structure (Bates and Watts, 1988; Seber and Wild, 1989; Marsili-Libelli et al., 2003), while C_{FIM}, being based on a linear approximation, does not contain this term. Because the curvature effect vanishes in the neighbourhood of $E(\hat{P})$, comparing the two confidence regions yields a measure of the estimation reliability, because if the two regions diverge, this implies that the search terminated at a point where the effect of the curvature is still significant, and therefore this cannot be the real minimum of $E(P)$ (Marsili-Libelli et al., 2003). Conversely, if the two regions provided by C_{FIM} and C_H coincide, this means that the curvature effect is negligible, and the identification can be considered reliable. In this case, these regions also coincide with the exact confidence region determined on the basis of the error surface (Lobry and Flandrois, 1991). It should be recalled, however, that the curvature, amplifying the estimation errors, is an indicator of a failure to reach the minimum, but it is not influenced by model inadequacy and residual characteristics. Its vanishing merely indicates that the residuals are orthogonal to the response surface in the neighbourhood of $\hat{P}$, but it does not attempt to characterize them, for example, whether they are normally distributed and uncorrelated. This test is therefore an assessment of the quality of the optimization, and *not* of the model structure. Though visual inspection of the two regions may often suffice in judging their coincidence, numerical tests have been developed for their exact definition (Checchi and Marsili-Libelli, 2005; Marsili-Libelli and Checchi, 2005; Checchi et al., 2007).

2.4.3.5 Parameter Validation for the Monod Kinetics

To demonstrate the above validation method based on the coincidence or divergence of the confidence ellipsoids defined by Equation 2.83, the parameters of the Monod kinetics were again estimated using noisy data and starting the search from two differing initial guesses, as described in detail in Marsili-Libelli et al. (2003). The initial and final parameter values are listed in Table 2.1, and the final estimates in the $\left(\mu_{max}, K_s\right)$ plane are shown in Figure 2.40 for both cases: case 1 (successful estimation) and case 2 (faulty estimation). In real life, where the 'real' parameter values are unknown, there would be no way of knowing whether the results are reliable. For this reason, the previous validity test is used, and the confidence ellipses with the two methods are compared. Figure 2.41 shows the coincidence of the ellipses, denoting that the estimated parameters

---

TABLE 2.1

Estimation of the Monod Kinetics from Noisy Data ($\sigma = 0.05$) with Differing Simplex Starting Points

	$K_s \left(\text{mgCODL}^{-1}\right)$	$\mu_{max}\left(\text{h}^{-1}\right)$	Y	$b_h(\text{h}^{-1})$	**Starting Point**	$E(\hat{P})$
True values	20.000	0.500	0.500	0.030		
Case 1	19.0997	0.483	0.525	0.0303	$[22.0\ 0.5\ 0.37\ 0.04]^{\mathrm{T}}$	0.265
Case 2	31.605	0.711	0.485	0.0289	$[33.3\ 0.8\ 0.57\ 0.08]^{\mathrm{T}}$	0.548

Source: Marsili-Libelli, S. et al., *Ecol. Model.*, 165, 127–146, 2003.

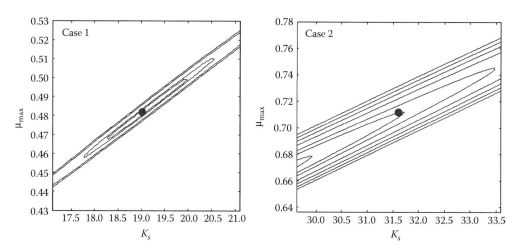

FIGURE 2.40 Contour portrait of $E(\mathbf{P})$ and the importance of a good search initialization. The dot indicates the location of the minimum $\hat{P}$ reached by the optimization algorithm for the two simplex initialization points of Table 2.1. Only in case 1 the true minimum is reached. (Redrawn with permission from Marsili-Libelli, S. et al., *Ecol. Model.*, 165, 127–146, 2003.)

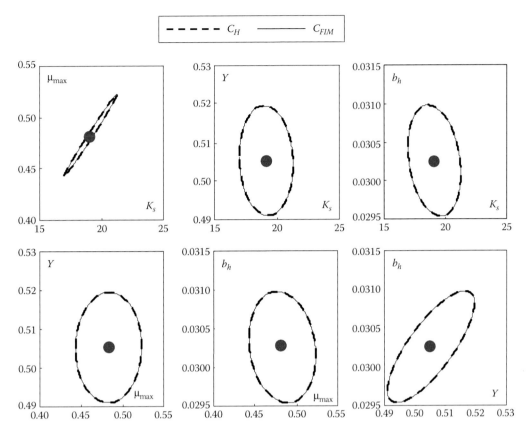

FIGURE 2.41 Confidence ellipsoids in the case 1 estimation. The coincidence of the confidence regions computed with the Hessian (dashed line) and the FIM matrix (solid line) indicates that the simplex converged to the 'real' parameters (dot). (Redrawn with permission from Marsili-Libelli, S. et al., *Ecol. Model.*, 165, 127–146, 2003.)

are reliable, whereas Figure 2.42 shows a considerable divergence, which should be interpreted as a warning that the final estimates are unreliable and deserve further investigation. For the 'good' estimates of case 1, the confidence intervals of the individual parameters are then computed by Equation 2.89 and listed in Table 2.2.

2.4.4 A MATLAB EXERCISE: PARAMETER ESTIMATION OF THE S&P MODEL

To wrap up the previous notions, let us again consider the simple S&P model consisting of two river reaches, each with an upstream point source of biodegradable pollutant. The model (2.32) will be used again, but now the exercise is aimed at the selection of differing sampling points along the DO trajectories and see how the differing sampling alternatives influence the estimation of the model parameters (K_b, K_c). But first, let us consider how the selection of the variables included in the error functional changes the nature of the estimation: if both model variables (BOD and DO) are used, then Figure 2.43a shows that the parameter correlation, giving rise to the 'narrow valley' shape, increases. In fact, the $E_{BOD,DO}$ contours are more elongated than those produced by E_{DO}. Further, Figure 2.43b shows that BOD gives a significant contribution, but being more difficult to measure than DO, it will not be much missed in the subsequent estimation, based on DO alone.

Two interconnected MATLAB scripts were designed for this exercise, and their use is illustrated in the flowchart of Figure 2.44. The software described here can be retrieved from the companion software bundle, in the subfolder \Exercises\Chapter_2\Calibrate_S&P. The first module to be used

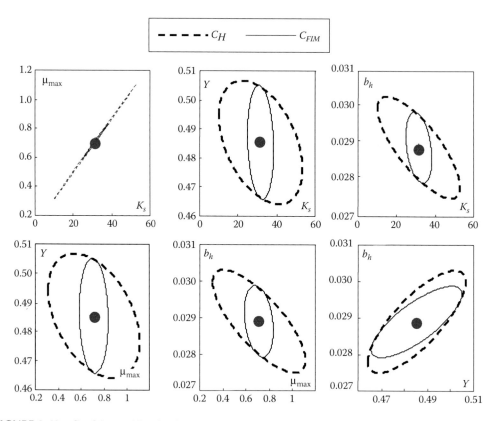

FIGURE 2.42 Confidence ellipsoids in the case 2 estimation. The divergence between the two confidence regions indicates that the simplex did not converge to the 'real' parameters (dot). (Redrawn with permission from Marsili-Libelli, S. et al., *Ecol. Model.*, 165, 127–146, 2003.)

TABLE 2.2

Values and Confidence Intervals of the Correctly Estimated Monod Kinetic Parameters (Case 1 in Table 2.1)

	$K_s \left(\text{mg COD L}^{-1}\right)$	$\mu_{\max}\left(\text{h}^{-1}\right)$	Y	$b_h\left(\text{h}^{-1}\right)$
Case 1	19.09968 ± 1.72181	0.48298 ± 0.03212	0.50524 ± 0.01139	0.03025 ± 0.00057

Source: Marsili-Libelli, S. et al., *Ecol. Model.*, 165, 127–146, 2003.

---

is Fisher_contour_SP_DO.m. It enables the user to simulate a sampling campaign by clicking with the mouse around the highest sensitivity spots (right-click to exit the sampling). Then, the FIM and the estimation bounds are computed based on these samples. All the data and the computed quantities (error surface, FIM, covariance matrix and parameter bounds) are then saved for the subsequent estimation, performed by the other module (Cal_SP_DO.m), which also validates the estimates by applying both the non-parametric and parametric methods.

Figure 2.45 shows the estimation results of the S&P model with a noisy data set (DO_sag_noise.mat) concentrated in the high-sensitivity zones. Given the data and the model, the FIM can be computed before the estimation, so that the expected confidence intervals for the two parameters are known in advance. In this case, these values are $\left(\delta K_b = \pm 0.01059; \delta K_c = \pm 0.011368\right)$, as shown

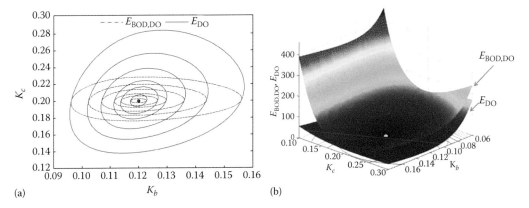

(a)   (b)

FIGURE 2.43   Contour profile (a) and 3D visualization (b) of the error functionals for the calibration of the Streeter & Phelps model using either both state variables BOD and DO (dashed lines) or only DO (solid lines). The counterintuitive result is that using only DO measurements, the 'narrow valley' problem is alleviated, though the gradient decreases. The 3D surface (b) shows that, although BOD gives a considerable contribution, it narrows the bottom of the functional.

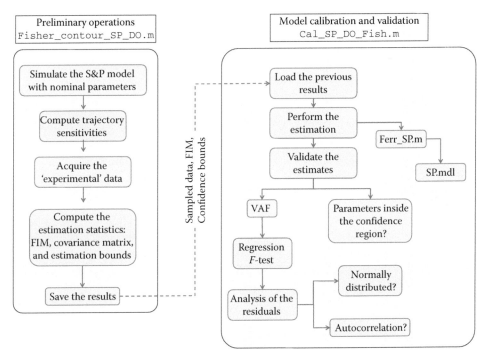

FIGURE 2.44   Software organization of the MATLAB scripts for the complete estimation of the Streeter & Phelps model, in the subfolder \Exercises\Chapter_2\Calibrate_S&P.

in the left part of Box 2.4. After performing the actual model calibration, the results confirm that the estimated parameters are indeed inside the expected confidence intervals. In fact,

$$\delta \hat{K}_b = 0.0028218 < 0.01059$$

$$\delta \hat{K}_c = -0.0037711 > -0.011368$$

(2.90)

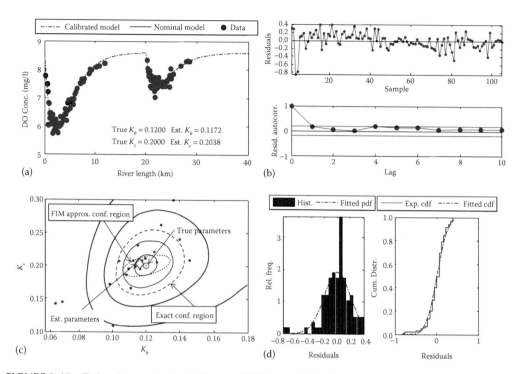

FIGURE 2.45 Estimation results for the Streeter & Phelps model using the data file DO _ sag _ noise. mat. In (a), the fitted model response is compared with the 'real' response, showing their coincidence. In (c), the dashed contour corresponds to the 95% exact confidence region and includes the approximate confidence region based on the FIM. The estimated parameters are well inside both regions. The graphs in (b) assess the residuals, showing the lack of autocorrelation, while in (d) the Gaussian cdf is well approximated by the experimental cumulative distribution.

The residual analysis in Figure 2.45b and d confirms that they are uncorrelated and normally distributed (both the KS and he Lilliefors test confirm this), and the F-test on the regression line in Figure 2.47a also confirms that the null hypothesis (regression line coinciding with the 1:1 line) cannot be rejected. The estimated parameters are well within the 95% confidence contour of the error functional, as shown in Figure 2.45c.

Figure 2.46a shows another example of the S&P model identification, but this time a highly biased data set (DO _ bias.mat) was used, where it is assumed that the oximeter has a strong bias, so that its readings are consistently lower than the actual DO value. In addition, it has a low accuracy (high divergence with true values) but a high precision (little measurement dispersion). The preliminary computation of the FIM yields surprisingly lower confidence bounds, due to the high precision $\left(\delta K_b = \pm 0.0055481; \delta K_c = \pm 0.0065491\right)$ that are not satisfied by the subsequent estimation; in fact, the differences between real and estimated parameters are

$$\delta \hat{K}_b = -0.027927 < -0.0055481$$

$$\delta \hat{K}_c = 0.046436 > 0.0065491$$

(2.91)

Also, Figure 2.46c shows that the estimated parameters are just on the border of the 95% confidence contour of the error functional. The residual analysis in Figure 2.46b indicates a high autocorrelation, as a consequence of a biased model, whereas the F-test of Figure 2.47b yields an extremely

---

BOX 2.4 STREETER & PHELPS MODEL CALIBRATION: COMPARISON OF THE ESTIMATION RESULTS WITH DATA FILES

File name: DO_sag_noise

Preliminary data analysis

Var. of DO meas. 0.05017
Estimation confidence band for Kb = ± 0.01059
Estimation confidence band for Kc = ± 0.011368

--- Estimation Results ---

Variance Accounted For 92.6251 %

Kolmogorov - Smirnov Test: 0
Normality hypothesis accepted at 95 %
Lilliefors Test: 0
Normality hypothesis accepted at 95 %

Residual analysis
Residual mean value 0.0043753
Residual variance 0.042787

Estimated Kb = 0.11718
Estimated Kc = 0.20377
Estimation error for Kb 0.0028218 equal to 2.3515 %
Estimation error for Kc − 0.0037711 equal to 1.8856 %

File name: DO_bias

Preliminary data analysis

Var. of DO meas. 0.032087
Estimation confidence band for Kb = ± 0.0055481
Estimation confidence band for Kc = ± 0.0065491

--- Estimation Results ---

Variance Accounted For 94.6089 %

Kolmogorov-Smirnov Test: 0
Normality hypothesis accepted at 95 %
Lilliefors Test: 1
Normality hypothesis rejected at 95 %

Residual analysis
Residual mean value − 0.67535
Residual variance 0.029617

Estimated Kb = 0.14793
Estimated Kc = 0.15356
Estimation error for Kb − 0.027927 equal to 23.2726 %
Estimation error for Kc 0.046436 equal to 23.2178 %

---

high F value, confirming that this identification should be rejected as a consequence of the poor data quality.

2.4.4.1 A Quick Guide to the Related Companion Software

The \Exercises\Chapter_2\Calibrate_S&P folder contains the set of MATLAB/Simulink functions demonstrated in the previous section. It is warned that the S&P model is very simple and should be viewed as an educational aid only. Here is how to use the two main scripts:

- `Fisher_controur_SP_DO.m` allows the user to generate a model response by inputting the two-reach river characteristics in the appropriate dialogue. It then displays a graph with the model response in terms of BOD and DO along the two reaches. In this graph, the user can select data by left-clicking on the DO curve (it is assumed that only DO data are acquired). To terminate the data selection process, right-click anywhere in the graph window. Remember that the last right-clicked point will not be included in the data set. It is advisable to produce several data files (DO*.mat) before proceeding to the calibration, to have several sets of usable data. At the end of the data input session, several graphs are produced:
  - A contour portrait of the error functional centred on the (K_b, K_c) couple used to generate the simulation.
  - A 3D surface of the same error functional, again centred on the (K_b, K_c) couple used to generate the simulation.
  - The FIM, computed on the basis of the sensitivity functions and the data error variance, defined as the difference between the clicked points and the corresponding simulation.

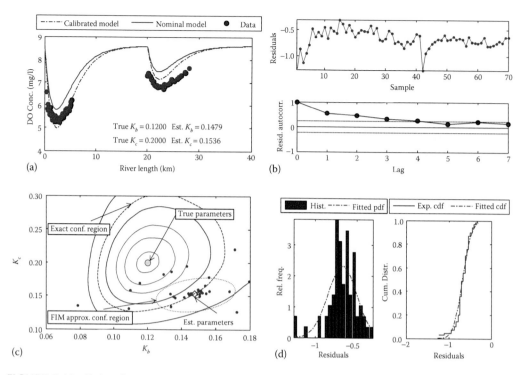

FIGURE 2.46 Estimation results for the Streeter & Phelps model using the data file `DO_bias.mat`. In (a), the fitted model response is compared with the 'true' response, showing their disagreement. In (c), the dashed contour indicates the 95% exact confidence region, while the dashed ellipse represents the approximate confidence region based on the FIM. The two regions are significantly divergent, with the estimated parameters placed just on the boundary of the exact confidence region. The graphs in (b) assess the residuals, showing a strong autocorrelation, while the Gaussian cdf is hardly approximated by the experimental cumulative distribution.

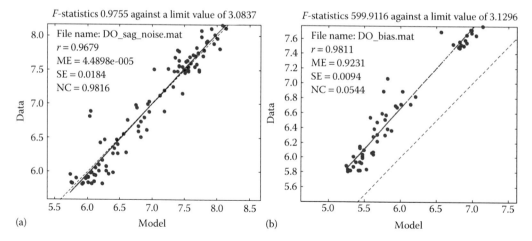

FIGURE 2.47 *F*-test on the regression lines performed for the calibration results of Figure 2.45 in (a) and of Figure 2.46 in (b), against the 1:1 perfect correspondence (black dashed line). In the (a) case the model can be accepted, while in the (b) case the model calibrated with the `DO bias.mat` data can be confidently rejected. Consider that in both cases the *r* coefficient has a (deceptively) very high value.

Based on the FIM, the expected parameter estimation bounds are computed and all the data are saved in a.mat file (starting with DO suffix) in the following order:

- DO _ sperim: clicked 'experimental' data
- DO _ Param: all the parameters needed to reproduce the simulation
- E, Par1, Par2: matrices to reproduce the contour and 3D portrait of the error functional
- Kb _ nom, Kc _ nom: nominal values of the parameters K_b and K_c used in the simulation

- Cal _ SP _ DO _ Fish.m is to be launched next: it performs the calibration of the two model parameters (K_b, K_c). First, a DO*.mat file containing all the information generated by Fisher _ controur _ SP _ DO.m as described above must be loaded by selecting it from a menu. After this file is loaded in the Workspace, the user is required to provide suitable initial values the (K_b, K_c) couple. It is suggested to try several differing starting estimates, as discussed in Figures 2.41 and 2.42. Then, the estimation procedure is started and its progress can be followed by observing the simulated system response with the current parameters values and the location of the current search point in the (K_b, K_c) space. Upon convergence, this module produces the following results:
  - The 'real' and calibrated model responses are compared with the data.
  - The estimated parameter values are displayed and compared with the 'real' ones.
  - The estimated (K_b, K_c) couple is displayed as a diamond in the contour figure.
  - The VAF is computed.
  - The estimation residuals are tested for whiteness with the autocorrelation test.
  - Their distribution is tested for normality through the KS and the Lilliefors tests.

2.4.5 Another MATLAB Exercise: Parameter Estimation of the Monod Model

This exercise illustrates the difficulties of estimating the parameters of the Monod kinetics, described by Equation 2.18. The relevant software can be retrieved from the folder \Exercises\ Chapter _ 2\Calibrate _ Monod of the companion software. Unlike the S&P model, the difficulties arise mainly from the high correlation among parameters, particularly that of the couple (μ_{max}, K_s), as amply described in Section 2.2.1. This exercise follows the same path as the previous one, but there are some differences. The estimation will be limited to the 'critical' parameter couple (μ_{max}, K_s), while the other parameters and the initial conditions are considered known. It is left to the reader to extend the identification to the four model parameters and to the initial conditions (hint: extend the parameter vector in Ferr _ Monod.m).

The workflow, shown in Figure 2.48, is similar to that illustrated for the S&P model identification. First, the Monod batch model is simulated with given parameters, and the trajectory sensitivities for both variables (substrate and biomass) are computed. Then, a certain number of data (N_{exp}) are uniformly sampled in the high-sensitivity interval, which can be extended to the whole simulation horizon. With this information the measurement accuracy matrix Q, as per Equation 2.9, and the FIM are computed as follows:

$$Q = \begin{bmatrix} \dfrac{1}{\sigma_S^2} & 0 \\ 0 & \dfrac{1}{\sigma_X^2} \end{bmatrix} \tag{2.92}$$

where σ_S^2 and σ_X^2 are the (constant) variances of the substrate and biomass samples, equal to the added measurement noise. The FIM matrix is computed using Q of Equation 2.92 and the sensitivity matrix S evaluated on the basis of the sampled data $\left(S_k, X_k, k = 1, \ldots, N_{exp} \right)$

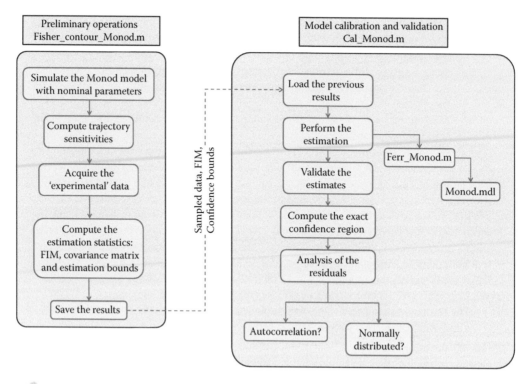

FIGURE 2.48 Workflow for the Monod kinetics identification, considered in the Example 2.4.5. The pertinent software can be found in the folder \Exercises\Chapter_2\Calibrate_Monod.

$$S_k = \begin{bmatrix} \dfrac{\partial S_k}{\partial \mu_{max}} & \dfrac{\partial S_k}{\partial K_s} \\[2ex] \dfrac{\partial X_k}{\partial \mu_{max}} & \dfrac{\partial X_k}{\partial K_s} \end{bmatrix} \tag{2.93}$$

as

$$\mathbf{FIM} = \sum_{k=1}^{N_{exp}} S_k^T \cdot Q \cdot S_k$$

$$= \begin{bmatrix} \dfrac{1}{\sigma_S^2}\sum_{k=1}^{N_{exp}}\left(\dfrac{\partial S_k}{\partial \mu_{max}}\right)^2 + \dfrac{1}{\sigma_X^2}\sum_{k=1}^{N_{exp}}\left(\dfrac{\partial X_k}{\partial \mu_{max}}\right)^2 & \dfrac{1}{\sigma_S^2}\sum_{k=1}^{N_{exp}}\left(\dfrac{\partial S_k}{\partial \mu_{max}}\cdot\dfrac{\partial S_k}{\partial K_s}\right) + \dfrac{1}{\sigma_X^2}\sum_{k=1}^{N_{exp}}\left(\dfrac{\partial X_k}{\partial \mu_{max}}\cdot\dfrac{\partial X_k}{\partial K_s}\right) \\[3ex] \dfrac{1}{\sigma_S^2}\sum_{k=1}^{N_{exp}}\left(\dfrac{\partial S_k}{\partial \mu_{max}}\cdot\dfrac{\partial S_k}{\partial K_s}\right) + \dfrac{1}{\sigma_X^2}\sum_{k=1}^{N_{exp}}\left(\dfrac{\partial X_k}{\partial \mu_{max}}\cdot\dfrac{\partial X_k}{\partial K_s}\right) & \dfrac{1}{\sigma_S^2}\sum_{k=1}^{N_{exp}}\left(\dfrac{\partial S_k}{\partial K_s}\right)^2 + \dfrac{1}{\sigma_X^2}\sum_{k=1}^{N_{exp}}\left(\dfrac{\partial X_k}{\partial K_s}\right)^2 \end{bmatrix} \tag{2.94}$$

The parameter covariance matrix is approximated by inverting the FIM of Equation 2.94, from which the confidence intervals for the individual parameters μ_{max} and K_s are computed according to Equation 2.89. The contour of the error functional can also be traced by repeated simulations with perturbed parameters (μ_{max}, K_s), as already shown in Figure 2.10 or 2.27 in the noisy data case.

With this preliminary information, the identification procedure continues according to the second box in the flow diagram of Figure 2.48. First, the sampled data and the FIM are loaded, then the simplex optimization is launched and the optimal parameters are assessed. The main difference between the Monod model flowchart of Figure 2.48 and the corresponding S&P workflow of Figure 2.44 is in the sensitivity of the Monod model to the number and quality of the data. To highlight this aspect, the data preparation module (`Fisher _ contour _ Monod.m`) has two data-generation options, controlled by the flag `Im` in line 104: the samples to be used in the estimation can be limited to the high-sensitivity interval (`Im=1`) or can be taken uniformly over the entire simulation horizon (`Im=2`). Specifically, in the first option the number of data specified by the user $\left(N_{exp}\right)$ are uniformly sampled only in the interval in which the sensitivities are greater than 40% of their maximum values. Through the initial menu, the user can also specify the variance of the measurement noise. The sampled data, nominal parameters, error function and FIM can be saved in a data file to be used by the subsequent `Cal _ Monod.m` script, which performs the actual estimation and validation.

As anticipated, the identification of the Monod kinetics presents several difficulties, due to the high parameter correlation. For this reason, it will be now shown that it also requires a large number of accurate data to achieve a successful identification. Table 2.3 compares the results obtained with data affected by a noise of varying intensity, in the case of uniform or concentrated sampling. The first two rows indicate that with a large number of samples (500) and no noise a perfect identification can be achieved. The accuracy of the estimates degrades as the noise variance increases. As predicted, the estimation error is always contained in the confidence bands estimated with the FIM. However, as Figure 2.49 shows, this condition considers only one parameter at a time, represented by each side of the dashed rectangle, while the confidence region considers the *joint* confidence of the two parameters simultaneously. Hence, having the estimates inside the FIM confidence region is a much stronger constraint than satisfying the single-parameter error bound. Two other considerations can be drawn by examining Figure 2.49: first, the simplex search develops mainly along the FIM confidence region; secondly, there is a considerable divergence between the exact and FIM confidence regions, confirming the difficulty of this exercise. One last consideration can be drawn by comparing the two confidence regions in Figure 2.49: the estimated parameter couple is certainly inside the ellipse defined by the FIM, but it is not contained in the exact confidence region, which is not based on the linear approximation.

TABLE 2.3

Estimation Results for the Monod Model Identification Exercise, Section 2.4.5

Data File Name	Estimated Confidence Band and Actual Estimation Error for μ_{max}		Estimated Confidence Band and Actual Estimation Error for K_s		Estimated Parameter Values	
	$\pm\delta\mu_{max}$	$\varepsilon_{\mu max}$	$\pm\delta K_s$	ε_{K_s}	$\hat{\mu}_{max}$	$\hat{K}_s$
Monod_500_0_All	$\pm 1.5184 \times 10^{-8}$	4.3612×10^{-8}	$\pm 1.2746 \times 10^{-6}$	3.5779×10^{-5}	0.45000	30.0000
Monod_500_0_HS	$\pm 1.1190 \times 10^{-7}$	-3.3716×10^{-7}	$\pm 9.4296 \times 10^{-6}$	-2.7848×10^{-6}	0.45000	30.0000
Monod_500_01_All	$\pm 1.7282 \times 10^{-4}$	3.4395×10^{-5}	$\pm 1.4507 \times 10^{-2}$	3.3251×10^{-3}	0.44997	29.9967
Monod_500_01_HS	$\pm 1.1673 \times 10^{-4}$	1.2141×10^{-4}	$\pm 9.8365 \times 10^{-3}$	1.0351×10^{-2}	0.44988	29.9896
Monod_50_01_All	$\pm 4.7423 \times 10^{-4}$	3.3686×10^{-4}	$\pm 9.808 \times 10^{-2}$	2.8363×10^{-2}	0.44966	29.9716
Monod_50_01_HS	$\pm 5.1921 \times 10^{-4}$	1.2782×10^{-4}	$\pm 4.3746 \times 10^{-2}$	1.0945×10^{-2}	0.44987	29.9891
Monod_50_10_All	$\pm 5.0478 \times 10^{-2}$	-3.7764×10^{-2}	± 4.2373	-3.1075	0.48776	33.1075
Monod_50_10_HS	$\pm 4.5277 \times 10^{-2}$	-1.1024×10^{-2}	± 3.815	-0.9053	0.46103	30.9053

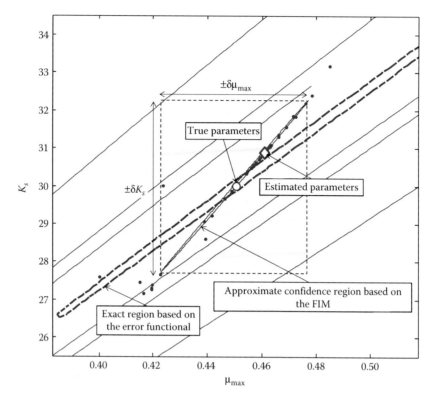

FIGURE 2.49 Estimation results for the Monod model exercise of Section 2.4.5, obtained with the data file Monod_50_10HS.mat. There is a considerable divergence between the confidence region based on the error functional (exact) (thick dashed line) and that based on the FIM (thin solid line). It is interesting to notice that the estimated parameters are in the FIM region, but just outside the exact region. Further, the parameter confidence intervals based on the FIM result in the grey dashed rectangle, which is much wider than the FIM region.

2.5 PEAS: A COMPUTER PACKAGE FOR PARAMETER ESTIMATION OF ECOLOGICAL MODELS

The previous parameter estimation and validation analysis for environmental system was condensed into a MATLAB toolbox named PEAS for Parameter Estimation Accuracy Software (Checchi et al., 2007), which performs both the operations described in the previous sections using a user-friendly graphical interface to minimize the required programming. The user is required to specify the model structure using the MATLAB/Simulink syntax, enter the experimental data, provide an initial parameter guess and select an estimation method. PEAS provides several model assessment methods, in addition to parameter estimation, such as error function plotting, trajectory sensitivity and Monte Carlo analysis, which are useful to assess the suitability of the experimental data to the estimation problem. After the parameters have been estimated, the reliability of the identification is assessed, approximate and exact confidence regions are computed, and a confidence test is produced. The Monte Carlo analysis is available for approximate accuracy assessment whenever the model structure prevents the application of the confidence regions method.

The PEAS toolbox is available for free on request to the author on an 'as is' basis. A minimum of MATLAB programming skill is required, and in the case of a complex model, C programming of the Simulink S-function is suggested, but not strictly required.

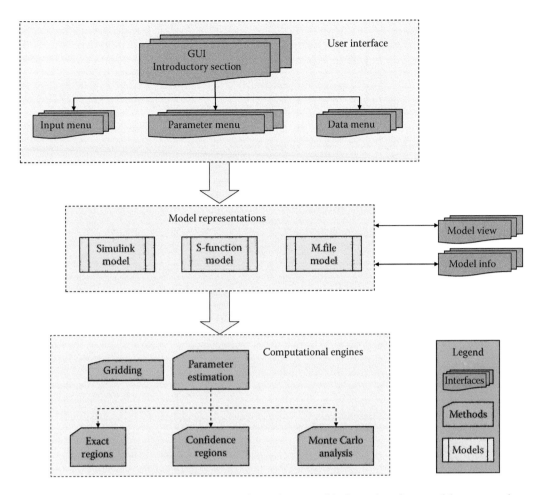

FIGURE 2.50 PEAS is composed of three main sections: graphical user interface, model representations and computational engines. The GUI frontend lets the user define the problem and the experimental data, then the model is specified and finally the selected methods are invoked. (Reproduced with permission from Checchi, N. et al., *Environ. Model. Softw.*, 22, 899–913, 2007.)

A few features of the PEAS toolbox are now reviewed. Figure 2.50 shows its software organization, which is composed of a graphic user interface, a modelling section and the computational engines. The user is taken through this logical sequence from top to bottom by first defining a model, together with its simulation specifications, then performing the estimation, after selecting the method, accuracy and weights, and finally assessing the results.

The user can define the problem and the experimental data through the main GUI, shown in Figure 2.51, after specifying the model either as a Simulink model (*.mdl), an M-file script (*.m) or as a compiled-C in either the *.mexw32 or *.mexw64 version, depending on whether you have a 32-bit or a 64-bit operating system. Additional interfaces are provided for model viewing. Then, the estimation algorithm is selected, choosing between a modified simplex with optimized expansion, a QN method, the Levenberg–Marquardt method (Himmelblau, 1972; Press et al., 1986; Marsili-Libelli, 1992) or a combination of simplex and QN, where the former is the starting algorithm, and the latter is initialized with the simplex result. After the estimation is completed, three-parameter estimation assessment tools are available: computation of exact and approximate confidence regions, shown in Figure 2.52, and Monte Carlo analysis, shown in Figure 2.53.

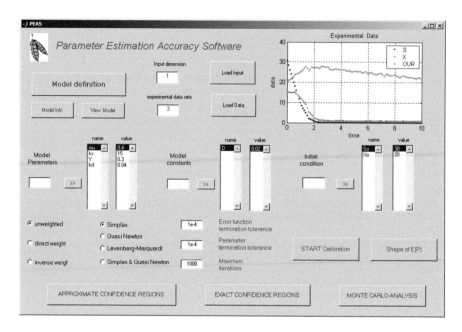

FIGURE 2.51 PEAS opening screen and initial interface, prompting the user to define the model and specify the estimation conditions. The example shown here refers to the Monod kinetics. (Reproduced with permission from Checchi, N. et al., *Environ. Model. Softw.*, 22, 899–913, 2007.)

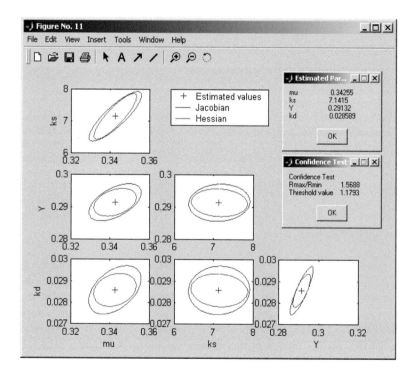

FIGURE 2.52 PEAS can perform the estimation reliability test by comparing differing confidence ellipsoids, as described in Section 2.4.3.4. In this example, the estimation fails to satisfy the ellipsoids coincidence. This behaviour was caused by an improper choice of the initial parameter guess and resulted in major estimation errors. (Reproduced with permission from Checchi, N. et al., *Environ. Model. Softw.*, 22, 899–913, 2007.)

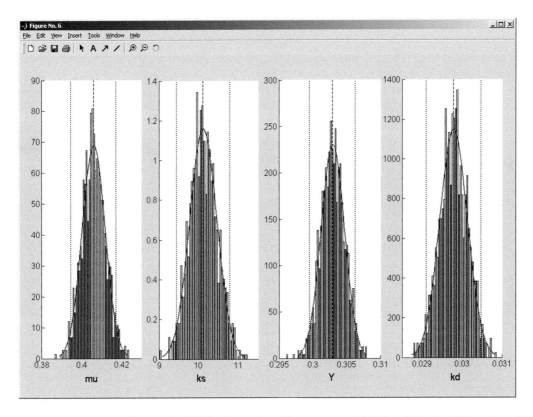

FIGURE 2.53 Monte Carlo analysis of the four estimated parameters of the Monod kinetics. The estimated values are represented by the dashed lines, and their confidence intervals by the dotted lines. (Reproduced with permission from Checchi, N. et al., *Environ. Model. Softw.*, 22, 899–913, 2007.)

3 Analysis of Environmental Time Series

True genius resides in the capacity for evaluation of uncertain, hazardous, and conflicting information.

Winston Churchill

3.1 INTRODUCTION

By time series, we mean a collection of homogeneous time-indexed data. The most important purpose of data is to carry information, and it is our prime goal to extract the useful information by separating it from unwanted interferences. In the context of environmental data, these data sets may be data either with which a model is identified (see Chapter 2) or which serve as a model input in simulations or scenario generation. In either case, these data cannot be used on an 'as is' basis, because there may be several features—or the lack of them—that make them unsuitable for direct use. The data are generally affected by noise due to sampling inaccuracies—the extreme case being the 'outliers', that is, data well outside their reasonably expected range—or there may be missing data. The sampling process itself may alter the data to a significant extent. It should always be remembered that in the environmental sciences, the practice of data collection is often troublesome and costly, which partly justifies their scarcity and the need to make the most of what we have available.

For these reasons, data treatment prior to their use in modelling and analysis is mandatory, both to improve their quality and to bring to the fore the information embedded therein. In this chapter, the techniques of data analysis will be reviewed with the aim of separating the informative content from the noise that affects it, supplementing missing data, and in general, making data more reliable for modelling purposes.

A time series is composed of a collection of samples taken at discrete time intervals from an intrinsically continuous process. Environmental time series may be acquired either manually or via automatic equipment. In the former case, the sampling involves the action of an operator and subsequent laboratory analysis, whereas in the latter case, unattended sampling stations are used. The first sampling mode is usually reserved for chemical parameters that require sophisticated laboratory processing, while automatic sampling can be used for comparatively simpler parameters, such as temperature, pH, flow and level. Recently the spectrum of chemical parameters that can be directly acquired by automated sampling stations has expanded to include specific ions such as $NH_4^+, NO_2^-, NO_3^-, PO_4^{2-}$, though the cost, reliability and maintenance issues still prevent their widespread use.

Manual data are often sampled at uneven rates, are relatively scarce, present a higher rate of missing data and need to be validated, whereas data coming from automatic monitoring stations are generally more abundant, sampled at regular intervals and are generally validated at the source, at least at a preliminary level, if the sensors are properly maintained.

Environmental data analysis can be divided into three main phases: data examination, filtering, and usage.

Data examination is based on a synthetic event description, to understand how the data are created and what is their meaning in the current context. Then, the data are preliminarily assessed in terms of their deterministic and random components, and the grossly inconsistent samples are discarded.

Data filtering performs the removal of the previously detected unwanted features, such as the removal of artefacts or of the stochastic component (denoising), together with outlier detection and removal. The data are then validated, and a more in-depth analysis is started with the aim of pattern detection and information extraction.

Data usage of the processed time series may range from forecasting future behaviour to synthetic time-series generation to be used as inputs in model simulation and scenario generation.

3.2 TIME DOMAIN VERSUS FREQUENCY DOMAIN OF ENVIRONMENTAL TIME SERIES

Figure 3.1 shows the basic signal decomposition into its deterministic and stochastic components. The former can be further decomposed into several components that differ mainly by their rate of change, with the *trend* being the slowest component, while the other components have increasingly higher rates. The decomposition of Figure 3.1 can be explained as follows:

Deterministic part: It contains the signal information and is, by definition, its reproducible part. It includes trend, cyclic and seasonal components, with the following meanings:
    Trend: A long-term variation (very low frequency), which can be generally represented by a linear relation.
    Cyclic components: A collection of periodic fluctuations, which can be represented by a collection of sinusoids of multiple frequencies.
    Seasonality: A particular cyclic component with a seasonal periodicity.
Stochastic part: This is the totally unpredictable (noisy) component of the signal, unless it has some inner 'autocorrelation' (see Chapter 2), which can be represented by an autoregressive (AR) stochastic model (see Chapter 1).

3.2.1 FREQUENCY ANALYSIS OF SAMPLED TIME SERIES

When the signal is examined in its native time domain, it is very difficult, if not impossible, to extricate the previously described components, whereas a *frequency analysis* will naturally separate

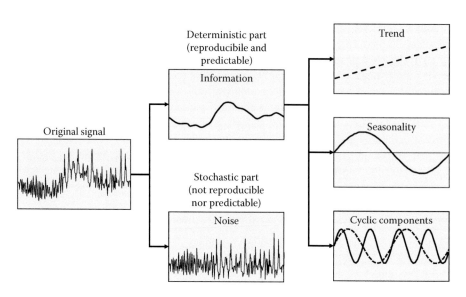

FIGURE 3.1 Decomposition of an environmental signal into its most significant components.

the components in the various frequency bands. The mathematical transformation that converts a time function $s(t)$ into a *frequency function* $S(f)$, and vice versa, was introduced by the French mathematician Jean B. Joseph Fourier (1768–1830), and is referred to as the *continuous* Fourier transform (FT) (Oppenheim and Schafer, 2010; Lyons, 2011), defined as follows:

$$S(f) = \int_{-\infty}^{\infty} s(t) \cdot e^{-j2\pi ft} dt \quad s(t) = \frac{1}{2\pi} \int_{-\infty}^{\infty} S(f) \cdot e^{j2\pi ft} df \tag{3.1}$$

where:
$j = \sqrt{-1}$ is the imaginary unit
f (Hz) is the frequency, expressed in hertz (periods per second) or radians per second (rad/s)

In practice, we shall be dealing not with continuous signals, but with a collection of samples acquired at T_s intervals. As we shall see, the sampling process produces a fundamental alteration in the frequency spectrum of the original signal. In fact, we are no longer dealing with a continuous signal $s(t)$, but with a discrete-time signal consisting of a collection of samples $s*(t) = \{s(kT_s)\}$, as shown in Figure 3.2, so that if $S(f)$ is the FT of the original continuous signal, the frequency content of the sampled signal $S*(f)$ is made up of infinite replicas of $S(f)$, spaced by multiples of the sampling frequency $f_s = 1/T_s$ (see, e.g., Oppenheim and Schafer, 2010)

$$S*(f) = \frac{1}{T_s} \sum_{k=-\infty}^{\infty} S(f + kf_s) \tag{3.2}$$

3.2.1.1 Aliasing

The FT of a sampled signal is generally referred to as the discrete FT (DFT), and Equation 3.2 describes how the original signal spectrum $S(f)$ is changed into $S*(f)$ as a consequence of the sampling process. In this context, it is not necessary to delve into the mathematical details of the FT and into the derivation of Equation 3.2, for which excellent textbook treatments are available (Oppenheim and Schafer, 2010; Lyons, 2011). It is, however, important to understand the role of the sampling frequency f_s, which, if wrongly selected, may considerably alter the frequency content of the original signal. Figure 3.3 provides a graphical visualization of Equation 3.2 showing the sidebands originated by sampling, with more bands extending from $-\infty$ to $+\infty$.

So long as the sampling frequency f_s is larger than twice the maximum frequency of the signal baseband, that is, $f_s > 2f_{max}$, the sidebands are well separated, so that the baseband can be recovered with a low-pass filter with bandwidth f_{max}, as shown in Figure 3.3a. However, if the sampling frequency is lower than twice f_{max}, that is, $f_s < 2f_{max}$, the sidebands will contribute to the frequency content *inside the baseband* (Figure 3.3b), so that its frequency content is permanently altered as

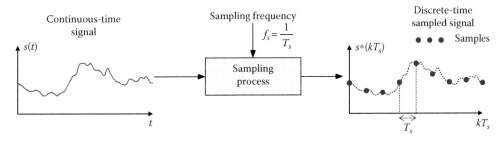

FIGURE 3.2 Schematization of the sampling process, which extracts discrete samples $s*(kT_s)$ from the continuous-time signal $s(t)$ spaced by T_s time intervals.

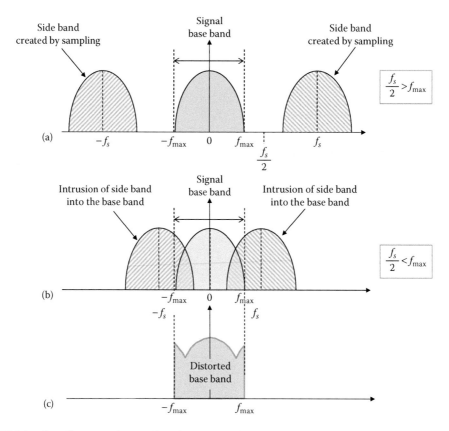

FIGURE 3.3 Sampling a continuous-time signal results in a proliferation of its original frequency spectrum with infinite replicas centred at multiples of the sampling frequency f_s. If the sampling frequency is larger than twice the maximum baseband frequency (a), then the baseband of the signal is unaffected, but if this condition is not satisfied, the sidebands invade the baseband (b) altering its frequency content (c).

shown in Figure 3.3c, and no recovery of the original signal is possible by subsequent low-pass filtering. This alteration is termed *aliasing*, meaning that a *frequency alias* creeps into the baseband causing distortion. To avoid aliasing, the sampling frequency should satisfy the so-called Nyquist constraint

$$f_{max} < \frac{f_s}{2} = f_N \tag{3.3}$$

which guarantees that sampling will not produce aliasing, by keeping the sidebands separate, as in Figure 3.3a. In general, the signal bandwidth f_{max} is not known in advance, and Equation 3.3 is often used in reverse by pre-filtering of the signal with an *anti-aliasing* low-pass filter that limits the signal bandwidth to a prescribed upper limit $f_{max} \ll f_N$ so that Equation 3.3 is amply satisfied and the subsequent sampling does not produce aliasing. The other possibility is to act on the opposite end of the inequality $f_{max} \ll f_N$ by increasing the sampling frequency. This approach, however, may have some drawbacks, such as the need for more sophisticated data acquisition systems or the increased dimension of the collected data, with obvious storage and handling problems.

   To recast the *aliasing* problem in the more familiar time domain, Figure 3.4 shows two differing samplings of the same sinusoid with frequency $f = 1.333\,\text{Hz}$. In the first case of Figure 3.4a, the sampling frequency $f_s = 10\,\text{Hz}$ is well above the Nyquist limit and no aliasing occurs

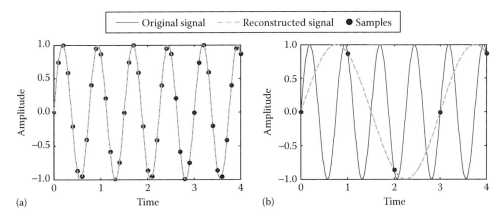

FIGURE 3.4 Sampling and reconstruction of a sinusoid using a sampling frequency satisfying the Nyquist constraint (a), and in violation of the Nyquist limit, causing aliasing (b), causing the original signal to be mistaken for an (erroneous) low-frequency sinusoid. This graph was obtained with the MATLAB script Ex_Aliasing.m in the ESA_Matlab\Exercises\Chapter_3\Time_Series folder.

$$
\left. \begin{array}{l} f = 1.333\,\text{Hz} \\ f_s = 10\,\text{Hz} \end{array} \right\} \; \frac{f_s}{2} > f \Rightarrow \text{No aliasing} \tag{3.4}
$$

but if the sampling frequency is lowered to $f_s = 1\,\text{Hz}$, then aliasing is produced. In fact,

$$
\left. \begin{array}{l} f = 1.333\,\text{Hz} \\ f_s = 1\,\text{Hz} \end{array} \right\} \; \frac{f_s}{2} < f \Rightarrow \text{Aliasing!} \tag{3.5}
$$

and the alias frequency f_a can be computed as

$$
f_a = \left(f + \frac{f_s}{2} \right) \bmod (f_s) - \frac{f_s}{2} = 0.333\,\text{Hz} \tag{3.6}
$$

The result of this 'under-sampling' is that an unaware user will (erroneously) believe that the data were sampled from the dashed sinusoid in Figure 3.4b with frequency $f_a = 0.333\,\text{Hz}$ given by Equation 3.6.

3.2.1.2 Feasible Frequency Range and Sampling Frequency

The other important aspect of Equation 3.3 is that once f_s is selected, the frequency analysis can be confined to the Nyquist limit frequency $f_N = f_s/2$ because beyond this limit, the frequency spectrum is made of infinite (and useless) replicas of the baseband. In practice, sampled environmental time series comes in the form of data sets whose sampling period is determined by the people who actually collected the data and over which the system analyst has no control. If, for example, we know that a certain set of data was sampled 10 min apart, the frequency analysis can be extended at most up to $f \leq 3\,1/\text{h}$. In fact,

$$
T_s = \frac{10}{60} = \frac{1}{6}\text{h} \rightarrow f_s = 6\,1/\text{h} \rightarrow f \leq 3\,1/\text{h} \tag{3.7}
$$

so the feasible frequency range is 0–3 1/h. Further, the computer algorithms performing the DFT are generally known as fast FT (FFT) and solve the problem of having to process a huge number of coefficients. If N data were processed in the conventional way, the number of required arithmetical operations would be in the order of N^2, whereas an FFT algorithm can do it in $N \cdot \log(N)$ operations. Almost all FFT algorithms are happier if the number of data to be processed is a power of 2. There is an abundant specialized literature about the FFT algorithms and their computer implementation (Duhamel and Vetterli, 1990; Frigo and Johnson, 2005).

To clarify how the sampling frequency may change the appearance of environmental data, Figure 3.5 shows a week-long portion of a dissolved oxygen (DO) time series, automatically sampled at 1 h intervals, in the Arno River upstream of the city of Florence (courtesy of the Regional Agency for Environmental Protection, ARPAT).

The three graphs in the left column show the same portion of data sampled at 1, 2, and 4 h, respectively, whereas the graphs in the right column show the corresponding frequency analysis. The periodic components at 24 and 12 h are indicated by the dashed vertical lines. Each frequency graph is limited to frequencies up to $f_N = f_s/2$, and as the sampling period T_s is increased, the graphs appear 'stretched' to the right because the frequency range is decreased. The corresponding time graphs on the left appear increasingly 'chunky' as the distance between samples (indicated by dots) increases. If the sampling period were increased beyond the 12 h or the 24 h periodicity, some important circadian characteristics would be lost. In general, the guidelines for selecting T_s may be summarized as follows:

A short T_s implies

- Expensive sampling equipment and procedures
- A large amount of data
- Capture of possible important high-frequency phenomena
- Redundant or poorly informative data
  - Noise samples (containing no information)
  - 'Boring data' (few significant variations, little information)

A long T_s implies

- Economical sampling equipment and procedures
- A small amount of data
- Possible information loss due to
  - Cut-off of potentially important high-frequency $(f > f_N)$ phenomena
  - Aliasing → Baseband distortion

3.2.1.3   Signal Components: Information, Artefacts, and Noise

The example of Figure 3.5 shows how the frequency components of the signal are distributed in the range $(0-f_s/2)$, and we are now in a position to reconsider the signal decomposition of Figure 3.1 from a frequency viewpoint. Figure 3.6 shows that, generally, the information-rich part is situated in the low-frequency part of the spectrum. Seasonality and cyclic components usually have a well-defined frequency range (e.g., hours, days, months), so they normally appear in the spectrum as fairly distinct lines, like the 12 h and 24 marks in Figure 3.5. The noise (random) component is usually confined to the high-frequency end of the spectrum. Artefacts, usually man-made non-periodic disturbances due to mechanical or electrical interferences, are somewhere in the middle between cycles and noise, because, being short lived, they contain high-frequency components. This frequency partition is important in view of signal conditioning, which tends to preserve the meaningful part of the signal and discard the non-informative components. In principle, a low-pass filter could do the job, but given the frequent superposition of the components, we shall need more sophisticated algorithms to 'separate the wheat from the chaff'.

3.2.1.4   MATLAB Implementation of Frequency Analysis

We now turn to describing the MATLAB® programming behind the graphs presented in Figure 3.5 and more generally to illustrate the frequency analysis of a time series in this platform. The FFT is

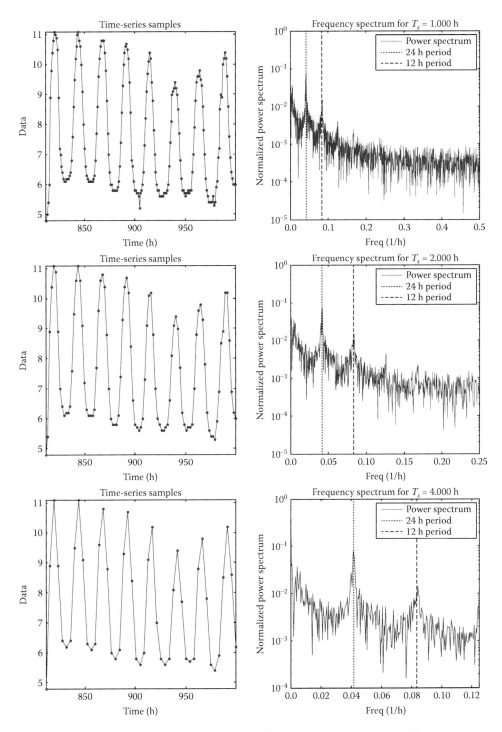

FIGURE 3.5 A dissolved oxygen time series in Arno River upstream of the city of Florence, sampled at differing frequencies. Of course, only sampling frequencies lower than the original one can be used. The same portion of data is shown in the left graph and the corresponding frequency spectrum is shown in the right graphs, which extend up to the Nyquist frequency $f_N = f_s/2$. The 24 and 12 h period lines are traced in each frequency spectrum as vertical dashed lines. In the left graphs the dots represent the samples, connected with straight lines just for visual representation. These graphs were obtained with the MATLAB script `Ex _ FFT _ Sampling _ Time.m` in the `ESA _ Matlab\Exercises\Chapter _ 3\Time _ Series` folder.

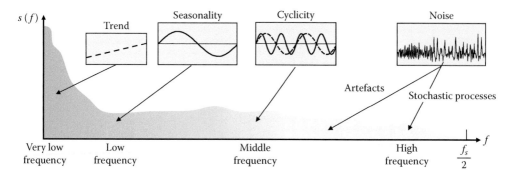

FIGURE 3.6 Frequency allocation of the signal components of Figure 3.1.

a basic MATLAB primitive (`fft`) and could, in principle, be used 'as is', but the correct graphical data presentation requires an additional bit of programming, which is listed in Box 3.1. After loading the data set, which includes two distinct quantities (time and value), the sampling interval T_s is computed as the average time interval between consecutive samples (it is hoped that the sampling is uniform). The length of data is computed and the largest power of 2 subset is extracted. In fact, working with data sets of exactly 2^L expedites the computation. Otherwise, the original data set would be padded with zeros until some 2^L length is reached, prior to processing by the `fft` function. The FFT produces a set of coefficient Y, which is a vector of the same length as the 2^L data portion. Because they cover the symmetrical spectrum around the origin, only half of them need to be plotted. The frequency range $(0–f_s/2)$ for the plot is created with the command

$$f = Fs/2*linspace(0,1,N\_samples/2+1); \qquad (3.8)$$

which shows that the FFT produces a *discrete* frequency spectrum (a *line* spectrum) with the components spaced by `Fs/N _ samples`. The single-sided normalized power spectrum is computed as

$$abs(Y(1:N\_samples/2+1)/abs(Y(1))). \qquad (3.9)$$

Because the spectral components `Y` are complex numbers, `abs(Y)` is equivalent to `sqrt(Y.*conj(Y))` where `conj` produces the conjugate of a complex number. Remember that the FFT produces a *bilateral* frequency spectrum, so in command (3.9) only half the coefficients need to be plotted, and hence the indexing of the vector `Y` of command (3.8). The frequency spectrum may be presented on a linear or log scale, depending on their spread and on which characteristics we are most interested in. The MATLAB script of Box 3.1 was used to produce the frequency analysis presented in Figure 3.5.

3.3 SIGNAL REGULARIZATION

Now that the basic features of the time series have been explored in both the time and frequency domains, we can set out to improve their characteristics and remediate possible flaws. The operations that can be performed include detrending, detection and removal of outliers, smoothing and replacement of missing data, usually accomplished in this order.

3.3.1 DETRENDING

With the decomposition of Figure 3.1 in mind, the first component to be removed is the trend. This operation can be done by subtracting from the data the fitted least-squares line (regression line).

**BOX 3.1   BASIC STRUCTURE OF THE MAIN SCRIPT
FOR THE FREQUENCY ANALYSIS**

*The FFT function computes the fast Fourier transform of the data. It is advisable
to select a power of 2 portion of the data for computational efficiency. The full
MATLAB script (*`Ex _ FFT _ sampling _ time.m`*) can be retrieved from
the* `ESA _ Matlab\Exercises\Chapter _ 3\Time _ Series` *folder.*

```
% load the data
g=uigetfile('*.mat','Load raw data');
eval(['load' g]);
% In the data file t is the time and q is the data set
% Ts is the sampling interval
% Define the kind of plot (linear or Log)
plotlog=1; % Select the linear plot
dt=diff(t); % Find the sampling intervals
Ts=mean(dt); % The sampling time is the average sampling intervals
Fs=1/Ts;     % Sampling frequency
% Frequency analysis
L=length(q); % Compute the time-series length
% Next power of 2 in the data length
N_samples = 2^(nextpow2(L)-1);
% Compute the FFT of the power-of-2-long data set
Y = fft(q,N_samples)/L;
% Compute the frequency range for plotting up to the Nyquist limit
f = Fs/2*linspace(0,1,N_samples/2+1);
switch plotlog
    case 1 % Linear plot
plot(f,abs(Y(1:N_samples/2+1)/abs(Y(1))))
    case 2 % Log plot
semilogy(f,abs(Y(1:N_samples/2+1)/abs(Y(1))))
end
% Downstream processing
% ........................................ . .
```

MATLAB provides the specific function `detrend` to do this job. Otherwise, this can be done with `polyfit` to compute the regression line coefficients and then subtract the linear contribution thus obtained from the data with the `polyval` function. The coefficients of the regression line are computed with the command `b = polyfit(x,y,1)`, where 1 indicates that we are fitting a line to the data, and the trend line is given by `y_trend=b(2)+b(1)*x`. If $b(1) \neq 0$, there is a data trend. The detrended data are then obtained by subtracting the trend line from the data, that is, `y_detrend=y-y_trend`. Both detrending methods are demonstrated in the MATLAB script of Box 3.2, and a sample detrended time series is shown in Figure 3.7. The detrended data have zero-mean value, and this can be useful when detrending several data sets before joining them to produce a longer stationary time series.

3.3.2   OUTLIERS DETECTION AND REMOVAL

Odd data may occur in any time series and, if overlooked, can disrupt all subsequent regularization operations. Therefore, these 'strange' data, usually referred to as *outliers,* should be dealt with prior to any further data processing, apart from detrending.

BOX 3.2   MATLAB SCRIPT FOR DATA DETRENDING

*The full MATLAB script (*Ex _ Detrending.m*) can be retrieved from the* ESA _ Matlab\Exercises\Chapter _ 3\Time _ Series *folder.*

```
%% load the data
g=uigetfile('*.mat','Load raw data');
eval(['load ' g]);
figure(1)
% Plot original data
plot(t,q,'.-')
hold on
% Perform detrending
qt=detrend(q);
plot(t,qt,':r')
%
%% Home-made detrending
% Compute regression coefficient
yp=polyfit(t,q,1);
% Compute regression line
yt=polyval(yp,t);
plot(t,yt,'--k','LineWidth',2)
% Subtract regression from the data
ydt=q-yt;
% Plot detrended data
plot(t,ydt,':g')
```

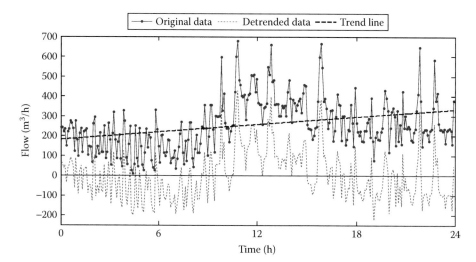

FIGURE 3.7   An example of detrended data, with zero-mean value. The trend line, which is subtracted from the original data, is also shown. This graph was obtained with the MATLAB script Ex _ FFT _ Detrending.m in the ESA _ Matlab\Exercises\Chapter _ 3\Time _ Series folder.

There are several definitions of *outliers* and consequently different ways to treat them, but first their very presence should be questioned. Are they just *wild* data produced by error or by some kind of arte-fact, or instead is there a reasonable explanation for their presence? Once their lack of significance has been assessed, there are several ways to eliminate them or bring them back into the fold of 'good' data.

An *outlier* is a sample that lies at an abnormal distance from other values in a coherent data set. This is, of course, a very subjective definition. A more objective definition may be formulated in statistical terms, for example, as the data lying outside the first and fourth quartile of the data population. This is a very severe restriction because it assumes that the time series is fairly stationary, with the well-behaved data remaining in the two central quartiles. This definition could be somewhat relaxed if we extend the tolerance band outwards and consider as outliers the data outside the bracket between, say, the 10th and the 90th percentiles. Figure 3.8 applies this definition to the same data of Figure 3.7, from which it appears that this definition may be unsuitable for data with a cyclic fluctuation. In fact, labelling as outliers the data with the smallest values at the beginning of the series is clearly unfair, although the data with peak values in the second half of the record do deserve to be classified as outliers.

Another possibility is to consider the data rate of change. Assuming that in nature, abrupt variations are rare and that data usually vary with a limited rate, the new outlier definition could be based on an abnormal rate of change, rather than on the value itself. So if the data rate of change Δs_k is computed as

$$\Delta s_k = \frac{s(t_{k+1}) - s(t_k)}{t_{k+1} - t_k} \quad k = 1, \ldots, N \tag{3.10}$$

then the percentile discrimination is applied to the Δs_k sequence rather than to the data, and the samples for which Δs_k falls outside the 10th–90th percentile bracket may be considered as outliers. This procedure is applied to the same data series of Figure 3.8, and the result is shown in Figure 3.9. The rate of change is checked for compliance with the 10th–90th percentile bracket, and the data whose rate of change is outside this band are marked as outliers, as shown in Figure 3.9c. It can be seen that this new definition is fairer on the data with a definite cyclic component and penalizes data with a large sudden variation from their neighbour.

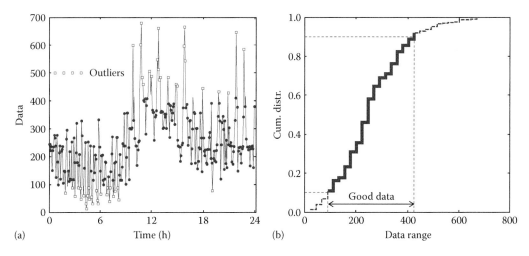

FIGURE 3.8 Definition of outliers on the basis of percentiles. The data falling outside the 10th–90th percentiles of the cumulative distribution (thick line in (b)) are considered as outliers (shown as hollow squares in (a)). Notice how inappropriate this definition may be for the initial data. These graphs were obtained with the MATLAB script Ex _ Outliers _ Perc _ Rates.m in the ESA _ Matlab\Exercises\ Chapter _ 3\Time _ Series folder.

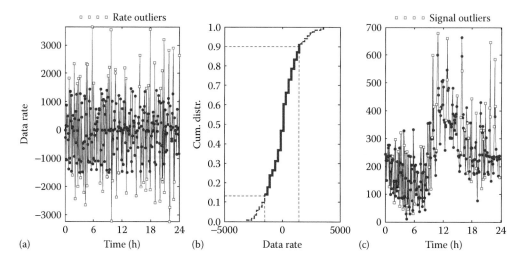

FIGURE 3.9 Outliers definition on the basis of maximum allowable rate. The data rate of change is shown in (a), with the squares indicating the values outside the 10th–90th percentile bracket (b). The resulting outliers are shown in (c). This method provides a fairer outlier detection, with a penalty on abnormal rate of change. These graphs were obtained with the MATLAB script Ex_Outliers_Perc_Rates.m in the ESA _ Matlab\Exercises\Chapter _ 3\Time _ Series folder.

Other outlier definitions are obviously possible and can be defined for specific contexts and needs. This issue will be reconsidered in conjunction with data smoothing, labelling as outliers the data with a significant departure from their smoothed counterpart.

3.3.3 Denoising through Approximating Splines Smoothing

After having assessed the data both in the time and frequency domains, and having removed the *wild* data as outliers, we now turn to the task of removing, or at least reducing, the noise affecting the data, which as we have seen, is generally confined to the high end of the frequency spectrum.

There are several good reasons why we may want to 'denoise' a time series. Noise affects data accuracy and may mask important information, essential for modelling, calibration, forecasting and so on. Further, noisy signals cannot be numerically differentiated, and this may become a serious limitation whenever the information we are seeking is contained in the data time derivative. Differentiation of a noisy signal amplifies the noise, with obvious disruption of the information contained in the data. Therefore, data smoothing prior to numerical differentiation is essential, and in this chapter, we shall consider two smoothing techniques, based either on spline approximation or on wavelet filtering.

Smoothing is a numerical technique to decrease the data 'roughness' and represents a simple and effective way of reducing the unwanted disturbance in the data, especially in anticipation of their numerical differentiation. It is based on the approximating properties of a family of functions named *splines* (De Boor, 2001), reminiscent of the bygone flexible ruler used to draw curves with china ink and tracing paper, that preserve derivative continuity. A spline is a numeric function that is piecewise defined by polynomials, each of which approximates the function in a specific portion of the domain. All these polynomials are connected to each other at specific points (knots) in a way that preserves derivative continuity up to a certain order. Figure 3.10 shows how a function $f(x)$ can be approximated by a collection of splines whose knots are placed at strategic positions along the domain. The placing of knots and the universal approximating properties of splines can be found in the specific literature (De Boor, 2001) and it will not be discussed here, being beyond the scope of this book.

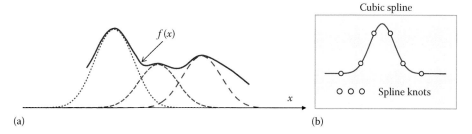

FIGURE 3.10   Approximation of a function $f(x)$ (solid curve) by local cubic splines (dashed curves) with suitably placed knots (a). A cubic spline is composed of seven polynomial segments connected at the knots as shown in (b).

We have already encountered the problem of *approximating* a function in a least-squares sense, that is, by minimizing the sum of squared errors. In fact, this has represented the main objective of Chapter 2 that dealt with model identification. Now, we require an additional property of the approximation: *smoothing*. Approximating splines smoothing is a trade-off between two conflicting objectives:

- *Information conservation*, which tends to place the approximation as close as possible to the data.
- *Smoothing*, which tends to distance the approximation from the (noisy) data.

An attractive way of reconciling these two opposite features is to define the objective function as a combination of the approximating and smoothing effects

$$p\sum_{k=1}^{N}\underbrace{\left(y(k)-ys(k)\right)^2}_{\text{approximation}}+(1-p)\underbrace{\int\lambda(t)\left|\frac{d^2ys(k)}{dt^2}\right|^2 dt}_{\text{smoothing}} \qquad (3.11)$$

and to seek its minimum with respect to the smoothing parameter $p \in (0,1)$. If $p = 1$, we get the *interpolating* spline, with no smoothing effect. As p is decreased from 1 towards 0, the emphasis in Equation 3.11 will gradually shift from approximation to smoothing until, for $p = 0$ the least-squares regression line is obtained. There is no general rule for selecting the right p value, which is strongly data dependent. The MATLAB function

$$ys = \texttt{csaps}(x,y,p) \qquad (3.12)$$

provides an 'adjustable' cubic spline in which the smoothing parameter p determines the relative importance of one term over the other and should be selected by the user. The output of Equation 3.12, ys, is the smoothed version of the original data y, whereas x is the independent variable; in our case, the time t, and p is the smoothing parameter. There are several factors that can assist in deciding the right amount of smoothing. Insufficient smoothing may leave too much noise in the data and thus fail its denoising purpose. On the other hand, too much smoothing may erode the information contained in the data, making them less informative. The frequency analysis can be used to compare the frequency spectra before and after smoothing, checking which frequencies have been attenuated, remembering the relative location of the signal components illustrated in Figure 3.6. Therefore, the smoothing is expected to have the highest impact on the medium-to-high frequencies. Figure 3.11 shows the smoothing of a flow time-series data sampled at 1 h intervals by varying the smoothing parameter *p*.

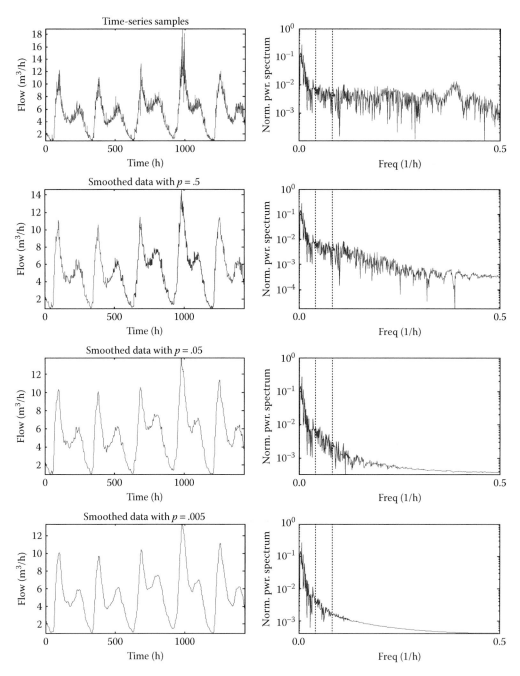

FIGURE 3.11   The effect of the smoothing parameter p on a noisy time series of flow data sampled at 1 h intervals. The time portraits (left column) show the progressive loss of the 'spiky' features as the smoothing is increased (p is decreased). This is confirmed by the frequency analysis (right column) where the high end of the spectrum is progressively attenuated. The dashed lines in the frequency portraits indicate the 24 and 12 h cycles. These graphs were obtained with the MATLAB script `Ex _ Spline _ Smooth.m` in the `ESA _ Matlab\Exercises\Chapter _ 3\Time _ Series` folder.

The progressive high-frequency loss as p is decreased is apparent in the frequency portraits (right column). It is up to the user to decide which portion of the spectrum can be cut out in the smoothed signal. Apart from improving the signal-to-noise ratio, smoothing may have two other applications: the replacement of missing data and the detection of outliers.

3.3.3.1 Replacement of Missing Data and Data Enhancement

Data in a time series may be missing as a consequence of a faulty equipment, or because some samples have been declared as outliers and removed. The approximating smoothing spline may be used to replace a limited amount of missing data with their corresponding approximation. Figure 3.12 shows a time series in which some data have been deleted. In the MATLAB file, this was accomplished by replacing the numerical values of the samples that we wanted to consider as missing with NaN (Not a Number). NaN's in a numeric data vector are simply ignored when the data are processed, and in particular, they are not plotted. In Figure 3.12, they are shown as hollow squares to indicate their position in the original file, but the approximating spline does not consider them when performing the data smoothing, for which only the remaining data (solid dots in the graph) are used. The reader can compare the relative position of the reconstructed data (filled squares) with respect to the original, but unknown, missing data (hollow squares). The solid line joining the known data becomes dashed when connecting reconstructed data.

The importance of a wise selection of the smoothing parameter p in missing data reconstruction is highlighted in Figure 3.13, in which the same portion of a data set including three missing samples was reconstructed with three differing choices for $p = .99$ (very close to the data, with little smoothing), $p = .5$ (medium smoothing) and $p = .01$ (strong smoothing). A large p value, favouring data fidelity, produces the closest approximation to the missing data, while the other two choices yield poor reconstruction. Further, the longer the data gap, that is, the number of consecutive missing data, the more uncertain will be the reconstruction, so this technique must be applied with great care and without excessive expectations. In fact, it is appropriate for replacing isolated missing data, but not for reconstructing more than a few consecutive missing data, for which specific techniques are available.

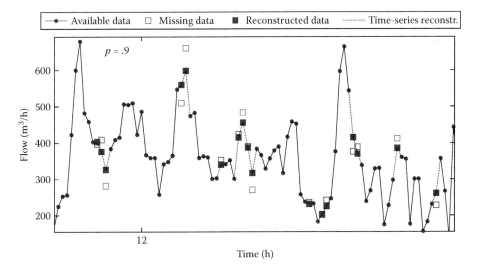

FIGURE 3.12 Missing data reconstruction using smoothing splines with $p = .9$. The hollow squares represent the missing data and the filled squares are their replacement obtained by spline smoothing. The line joining the data is dashed when it connects reconstructed data (filled squares). This graph was obtained with the MATLAB script Ex_Spline_Missing_Data.m in the ESA_Matlab\Exercises\ Chapter_3\Time_Series folder.

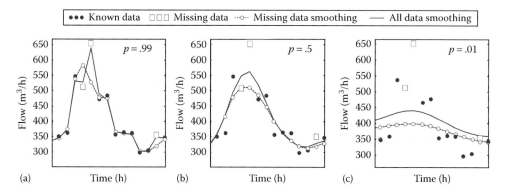

FIGURE 3.13 Differing choices for p produce very different reconstructions of the missing data. In (a), a p near unity produces a fairly close reconstruction of the three missing data (hollow squares). In (b), the medium smoothing ($p = .5$) yields regular data and good reconstruction of all but one missing data, while in (c), the very small value of p makes the reconstructed data too far from the originals.

An example of denoising through smoothing is shown in Figure 3.14, in which the DO circadian fluctuations are apparent, though partly masked by noise. The latter can be removed by spline smoothing, with the selection of p monitored by checking the frequency spectrum. It can be seen that the smaller the value of p, the more pronounced is the noise removal, without affecting the daily pattern, marked by the dashed line corresponding to the 24 h period.

Data enhancing via spline smoothing may be used when preparing input files for ecosystem simulation. Often, there are not enough experimental data to perform a long-range simulation, or the data are too sparse. In this case, applying an approximating smoothing spline to the data produces a data file with ample, well-behaved data, which can be used as input to the simulation model. Figure 3.15 shows the water quality data enhancement required to provide the inputs to the Massaciuccoli Lake simulation model (Giusti et al., 2011b) developed in Aquatox (Park and Clough, 2014).

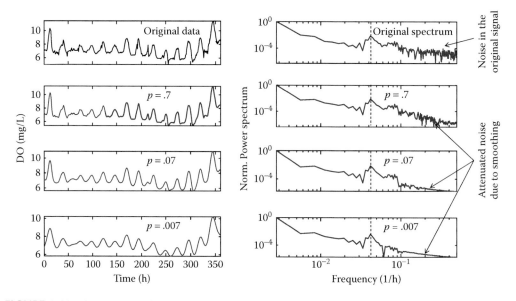

FIGURE 3.14 An example of FFT-assisted smoothing of a dissolved oxygen time series in the Orbetello lagoon (thin black line in the upper-left plot). The smoothing effect is clearly reflected in the frequency spectra, where the effect of noise attenuation is quite evident, without disrupting the basic time pattern. The dashed line in the spectra corresponds to the 24 h period, which is unaffected by smoothing.

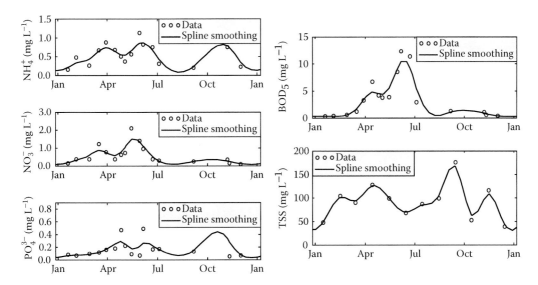

FIGURE 3.15 Data enhancement via smoothing splines to construct sufficient data for feeding a wetland water quality model. (Reprinted with permission from E. Giusti et al. 2011b, *Water Science & Technology*, 63, 2061–2070, IWA Publishing.)

3.3.3.2   Outliers Revisited

The use of smoothing approximations can lead us to reconsider the definition of outliers in the light of data regularization. In fact, we can now assess the consistency of the data by comparing the raw data with their smoothed version and consider as outliers the samples for which the difference between raw and smoothed values exceeds a certain threshold. So we are again considering the definition of outliers in terms of a derived quantity, as in Figure 3.9, which in this case is represented by the *residuals* defined as the difference between the original and the smoothed data. Figure 3.16 shows a new definition of outlier, represented by data whose difference with the smoothed data is outside the 10th–90th percentile bracket. Of course, this definition is heavily p-dependent as the number of outliers will tend to increase with the amount of smoothing. In this example, $p = .99$ was selected, so that the smoothing approximation (thick dashed line) is still reasonably close to the data. If a lower p is selected, then the number of outliers will increase considerably.

Because we are dealing with residuals, their characteristics may be regarded as a further check for the choice of p. In fact, we have seen in Chapter 2 the importance of checking the residual autocorrelation and normality, and we can now use these characteristics as further guidance in the selection of the smoothing parameters. Figure 3.17 shows the autocorrelation of the smoothed DO data of Figure 3.14 smoothed with $p = .7$. A lower p would have caused a proliferation of outliers, whereas, on the other hand, a 1-lag autocorrelation is acceptable in view of moderate smoothing.

3.3.4   Time-Series Synthesis

Separating the components of a time series may assist in the synthesis of more series, to enhance the input library of model simulation. As previously noted, the two main constituents of a time series are the deterministic pattern and the stochastic disturbance. We are now in a position to shape or alter each element and produce more synthetic time series by recombining them as shown in Figure 3.18. In this procedure, the deterministic part may be viewed as a kind of 'average' behaviour determined by the primary environmental agents, which can be predicted on the basis of a sufficiently comprehensive historical data set describing the typical behaviours to be expected

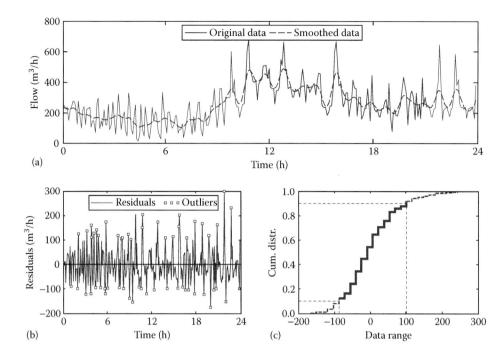

FIGURE 3.16 Outliers detection using a smoothing spline with $p = .99$. (a) compares the original and smoothed data. (b) plots their differences (residuals), while (c) shows their cumulative distribution, defining as outliers the data whose residuals lie outside the 10th–90th percentile bracket. These graphs were obtained with the MATLAB script Ex_Spline_Smooth_Outliers.m in the ESA_Matlab\Exercises\ Chapter_3\Time_Series folder.

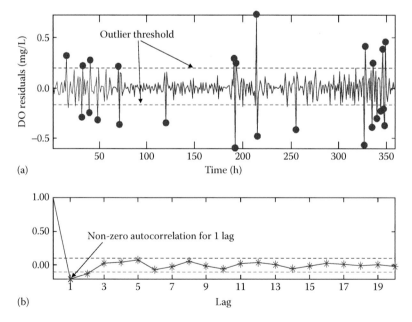

FIGURE 3.17 Residual autocorrelation analysis of the DO data of Figure 3.14 for $p = .7$. (a) shows the time plot of the residuals with the outliers limits set by the cumulative distribution, while (b) shows their autocorrelation. The non-zero 1-lag autocorrelation is acceptable vis-à-vis the limited number of resulting outliers. A smaller p would have produced a much larger number of outliers. The Lilliefors test suggests rejecting the normality hypothesis for the residuals distribution.

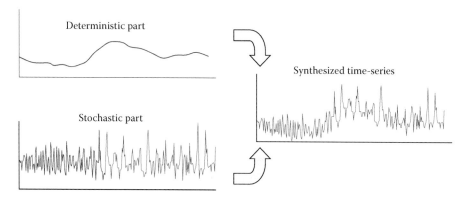

FIGURE 3.18   Time-series synthesis by combining the deterministic and stochastic components.

in that context. Conversely, the stochastic part is the random component, influenced by secondary environmental agents, and by its very nature it is not predictable from the past observations. Yet, this part too is generally context dependent in terms of autocorrelation and probability density function.

In defining the characteristics of the deterministic module of the time series, several aspects must be considered. The time series normally contains circadian periodicities due to the cyclic variability of the environmental variables on a daily/weekly/monthly/yearly basis. Examples of this are the daily periodicity of wastewater flow and organic load, both linked to the daily cycle of human activities, or the daily periodicity of DO in rivers and lakes due to the solar radiation, which stimulates photosynthesis. Other 'typical' periodic behaviours in domestic wastewater may be due to differing weekdays and weekend habits, whereas in natural waters, the DO daily patterns strongly differ in summer and winter. So it is important to define a criterion to decide under which circumstances a behaviour is considered representative, its period and the typical conditions under which it occurs. By observing a large number of similar behaviours, the typical features of this pattern can be isolated and will form the basis for the deterministic component of the synthetic time series.

As an example, Figure 3.19 (Marsili-Libelli, 2004) shows two sets of observed daily data from which representative patters have been extracted. In (a), 30 daily patterns of DO in the Orbetello lagoon were recorded during April 2001. The strong increase around noon is due to the intense algal activity, as will be better explained in Chapter 8, and the typical hourly behaviour (thick line) is obtained as the mean of the observations. In (b), five typical loading patterns were observed at the

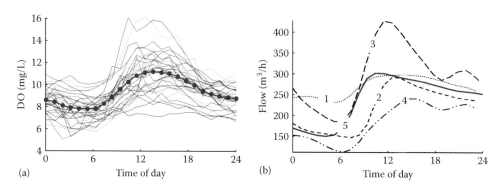

FIGURE 3.19   Typical patterns isolation in daily recurring behaviours. (a) shows the circadian evolution of dissolved oxygen in the Orbetello lagoon, whereas (b) shows five typical daily loading profiles of a wastewater treatment plants. (Reproduced with permission from Marsili-Libelli, S., *Ecol. Model.*, 174, 67, 2004.)

input to a small wastewater treatment plant treating domestic reject water. The five observed typical loading patterns have the following explanations:

1. Low load (summer)
2. Medium load with wide daily variability
3. High load with large daily variability
4. Low load with moderate daily variability
5. Medium load with average daily variability

While the extraction of the daily patterns is fairly simple, the identification of the stochastic component is more complex and hinges on the ability to identify the noise-generating stochastic process in terms of autocorrelation and probability density function. Figure 3.20 shows the four steps in the synthesis of environmental time series. First, the deterministic pattern is isolated. There may be a collection of possible patterns, such as the ones in Figure 3.20b for the daily wastewater loading profiles. The residuals (observation pattern) are then analysed for autocorrelation (Figure 3.20c) and distribution (Figure 3.20d). Then, more random data series can be generated with these characteristics and summed back to the pattern to obtain a synthetic time series. The resulting time series may look like the one shown in Figure 3.21.

3.3.4.1 Wind Time-Series Synthesis for Lagoon Simulation

Wind is the prime mover in shallow water, such as the Orbetello lagoon on the Tyrrhenian coast of central Italy (see the inset in Figure 3.22). In the development of a comprehensive lagoon water quality model (Giusti and Marsili-Libelli, 2006), the wind modelling was included as an input generator for long-term simulation to understand the role played by the wind in determining the water movement and the dispersal of seeds, the basis for submerged vegetation expansion.

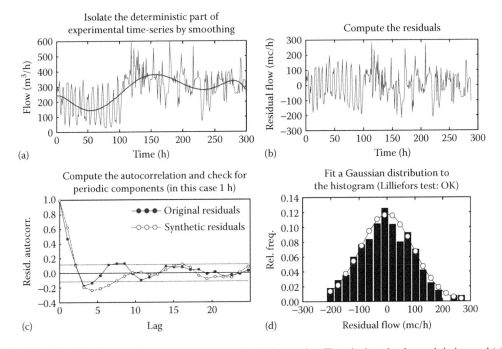

FIGURE 3.20 The four steps in constructing a synthetic time series. First, isolate the deterministic trend (a), then analyse the residuals (b–d) to generate more random patterns which can be summed to the basic deterministic trend (a) to produce an endless number of combinations.

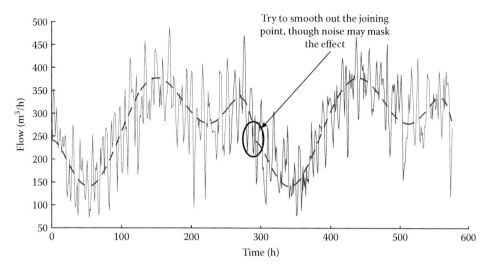

FIGURE 3.21 Two synthetic time-series modules are joined together to produce a longer record of synthetic data. There may be a joining problem that could be avoided by careful matching the initial and final derivatives, though the noise may partially mask the discontinuity.

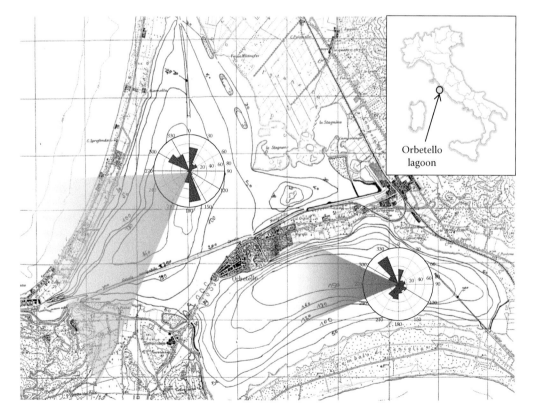

FIGURE 3.22 Wind statistics observed in the two ponds of the Orbetello lagoon. (Reproduced with permission from Giusti, E. and Marsili-Libelli, S., *Ecol. Model.*, 196, 379, 2006.)

The model was based on three years (2001–2004) of data acquired by two fixed stations placed in the middle of the two ponds that form the Orbetello lagoon. From those data, it was observed that the wind distribution in the two basins is substantially different, given the differing shielding provided by the Monte Argentario promontory on the sea side. Therefore, two wind models were set up, one for each pond, with correlated time series. The available data were used to form the reference daily sequence. These data were originally sampled at 10 min intervals, with about 20% of missing data. To fill the gaps, the most frequent value in the corresponding time slot was considered. Then, a synthetic daily value was extracted by considering the prevailing wind of the day as the one blowing for at least three consecutive hours with a constant direction and a speed of at least 0.5 m/s, which was assumed as the minimum wind speed required to produce a water movement. Several synthetic time series were obtained from these reference data by the following procedure:

1. The deterministic trend was obtained by spline smoothing.
2. The autocorrelation of the residuals was computed to determine the order of the AR model producing the stochastic component of the series.
3. The synthetic sequence was then obtained by adding the residuals generated by the AR models to the deterministic trend data.

Further, because a correlation was observed between the same-day wind speeds in the two basins, a correlated noise was used to drive the AR models and conserve this feature in the synthetic data. The observed wind series were then decomposed as shown in Figure 3.23, with the smoothed part forming the deterministic backbone of the series and the stochastic part being further analysed for autocorrelation and normal probability density function.

In fact, the synthesized stochastic residuals should have the same autocorrelation as their observed counterparts. The order of the stochastic model was assumed equal to the first non-zero residual, that is, 3, as shown in Figure 3.24, which compares the autocorrelation of the original and synthetic random time series, whereas the Gaussian distributions estimated from the wind samples are shown in Figure 3.25.

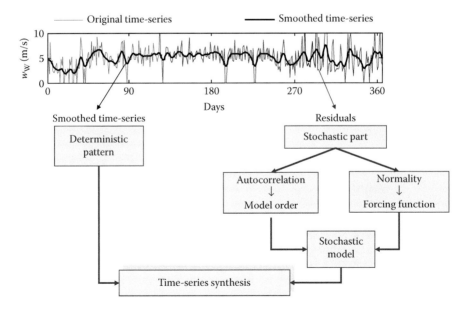

FIGURE 3.23   Extracting information from the observed wind data to construct a synthetic time series.

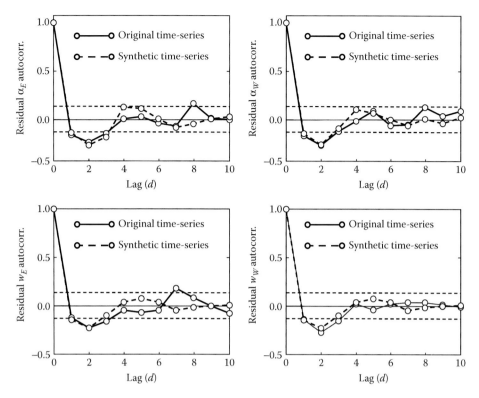

FIGURE 3.24 Autocorrelation of the observed and synthetic wind time series. (Reproduced with permission from Giusti, E. and Marsili-Libelli, S., *Ecol. Model.*, 196, 379, 2006.)

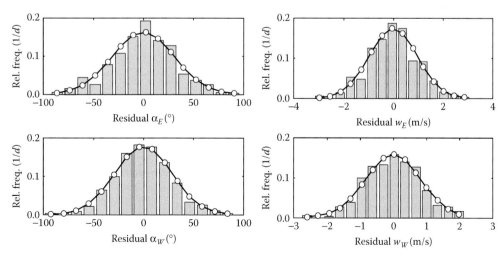

FIGURE 3.25 Fitting the experimental wind histograms obtained from the smoothed time series. The normal distribution hypothesis passed the Kolmorogov–Smirnov test. (Reproduced with permission from Giusti, E. and Marsili-Libelli, S., *Ecol. Model.*, 196, 379, 2006.)

The correlated AR models for the wind direction and speed in the two basins were estimated from the data using the Yule–Walker method (Friedlander and Porat, 1984), which minimizes the forward least-squares prediction error based on the estimated autocorrelation function. The following four AR models were obtained for wind angle and speed in the two East and West ponds:

Angle East $\tilde{\alpha}_E(i) = -0.31863 \cdot \tilde{\alpha}_E(i-1) - 0.37667 \cdot \tilde{\alpha}_E(i-2) - 0.27547 \cdot \tilde{\alpha}_E i(i-1) + \sigma_E^\alpha \cdot e_E(i)$

Angle West $\tilde{\alpha}_W(i) = -0.29189 \cdot \tilde{\alpha}_W(i-1) - 0.33872 \cdot \tilde{\alpha}_W(i-2) - 0.27405 \cdot \tilde{\alpha}_W(i-3) + \sigma_W^\alpha \cdot e_W(i)$

Speed East $\tilde{w}_E(i) = -0.28069 \cdot \tilde{w}_E(i-1) - 0.36070 \cdot \tilde{w}_E(i-2) - 0.28083 \cdot \tilde{w}_E(i-3)$

$$+ (1-v)\sigma_W^w \cdot e_W(i) + v \cdot \sigma_E^w \cdot e_E(i) \qquad (3.13)$$

Speed West $\tilde{w}_W(i) = -0.22924 \cdot \tilde{w}_W(i-1) - 0.29776 \cdot \tilde{w}_W(i-2) - 0.25496 \cdot \tilde{w}_W(i-3)$

$$+ (1-v)\sigma_W^w \cdot e_W(i) + v \cdot \sigma_E^w \cdot e_E(i)$$

$$i = 1,\ldots,365$$

The speed time series are driven by two independent normally distributed uncorrelated noise sources $e_E(t)$ and $e_W(t)$ with standard deviations σ_E and σ_W. To reproduce the observed same-day speed correlation between the two basins, a driving function was constructed combining the two noises with the correlation coefficient $v = 0.732$. This provided consistently related time series with a correlation very close to the observed value of 0.7. The synthetic time series were obtained by adding the smoothed deterministic part (trend data) to the pertinent stochastic model, to obtain

		trend data smoothing	stochastic component
Angle East	$\alpha_E(i) =$	$\alpha_E^s(i) +$	$\tilde{\alpha}_E(i)$
Angle West	$\alpha_W(i) =$	$\alpha_W^s(i) +$	$\tilde{\alpha}_W(i)$
Speed East	$w_E(i) =$	$w_E^s(i) +$	$\tilde{w}_E(i)$
Speed West	$w_W(i) =$	$w_W^s(i) +$	$\tilde{w}_W(i)$

$$(3.14)$$

A sample of a synthetic time series derived by the above method is shown in Figure 3.26, where it is compared to the original data.

3.4 WAVELET SIGNAL PROCESSING: ADAPTIVE COMBINATION OF TIME AND FREQUENCY ANALYSIS

The FT described in Section 3.2 performs a harmonic decomposition and reveals the frequency content of a time signal, but it has limitations: in addition to losing any time information, the signal is supposed to be stationary, meaning that the same frequency resolution is applied for any time portion of the signal. To overcome this limitation, in 1946, Gabor adapted the FT considering only a portion of the signal through a time window. This short-term FT (STFT) is a compromise between time and frequency, but its accuracy depends on the window width, which was the same for all frequencies, whereas many signals may require a more flexible approach, where the window can be adapted to the time and frequency features of the signal. Intense research spanning several decades has produced the wavelet transform (WT) that, adapting the time/frequency window to the signal characteristics, has shown great potential and applicability in many areas of science and engineering (Rioul and Vetterli, 1991; Vidakovic and Mueller, 1991; Heil, 1993; Strang, 1994; Graps, 1995;

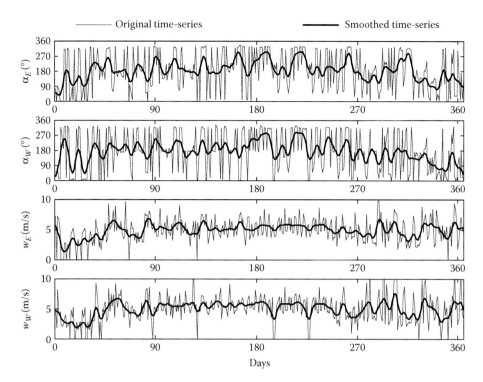

FIGURE 3.26 Smoothed wind time series obtained from the wind data over the period 2001–2004. The thin dashed lines represent the original data and the thick solid lines are the smoothed deterministic component, obtained with $p = .0924$. (Reproduced with permission from Giusti, E. and Marsili-Libelli, S., *Ecol. Model.*, 196, 379, 2006.)

Goswami and Chan, 2008; Walker, 2008; Misiti et al., 2010; Yajnik, 2012). The WT provides an adaptive window with long time-intervals to analyse low-frequency phenomena, and short time-intervals to analyse high-frequency phenomena. So the signal analysis is performed over a grid of time–frequency slots whose dimensions are adapted to the *local* signal characteristics. The qualitative difference between the FT and the WT is shown in Figure 3.27.

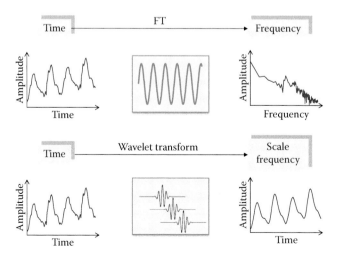

FIGURE 3.27 While the FT uses a set of constant sinusoids, the Wavelet Transform adapts the transformation to the signal local features through a bank of scalable wavelets.

As the name suggests, a wavelet ('little wave') is non-zero only for a limited time span, as opposed to the 'great wave' used by Fourier, the sine function that exists from $-\infty$ to $+\infty$. A function qualifies as a wavelet (ψ) if it exists only in the finite interval $(-T, T)$ and satisfies the following properties:

$$\text{zero mean value} \quad \int_{-\infty}^{\infty} \psi(u)\,du = \int_{-T}^{T} \psi(u)\,du = 0$$

$$\text{finite energy} \quad \int_{-\infty}^{\infty} \psi^2(u)\,du = \int_{-T}^{T} \psi^2(u)\,du = 1 \tag{3.15}$$

In addition to the literature already cited, a very good tutorial is available online (http://users.rowan.edu/~polikar/WAVELETS/WTtutorial.html). Many functions satisfy conditions (3.15), and the most popular wavelets are due to Haar, Morlet, Meyer and Daubechies, as surveyed by Heil (1993).

3.4.1 THE CONTINUOUS WAVELET TRANSFORM

Like splines, wavelets use the concept of local approximation, here in two dimensions (time and scale). A signal is decomposed into a combination of 'daughter' wavelets, originated from a 'mother' wavelet, which can be adapted to the *local* signal properties, both in amplitude and in frequency, through the two basic operations of *shifting* and *scaling*. Given a generic wavelet $\psi(a,b,t)$, where a and b represent the *scaling* and *shifting* factors, the continuous wavelet transform (CWT) is defined as the integral of the signal $s(t)$ multiplied by the scaled and shifted wavelet $\psi(t,a,b)$

$$C(a,b) = \frac{1}{\sqrt{a}} \int_{-\infty}^{\infty} s(t) \cdot \psi\left(\frac{t-b}{a}\right) dt \tag{3.16}$$

If the wavelet is stretched $(a > 1)$, it contains predominantly low frequencies and yields a global signal approximation including slowly changing, coarse features corresponding to low frequencies. Conversely, if it is shrunk $(a < 1)$, the compressed wavelet contains predominantly high frequencies, and yields a detailed portrait of the rapidly changing details in a small portion of the signal $s(t)$. Thus, high scales correspond to stretched wavelets, which consider in coarse details a large portion of the signal (the signal is zoomed out), whereas the low scales correspond to compressed wavelets, providing insight into a small portion of the signal, of which the fine details will be revealed (the signal is zoomed in). Through shifting and scaling, shown in Figure 3.28, the entire signal is covered in an adaptive way.

In principle, the CWT of a signal is performed through the following steps:

1. Select a wavelet (some examples of which are shown in Figure 3.29) and compare it to an initial portion of the signal.
2. Calculate a number (c) that adapts the wavelet to that signal section. The magnitude of c is a measure of the similarity between the signal portion and the wavelet.
3. Shift the wavelet forward and repeat steps 1 and 2 to cover the whole signal.
4. Repeat steps 1 through 3 with a different scale.
5. Repeat steps 1 through 4 for all scales.

The CWT produces a huge amount of coefficients, and the results are difficult to interpret. They are often presented in graphical form, as in Figure 3.30, where it is applied to a sine wave with a sudden frequency change.

The scales used in this example are up to the fifth power of two $(2^5 = 32)$. The FT would have been incapable of adapting to the frequency change at $t = 500$ and detect the exact instant when the break occurred, whereas the wavelet coefficients change to adapt to the discontinuity. In the

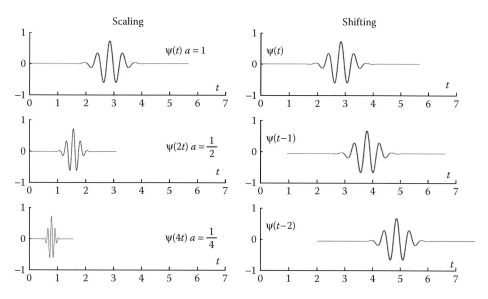

FIGURE 3.28 Graphical illustration of the two main wavelet operations: Scaling changes the frequency range of the wavelet, whereas Shifting changes the portion of the signal to which it is applied.

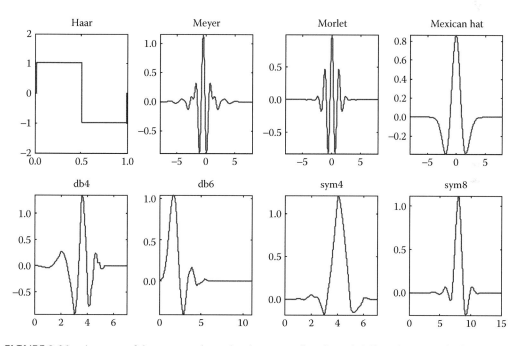

FIGURE 3.29 A survey of the most-used wavelet elementary functions. A full catalogue can be found in the MATLAB Wavelet Toolbox Help.

graphical representation of the coefficients, the time/scale graph associates a colour with the magnitude of the c coefficient, indicating that they are large at scales near the frequencies of the signal or near discontinuities. The CWT detects both the abrupt transitions and the steady oscillations in the signal. The central break in the middle of the signal affects the CWT coefficients at all scales. On the other hand, the maxima and minima of the steady-state sinusoid are represented by

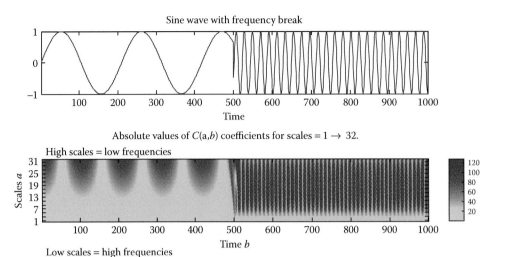

FIGURE 3.30 CWT of a sine wave with a sudden frequency change. The lower graph shows the absolute values of the CWT coefficients up to scale 31. The colour scale on the right gives an idea of the relative coefficient magnitudes. (Adapted from the MATLAB Wavelet Toolbox Demos.)

large CWT coefficients at high scales and are not apparent at small scales. Following this analysis, even hidden signal changes that are not as apparent as this break could be detected by observing changes in the coefficients.

3.4.2 THE DISCRETE WAVELET TRANSFORM

As clearly shown in Figure 3.30, the CWT is computationally demanding, and the amount of its coefficients is overwhelming. The analysis of environmental time series could be more easily carried out by restricting the scales to powers of 2, thus performing a *dyadic* signal decomposition, without loss of accuracy. The discrete wavelet transform (DWT) is defined by

$$s(t) = \sum_{j=-\infty}^{\infty} \sum_{k=-\infty}^{\infty} c_{jk} 2^{j/2} \psi\left(2^j t - k\right) \tag{3.17}$$

The wavelets $\psi\left(2^j t - k\right)$ are 2^j scaled and k shifted versions of the original wavelet ψ. The DWT decomposes the signal into *approximations* and *details*. The *approximations* are the high-scale, low-frequency components of the signal, whereas the *details* refer to the low-scale, high-frequency components. However, there is a problem in this decomposition: at each step the two signals (Approximations and Details) combined have twice the number of samples as the original signal. Therefore, to keep the sample number constant, *down-sampling* is applied, by retaining every other sample after the decomposition, so that half samples are available at double period. This is equivalent to doubling the sampling period at each doubling of scale, so at each step lower frequencies are involved. Figure 3.31 shows the dyadic decomposition tree of a signal. At each step an approximation/details pair is created, and the period doubles because of down-sampling. Likewise, the number of data on which the DWT is performed is halved. This process may be repeated for as many steps as required by the frequency range of interest and is limited either by the availability of samples or because the sampling period becomes too long.

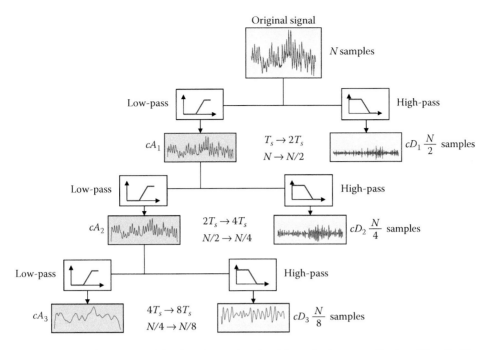

FIGURE 3.31 Dyadic decomposition into Approximation (through a low-pass filter) and Details (through a high-pass filter) to obtain the DWT of a signal. Because of down-sampling at each step, the period is doubled.

3.4.3 THE MATLAB WAVELET TOOLBOX

If we were to set up wavelet signal processing from scratch, we would be confronted with a formidable task. Luckily, the Wavelet Toolbox is a very comprehensive environment in which all the useful wavelet functions can be found. Abundant demos and a very detailed help make the toolbox very friendly and useful, even for the novice. Given a time series that we want to analyse, its DWT can be accomplished with a few, effective, commands. The first thing to decide is the kind of wavelet to use and the maximum level of decomposition. As to the first, several wavelets can be tested, but from experience, the Morlet or Meyer wavelets appear to give the best results in environmental data processing. Regarding the level of decomposition, it is not advisable to push the dyadic splitting too far, because of possible data shortage and lack of interest in the very long period of the last stages. A good compromise is generally to extend the decomposition for two or three steps, but not more. At this level, the noise components should have been sufficiently 'tamed'.

The multilevel decomposition tree of Figure 3.32 operates by filtering the signal (X) with either a low-pass (approximation) or a high-pass (detail) filter, followed by down-sampling. The actual wavelet decomposition is formed by the last approximation and all the details $W_{dec} \sim (A_n, D_1, D_2, ..., D_n)$. If the original signal has N samples, Wdec may contain at most $n = \log_2(N)$ levels.

The MATLAB command producing the DWT decomposition is

$$[\text{C,L}] = \texttt{wavedec}(\texttt{x,N,wname}); \tag{3.18}$$

where:

x is the data vector
N is the maximum level of decomposition
wname is a string containing the name of the wavelet

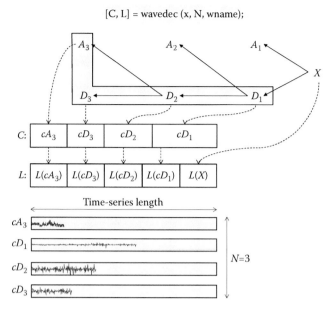

FIGURE 3.32 Vectors of coefficients produced by the wavelet decomposition function `wavedec`. The function returns the coefficients of the last approximation and of all the details up to the Nth level. The coefficients are plotted at the bottom of the figure, but in practice, they are hardly ever looked at.

The output data consist of two vectors, the coefficient vector C and their lengths L. Therefore, the output is not the processed signal, but the set of coefficients, both approximations and details, bundled in the vector C of Figure 3.32, and whose lengths are stored in the vector L. To view the coefficients, the approximations and the details, the following commands are required, after `wavedec` has been executed:

View the approx. coefficients	`cAi = appcoef (C,L,i);`
View the details coefficients	`cDi = detcoef (C,L,i);`
View the ith approximation	`Ai = wrcoef ('a',C,L,wname,i);`
View the ith detail	`Di = wrcoef ('d',C,L,wname,i);`

(3.19)

In commands (3.19), the index `i` refers to the level of the decomposition, and therefore $(i \leq L)$. It should be reminded that the exact number of coefficients in the decomposition is not accurately predictable, because the windowing effects of the algorithm is such that the coefficients at a given level are not exactly half those of the previous level.

The wavelet-filtered signal can be reconstructed by summing the *last* approximation A_n and *all* the details $(D_1, D_2, ..., D_n)$, that is,

$$s(t)_{\text{rec}} = A_n + \sum_{i=1}^{n} D_i$$

(3.20)

The corresponding MATLAB command is

$$S\_rec = \texttt{waverec(C,L,wname)};$$

(3.21)

where the input parameters have the same meaning as in commands (3.19). Therefore, the reconstruction hinges on the coefficient vector C and of course refers to the same wavelet (`wname`) used for the decomposition. The reconstructed signal has the same number of samples as the original signal.

3.4.4 Signal Denoising

Among the many application of wavelets, the one that is particularly important in this context is *denoising*, which is another way of improving the data quality, as we have already attempted with smoothing splines. Going back to the previous paragraph, one might wonder what is the use of performing a series of operations (decomposition + reconstruction) that eventually reproduces the original signal. The answer is that in the decomposition the signal was partitioned into different components, each of which could be separately manipulated before reassembling all of the parts to reconstruct the signal. Denoising does exactly that: it separately processes the parts where the noise is predominant before reconstructing the signal. The details, containing mainly high-frequency components, can be treated for noise removal by hard or soft thresholding, as will be shown in a moment, after which the signal is reconstructed by summing all the processed details $D_{i_{thr}}$ to the last approximation A_n. Thus the denoised reconstruction signal is similar to Equation 3.20 but with altered details, that is,

$$s(t)_{dn} = A_n + \sum_{i=1}^{n} D_{i_{thr}} \tag{3.22}$$

The details in Equation 3.22 differ from those in Equation 3.20 in that they have been 'thresholded' to reduce their noise contribution. It should be stressed that the thresholding process involves all the details, and not just those of the last level. The logical sequence of decomposition, thresholding of the details, and reconstruction is described in Figure 3.33, relative to the denoising of a pH signal, typically affected by noise. It shows the signal decomposition down to level 2 and the separate thresholding of the two details coefficients D_1 and D_2. The denoised signal is then the sum of the processed components, as in Equation 3.22.

Figure 3.34 shows two possible thresholds to be applied to the details coefficients: in the 'hard' case, all the details coefficients smaller than a given threshold δ are set to zero, whereas in the 'soft' case, the non-zero coefficient after hard thresholding are compressed toward zero by an amount equal to the threshold δ

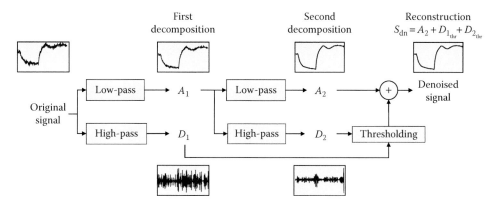

FIGURE 3.33 Logical sequence of the denoising process, applied here to the denoising of the pH signal in a sequencing batch reactor. The signal is first decomposed (down to level 2 in this example), and then the details coefficients are thresholded before reconstruction. In this way, the high-frequency part of the signal is conditioned, and the reconstructed signal is the sum of the last approximation (A_2) and the conditioned details. (Reproduced with permission from Marsili-Libelli, S., *Water Res.*, 40, 1095, 2006.)

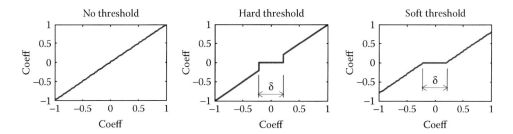

FIGURE 3.34 Thresholding methods to alter the relative importance of the details coefficients.

$$\text{Hard threshold} \quad D_{i_\text{ht}} = \begin{cases} D_i & \text{if} \quad |D_i| > \delta \\ 0 & \text{if} \quad |D_i| \le \delta \end{cases}$$

$$\text{Soft threshold} \quad D_{i_\text{st}} = \begin{cases} D_i - \delta & \text{if} \quad |D_i| > \delta \\ 0 & \text{if} \quad |D_i| \le \delta \end{cases}$$

(3.23)

The differing effects of hard or soft thresholding are shown in Figure 3.35.

MATLAB provides a specific denoising command `wden`, whose syntax and options are shown in Figure 3.36. It can operate directly on the signal (X) or on its decomposition ([C,L]), computed by a previous `wavedec` command. The other input parameters control the manner in which the threshold is computed and applied, in addition to the level of decomposition and the type of wavelet. As shown in Figure 3.36, there are four possible methods to compute the threshold:

1. `rigrsure`: Applies Stein's Unbiased Risk (SURE) and computes the threshold as $\delta = \sqrt{2\ln\left(N\log_2 N\right)}$

2. `heursure`: A heuristic variant of the `rigrsure`
3. `sqtwolog`: Universal threshold computation $\delta = \sqrt{2\log N}$, the one used to compute the thresholds in Figure 3.35
4. `minimaxi`: Minimum of maximum mean squared error

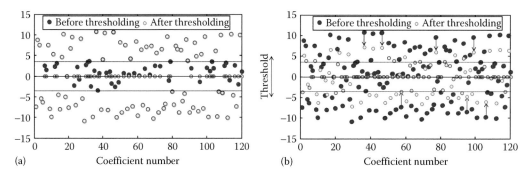

FIGURE 3.35 Differing effect of 'hard' (a) or "soft" (right) thresholding. In hard thresholding, the details inside the threshold band (filled dots) are set to zero, whereas those outside are unchanged (filled and hollow dots superimposed). In the soft thresholding (b), the details inside the band are equally set to zero, but those outside are moved inwards (see arrows) by a half threshold width. These graphs were obtained with the MATLAB script `Ex_Wavelet_Thresholding.m` in the `ESA_Matlab\Exercises\Chapter_3\Time_Series` folder.

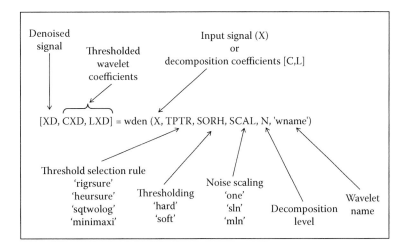

FIGURE 3.36 Syntax of the denoising command wden. It can operate on either the data (X) or their previous decomposition ([C,L]). The output data consist of the denoised signal and related coefficients.

Denoising assumes that the signal model is $s(t) = f(t) + \sigma N(0,1)$, where the 'true' signal $f(t)$ is affected by an additive zero-mean noise with a Gaussian distribution $N(0,1)$ and standard deviation σ. To estimate the actual noise level, wden offers three options for the multiplicative threshold rescaling parameter SCAL:

1. one: No rescaling
2. sln: Estimates σ based on first-level coefficients
3. mln: Estimates σ based on all levels

The best combination of denoising parameters must be adapted to the characteristics of the signal being processed. For example, Figure 3.37 shows a portion of the denoised DO in the Orbetello

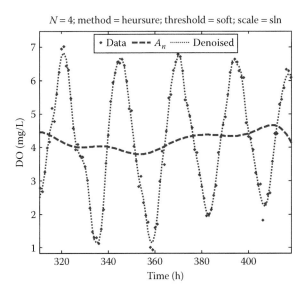

FIGURE 3.37 Details of the denoised DO signal with a four-level decomposition applied to a portion of the DO data of Figure 3.19. In this case the approximation A_4 would be totally unsuitable for denoising.

lagoon for five running days, using a level 4 decomposition with a `heursure` soft thresholding. As a general rule, if the decomposition is not very deep (two levels at most), then A_2 may provide a satisfactory denoised signal, but with deeper decompositions, the difference between A_n and the denoised signal defined by Equation 3.22 increases dramatically.

A final comparison of the four denoising method provided by the `wden` command are shown in Figure 3.38, where they are applied to a portion of DO recording in the Orbetello lagoon. It appears that some methods do not provide enough denoising because they follow the data too closely (particularly `rigrsure` and `minimaxi`), whereas `heursure` and `sqtwolog` are less distracted by wild data and provide a very robust denoising, especially `heursure`.

Do-It-Yourself addicted may not find in the `wden` command a level of control that they like. In this case, a more hands-on approach is provided by the group of commands shown in Figure 3.39. First, the signal is decomposed, and then details are processed with the `wthcoef` command,

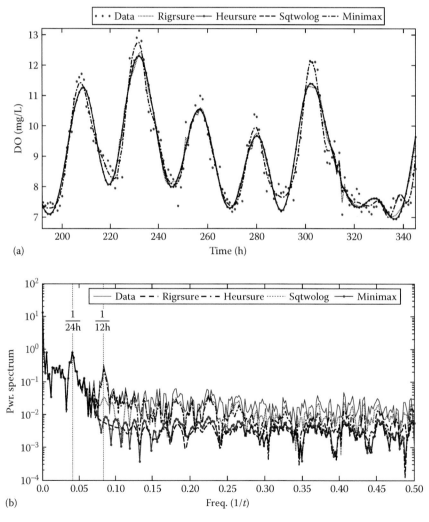

(a)

(b)

FIGURE 3.38 Comparison in time (a) and frequency (b) of the various denoising methods applied to a portion of the DO data of Figure 3.19. The heuristic method (`heursure`) appears to be the most robust, being least affected by 'wild' data and providing the highest attenuation in the high-frequency band. These graphs were obtained with the MATLAB script `Ex_Wavelet_Denoising_Comparison.m` in the `ESA_Matlab\Exercises\Chapter _ 3\Time _ Series` folder.

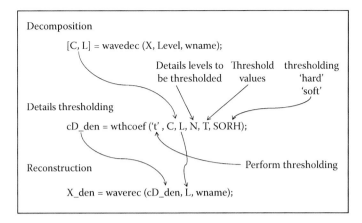

FIGURE 3.39 The Do-It-Yourself (DIY) version of denoising: after decomposing the signal, the command `wthcoef` enables direct control over which details should be the thresholded and how. The signal is eventually reconstructed with the manipulated details.

enabling full control over which details are to be thresholded and in which way. The denoised signal is eventually reconstructed using the processed details.

3.4.5 NUMERICAL DIFFERENTIATION OF NOISY SIGNALS

Often the information we want to extract from the data is embedded in their derivative rather than in the data themselves. On the other hand, differentiating noisy data may end in frustration because the noise will be amplified by the operation and produce totally unreliable results. Therefore, denoising is mandatory prior to differentiation if we are to produce meaningful results. The most-used numerical differentiation formulas with a finite increment h are (Abramowitz and Stegun, 1970; Hildebrand, 1974; Dahlquist and Bjorck, 2003)

$$
\begin{array}{ll}
A & \dfrac{dx}{dt} \cong \dfrac{x(t+h)-x(t-h)}{2h} \\[2mm]
B & \dfrac{dx}{dt} \cong \dfrac{1}{12}\Big[x(t-2h)-8x(t-h)+8x(t-h)-x(t+2h)\Big] \\[2mm]
C & \dfrac{d^2x}{dt^2} \cong \dfrac{x(t+h)-2x(t)+x(t-h)}{h^2} \\[2mm]
D & \dfrac{d^2x}{dt^2} \cong \dfrac{1}{12}\dfrac{-x(t-2h)+16x(t-h)-30x(t)+16x(t+h)-x(t+2h)}{h^2}
\end{array}
\tag{3.24}
$$

Formulas A and C in Equation 3.24 are simple centred differences, where B and D use two points on each side of the central point and are, therefore, more accurate. Numerical accuracy is largely determined by the prior signal conditioning, hence the need to remove as much noise as possible. For this reason, preliminary denoising is recommended, and wavelets can assist in this task better than splines, because the latter cannot control the information embedded in the approximation, whereas wavelets, with their approximation/details separation provide a more efficient signal/noise partitioning. It should be recalled that if differentiation is applied to the denoised signal, the increment h is equal to the sampling interval, but if it is applied to approximations, it should be considered that each approximation has a sampling interval, that is, $n \cdot h$, where n is the approximation level. This should be considered when applying Equation 3.24. Figure 3.40 demonstrates the importance of

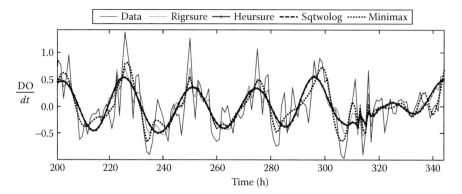

FIGURE 3.40 Numerical differentiation of a portion of the dissolved oxygen data of Figure 3.38. The differentiation of the raw data is totally useless, whereas the smoothest derivatives are provided by the `heursure` and `sqtwolog` methods that are almost superimposed in this example. This graph was obtained with the MATLAB script `Ex_Wavelet_Denoising_Comparison.m` in the `ESA_Matlab\ Exercises\Chapter_3\Time_Series` folder.

denoising prior to differentiation, and that trying this operation on the raw data is completely useless. The `heursure` and `sqtwolog` methods provide the smoothest differentiation.

An application of denoising wavelets to the pH signal is shown in Figure 3.41 following the block diagram of Figure 3.33. This signal is particularly noisy because it is measured with a high impedance probe, which may easily pick up all sorts of electromagnetic disturbances. On the other hand, when pH is used to control the switching in sequencing batch reactors, the important information is to be found in its rate of change, rather than in the pH value itself. If numerical differentiation is performed on the raw signal, the noise is amplified to such an extent that it makes the derived signal useless. Instead, as Figure 3.41b shows, the differentiation performed on the denoised signal is much smoother and can be used to infer the important process transitions (Marsili-Libelli, 2006).

3.5   PRINCIPAL COMPONENTS ANALYSIS

Principal Component Analysis (PCA) is a widely used statistical technique to extract relevant information hidden in the data. Because of its wide and deep coverage in the statistical literature (Dunteman, 1989; Jolliffe, 2002; Ringnér, 2008; Abdi and Williams, 2010), only the basic concepts

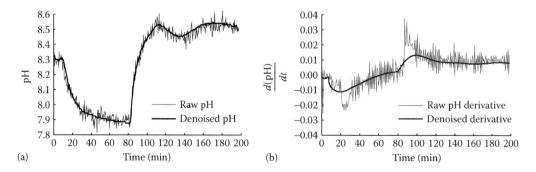

FIGURE 3.41 The pH signal is known to be noisy because it is measured with a high impedance probe, and therefore it is liable to pick up all sorts of disturbances. (a) compares the raw pH signal with its denoised counterpart, whereas (b) shows the same comparison for the derivatives. Notice how the differentiation amplifies the noise, hence the need to denoise prior to derivation.

will be recalled here, together with a summary of the main MATLAB commands to perform PCA. Given a data set $X \in \mathbb{R}^{N \times p}$, where N is the number of observations and p the number of features of each observation, PCA is a linear transformation with two objectives: feature selection and dimension reduction. The first objective is achieved through a change of the reference coordinates and a ranking of the features of X, whereas the second (optional) goal consists of reducing the number of features from p to $a < p$, controlling the related loss of information.

3.5.1 Feature Selection

The features (or variables) of a multidimensional data set in general are related to one another, in the sense that the variability of each one may influence the others. This interdependence, in addition to masking the full significance of each feature and hiding the information embedded in the data, represents a redundancy by partially replicating the information in several variables. A measure of redundancy in a multidimensional data set is the covariance matrix computed on the zero-mean data matrix x obtained from X by subtracting the average of each feature $(1,...,p)$

$$\bar{x}_j = \frac{1}{N} \sum_{i=1}^{N} X(i,j) \qquad j = 1,...,p \qquad (3.25)$$
$$x_j = X_j - \bar{x}_j$$

The diagonal elements of the (symmetrical) covariance matrix C are the variance of each feature, whereas the off-diagonal terms represent the mutual interaction between feature couples

$$C = \frac{1}{N-1} x^T \cdot x = \begin{bmatrix} \sigma_1^2 & \rho_{12} & \cdots & \rho_{1p} \\ \rho_{21} & \sigma_2^2 & \cdots & \rho_{2p} \\ \cdots & \cdots & \cdots & \cdots \\ \rho_{p1} & \rho_{p2} & \cdots & \sigma_p^2 \end{bmatrix} \quad \begin{array}{l} \sigma_j^2 = \dfrac{1}{N-1} x_j^T \cdot x_j \\[2mm] \rho_{jk} = \rho_{kj} = \dfrac{1}{N-1} x_j^T \cdot x_k \end{array} \quad j,k = 1,...,p \quad (3.26)$$

From the covariance matrix C of Equation 3.26, the correlation matrix can be obtained by normalizing C by the joint variances

$$R : r_{i,j} = \frac{C(j,k)}{\sqrt{C(j,j) \times C(k,k)}} \qquad j,k = 1,...,p \qquad (3.27)$$

The diagonal elements of R are all one, and the off-diagonal terms represent the correlation between feature couples. Normally, PCA is performed on zero-mean data, indicated with x in Equation 3.25, but often *standardized* data x_s are also used, meaning that in addition to having zero mean they should also have unit variance, that is,

$$x_s = \frac{X - \bar{x}}{\sigma} \qquad (3.28)$$

The implications of performing PCA on either zero-mean or standardized variables will be discussed later. However, the covariance matrix computed with the standardized variables is equal to the correlation matrix. The covariance and correlation matrices have an appealing graphical representation, at least in two dimensions, and their tracing follows the same procedure already described in Chapter 2. For this, consider a two-feature data set, that is, $p = 2$ and let (w_1, w_2) and (λ_1, λ_2) be the eigenvectors and the eigenvalues of C. Then, similar to the confidence ellipse considered in Chapter 2, the related ellipse centred at the origin can be traced according to

$$\begin{bmatrix} x_1 \\ x_2 \end{bmatrix} = \begin{bmatrix} w_{11} & w_{12} \\ w_{21} & w_{22} \end{bmatrix} \cdot \begin{bmatrix} 1.96\sqrt{\lambda_1} & 0 \\ 0 & 1.96\sqrt{\lambda_2} \end{bmatrix} \cdot \begin{bmatrix} \cos\varphi \\ -\sin\varphi \end{bmatrix} \quad \varphi \in (0, 2\pi) \tag{3.29}$$

The more elongated the ellipse, the higher the correlation, whereas two uncorrelated variables would produce a circle. The multiplying factor 1.96 means that if the data come from a Normal distribution, the ellipse will include on average 95% of its samples. As an example, Figure 3.42a shows 100 samples of a two-dimensional time series. The correlation between the two variables (x_1, x_2) is apparent from the covariance and correlation symmetrical matrices

$$C = \begin{bmatrix} 0.2359 & -0.3333 \\ -0.3333 & 0.6773 \end{bmatrix} \quad R = \begin{bmatrix} 1.0000 & -0.8340 \\ -0.8340 & 1.0000 \end{bmatrix} \tag{3.30}$$

In Figure 3.42b, the eigenvectors (w_1, w_2) of C determine the direction of the covariance ellipse, whose length is given by the eigenvalues (λ_1, λ_2).

Contrary to the orientation of the covariance ellipse, which is determined by its eigenvectors, the correlation matrix has an invariant 45° tilt, and its shape indicates the extent of the correlation between the two variables (ρ). In fact, the eigenvectors and eigenvalues of a 2×2 correlation matrix are

$$R_2 = \begin{bmatrix} 1 & \rho \\ \rho & 1 \end{bmatrix} \rightarrow w_1 = \begin{bmatrix} -\dfrac{\sqrt{2}}{2} \\ \dfrac{\sqrt{2}}{2} \end{bmatrix}; \quad w_2 = \begin{bmatrix} \dfrac{\sqrt{2}}{2} \\ \dfrac{\sqrt{2}}{2} \end{bmatrix}; \quad \begin{matrix} \lambda_1 = 1 - \rho \\ \lambda_2 = 1 + \rho \end{matrix} \tag{3.31}$$

so that by inspection, the correlation ellipse gives an immediate visual representation of the correlation between the two variables, as shown in Figure 3.43. A thick ellipse indicates a low correlation (a), while a lean ellipse denotes a high correlation (c).

3.5.1.1　The PCA Transform

If data redundancy is related to correlation, the linear transformation that we are seeking should change the reference basis of the data to eliminate the off-diagonal terms of C. As Figure 3.44 shows,

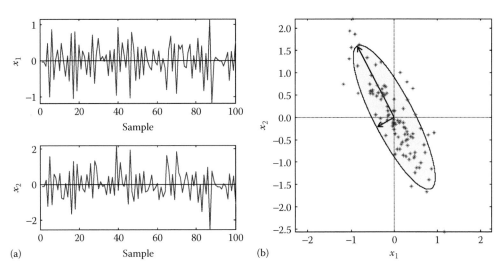

(a)　　　　　(b)

FIGURE 3.42　Two correlated time series. From the 100 samples shown in (a), it is very hard to detect any correlation, which instead is clearly shown by the correlation ellipse (b), whose shape is determined by the eigenvectors (black arrows) of C_x according to Equation 3.29.

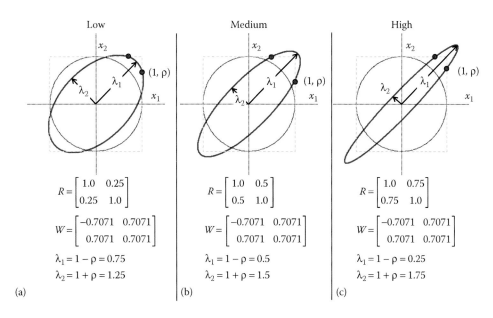

FIGURE 3.43   (a–c) Correlation ellipses show the degree of interdependence between variables. The ellipses are always oriented at 45°, and their eigenvalues are directly related to the correlation coefficient ρ_{12}. The dots indicate where the ellipse crosses the unit square.

(x_1, x_2) = original reference

(w_1, w_2) = variance maximizing reference (eigenvectors)

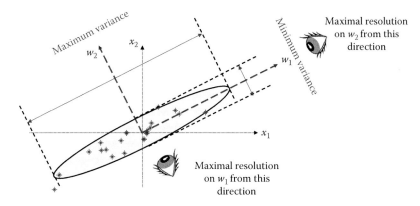

FIGURE 3.44   Graphical representation of the change of basis of Equation 3.32, providing maximum data resolution.

the axes of the new system of coordinates z are aligned with the principal axes of the covariance ellipse, given by the eigenvectors, so that the spread (variance) of each variable is maximal and perpendicular to all others.

The required similarity transformation is therefore defined by the eigenvectors and eigenvalues of C.

$$\begin{aligned} C &= W \cdot L \cdot W^T \\ L &= W^T \cdot C \cdot W \end{aligned} \rightarrow \begin{aligned} Z &= X \cdot W \\ X &= Z \cdot W^T \end{aligned} \quad \text{where} \quad \begin{aligned} W &= \begin{bmatrix} w_1, w_2, ..., w_p \end{bmatrix} \\ L &= \mathrm{diag}\left(\lambda_1, \lambda_2, ..., \lambda_p \right) \end{aligned} \tag{3.32}$$

In Equation 3.22, W^T is used in place of W^{-1} because the eigenvector matrix W is orthonormal, hence $W^T = W^{-1}$. In the new basis, the transformed data Z have a diagonal covariance matrix, thus solving the first problem of removing the redundancy among data features. However, PCA goes a step further and *ranks* the eigenvector matrix W according to the decreasing magnitude of the corresponding eigenvalue. Thus, the rearranged similarity matrix P becomes

$$P = \begin{bmatrix} p_{11} & p_{12} & \cdots & p_{1p} \\ p_{21} & p_{22} & \cdots & p_{2p} \\ \cdots & \cdots & \cdots & \cdots \\ p_{p1} & p_{p2} & \cdots & p_{pp} \end{bmatrix} \tag{3.33}$$

$$\uparrow \qquad \uparrow \qquad\qquad \uparrow$$

$$\lambda_1 > \lambda_2 \quad > \cdots > \quad \lambda_p$$

where the columns of P are still the eigenvectors of C, but now they are sorted according to the decreasing magnitude of the corresponding eigenvalue. So, the PCA transform is obtained from Equation 3.32 by substituting the sorted matrix P in place of W, that is,

$$Z = X \cdot P$$
$$X = Z \cdot P^T \tag{3.34}$$

Like W, P too is orthonormal, meaning that its columns are mutually orthogonal and have unit length, that is,

$$p_i^T \cdot p_j = 0 \quad i \neq j$$

$$\sum_{k=1}^{p} p_{i,k}^2 = 1 \quad i = 1,\ldots,p \tag{3.35}$$

$$P^{-1} = P^T$$

Figure 3.45 summarizes the procedure for obtaining the principal components (PCs) from the given data set. After subtracting the average from each feature, the covariance matrix C is computed. Its eigenvectors, once rearranged according to the decreasing magnitude of the corresponding eigenvalue, form the basis of the new reference system that provides maximum data resolution for the transformed data Z. In Figure 3.45, the typical PCA jargon is added, whereby the eigenvectors are called *loadings,* the eigenvalues are called *latent variables* and the transformed variables Z are referred to as *scores.*

Thus, we can summarize the main aspects of PCA: it is a linear data transformation that eliminates the data redundancy (represented by covariance) and at the same time maximizes the information (represented by variance). In ranking the eigenvectors, it also indicates to what extent each original variable contributes to the new ones. The columns of the P matrix (or *loadings*) are obtained through an orthonormal transformation with these properties:

- Each PC is a linear combination of the original variables.
- The PCs are mutually orthogonal, that is, they are uncorrelated (the redundant information is eliminated).
- The covariance matrix of the transformed data is diagonal and its eigenvalues are equal to the variance of each feature, that is,

$$\text{cov}(Z) = \frac{1}{N-1} Z^T \cdot Z = \frac{1}{N-1} \text{diag}(\lambda_1, \lambda_2, \ldots, \lambda_p) = \frac{1}{N-1} \text{diag}(\sigma_1^2, \sigma_2^2, \ldots, \sigma_p^2). \tag{3.36}$$

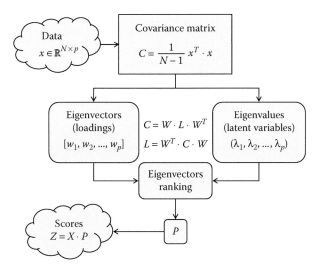

FIGURE 3.45   The PC transformation consists of computing the similarity matrix **P**, which makes the covariance matrix diagonal and reordering its columns (eigenvectors) according to the relative magnitude of the corresponding eigenvalues. The typical PCA nomenclature is added for completeness.

3.5.1.2   The Biplot

Once the original data set has been transformed according to Equation 3.34, we can look into the new data in search of fresh information. Each column of the **P** matrix contains the coefficients of one PC, sorted by decreasing magnitude of the corresponding eigenvalue. They represent the coefficients of the linear combinations of the original variables that generate the PCs. For example, the elements of first column show how much each original variable contributes to the first PC, that is,

$$p_1 = \begin{bmatrix} p_{11} \\ p_{21} \\ ... \\ p_{p1} \end{bmatrix} \begin{matrix} \leftarrow \text{ contribution of } x_1 \text{ to PC}_1 \\ \leftarrow \text{ contribution of } x_2 \text{ to PC}_1 \\ ... \\ \leftarrow \text{ contribution of } x_p \text{ to PC}_1 \end{matrix} \tag{3.37}$$

and the same applies to the other columns $\left(p_2, p_3,..., p_p \right)$. The *biplot* is the graphic equivalent of Equation 3.37 as it visualizes each variable's contribution to the PCs, and unfortunately it is limited to 2D and 3D visualizations. The axes of the biplot are the first two or three PCs, the columns of **P** are represented as unit vectors and the scores (observations transformed in the PC space) are represented as dots. Figure 3.46 shows a two-dimensional biplot where the axes represent the PCs (columns of **P**). It visualizes the magnitude and sign of each variable's contribution to the first two or three PCs, and how each observation is represented in terms of those components. The dispersion of the scores is equal to their standard deviation, which is given by the square root of the corresponding eigenvalue.

3.5.1.3   The Scree Plot

Another graphical aid providing insight into the PCA is the *scree* plot (PCA jargon again), which is generally composed of two plots: one showing the amount of data variability explained by each PC, and the other showing the magnitude of the sorted eigenvalues. Given the ranking of the columns of **P** and of the corresponding eigenvalues, the contribution of each PC and the cumulative sum can be computed as

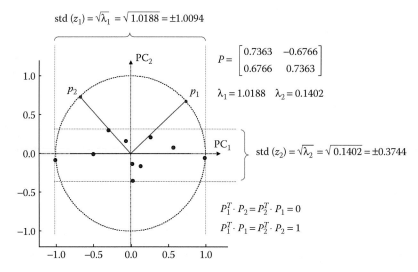

FIGURE 3.46 A 2D biplot showing the relative position of the two eigenvalues (latent vectors) and the relative spread of the scores (dots) inside the bands delimited by the standard deviations, which are determined by the eigenvectors (latent values) of the PCA matrix *P*.

$$q_i = \frac{\lambda_i}{\lambda_1 + \lambda_2 + \cdots + \lambda_p} \quad i = 1,\ldots,p$$

$$v_i = \sum_{k=1}^{i} q_k \quad k = 1,\ldots,p$$

(3.38)

An example is shown in Figure 3.47. The individual (bar) and cumulative (line and dot) explained variance (a) specifies how much each PC contributes to the total data information, whereas the ranked magnitude of the eigenvalues (b) indicates which components of the data are expanded $(\lambda > 1)$ or compressed $(\lambda < 1)$.

3.5.1.4 PCA on Correlation or Covariance?

PCs may be computed using either the covariance or the correlation matrix. The covariance PCA uses the zero-mean data *x* computed with Equation 3.25, whereas the correlation PCA operates on *standardized* data defined by Equation 3.28. If the variables have comparable magnitudes, either approach can be used, but if they differ by some order of magnitude the method should be selected with care. In fact, it should be reminded that in the correlation PCA all the variables are equally important because they all are standardized to have unit variance, whereas in covariance PCA, each variable retains its original variance. If the variables differ in magnitude, then correlation PCA may be advisable to make all the variables comparable; otherwise the largest one will bias the results. However, there is no definitive answer as to which data to use (zero-mean or standardized) and the decision should be taken on a case-by-case basis. Let us consider the pros and cons of each choice:

1. Covariance PCA:
   a. Maintains the natural proportions among variables
   b. It is sensitive to units, emphasizing the importance of the largest variables
2. Correlation PCA:
   a. Operating on standardized data it is insensitive to units
   b. The results bear no relation with covariance

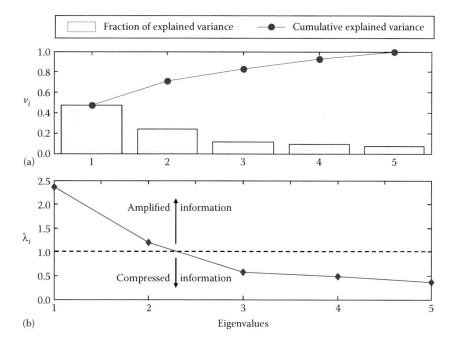

FIGURE 3.47 Scree plot showing the fraction of explained variance (a) for each ranked eigenvalue (b). Any eigenvalue greater than 1 amplifies the information along its PC.

As an example, consider Figure 3.48, which compares the covariance ellipses and the biplots of a 100 samples record of two correlated variables (x_1, x_2) with $x_2 \sim 20 \cdot x_1$. Performing the PCA on the zero-mean data (a), rather than on standardized data (b), produces totally different results. In fact, the largely differing orders of magnitudes in case (a) are such that x_2 totally dominates the analysis. The covariance and the PCA matrices in the two cases are

$$
\begin{aligned}
&\text{(A)} \quad C_A = \begin{bmatrix} 0.2572 & -7.2595 \\ -7.2595 & 293.2540 \end{bmatrix} \quad P_A = \begin{bmatrix} -0.0248 & 0.9997 \\ 0.9997 & 0.0248 \end{bmatrix} \quad \begin{aligned} z_1^{(A)} &= -0.0248 \cdot x_1 + 0.9997 \cdot x_2 \\ z_2^{(A)} &= 0.9997 \cdot x_1 + 0.0248 \cdot x_2 \end{aligned} \\
&\text{(B)} \quad C_B = \begin{bmatrix} 1.0000 & -0.8359 \\ -0.8359 & 1.0000 \end{bmatrix} \quad P_B = \begin{bmatrix} 0.7071 & 0.7071 \\ 0.7071 & 0.7071 \end{bmatrix} \quad \begin{aligned} z_1^{(B)} &= 0.7071 \cdot x_1 + 0.7071 \cdot x_2 \\ z_2^{(B)} &= 0.7071 \cdot x_1 + 0.7071 \cdot x_2 \end{aligned}
\end{aligned} \tag{3.39}
$$

The dominance of x_2 clearly shows in (a,) whereas in (b) the two components have comparable magnitude. A similar difference emerges from the comparison of the biplots, where (c) shows that x_2 totally dwarfs the PC_2, whereas (d) shows a more balance contribution of both variables to the two PCs. The point made by this example is that if one feature is considerably larger than the others, it will dominate the PCA by being itself a good approximation of PC_1 (see Figure 3.48c), and the other features may be almost lost in comparison. On the other hand, standardization tends to level out the differences, so it should be applied with care. Cases where a single feature dominates over the others should be avoided, though no general guideline can be set and the decision on whether the data should be standardized must be taken on a case-by-case basis.

3.5.2 DATA REDUCTION AND RECONSTRUCTION

Now, we turn to the second goal of PCA, considering the possibility of reducing the complexity of the data set without losing too much information. Needless to say, we have to find a compromise between the amount of information we are prepared to discard and the advantage of dealing with less data.

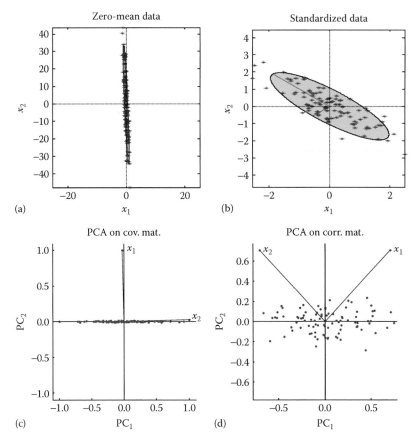

FIGURE 3.48 Comparison of PCA using zero-mean or standardized data with significantly different orders of magnitude. The covariance ellipses are shown (a) for the zero-mean data, which shows an overwhelming dominance of the largest variables (x_2), as opposed to (b) where the standardization removes this dominance. The same happens with the biplots, where in (c) the PC_1 is totally dominated by x_2, whereas the biplot in (d) yields a more balance portrait.

By inspection of the scree plot of Figure 3.47, a reduction in the number of features is possible if we limit the PC transform to the components for which the corresponding eigenvalue is larger than a prescribed threshold. In principle, $\lambda = 1$ should be the natural yardstick between information amplification $(\lambda > 1)$ and compression $(\lambda < 1)$, but it is advisable to make a more conservative choice and retain the columns of P for which $\lambda > 0.7$ (Dunteman, 1989; Jolliffe, 2002). The selection of the maintained components can be cross-checked with the cumulative explained variance given by Equation 3.38. Normally, a valid reduction should conserve at least 80% of the total variability. Data reduction is performed by partitioning the P matrix between the retained and discarded parts

$$P = \left[P_a \mid P_{p-a} \right],$$
$$Z_a = X \cdot P_a$$

(3.40)

and using only the left sub-matrix P_a, which contains the retained eigenvalues of the PC transformation to obtain the reduced transformed data Z_a. The selection of a suitable $a < p$ is assisted by the scree plot of Figure 3.47, which indicates the loss of information related to the reduction. The drawback of this operation is that when we retrieve the original data by inverse PCA transform

$$X_a = Z_a W_a^T$$
$$X = Z_a W_a^T + E \tag{3.41}$$

the data retrieved with the reduced PCA (X_a) will differ from the original data by the residual matrix E, which represents the contribution of the discarded components. As an example, consider a data set composed of three variables: per cent saturation (%) of DO, pH and temperature (°C). These data were acquired by an automatic sampling station placed along the Arno River (Italy) upstream of the city of Florence in September 2006. Figure 3.49 shows the reconstruction accuracy by using all the three PCs ($a = 3$), the first two PCs ($a = 2$) or just the first PC ($a = 1$). Obviously, the fewer the PCs, the lower is the accuracy.

The scree plot of Figure 3.50 shows that one PC would yield a sufficiently accurate reconstruction, because only λ_1 is above the 0.7 threshold and the explained variance would be in excess of 90%.

3.5.3 CONSISTENCY STATISTICS

PCA is often used to check the consistency of a data set, and for this reason, it is at the basis of many fault detection algorithms. In this context, it has been widely applied to process monitoring and to wastewater treatment processes (Schraa et al., 2006; Alferes et al., 2013; Nagy-Kiss and Schutz, 2013).

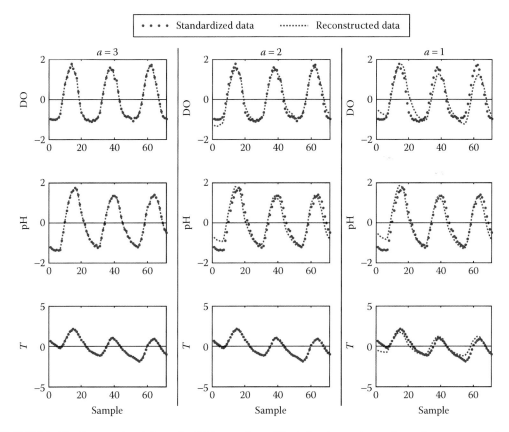

FIGURE 3.49 Data retrieval from the PCA transform. The left column shows the perfect reconstruction obtained using the full PCA space ($a = 3$), whereas the other two columns show the loss of reconstruction accuracy when a subspace of decreasing dimension ($a = 2$ or $a = 1$) is used. These graphs were obtained with the MATLAB script Ex_PCA_Reconstr_Reduced.m in the ESA_Matlab\Exercises\Chapter_3\PCA folder.

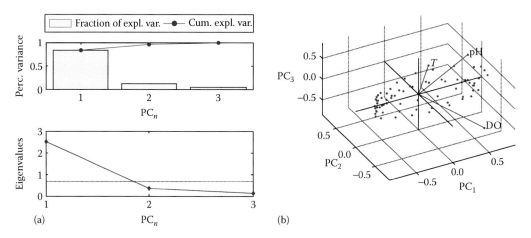

FIGURE 3.50 Scree plot (a) and biplot (b) of the Arno River data used in Figure 3.49.

3.5.3.1 Hotelling's T^2 and Q Statistics

The typical consistency measure used with PCA is Hotelling's T^2 (Hotelling, 1931, 1947), which may be viewed as a generalization of the Student-t statistics used in hypothesis testing. When PCA is applied to a given set of data, assumed as the reference data set, we may be interested in knowing whether some later sample originating from the same process can still be explained by that model, or if a significant divergence can be detected. Given $X \in \mathbb{R}^{N \times p}$, suppose that a PCA model $Z = X \cdot P$ was computed. Then Hotelling's T^2 is defined as

$$T^2 = Z^T \cdot L^{-1} \cdot Z \tag{3.42}$$

with $L = \text{diag}\left(\lambda_1, \lambda_2, \ldots, \lambda_p\right)$ given by Equation 3.32. We can use T^2 as a hypothesis testing variable, with the null hypothesis H_0 being that the data can be explained by the given PCA. For this, the T^2 value can be related to an F-statistics (Hotelling, 1947; Dunteman, 1989; Mason et al., 2001; Jolliffe, 2002) to set an acceptability threshold

$$T^2_{\lim} = \frac{p(N-1)}{N-p} F^{\alpha}_{N-p,p} \tag{3.43}$$

where $F^{\alpha}_{N-p,p}$ is the Fisher statistics with $N - p$ and p degrees of freedom and $100(1 - \alpha)$ confidence level. T^2 is unaffected by standardization, so it yields the same results when applied to the zero-mean data x or to the standardized data x_s.

If a dimension reduction has been made, the T^2 will record anomalies in the subspace of the retained components $\left(\text{PC}_1, \text{PC}_2, \ldots, \text{PC}_a \text{ with } a < p\right)$ but not on the excluded ones $\left(\text{PC}_{a+1}, \ldots, \text{PC}_p\right)$. Any anomaly in this subspace will be detected by the Q-statistics, defined as the sum of squared residuals of the active PCs (Russell et al., 2000; Villegas et al., 2010)

$$Q = x^T \left(I - L \cdot L^T\right) x \tag{3.44}$$

In a similar manner to the T^2, the Q-statistics too should be compared to a threshold value given by

$$Q_{\lim} = \theta_1 \left[1 + \frac{h_0 c_{\alpha} \sqrt{2\theta_2}}{\theta_1} + \frac{\theta_2 h_0 (h_0 - 1)}{\theta_1^2}\right]^{(1/h_0)} \tag{3.45}$$

where c_{α} is the $(1 - \alpha)$ quantile of the normal distribution and the various quantities in Equation 3.45 are defined as

$$\theta_1 = \sum_{i=a+1}^{p} \lambda_i \quad \theta_2 = \sum_{i=a+1}^{p} \lambda_i^2 \quad \theta_3 = \sum_{i=a+1}^{p} \lambda_i^3 \quad h_0 = 1 - 2\frac{\theta_1\theta_2}{3\theta_3^2} \tag{3.46}$$

The geometric interpretation of T^2 and Q is shown in Figure 3.51.

3.5.3.2 Contribution Variables

Hotelling's T^2 and the Q statistics are global indicators that certain data diverge from the majority of the data set, but they do not single out which variable of the original data X is responsible for the anomaly. To get a further insight into the fault, the *contribution variables* (CVs) are a tool that may shed more light into the analysis. They are defined as

$$\mathrm{CV} = z \cdot \mathbf{L}^{-1/2} \cdot \mathbf{P}^T \tag{3.47}$$

which is similar to the inverse PCA of Equation 3.34, so the CVs are consistent with the original reference system X. This allows the user to immediately grasp its meaning without having to go through the inspection of the single variables.

Figure 3.52 shows how the T^2 and the CVs react to an abrupt change in the data. The same data from the Arno River shown in Figure 3.49 are used here, but in this record a malfunction of the pH probe was simulated at about 30 h, represented by a short signal drop (Figure 3.52a). The dip in the PCA scores (Figure 3.52b) is hardly discernible, but the jump in T^2 is significant (Figure 3.52c). This is, however, a global indication of a substantial divergence from the normal behaviour, while the CVs (Figure 3.52d) clearly single out pH as the faulty variable.

3.5.4 PCA AND MATLAB

MATLAB Statistics toolbox provides full PCA capability. The main commands and their relationships are summarized in Figure 3.53. The PCA can be obtained by directly processing the data matrix X consisting of N observations (rows) and p variables (columns) with the command

$$\left[\mathrm{P,score,L,T2}\right] = \mathrm{princomp}\left(\mathrm{X}\right) \tag{3.48}$$

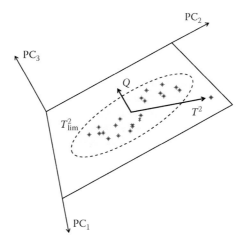

FIGURE 3.51 A geometric representation of the T^2 and Q statistics. It is assumed that the three-dimensional original space has been reduced to a two-dimensional subspace. Hotelling's T^2 shows when an observation diverges from the 'normal' set (asterisks), delimited by T_{lim}, in the retained PC$_1$–PC$_2$ subspace, whereas the Q statistics shows when a residual, in this case along the PC$_3$ component, exceeds the threshold limit Q_{lim}.

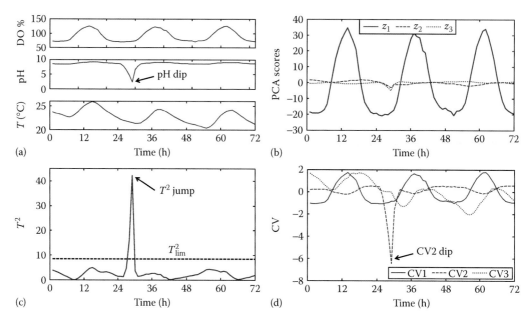

FIGURE 3.52 Detecting an anomaly in the data through T^2 and CV. Around 30 h pH has a dip (a), which is recoded as a T^2 jump (c). At the same time, the CVs single out CV2 (pH) as the responsible variable (d). This backtracking would not have been possible with the scores, two of which exhibit a small dip (b). These graphs were obtained with the MATLAB script `Ex _ PCA _ Rosano _ T2 _ CV.m` in the `ESA _ Matlab\ Exercises\Chapter _ 3\PCA` folder.

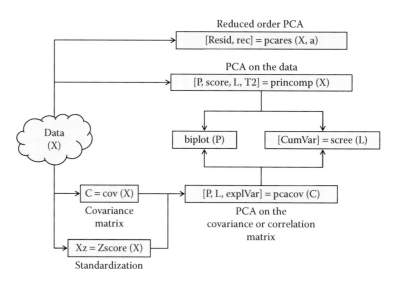

FIGURE 3.53 Main MATLAB commands to perform the PCA.

The column mean values are automatically subtracted from X, making the data zero-mean before processing, but their relative magnitudes are unaffected. If it is desired to compute the PCA on standardized data, they should be pre-processed using the `zscores` command and then either `princomp` or `pcacov` commands can be used, with the latter operating directly on the covariance or correlation matrix. Going back to command (3.48), its output consists of the PCA matrix P, referred to as `coeff` in MATLAB, while the transformed data are referred to as `score`, while L is the vector of the eigenvalues already sorted in descending order, and T2 is Hotelling's indicator of the distance of each observation from the 'core' of the observations, as discussed in Section 3.5.3.1. If PCA on the covariance or correlation matrix is pursued, then the following command should be used:

$$[\mathrm{P,L,explVar}] = \mathrm{pcacov}(\mathrm{C}).\qquad(3.49)$$

In this case, the scores are not included among the output variables, because pcacov does not operate on the data. The PC matrix (P) and the eigenvalues (L) are the same as in command (3.48), but in this case the third output explVar is a vector containing the percentage of the variance explained by each PC, starting with the largest (PC_1). In practice, this is equivalent to the cumulative explained variance of the scree plot of Figure 3.47a. The sample code listed in Box 3.3 shows the basic commands to perform the PCA on a data set.

After loading the data, the transform matrix (P), the transformed scores (Z) and the eigenvalues (L) can be obtained with the single command princomp, which also provides Hotelling's T^2. The results of the PCA can be visualized with the scree plot, in view of a possible dimension reduction, or with the biplot, which indicates how each original variable is reflected in the PCs. The last lines show how the scores (Z) can be reconverted back into the original data (X) by computing the inverse transformation of Equation 3.34. A comparison is made in Figure 3.54 between the full data recovery ($R_2 X$) and the first-order approximation ($R_1 X$) of a random time series. While a perfect data reconstruction is obtained with R_2, the retrieval provided by R_1 is, as expected, not nearly as good.

BOX 3.3   ESSENTIALS OF PCA COMPUTATION

```
load Data
%% Perform PCA transform
[P,Z,L,T2]=princomp(X);
% Draw the scree plot
scree(L,2)
% Draw the biplot
figure(3)
vl={' x_1',' x_2'}; % Set the labels
biplot(P,'Scores',Z,'VarLabels',vl)
% --------------------------------
%% Draw Hotelling's T^2
plot(T2)
hold on
% Draw the Tlim line for comparison
T_lim=(p*(N-1)/(N-p))*finv(0.95,p,N-p);
plot([0 length(T2)],[T_lim T_lim],'--k')
% --------------------------------
%% Retrieve the data by
%  inverse PCA transform
figure(5)
% Inverse PCA transform
X1=Z*P'; % Full order data retrieval
% compare with inverse reduced
% reconstruction of order 1
% R1 = first order data retrieval
[resid,R1]=pcares(X,1);
% --------------------------------
```

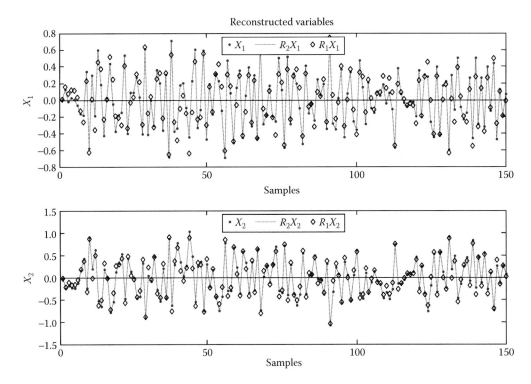

FIGURE 3.54   Comparison of full and reduced order reconstruction of the observations, obtained with the last few lines of the sample code in Box 3.3. The complete MATLAB script is `Ex _ PCA _ Reconstruction _ Random _ Data.m` in the `ESA _ Matlab\Exercises\Chapter _ 3\PCA` folder.

3.5.5   PCA Case Studies

Two applications are now discussed in which the use of PCA can assist in extracting information from the data, which could not be otherwise apparent. The broad applicability of PCA is also demonstrated by applying this method to two widely differing time series: the turbidity trend in a potabilization process, and the detection of faults in a wastewater treatment process.

3.5.5.1   Turbidity Analysis in a Potabilization Process

The recording of Figure 3.55 shows the main parameters of the raw water drawn from the Arno River to be processed by the potabilization plant of the municipality of Florence (Italy). The data span the first half of 2013 and include the abstraction flow from the river, the raw water temperature and its turbidity. The flocculant dosage and the output turbidity after the clarification stage, representing the process output, are also recorded. The main goal of the clarification-flocculation process is to cope with the input turbidity peaks, indicated by the shaded areas, and limit the variability of the output turbidity as much as possible prior to the sand filtering, which represents the next process stage, and which are very sensitive to input turbidity variations.

The PCA may assist in establishing relations between the process inputs and its output, and indicate which inputs are most responsible for the output variations. Given the large magnitude differences of the variables, the data are standardized prior to PCA processing. Then, the scree plot in Figure 3.56a indicates that the feature dimensions could be reduced from five to three. Further, the biplot of Figure 3.56b shows to what extent the process variables contribute to the first two PCs and how they are correlated to one another. In particular, the input turbidity ($Turb_{in}$) and the flocculant dosage (Floc) are strictly related, but the river flow also plays an important role, and its influence is almost equally shared between PC_1 and PC_2, whereas the temperature is almost entirely

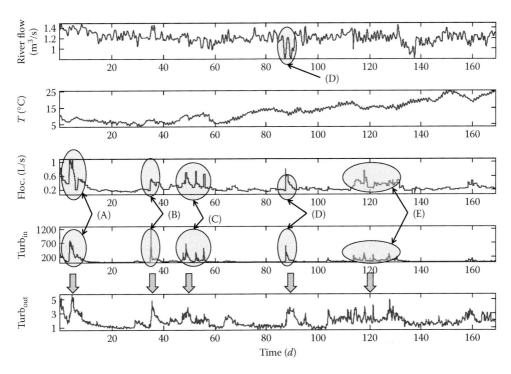

FIGURE 3.55   Turbidity case study: the four input variables that influence the output turbidity in a potabilization process are recorded for the period January–June 2013. The major turbidity events are highlighted by the shaded areas and they cause an increase in the output turbidity.

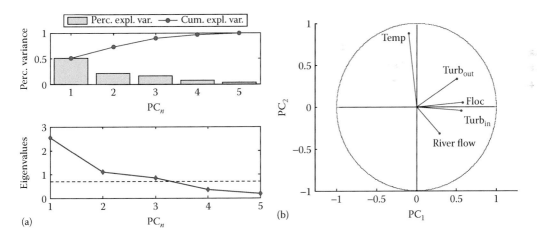

FIGURE 3.56   Scree plot (a) and 2D biplot (b) of the turbidity recording of Figure 3.55.

represented by PC_2. The positive correlation between river flow and $Turb_{in}$ indicates that turbidity obviously increases during high-flow events.

Figure 3.57 examines the consistency of the variables trend: Hotelling's T^2 indicates when the process behaviour globally diverges from its normal condition by exceeding the threshold defined by Equation 3.43. It can be seen that the T^2 peaks correspond to the turbidity events induced by a flood condition, but the CVs provide a further insight into each event. For example, the input turbidity is the main feature responsible for events (A)–(C) and (E), but in event (D) the flow is also involved.

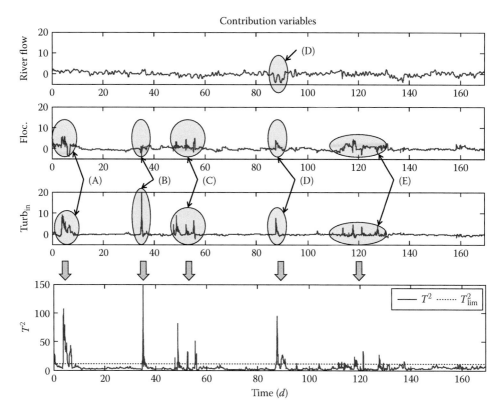

FIGURE 3.57 Consistency indicators (CV and T^2) for the turbidity recording of Figure 3.55. The shaded areas indicate which variable was mainly responsible for the anomalies globally indicated by T^2.

This information may be used for process control, for example, by trimming the flocculant dosage according to the kind of turbidity event.

3.5.5.2 Fault Detection and Isolation in a Wastewater Treatment Process

The use of PCA for fault detection and isolation (FDI) is well established, and important engineering results have been achieved thanks to this technique (Lennox and Rosen, 2002; Yoo et al., 2002, 2006; Rosen et al., 2003; Schraa et al., 2006; Corominas et al., 2009, 2011). The case study presented here is fully documented elsewhere (Baggiani and Marsili-Libelli, 2009), and only the fault detection motivation and results are summarized here. The monitoring algorithm was implemented on a full-scale activated sludge plant in Tuscany (Italy) with a capacity of 88,600 PE treating a mix of domestic sewage and septic tank discharges. The process was conceived for biological nutrient removal by fitting a pre-denitrifying tank followed by three parallel-fed oxidation tanks and three secondary settlers. Nutrient removal requires sophisticated ion-specific probes, which need constant maintenance and monitoring. For this reason, PCA-based diagnostic software was set up to process the probe signals in real time and discriminate between sensor failure and process anomalies. The process data consisted of ammonium-N (NH_4^+) and oxidized-N $\left(NO_{X_{OX}}\right)$ measurements at the outlet of the oxidation tanks and one additional oxidized nitrogen measurement at the output of the anoxic denitrification stage $\left(NO_{X_{DEN}}\right)$. In this project, PCA was used to detect the departure of operational data from the correct performance, previously defined on the basis of an initial record of 'normal' data, assumed as a yardstick, and to be updated during real-time operation. The faults were detected by checking the global statistics T^2 and Q against their thresholds, together with the CVs, to indicate which sensor was responsible for each fault. The innovative aspect of the project is in the real-time updating of the reference data set, which is constantly refreshed with new data, provided they are proved to represent a 'correct performance' indication. The discrimination procedure whereby new data are accepted or rejected into the reference data set is fully described in Baggiani and Marsili-Libelli (2009) and is based on a T^2 check, setting a stricter acceptance limit well below T_{lim}.

Some examples are now presented, drawn from the plant operational record during the experimentation phase (September 2007–June 2008) implemented in collaboration with the Information Technology department of the responsible integrated water cycle manager (Acque SpA), to illustrate the performance of the FDI procedure. Figure 3.58 shows how the FDI reacts to a malfunction of the ammonium-N probe (NH_4^+), which at day 7.2 in the recording fails to operate properly. Both T^2 and Q jump out of the plot to slowly return later, though neither of these indicators return below their thresholds because the faulty sensor was not promptly repaired. The problem with a delayed repair of the fault is that bad data continue to be produced, and obviously they are not accepted into the reference data set of the FDI. In this way, there is a risk of the database becoming impoverished, or at least becoming outdated for lack of fresh data. The contribution plot, shown in the lower part of the figure, indicates that the NH_4 sensor was responsible for the malfunction. It could be argued that a simple checking of the NH_4 sensor output could do the job equally well, but the FDI algorithm is conceived to detect more complex faults and, thanks to the contribution plots, can assist the operator with a centralized information about the state of the plant, by immediately indicating

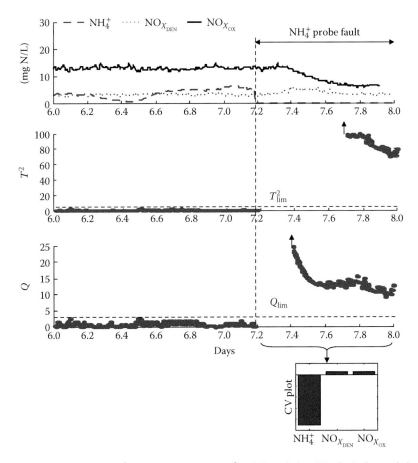

FIGURE 3.58 Detection of an NH_4^+ sensor fault through T^2 and Q statistics. The fault diagnosis is confirmed by the contribution plots, which indicate that the NH_4^+ sensor caused the fault. The CV bars refer to the time when the peak T^2 occurred. (Reproduced with permission from Baggiani, F. and Marsili-Libelli, S., *Water Sci. Technol.*, 60, 2949, 2009.)

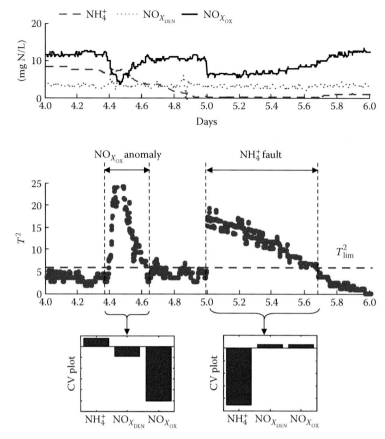

FIGURE 3.59 Discrimination between differing faults through nitrogen measurements and T^2 statistics and CVs, which play a fundamental role in discriminating the nature of the two faults. (Reproduced with permission from Baggiani, F. and Marsili-Libelli, S., *Water Sci. Technol.*, 60, 2949, 2009.)

the faulty element. This aspect becomes important when many sensors are connected to the FDI and concurrent failures must be resolved.

Figure 3.59 shows the discriminating capacity of the CVs (Section 3.5.3.2) in addition to the global indicator T^2 and Q. In this instance, the first warning was caused by a sudden decrease of the oxidized nitrogen measurement in the oxidation tank $\left(NO_{X_{OX}}\right)$, but this decrease was not large enough to be diagnosed as a sensor fault. Rather, it could be ascribed to an excess ammonia loading. The CV plot clearly indicates which sensor was mainly responsible for the alarm. The second warning, instead, was clearly recognized by the CVs as caused by the ammonia sensor (NH_4^+), whose output remained near zero for nearly 14 h, after which the sensor was repaired and resumed normal operation. In response to this, the T^2 returns below the T_{lim}^2 threshold as soon as the fault was fixed, contrary to the previous example of Figure 3.58.

4 Fuzzy Modelling of Environmental Systems

Neurosis is the inability to tolerate ambiguity.

Sigmund Freud

4.1 INTRODUCTION TO THE FUZZY LOGIC

In the previous chapters, we have dealt with 'normally' defined concepts and measurements, in the sense that each concept was unequivocally defined by its attributes. We have used, therefore, a single-valued logic associated with a clear-cut partition of the reality, where a concept is either true or false. This clear-cut logic is a legacy of ancient positivistic societies, such as the Romans, for whom *tertium non datur* (there is not a third option). In contrast, our flexible, or hypocritical, society tolerates, and even invites, ambiguity. Sometimes, ambiguity is even necessary for our survival, or at least for our peace of mind.

In this chapter, we move away from the 'conventional' logic and consider a *multi-valued* logic in which the boundary between true and false becomes blurred—hence the term 'fuzzy'—and a concept can be anything between the extremes of true or false, with its *degree of truth* varying smoothly from zero (totally false) to one (totally true). Fuzzy logic was conceived to handle such uncertain situations. It was first introduced by the Iranian-born mathematician Lofti Zadeh (Zadeh, 1965; Yager et al., 1987) and has been a hot issue in science and engineering ever since, spawning a vast literature in mathematics (Klir and Folger, 1988; Pedrycz, 1995, 1996; Patyra and Mlynek, 1996; Jamshidi et al., 1997; Pelletier, 2000; Hajek, 2010). Moreover, it has been successfully applied to such diverse fields as industrial engineering (Sugeno, 1985; Hirota, 1993; Lu, 1996) and environmental sciences (Ganoulis, 1994, 2009; Bàrdossy and Duckstein, 1995; Marsili-Libelli and Cianchi, 1996). It also paved the way for advances in control engineering (Pedrycz, 1993; Kandel and Langholz, 1994; Yager and Filev, 1994; Nguyen et al., 1995; Ross, 1995; Passino and Yurkovic, 1998; Abonyi, 2003), artificial intelligence, pattern recognition and machine learning (Yager and Zadeh, 1994; Jang et al., 1996; Roger and Sun, 1997; Kecman, 2001; Theodoridis et al., 2010; Marsland, 2015; Padhi and Simon, 2015).

According to fuzzy logic, a concept is hardly ever *completely true* or *completely false*, but it is rather somewhere in between these two extremes. Further, no concept exists *per se* but is defined in comparison to a reference concept (the *prototype*). A simple example of this difference is shown in Figure 4.1, in which instead of a *singleton* number defining the number x^*, as we instinctively use in everyday logic, the concept of *degree of similarity* (or *degree of truth* or *degree of membership*) is introduced, defined by the membership function (mf) $A : \mu_a(x)$ centred around the reference (*prototype*) value $\bar{x}$, so that the mf $\mu_a(x^*)$ replaces the value x^* itself. Thus, what is important in fuzzy logic is not the independent value of a concept, but its similarity to a predefined prototype.

Another example is given in Figure 4.2, considering the statements 'Real numbers smaller than 20' and 'Temperature between 15 and 25°C' in the crisp and fuzzy contexts. The crisp logic makes a hard distinction by classifying each entry as either true (1, black) or false (0, white), whereas the fuzzy logic ranks the truth of each statement, depending on the relative similarity with the reference concept.

What might appear as a mere complication when dealing with simple numbers, does, in fact, open vast possibilities for including non-numerical information in the logical process. In broad terms, fuzzy logic could be defined as an *approximate reasoning method dealing with vaguely defined concepts*. Figure 4.3 shows how a concept (temperature) can be defined using linguistic

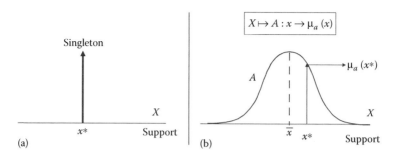

FIGURE 4.1 The basic difference between crisp and fuzzy logic is representing the number x^*. In the crisp logic (a) x^* is represented by its numerical value, whereas in fuzzy logic (b) it is defined by its similarity to a given concept $\bar{x}$.

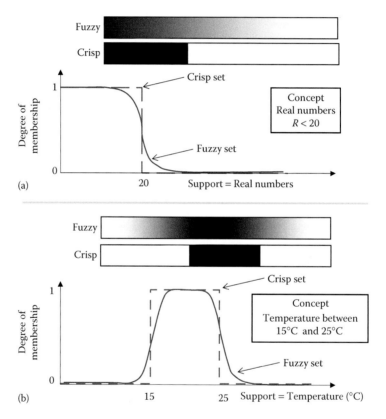

FIGURE 4.2 Differing representations of the same concept in either crisp or fuzzy logic. While crisp logic uses only the 'equal' and 'not equal' concepts (represented by the dashed lines), fuzzy reasoning is based on the concept of similarity with respect to the concept represented by the solid curves. In this way, the crisp true/false duality is softened by the membership grade that can vary continuously between 0 (no similarity) to 1 (full similarity).

terms (low, medium and high) and translated into precise mathematical terms through a set of suitably selected reference mfs used as prototypes. The novel aspect with respect to crisp logic is that a concept may 'straddle' differing, if not conflicting, definitions (e.g., low temperature vs. medium temperature), a situation totally alien to crisp logic.

Another 'conventional wisdom' belief to be dispelled is that fuzzy logic is a newly baked concoction for probability. In the opening editorial of the IEEE Transaction on Fuzzy Systems, Bezdek (1993) proposed a nice example to clarify the difference between fuzziness and probability, which is illustrated in Figure 4.4. In Bezdek's example, the 'weary traveller' is confronted with two bottles

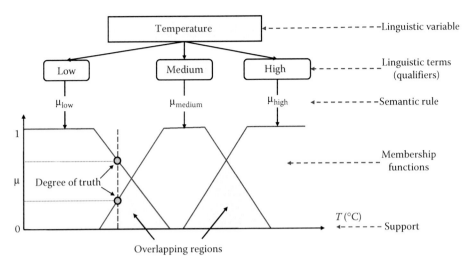

FIGURE 4.3 The full logical path of translating a linguistic concept (temperature) into a set of mfs. The novel aspect of fuzzy logic is the partial overlapping of the mfs (shaded areas), indicating that differing concepts are not necessarily mutually exclusive.

FIGURE 4.4 The 'weary traveller' coming to terms with the fundamental difference between fuzziness and probability. (From Bezdek, J. C., *IEEE Trans. Fuzzy Syst.*, 1, 1–5, 1993.)

containing an unknown liquid. Bottle A is labelled 'Drinkable with *fuzziness* = 0.9', whereas the label on bottle B reads 'Drinkable with *probability* = 0.9'. Which bottle should he drink from? Let us elaborate: in bottle A, there is *for sure* some liquid that resembles drinking water to a considerable degree (90%). It may not be mountain spring water but it is drinkable nonetheless. What is clear is that the statement 'Drinkable with *fuzziness* = 0.9' means that the liquid inside *has been tested* and found drinkable to a high degree. Then *as a result* of the test that label has been issued.

Conversely, the label on bottle B sounds a bit spooky. It says that the bottle contains some liquid that *over a large number of trials* of similar bottles proved to be drinkable in 90% of the cases. This is a statement based on statistical testing on many similar samples, but no assurance is given for *that* particular sample, for which there is one chance out of ten that it could be anything from dreadful to poisonous. Obviously, Bezdek concludes that it is advisable to drink from the bottle with the fuzzy label, and how could anyone disagree with him? This entertaining and instructive example underlines the different kind of information provided by fuzziness (proven similarity of concepts) as opposed to probability (statistical averaging). In other words, fuzziness identifies a *certain a posteriori quality*, whereas probability represents an *uncertain a priori expectation*. Needless to say, after opening bottle B, its probability either jumps to 1 or plummets to 0, but—alas—too late.

4.1.1 Fuzzification

The previous examples indicate that to enter the fuzzy world, a concept needs to be *fuzzified*, meaning that its value or its semantics must be converted from *absolute* to *relative* by comparison to a set of reference concepts, mathematically represented by the mfs. Figure 4.5 illustrates the conversion of the real number x^* from crisp to fuzzy, producing a set of mfs:

$$x^* \xrightarrow{\text{fuzzification}} \begin{bmatrix} 0 & \mu_2(x^*) & \mu_3(x^*) & 0 \end{bmatrix} \tag{4.1}$$

Equation 4.1 refers to the example of Figure 4.5, where only μ_2 and μ_3 yield a non-zero degree of membership, while μ_1 and μ_4 are not involved. Fuzzification therefore implies substituting a vector of degrees of membership to the original number or concept.

4.1.1.1 Possible Analytical Forms of mfs

There is great freedom in shaping the mfs and the MATLAB® Fuzzy Toolbox provides a wide choice of predefined functions, as shown in Figure 4.6. Each of them has a number of 'handles' (numerical parameters) that can be manipulated to change its shape, as exemplified in Box 4.1 for the gauss2mf mf, an asymmetrical Gaussian function with four shape parameters, which represent the variance and mean of each half Gaussian (left and right of the middle).

4.1.2 Basic Properties of Fuzzy Sets

The basic properties of fuzzy sets are summarized in Table 4.1. The elementary set operations of union and intersection are shown in the first row of Figure 4.7, whereas the second row shows a result quite unlike the crisp sets: the union of a fuzzy set A with its complement $\left[\overline{A} : 1 - \mu_a(x) \right]$ is not the universe, and the intersection is not empty. The last row shows the relationship between union and intersection fuzzy sets.

4.1.3 Logic Connectives: T-Norms and S-Norms

In Figure 4.7, the fuzzy union and intersection were implemented with the *min* and *max* operators, but this is not the only choice. A whole class of alternate operators can be defined, provided they

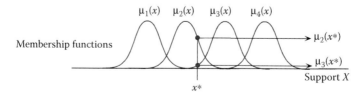

FIGURE 4.5 Fuzzification of the crisp concept x^* means computing its degree of membership with respect to the set of mfs $(\mu_1, \mu_2, \mu_3, \mu_4)$.

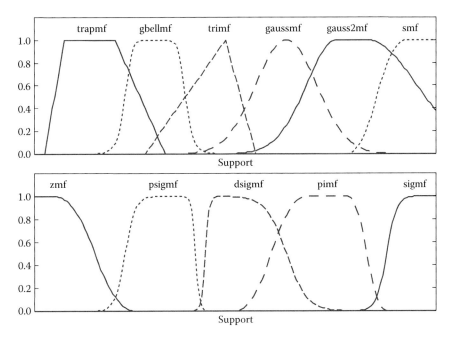

FIGURE 4.6 Mfs available in the MATLAB Fuzzy Toolbox.

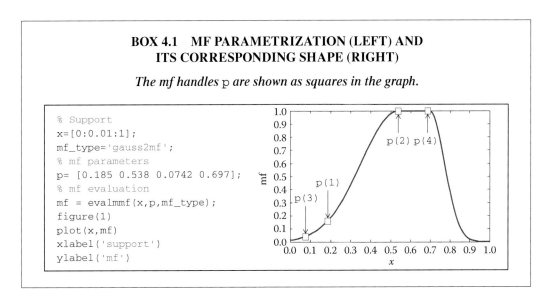

satisfy the properties of S-norms for the 'or' ($\vee$) connector and T-norms for the 'and' ($\wedge$) connector, respectively, as shown in Table 4.2.

4.2 BUILDING A FUZZY INFERENCE SYSTEM

We are now going to use the fuzzy sets just defined to implement an automated fuzzy inference system (FIS), capable of deducing a logical result on the basis of the given premises. The deductive logic is based on the *modus ponens* rule of inference. It represents a widely accepted method for the construction of deductive reasoning, whereby given an *antecedent* and a *consequent,* the truth of the latter is determined by the truth of the former. In crisp logic, the antecedent may only be either

TABLE 4.1

Basic Fuzzy Sets Properties

Property	Definition
Normality: a fuzzy set is said to be normal if there exists at least one element whose membership is equal to one.	$\exists x \mid \mu(x) = 1$
Height: largest membership grade of any element in the set A.	$h(A) = \max_{x} \mu(x)$
Support: the crisp subset X in the domain of x for which all mfs are nonzero.	$\text{Supp}(A) = \left\{ x \mid \mu(x) > 0 \text{ and } x \in X \right\}$
Core: crisp subset of A containing the elements with membership equal to one.	$\text{Core}(A) = \left\{ x \mid \mu(x) = 1 \text{ and } x \in X \right\}$
Containment: given two fuzzy sets A and B, A is said to be *contained* in B if	$B \subset A \mid \mu_B(x) > \mu_A(x) \text{ for } \forall x \in X$
Equality: fuzzy sets A and B are said to be equal if	$A \subset B \text{ and } B \subset A \Rightarrow \mu_B(x) = \mu_A(x) \text{ for } \forall x \in X$
Cardinality: extends the notion of conventional sets. For fuzzy sets with n elements, it is defined as the sum of membership grades.	$\text{Card}(A) = \sum_{i=1}^{n} \mu_A(x_i)$
α-cut: subset of A formed by the elements whose membership is greater than α.	$A_\alpha = \left\{ x \in X \mid \mu_A(x) \geq \alpha \right\}$

Fuzzy union (OR)

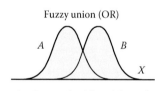

$A \vee B : \max \{\mu_A(x), \mu_B(x)\ x \in X\}$

Fuzzy intersection (AND)

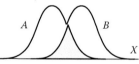

$A \wedge B : \min \{\mu_A(x), \mu_B(x)\ x \in X\}$

Fuzzy union with complement

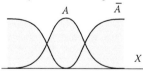

$X \neq A \vee \overline{A} : \max \{\mu_A(x), 1 - \mu_A(x)\ x \in X\}$

Fuzzy intersection with complement

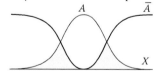

$\varnothing \neq A \wedge \overline{A} : \min \{\mu_A(x), 1 - \mu_A(x)\ x \in X\}$

If $C = A \vee B$ and $D = A \wedge B$, then $\begin{cases} D \subset C \\ A \subset C \text{ and } B \subset C \\ D \subset A \text{ and } D \subset B \end{cases}$

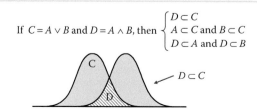

$D \subset C$

FIGURE 4.7 Logical operations on fuzzy sets. The shaded areas represent the result of the operation.

TABLE 4.2

Definition of T-Norms and S-Norms to Implement the Fuzzy Logical Operations of Union and Intersection

T-Norm = AND ($\wedge$)		S-Norm = OR ($\vee$)	
$T\left(\mu_A(x),\mu_B(x)\right)=\mu_A(x)\wedge\mu_B(x)$		$S\left(\mu_A(x),\mu_B(x)\right)=\mu_A(x)\vee\mu_B(x)$	
Boundedness (*The smallest prevails*): $T(0,0)=0; T(a,1)=T(1,a)=a$		Boundedness (*The largest prevails*): $S(1,1)=1; S(a,0)=S(0,a)=a$	
Monotonicity: $b\leq c\Rightarrow T(a,b)\leq T(a,c)$		Monotonicity: $b\leq c\Rightarrow S(a,b)\leq S(a,c)$	
Commutativity $T(a,b)=T(b,a)$		Commutativity $S(a,b)=S(b,a)$	
Associativity: $T\left(a,T(b,c)\right)=T\left(T(a,b),c\right)$		Associativity: $S\left(a,S(b,c)\right)=S\left(S(a,b),c\right)$	
Possible T-norms		Possible S-norms	
Standard intersection	$T(a,b)=\min(a,b)$	Standard union	$S(a,b)=\max(a,b)$
Algebraic product	$T(a,b)=a\times b$	Probabilistic sumt	$S(a,b)=a+b-a\times b$
Lukasiewicz intersection	$T(a,b)=\max(0,a+b-1)$	Lukasiewicz union	$S(a,b)=\min(1,a+b)$

true or false, thus implying that the consequent is also either true or false, but in fuzzy logic, things get a lot more interesting because the *degree of truth* of the consequent is determined (actually upper limited) from the *degree of truth* of the antecedent, of which it cannot be greater. In other words, the consequent cannot be 'truer' than its premises. Thus, fuzzy logic provides an attractive way to deal with partially true statements, which can be composed with the logical operators *or* and *and* to yield the degree of truth of the implied concept.

4.2.1 FUZZY INFERENCE

In strictly logical terms, given a fuzzy *antecedent* $x \in X$ and a fuzzy *consequent* $y \in Y$ the *modus ponens* inferential logic rule R can be stated as

$$R: \text{if}\left(x \text{ is } A\right) \text{then}\left(y \text{ is } B\right) \quad R: X \times Y \rightarrow [0,1] \tag{4.2}$$

where A and B are fuzzy sets defined by their mfs, so that the linguistic statements $\left(x \text{ is } A\right)$ and $\left(y \text{ is } B\right)$ actually produce the degrees of membership $\mu_a(x)$ and $\mu_b(y)$ on their supports X and Y. Therefore, the implication R operates over the Cartesian product of the two spaces (antecedent X) and (consequent Y). Far from being a trivial tautology, Equation 4.2 states that the degree of truth of A determines the upper bound for the degree of truth of B, according to the logical operators previously defined. In particular, the implication *then* is normally implemented with a T-norm, representing a logical *and* ($\wedge$). Substituting the mfs in Equation 4.2 yields the practical rule of inference:

$$R: \mu_a(x)\wedge\mu_b(y) \rightarrow \begin{cases} \min\left(\mu_a(x),\mu_b(y)\right) & \text{min operator} \\ \mu_a(x)\times\mu_b(y) & \text{product operator} \end{cases} \tag{4.3}$$

Now we encounter a new problem: x and y are defined over two independent supports X and Y, so the relation of Equation 4.3 should be defined over the Cartesian product with a considerable increase of the computational complexity $\mathbb{R} \rightarrow \mathbb{R}^2$. However, we do not need to carry on the whole two-dimensional relation R because this is actually computed only for the current value $x*$ of the antecedent. Therefore, instead of the whole Cartesian product, only the 'slice' for $x = x*$ is required, and relation (4.3) simplifies into

$$R : \mu_a(x*) \wedge \mu_b(y) \rightarrow \begin{cases} \mu_b(y \mid x*) = \min(\mu_a(x*), \mu_b(y)) & \text{min operator} \\ \mu_b(y \mid x*) = \mu_a(x*) \times \mu_b(y) & \text{product operator} \end{cases} \quad (4.4)$$

Figure 4.8 shows three possible representations of the fuzzy implication (4.2) for the two most commonly used operators (min and product). Figures 4.8a and b show the full extent of the implication represented by the relational operator (*min* in a and *prod* in b) over the entire domains X and Y, whereas Figures 4.8c and d show the 'slicing' of the implication surface for $x = x*$. In these cases, the implication reduces to the one-dimensional shaded area. In the last row, Figures 4.8e and f show the same one-dimensional implications deduced from the previous two-dimensional plots. This last representation will be the preferred one to describe the implication in the following.

The implication of Equation 4.2 can be extended to include more than one antecedent and differing connectors. The general rule, however, is that the overall degree of truth of the antecedents determines the degree of truth of the consequent. An example of how a composite antecedent might look like is

$$\text{if } (x_1 \text{ is } A_1) \text{ and } (x_2 \text{ is } A_2) \text{ or } (x_3 \text{ is not } A_3) \quad (4.5)$$

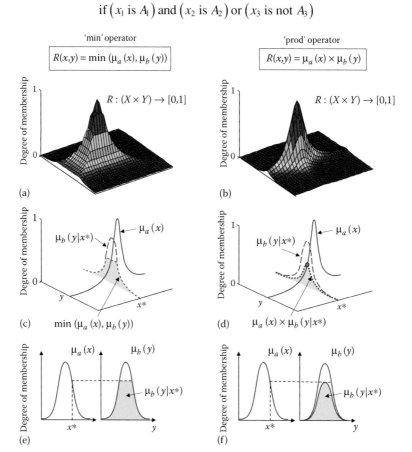

FIGURE 4.8 Three different aspects of the fuzzy implication for the two most common operators (min and prod). Figures (a) and (b) show the whole implication surface resulting from the Cartesian product. Figure (c) and (d) show that the whole antecedent domain can be reduced to the single $x*$ antecedent value. Figures (e) and (f) present the same results, but with independently plotted supports x and y. The shaded areas represent the mf resulting from the implication.

whose computational version in terms of mfs is

$$\mu_{a_1}(x_1) \underbrace{\wedge}_{\text{AND}} \mu_{a_2}(x_2) \underbrace{\vee}_{\text{OR}} \underbrace{\left[1 - \mu_{a_3}(x_3)\right]}_{\text{NOT}} \tag{4.6}$$

The resulting degree of truth of Equation 4.6 determines the highest degree of truth of the consequent. Thus, the general ith rule R_i would look like

$$R_i : \text{if } \underbrace{\left(x_1 \text{ is } A_{1,i}\right) \text{ and } \left(x_2 \text{ is } A_{2,i}\right) \text{ and } \dots \text{ and } \left(x_n \text{ is } A_{n,i}\right)}_{\text{Antecedents}} \text{ then } \underbrace{\left(y \text{ is } B_i\right)}_{\text{Consequent}} \tag{4.7}$$

which, using T-norms as connectives, translates into

$$\mu_{1,i}(x_1) \wedge \mu_{2,i}(x_2) \wedge \dots \wedge \mu_{n,i}(x_n) \wedge v_{b_i}(y) \tag{4.8}$$

Figure 4.9 provides a graphical representation of a rule with two antecedents (x_1 and x_2) and triangular mfs. The two examples differ in the choice of the 'then' operator, whereas the 'min' is used in both cases as the 'and' connector, which means that the antecedent with the smallest degree of truth prevails in the implication.

4.2.2 Fuzzy Reasoning à la Mamdani

Fuzzy reasoning is based on a set of implications (rules) where each consequent output is determined by the combination of the antecedents, as in Equation 4.7. This structure of fuzzy reasoning, which will now be discussed, was introduced by Mamdani (Procyk and Mamdani, 1979). In general, a complete FIS will involve m rules like Equation 4.7, which can be combined with the new connective *else*, usually implemented as a logical 'or' ($\vee$) because each rule represents an alternative outcome of the reasoning, albeit in a non-exclusive fuzzy sense. The various rules thus combined yield the FIS:

$$R_i : \text{if } \left(x_1 \text{ is } A_{1,i}\right) \text{ and } \left(x_2 \text{ is } A_{2,i}\right) \text{ and } \dots \text{ and } \left(x_n \text{ is } A_{n,i}\right) \text{ then } \left(y_i \text{ is } B_i\right) \quad i = 1, \dots, m \tag{4.9}$$

In Equation 4.9, each rule is composed of n antecedent variables $\left(x_1, x_2, \dots, x_n\right)$ each of which is fuzzified with q mfs $\left(A_{i,1}, A_{i,2}, \dots, A_{i,q}\right)$ according to the fuzzification rule:

$$\left(x_i \text{ is } A_{i,k}\right) \rightarrow \mu_{i,k}(x_i) \quad i = 1, \dots, n;$$
$$\left(y_i \text{ is } B_i\right) \rightarrow v_i(y_i) \quad k = 1, \dots, q. \tag{4.10}$$

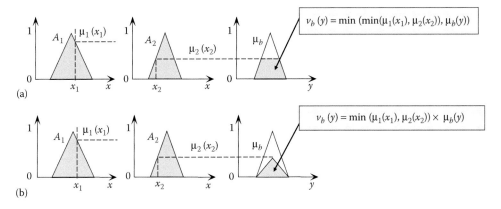

FIGURE 4.9 Graphical representation of a fuzzy implication with two antecedents, using differing 'then' operators: 'min' in case (a) and 'prod' in case (b). In both cases, the 'and' between antecedents were implemented with a 'min' operator, therefore the antecedent with the smallest degree of truth prevails.

The maximum number of different rules m that can be written given n antecedents each defined with q mfs is therefore

$$\text{Maximum number of rules: } m = q^n. \qquad (4.11)$$

Further, assuming that the rules of Equation 4.9 are composed with an *else* operator, which has the S-norm properties, then the set of rules can be written in a compact form as

$$R : \bigcup_{i=1}^{m} \left(\bigcap_{j=1}^{n} \mu_{j,i}(x_i) \wedge v_i(y_i) \right). \qquad (4.12)$$

The maximum number of rules computed by Equation 4.11 is a purely combinatorial result, which in practice may be redundant. In fact, several rules are likely to be discarded as being contradictory or unrealistic. Though rules can be directly defined by inspection, if the relevant information is available, a more systematic approach to rule writing may consist of writing down the whole set of possible combinations, according to Equation 4.11, and later pruning the lot by removing the inconsistent rules. A further refinement can then be made by running the FIS with the available data and discarding the rules that are activated in less than a given percentage of instances, for example, less than 5%.

4.2.2.1 Fuzzy Inference à la Sugeno

An alternative structure of an FIS was later proposed by Takagi and Sugeno (1985), who supposed that the consequent could be a deterministic (crisp) quantity, thus modifying Equation 4.9 into

$$R_i : \text{if } (x_1 \text{ is } A_{1,i}) \text{ and } (x_2 \text{ is } A_{2,i}) \text{ and}, \dots, \text{ and } (x_n \text{ is } A_{n,i}) \text{ then } y_i = b_i \quad i = 1, \dots, m. \qquad (4.13)$$

Compared with the previous Mamdani definition, Equation 4.13 may appear as a setback. However, this approach paved the way to incredible developments, given the freedom to define the consequents in the most diverse ways, from a numerical constant, as in Equation 4.13, to anything from a linear function to full dynamical systems that could be blended by the antecedents degree of truth to represent any complex system behaviour in differing regions of operation.

Figure 4.10 summarizes the possible operators to be used in building an FIS in either Mamdani or Sugeno logic. The intra-rule connectors are normally of the 'and' kind (T-norms), whereas the

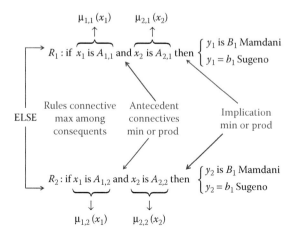

FIGURE 4.10   Summary of possible operators used in building an FIS.

rules are linked by 'else' statements implemented as 'or' connectors (S-norms). The 'then' connector is always implemented as a T-norm.

4.2.3 Defuzzification

The result of the FIS of Equations 4.9 or 4.13 is either a fuzzy set (Mamdani) as shown in Figure 4.11 or a set of constants $(v_i b_i)$ in the Sugeno logic of Figure 4.12.

In either case, the outcome of the FIS is not compatible with 'human' logic, which requires a crisp output. The FIS result must then be *defuzzified* to yield a unique result. In other words, defuzzification designates the 'most representative' crisp equivalent of the fuzzy reasoning outcome. In the Mamdani logic, there are several ways to compute the defuzzified equivalent of the output (y_{defuz}).

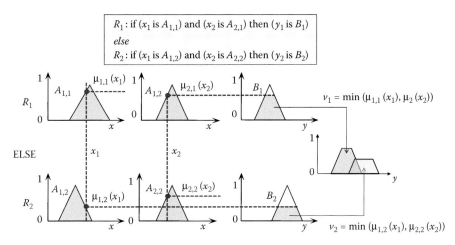

FIGURE 4.11 Composition of two Mamdani fuzzy inference rules (R_1 and R_2) producing the consequent mf shown at the far right. Both the 'and' and the 'then' connectors were implemented with the 'min' operator, whereas 'max' was used for the 'else' connector.

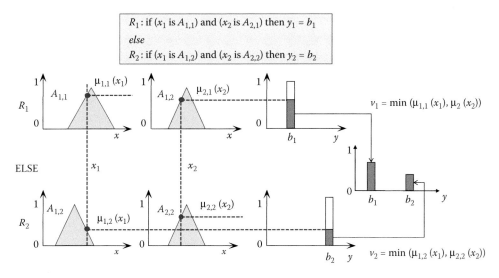

FIGURE 4.12 Composition of two Sugeno fuzzy rules (R_1 and R_2) resulting in the consequent bars at the far right. Both the 'and' and the 'then' connectors were implemented with the 'min' operator, whereas 'max' was used for the 'else' connector.

The most popular of them is the 'centre-of-gravity' (COG), or *centroid,* method, which considers the output fuzzy set as a geometrical figure whose value on the *y*-axis can be determined as

$$y_{\text{defuz}} = \frac{\int_Y y \cdot v(y) dy}{\int_Y v(y) dy} \tag{4.14}$$

where the integral is extended to the whole support *Y,* as shown in Figure 4.13a. Numerical integration poses some practical problems, because the shape of the output fuzzy set is not known in advance and changes with the inputs as the FIS evolves, but it can be approximated by a discrete set of n_z values, as shown in Figure 4.13b, so that the integral (4.14) may be replaced by a summation

$$y_{\text{defuz}} = \frac{\sum_{i=1}^{n_z} y_i \cdot v(y_i)}{\sum_{i=1}^{n_z} v(y_i)} \tag{4.15}$$

In the Sugeno case, this approximation is not necessary because the FIS produces a set of *discrete* outputs, represented in Figure 4.13c by a collection of bars, placed at the consequent output values (b_i), each of height v_i. Therefore, the y_{defuz} value is computed as a weighted sum similar to (4.15),

$$y_{\text{defuz}} = \frac{\sum_{i=1}^{m} b_i \cdot v_i}{\sum_{i=1}^{m} v_i} \tag{4.16}$$

with the summation extended to the number of consequents *m.*

In addition to the COG method, the MATLAB Fuzzy Toolbox provides several alternate defuzzification methods in the Mamdani case, as shown in Figure 4.14. The bisector method divides the output mf into two parts of equal area. If trapezoidal mfs are used, the output is likely to have a flat top, in which case any of the three methods (som, mom, lom) can be used. They would yield y_{defuz} as the smallest, middle or maximum value of the highest flat part of the mf.

Regardless of the method used for the defuzzification, the structure of the FIS is summarized in Figure 4.15. First, the antecedent variables are fuzzified to produce their degrees of truth with respect to the predetermined mfs. They yield the degree of activation of each rule from which the final output is obtained by defuzzification.

4.2.3.1 Fuzzification/Defuzzification Errors

The processes of fuzzification and defuzzification (fuz/defuz) begin and end the FIS and as such are the two communication ports between the crisp and the fuzzy worlds. They are non-linear data transforms that might introduce some error in this two-way conversion process. To compare the performance of the two fuzzy logic approaches (Mamdani and Sugeno), Figure 4.16 shows a

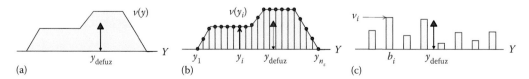

FIGURE 4.13 Mamdani and Sugeno defuzzification methods. In the Mamdani theoretical case (a) the COG y_{defuz} can be obtained by integration of the continuous mf. In practice (b) this is reduced to n_z values and the integral is approximated by a weighted summation. In the Sugeno case (c) the output is naturally a collection of discrete values and y_{defuz} is computed by the weighted sum of the consequent values.

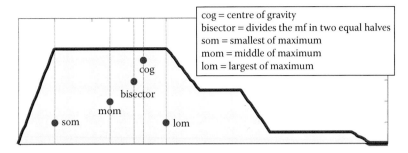

FIGURE 4.14 Defuzzification methods available in the MATLAB Fuzzy Toolbox.

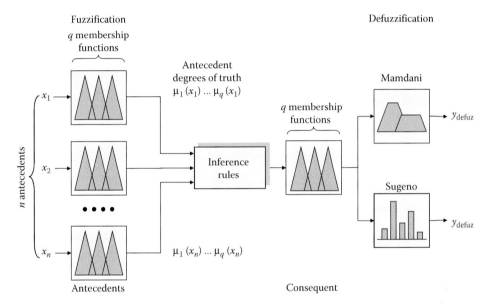

FIGURE 4.15 Structure of a Mamdani or Sugeno FIS. First, the antecedent variables are fuzzified to produce their degrees of truth with respect to the predefined mfs. The inference rules combine this information to produce the degree of truth of the consequents, from which the final output is obtained by defuzzification.

simple logic that converts a number (x) into its fuzzy equivalent and then back into a crisp quantity. In the Mamdani case, the same mfs are used in both the antecedent and the consequent. Because the two quantities are in the same principle, any difference is due to errors introduced by the fuz/defuz operations. In this example, we used triangular mfs that overlap at the middle value (0.5). It is known (Pedrycz, 1995) that this mfs arrangement yields the best results in terms of fuzzification errors. Nevertheless, some errors appear in Figure 4.16b due to the fact that the first and last triangular mfs do not have the same symmetrical shape of the internal ones and therefore have a lesser weight in the defuzzification. Conversely, the Sugeno logic does not introduce any fuz/defuz error, as can be seen in Figure 4.16c.

4.3 THE MATLAB FUZZY TOOLBOX

The MATLAB Fuzzy Toolbox provides all the basic functionalities required to build an FIS and much more. Its features will now be illustrated.

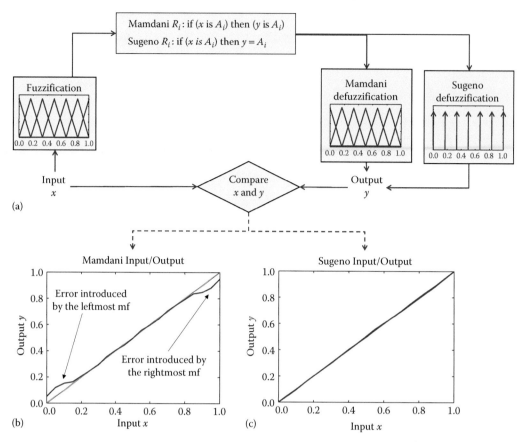

FIGURE 4.16 Comparison of fuzzification/defuzzification errors in the Mamdani and Sugeno fuzzy logic shown in (a). The Mamdani errors (b) near the boundaries of the domain are caused by the different shape of the external mfs, which are half the size of the central ones. By contrast, the Sugeno logic (c) does not introduce any such errors.

4.3.1 THE FIS EDITOR

The best way for the beginner to become familiar with the world of fuzzy reasoning is to move the first steps within the Fuzzy Editor in the MATLAB Fuzzy Toolbox, where a complete FIS can be set up without writing any code. The editor consists of the five interconnected views shown in Figure 4.17, whose capabilities are now briefly described in the order in which they are normally used. The following discussion, and related screen shots, refers to a simple non-technical example drawn from the fuzzy toolbox demos, named `tipper2`, which suggests how to tip the waiter at a restaurant. The decision is based on two factors (antecedents): quality of food and quality of service, and the consequent is obviously the tip amount.

a. *FIS definitions*: This is the first window to appear when the editor is invoked by typing `fuzzy` in the command window. If the user wants to edit a previously defined FIS, this can be retrieved by typing `fuzzy <fisname>`, where `<fisname>` is an existing `.fis` file. In the upper part of the window, the FIS structure is depicted, with the antecedents on the left, the rules in the middle (specifying whether they follow a Mamdani or a Sugeno logic) and the consequent on the right. In the lower part of the window, the user can specify the kind of connectors and the defuzzification method. By double-clicking on an antecedent icon, the control is transferred to the membership editor.

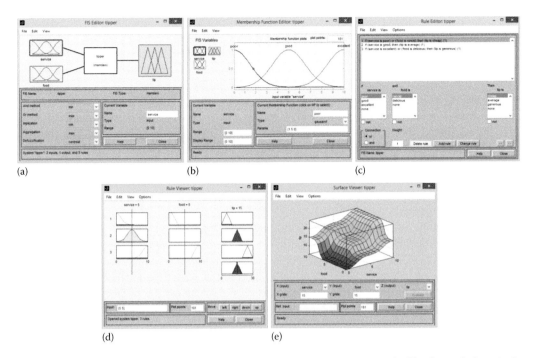

FIGURE 4.17   The five views of the Fuzzy Editor for the `tipper2.fis` example. The three windows in the upper row refer to the design environment: (a) FIS definition; (b) membership editor; (c) rule editor; while the lower row shows the two windows where the FIS can be tested: (d) rule viewer, and (e) surface viewer.

b. *Membership editor*: In this window, the mfs of each antecedent can be defined (type, support, etc.). Each mf can be assigned a verbal label, which will be used in the rule definition. Each mf can be manually changed by clicking and dragging the 'handles' in the graph, whose numerical values are displayed in the window below. Once all the mfs have been defined, the user can switch to the rule editor window

c. *Rule editor*: By default, the inference rules are specified in a 'verbose' way, that is, using the syntax used in Equations 4.9 for the Mamdani logic or in Equations 4.13 for the Sugeno logic, and using the labels defined in the previous window (b). Each rule can be set up simply by clicking on the antecedent labels, selecting the connectors, and finally clicking on 'add rule'. Existing rules can be changed by reformulating the rule and then clicking 'change rule'. An unwanted rule can be discarded by highlighting the rule and clicking 'delete rule'. All rules, by default, have a weight of 1, but the user may decide to reduce the importance of a rule by inputting a smaller weight, for example, 0.5, meaning that the maximum degree of truth of that rule will be 0.5 instead of 1.

   With reference to the tipper example, let us examine the rules and see how they make sense. Here are the rules:

$$1: \text{if} \left(\text{service is poor}\right) \text{or} \left(\text{food is rancid}\right) \text{then} \left(\text{tip is cheap}\right)$$

$$2: \text{if} \left(\text{service is good}\right) \text{then} \left(\text{tip is average}\right)$$

$$3: \text{if} \left(\text{service is excellent}\right) \text{or} \left(\text{food is delicious}\right)$$

$$\text{then} \left(\text{tip is generous}\right)$$

(4.17)

Contrary to common practice, the antecedents of rules 1 and 3 in Equation 4.17 are linked with an *or* connector, whereas *and* is the most used connector. The second rule has only one antecedent. The rules (4.17) are written in what MATLAB calls a *verbose* format, using the linguistic labels defined in the membership editor of Figure 4.17b, but other more synthetic formats are available, such as symbolic or indexed.

Having going through the three previous steps, the FIS is now fully specified and can be tested. The FIS editor provides two graphical tools (*viewers*) to check the behaviour of the inference system just set up: the rule viewer and the surface viewer.

d. *Rule viewer*: This interactive window shows the FIS output when the vertical lines are clicked and dragged over each antecedent support. The degrees of activation of all the mfs are shown in colour and a thick bar shows the defuzzified output value. By spanning the entire antecedent supports, the user can check the output changes and get an idea of the overall FIS working.

e. *Surface viewer*: While the previous rule viewer provides an output corresponding to a specific selection of the antecedent values, the surface presents a global view of how each combination of antecedents yields the FIS output. If there is only one antecedent, the graph reduces to a curve relating the input (antecedent) to the output (defuzzified consequent). If there are two antecedents, the three-dimensional graph shows the surface relating them to the defuzzified output. If there are more than two antecedents, the user can select the pair of antecedents to be displayed.

4.3.2 STRUCTURE OF THE FIS FILE

Once the FIS has been designed and tested, it can be saved either to the workspace for immediate use or to the disk for permanent storage in a file with the extension `.fis`. In this way, MATLAB recognizes FIS file as a *structure* organized as follows:

$$
\begin{array}{rl}
1 & \text{name: 'tipper'} \\
2 & \text{type: 'mamdani'} \\
3 & \text{andMethod: 'min'} \\
4 & \text{orMethod: 'max'} \\
5 & \text{defuzzMethod: 'centroid'} \\
6 & \text{impMethod: 'min'} \\
7 & \text{aggMethod: 'max'} \\
8 & \text{input: [1x2 struct]} \\
9 & \text{output: [1x1 struct]} \\
10 & \text{rule: [1x3 struct]}
\end{array}
\tag{4.18}
$$

The first seven lines of the structure (4.18) define the name and type of the FIS together with the inference method, then lines 8–10 declare the dimensions of the input, the output and the rules which are themselves structures nested in the overall FIS structure. Their detailed content is shown in Box 4.2, where each input/output element is defined by name, range (support) and number of mfs, whose parameters are also defined in terms of label, type of mf and 'handles', that is, parameters controlling its shape relative to the support. The rules are defined in the so-called indexed

---

BOX 4.2 DETAILED STRUCTURE OF THE 'TIPPER.FIS' SUB-STRUCTURES DEFINING THE INPUTS, THE OUTPUT AND THE RULES

```
[Input1]
Name='service'
Range=[0 10]
NumMFs=3
MF1='poor':'gaussmf',[1.5 0]
MF2='good':'gaussmf',[1.5 5]
MF3='excellent':'gaussmf',[1.5 10]

[Input2]
Name='food'
Range=[0 10]
NumMFs=2
MF1='rancid':'trapmf',[0 0 1 3]
MF2='delicious':'trapmf',[7 9 10 10]

[Output1]
Name='tip'
Range=[0 30]
NumMFs=3
MF1='cheap':'trimf',[0 5 10]
MF2='average':'trimf',[10 15 20]
MF3='generous':'trimf',[20 25 30]

[Rules]
1 1, 1 (1): 2
2 0, 2 (1): 1
3 2, 3 (1): 2
```

---

form that requires some explanation: the numbers before the comma refer to the antecedents, whereas the one after the comma points to the consequent. The number in parentheses is the rule weight (generally 1) and the rightmost number refers to the kind of antecedent connector (1 for *and*, 2 for *or*). The correspondence between the verbose and indexed formats of the first rule is shown in Table 4.3.

TABLE 4.3

Correspondence between the 'Verbose' and the 'Indexed' Formats for Rule Representation in the MATLAB Fuzzy Editor

	Antecedent 1	Antecedent 2	Consequent	Weight	Connector
Verbose format	(service is poor) or	(food is rancid) then	(tip is cheap)	(1)	
Indexed format	1	1	1	(1)	2
Symbol explanation	Input 1, mf1	Input 2, mf1	Output, mf1	rule weight	antecedent connector OR

The example refers to the first rule of the `tipper2.fis`, Equation 4.17.

TABLE 4.4

Some Frequently Used FIS Editing Commands, Defined as Fields in the FIS Structure

Command	Purpose
`a=newfis('fis_name')`	Defines a new FIS
`a.input(1).range=[min max];`	Defines the attributes of each mf of each antecedent
`a.input(1).mf(1).name='mf_name';`	(to be repeated for all the inputs and for each mf)
`a.input(1).mf(1).type='mf_type';`	
`a.input(1).mf(1).params=[par1 par2 …];`	
`a.output(1).name='';`	Defines the attributes of the consequent
`a.output(1).range=[0 30];`	(to be repeated for each mf)
`a.output(1).mf(1).name='cheap'`	
`a.output(1).mf(1).type='trimf';`	
`a.output(1).mf(1).params=[0 5 10]`	
`a=readfis('fis_name')`	Loads the FIS 'fis_name' into the workspace by creating the corresponding structure a
`writefis(a,'fis_name')`	Writes the workspace fis 'a' into a.fis file named 'fis_name'
`showrule(a)`	Lists the rules in the previously loaded 'a' FIS
`ruleedit(a)`	Edits the rules in the previously loaded 'a' FIS
`a = addrule(a, ruleList)`	Adds rules contained in the matrix 'ruleList' to the previously loaded 'a' FIS (one rule per row)
`plotmf(a, varType, varIndex)`	Plots the required mf contained in the previously loaded 'a' FIS, with `varType` representing the 'input' (antecedent) or 'output' (consequent) variables, and `varIndex` referring to the mf number.
`Fuzzy Logic Controller`	Inserts an FIS into a Simulink model.

The Fuzzy Editor is very handy for beginners, but it may become a limitation for the advanced user, who will find it more convenient to build an FIS using MATLAB commands, the most common being listed in Table 4.4. Using these commands, the user can directly write a script performing all the required operations to set up an FIS, which can be eventually stored as a `.fis` file onto the hard disk with the `writefis` command and *not* the `save` command reserved for saving data files.

4.3.3 An Example of Sugeno FIS: Function Approximation

The following example demonstrates how a Sugeno FIS can be set up to approximate a set of data in the least-squares sense, as described in Chapter 2. Let us assume that the data are available as $N(x,y)$ couples over a support $(x = 0,...,10)$. We want to approximate the function $y = f(x)$ by a Sugeno FIS consisting of the following three simple rules having three straight lines as consequents:

$$R_1: \text{if } (x \text{ is Small}) \text{ then } y_1 = a_1 \cdot x + b_1$$

$$R_2: \text{if } (x \text{ is Medium}) \text{ then } y_2 = a_2 \cdot x + b_2 \tag{4.19}$$

$$R_3: \text{if } (x \text{ is Large}) \text{ then } y_3 = a_3 \cdot x + b_3$$

where the three mfs labelled Small, Medium and Large can be suitably defined to cover the support. The FIS output is then obtained by defuzzification, according to Equation 4.16, as

$$y_{\text{fit}} = \frac{\sum_{i=1}^{3} y_i \cdot \mu_i}{\sum_{i=1}^{3} \mu_i} \tag{4.20}$$

where μ_i is the degree of truth of each antecedent. Assuming that the antecedent mfs have been suitably defined, the adjustable parameters are the coefficients $(a_1, b_1, a_2, b_2, a_3, b_3)$ of the linear consequents. Thus, we have an optimization problem similar to the one considered in Chapter 2, which can be defined as

$$\min_{(a_i, b_i)} \sum_{k=1}^{N} \left(y(k) - y_{\text{fit}}(k) \right)^2 \tag{4.21}$$

Figure 4.18 shows the antecedent mfs, each defined by four parameters, and the data to be approximated. The approximation y_{fit} is then obtained by blending the three linear consequents through defuzzification (4.20).

In this optimization problem, the six consequent coefficients of Equation 4.19 have to be adjusted. Their initial values can be guessed by computing the regression lines in the intervals where each line best approximates the data, as illustrated in Box 4.3 and in Figure 4.19a, while the mfs are defined in (b). Following the general procedure of Chapter 2 for setting up the optimization procedure, an error function is defined (Ferr _ fun) to compute the sum of squared errors (sse) as a function of the consequent parameters (Par), as shown in Box 4.4. The degrees of truth of each rule (Mu) are passed as external parameters via the global statement, together with the data (x, y), and the data dimension (n) required to compute the sse.

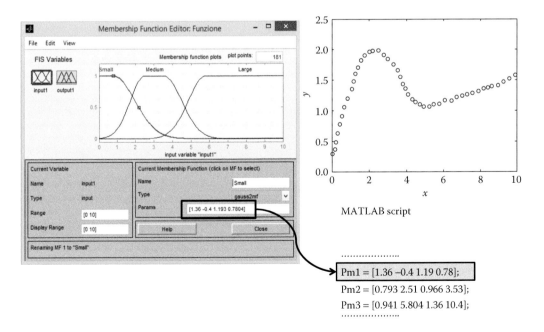

FIGURE 4.18 The antecedent mfs can be defined in the fuzzy editor to cover the three main sections of the data shown in the right graph. A separate approximating line is associated with each section. The mf parameters shown in the box are copied into the MATLAB script.

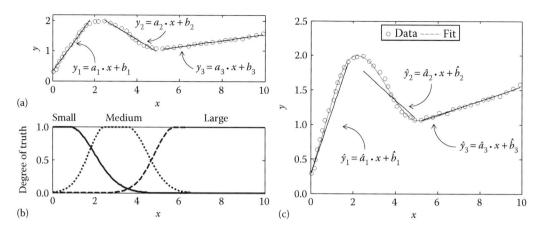

FIGURE 4.19 Piece-wise linear approximation of the data (a) and antecedent mfs (b). The optimized consequents and the resulting data approximation are shown in (c). These graphs were obtained with the MATLAB script `Function_Approx_Sugeno.m` in the `ESA_Matlab\Exercises\Chapter_4\Function_Approx` folder.

BOX 4.3   GENERATING THE INITIAL VALUES (PAR) OF THE REGRESSION COEFFICIENTS IN THE SUB-DOMAINS OF X

The parameters are grouped in the matrix `Par`, *where each row corresponds to the coefficients of each approximating line.*

```
% First line
i1=find(x<1.8);
    p1=polyfit(x(i1),y(i1),1);
    y1=polyval(p1,x(i1));
% Second line
i2=find((x>2.2)&(x<5));
    p2=polyfit(x(i2),y(i2),1);
    y2=polyval(p2,x(i2));
% Third line
i3=find(x>5);
    p3=polyfit(x(i3),y(i3),1);
    y3=polyval(p3,x(i3));
% Initial parameter matrix
Par=[p1;p2;p3];
```

After computing the individual model outputs (ym), the fitted output (y_fit) is obtained by defuzzification. The sum of squared errors (sse) is computed in the last line. In the main script, the optimization procedure, supervised by Simplex, is then invoked as ParBest=Simplex ('Ferr_fun',Par, options), and the results are shown in Figure 4.19c, where it can be seen that a good data fit is obtained, although the final linear consequents are fairly different from the initial ones.

In particular, the middle line, corresponding to the medium antecedent, looks 'wrong', but it should be recalled that the overall model output is a blending of the three lines through defuzzification (4.20). The FIS parameters (antecedent mfs, initial and optimized consequent coefficients) are shown in Table 4.5.

BOX 4.4   DEFINITION OF THE ERROR FUNCTION TO OPTIMIZE THE CONSEQUENT COEFFICIENTS IN THE FUNCTION APPROXIMATION EXERCISE FUN_UP_DOWN

```
function sse=Ferr_fun(Par)
global Mu x y n
ym=zeros(n,3);
% Compute individual models
for w=1:3
ym(:,w)=Par(w,1)*x+Par(w,2);
end
% Compute model output by defuzzification
y_fit= sum(ym.*Mu',2)./sum(Mu',2);
% Compute sum of squared errors
sse=sum((y-y_fit).^2)/n;
```

TABLE 4.5

Antecedent and Consequent Parameters for the Exercise of Section 4.3.3

Antecedent Parameters	Initial Consequent Parameters	Optimized Consequent Parameters
Pm1=[1.36 - 0.4 1.19 0.78];	0.9752 0.3067	0.9166 0.2684
Pm2=[0.793 2.51 0.966 3.53];	-0.4287 3.0951	-0.2746 2.4548
Pm3=[0.941 5.804 1.36 10.4];	0.1031 0.5197	0.1027 0.5243

As a concluding remark on this exercise, it should be considered that the antecedent mfs were manually adjusted by tweaking their 'handles' in the membership window of the fuzzy editor, as shown in Figure 4.18 and were not changed during the optimization. For this reason, the mfs (Mu) are passed as parameters to the function Ferr _ fun once and for all, with a considerable saving in computational time. If the antecedent parameters had been included in the estimation, this would not have been possible, and there would have been $3 \times 4 + 3 \times 2 = 18$ parameters to optimize. This extension is left to the reader as an exercise, though it can be anticipated that this example will be reconsidered soon by using a much more powerful method (ANFIS) to obtain a more accurate approximation of the data by estimating all the above 18 parameters.

4.4   FUZZY MODELLING

The example of the previous section considered the fuzzy approximation of a simple static model $y = f(x)$, but we are primarily interested in *dynamical* models, whose behaviour by definition depends on their past history, as amply treated in Chapter 1. Given the implication structure of an FIS, we shall consider *discrete-time dynamical* models written in the following form:

$$y_k = f\left(y_{k-1}, y_{k-2}, \ldots, y_{k-n_y}, u_{k-\delta}, \ldots, u_{k-\delta-n_u}\right) \tag{4.22}$$

where:

$(n_y + n_u)$ is the model order, given by the number of past input and outputs needed to compute the current output

$\delta \geq 1$ is the input/output delay, meaning that the most recent input $(u_{k-\delta})$ influencing the current output (y_k) has entered the system δ samples earlier

No special assumption is made to the model structure, which may be non-linear both in the past inputs and outputs, and in the parameters.

4.4.1 Mamdani versus Sugeno Models

The general model structure of Equation 4.22 can be represented by either a Mamdani or a Sugeno FIS. In both cases, the antecedents are defined as a collection of mfs, whereas the consequent may be another fuzzy set (Mamdani) or a deterministic quantity (Sugeno) ranging from a constant to a specific dynamical model, which best describes the system behaviour in a certain operating region. In the Mamdani case, the model structure can be described by the generic rule

$$R_i : \text{if } \underbrace{\left(y_{k-1} \text{ is } Y_1^{(i)}\right) \dots \text{ and } \dots \left(y_{k-n_y} \text{ is } Y_{n_y}^{(i)}\right)}_{\text{output past samples}} \text{ and } \underbrace{\left(u_{k-\delta} \text{ is } U_\delta^{(i)}\right) \dots \text{ and } \dots \left(u_{k-\delta-n_u} \text{ is } U_{n_u}^{(i)}\right)}_{\text{input past samples}},$$
$$\text{then } \left(y_k^{(i)} \text{ is } Y_0^{(i)}\right) \tag{4.23}$$

where $\left(Y_1^{(i)}, \dots, Y_{n_y}^{(i)}, U_\delta^{(i)}, \dots, U_{n_u}^{(i)}, Y_0^{(i)}\right)$ are suitably defined mfs and $\delta \geq 1$ is the input–output delay. As already pointed out by Equation 4.11 that computes the number of possible rules, the set of rules (4.23) may quickly become explosive, because if q mfs are used to define each antecedent, the maximum number of rules m is

$$m = q^{\left(n_y + n_u\right)} \tag{4.24}$$

which suggests that the model order $\left(n_y + n_u\right)$ should be kept as low as possible. The general structure of a Mamdani fuzzy dynamical model is shown in Figure 4.20.

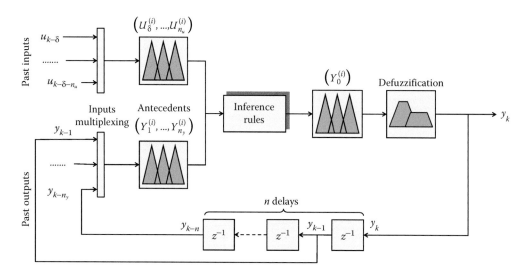

FIGURE 4.20 General structure of a Mamdani fuzzy dynamical model. The dynamic nature of the model is reflected in the delay chain, used to store the past model outputs. A similar arrangement can be set up for storing past input data.

In the Sugeno logic, the fuzzy model (4.23) becomes

$$R_i : \text{if } \underbrace{\left(y_{k-1} \text{ is } Y_1^{(i)}\right) \dots \text{ and } \dots \left(y_{k-n} \text{ is } Y_{n_y}^{(i)}\right)}_{\text{output past samples}} \text{ and } \underbrace{\left(u_{k-\delta} \text{ is } U_\delta^{(i)}\right) \dots \text{ and } \dots \left(u_{k-\delta-n_u} \text{ is } U_{n_u}^{(i)}\right)}_{\text{input past samples}} \quad (4.25)$$

$$\text{then } y_k^{(i)} = f_i\left(y_{k-1},\dots,y_{k-n_y},u_{k-\delta},\dots,u_{k-\delta-n_u}\right)$$

where the consequents $f_i\left(y_{k-1},\dots,y_{k-n_y},u_{k-\delta},\dots,u_{k-\delta-n_u}\right)$ can be any deterministic function, ranging from a constant (singleton) to a local linear model such as

$$y_k^{(i)} = a_1^{(i)} \cdot y_{k-1} + \dots + a_n^{(i)} \cdot y_{k-n} + b_1^{(i)} \cdot u_{k-\delta} + \dots + b_{n_u}^{(i)} \cdot u_{k-\delta-n_u} \quad (4.26)$$

that approximates the system behaviour in a specific operating region. Its structure is shown in Figure 4.21.

The common feature of the two fuzzy models (4.23) and (4.25) is the delay chain needed to store the past model output to be used as antecedents as the system evolves in discrete-time steps. As already pointed out in Chapter 1, memory is the essential feature of dynamical models and the delay chain is essential to 'remember' the model past outputs. Another common feature of fuzzy models is their intrinsically discrete-time nature, whereby the model output is evaluated at T_s time intervals.

4.4.2 An Example of a Heuristic Mamdani Fuzzy Model

We are now going to set up a fuzzy model of the pressure tank system, previously considered in Section 1.3, using the Mamdani fuzzy logic. Assuming that we do not know the laws of hydraulics, the model is set up by just looking at the input–output data, thus adopting the data-driven approach. The mechanistic model developed in Section 1.3 will be used only to generate the data

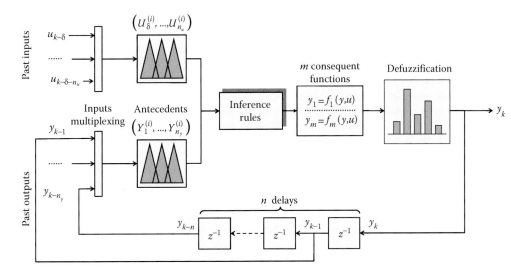

FIGURE 4.21   General structure of a Sugeno fuzzy dynamical model. As in the previous case, the dynamic nature of the model is reflected in the delay chain, used to store the past model outputs. A similar arrangement is can be set up for storing past input data.

for training and validation, and as a comparison. The pressure flow equation, recalled here for convenience, is

$$A\frac{dh}{dt} = F - k_v\sqrt{h} \tag{4.27}$$

where:

A is the tank cross section
k_v is the friction coefficient of the outlet
F is the input flow
h is the height of the liquid in the tank

The input–output data used to train the fuzzy model are shown in Figure 4.22, together with the deterministic model (4.27).

To build the fuzzy model, the following operations are sequentially performed:

1. Definition of the 'meaningful' regions of the data for support delimiting.
2. Definition of the number and type (triangular, Gaussian, etc.) of the mfs.
3. Definition of the model structure (number of past inputs and outputs) and of its rules of inference.

Point 1 reminds us that the model will 'learn' what we 'teach' it, in other words, if we train the model using data in a given range, it will be lost when confronted with data outside this region. Consequently, we must check that our training data are exhaustive and span the entire domain of operation.

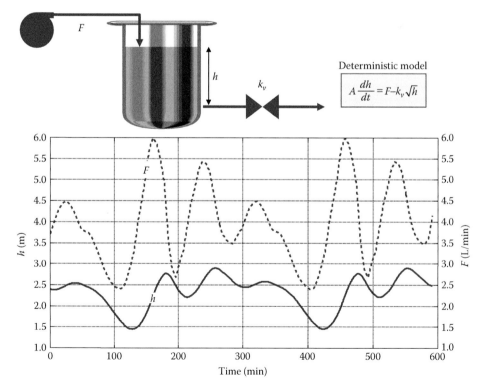

FIGURE 4.22 Input–output data from the tank with pressure output flow, used to set up the fuzzy model. The upper-right inset shows the deterministic model derived from elementary hydraulics (see Section 1.3).

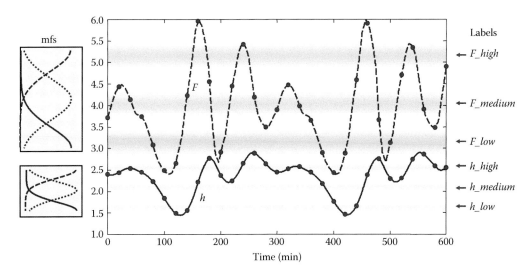

FIGURE 4.23 Covering the data with a suitable collection of mfs (left) corresponding to the linguistic labels shown on the right side. The darkest shaded bands show where each membership is close to one. The dots represent the sampled data.

Point 2 requires that we cover the data range with a 'complete' set of mfs, meaning that over the whole support at least one mf will yield a non-zero degree of truth. Then, the mathematical form of the mfs can be selected and each of them may be given a linguistic label.

Point 3 regards the model structure, in the sense that we should decide the number of past inputs and outputs that contribute to the current output. Remembering Equation 4.24 and to avoid an explosive number of rules, the model order should be kept as low as possible. In this example, it was decided to include the last two inputs and outputs, so that the general model structure of Equation 4.26 now becomes

$$h_k = f\left(h_{k-1}, h_{k-2}, F_{k-1}, F_{k-2}\right) \tag{4.28}$$

We also assume that there is no additional delay in the model input $(\delta = 1)$. The actual relationship $f(\bullet)$ of Equation 4.28 is of no concern here, as it will be approximated by the fuzzy rules.

Assuming that three mfs have been defined as in the left column of Figure 4.23, we now propose a heuristic method to define the rules, having set the model structure as in Equation 4.28.

Let us consider a two-sample wide moving window, as shown in Figure 4.24, encompassing sampling the instants $(k, k-1, k-2)$. By sliding the window (this could be actually done by tracing it on a transparency film) over the data, the relative position of the input and output samples define the rule for the current position of the window. It should be recalled that given three mfs and four antecedents, the maximum number of independent rules is a staggering $3^4 = 81$, but this method is not likely to generate the whole lot, nor we will need all of them.

By scanning the data set of Figure 4.24, the rules listed in Table 4.6 were generated and inserted in the rule editor of Figure 4.17c, together with the mfs definitions. Then the FIS can be saved to disk and incorporated in the Simulink® diagram of Figure 4.25 using the Simulink controller block.

The communication between the fuzzy editor and the Simulink diagram is shown in Figure 4.26. The FIS named Tank.fis, previously saved in the hard disk, is retrieved by the launch MATLAB script Go _ Fuzzy _ Tank.m with the readfis command (not load!) and assigned to a structure named Tank, which must be written in the input mask of the Fuzzy Logic Controller Simulink block. The FIS and the Simulink structure may have differing names, but the name of the structure in the Simulink block must be the same as the one retrieved with the readfis command as detailed in Figure 4.26.

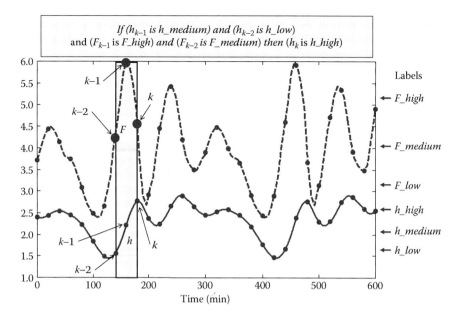

> *If (h_{k-1} is h_medium) and (h_{k-2} is h_low)*
> *and (F_{k-1} is F_high) and (F_{k-2} is F_medium) then (h_k is h_high)*

FIGURE 4.24 Heuristic rule definition by sliding a two-sample moving window (shaded area) over the data. The relative positions of input and output samples in the window define the current rule (upper box). Each time the window is advanced by one step a new rule can be defined by inspection.

TABLE 4.6

Heuristic Rules for the Tank Fuzzy Model

1. If (h1 is h_medium) and (f1 is f_high) and (f2 is f_medium) and (h2 is h_low) then (h is h_high)
2. If (h1 is h_medium) and (f1 is f_high) and (f2 is f_low) and (h2 is h_medium) then (h is h_high)
3. If (h1 is h_low) and (f1 is f_high) and (f2 is f_low) and (h2 is h_low) then (h is h_low)
4. If (h1 is h_medium) and (f1 is f_high) and (f2 is f_medium) and (h2 is h_low) then (h is h_high)
5. If (h1 is h_medium) and (f1 is f_medium) and (f2 is f_low) and (h2 is h_medium) then (h is h_low)
6. If (h1 is h_high) and (f1 is f_medium) and (f2 is f_high) and (h2 is h_high) then (h is h_high)
7. If (h1 is h_high) and (f1 is f_high) and (f2 is f_medium) and (h2 is h_high) then (h is h_high)
8. If (h1 is h_medium) and (f1 is f_low) and (f2 is f_medium) and (h2 is h_high) then (h is h_low)

The performance of the fuzzy model is shown in Figure 4.27, where it is compared to the deterministic model that it was supposed to mimic.

Though the model agreement between the fuzzy output (*h_fuzzy*) and the real height (*h*) is not very good, some useful conclusions may be drawn from this naive exercise, whose pros and cons are now assessed:

Pros:
- It is not necessary to know the system structure, nor its internal (mechanistic) working.
- The data-driven model is derived only on the basis of observed behaviours.

Cons:
- No previous knowledge is available to assist in the definition of the model structure (e.g., its order, number and shape of mfs, number and nature of the rules).
- The model validity is limited to the range of the observations. Having trained the model with flow in the 2–6 L/min range, how would the model behave if the flow were increased to 10 L/min? The present training provides no rule for this value, so the model could not reproduce that behaviour.

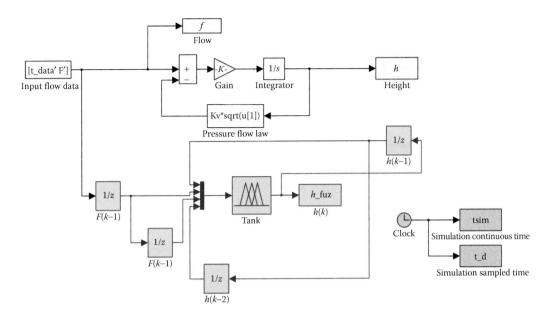

FIGURE 4.25 Simulink diagram implementing the tank fuzzy model (`Tank_fuzzy_mamdani.mdl` in the `ESA_Matlab\Exercises\Chapter_4\Heuristic_Fuzzy_Tank` folder). The fuzzy model is represented by the blue blocks, including the fuzzy logic controller implementing the FIS. The Mux at its input bundles the antecedent data in the order in which the rules are defined (see Table 4.6). The upper diagram in grey is the deterministic model for comparison.

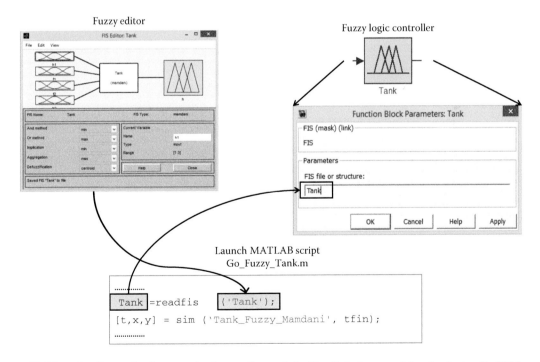

FIGURE 4.26 Connection between the fuzzy editor and the Simulink diagram. After developing the FIS in the fuzzy editor, the FIS named `Tank.fis` is stored in the hard disk, to be retrieved by the launch MATLAB script `Go_Fuzzy_Tank.m` with the `readfis` command and assigned to a structure named `Tank`, which is then written into the input mask of the Fuzzy Logic Controller Simulink block. The FIS and the Simulink structure may have differing names.

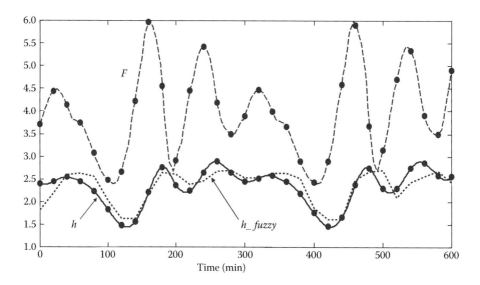

FIGURE 4.27 Fuzzy tank model response (dotted line) compared to the training data (solid line) obtained with the Simulink model of Figure 4.25. This graph was obtained with the MATLAB script `Go _ Fuzzy _ Tank.m` in the `ESA _ Matlab\Exercises\Chapter _ 4\Heuristic _ Fuzzy _ Tank` folder.

Given the limits of this heuristic approach, the question arises as to whether there are systematic methods to develop a valid fuzzy model (dynamical or otherwise). The answer is fortunately 'yes', as will be explained in the next section.

4.4.3 ANFIS: A SYSTEMATIC MODELLING APPROACH

Those who were disappointed by the previous results will be relieved to learn that several systematic approaches are available to obtain *good* fuzzy models from the data, thus retaining the advantage of avoiding the heavy theoretical involvement required by mechanistic modelling and still produce a model capable of accurately reproducing the observed behaviour.

One such approach is the so-called ANFIS, for Adaptive Neuro-Fuzzy Inference System, which was originally proposed by Jang (1993) and was later used, among other applications, to model wastewater treatment processes (Tay and Zhang, 1999, 2000), drinking water processes (Wu et al., 2014) and, in general, in the context of soft-sensors development (Dürrenmatt and Gujer, 2011; Haimi et al., 2013). ANFIS is a composite neuro-fuzzy structure that implements Sugeno-type fuzzy logic in the form

$$R_i : \text{if } \left(x_1 \text{ is } A_1^{(i)} \right) \dots \text{ and } \dots \left(x_n \text{ is } A_n^{(i)} \right) \text{ then } f_i = \boldsymbol{a}_i^T \cdot \boldsymbol{x} + b_i \qquad (4.29)$$

where $\left(\boldsymbol{x} \in \mathbb{R}^n \right)$ is the vector of inputs and $\left(\boldsymbol{a}_i, b_i \right)$ are the coefficients of the linear Sugeno consequents. The novelty of ANFIS with respect to the previous Sugeno model of Equation 4.25 is that this algorithm can adapt *both* the antecedent mfs *and* the linear consequents using a hybrid learning algorithm based on a combination of the least-squares and the backpropagation methods to minimize the sum of squared differences between the ANFIS output and a training data set. The general ANFIS structure is shown in Figure 4.28: the inputs $\left(\boldsymbol{x} \in \mathbb{R}^n \right)$ are fuzzified and the degree of truth is evaluated as the product of the individual degrees of truth of the antecedents:

$$w_i = \prod_i \mu_i \left(\boldsymbol{x} \right) \qquad (4.30)$$

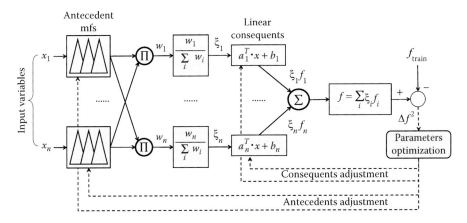

FIGURE 4.28 General structure of ANFIS. The dashed lines indicate the adjustment paths of both the antecedents and the consequents, depending on the squared error between the ANFIS output (f) and the training data (f_{train}).

Then, each degree of truth is normalized with respect to their sum

$$\xi_i = \frac{w_i}{\sum_i w_i} \qquad (4.31)$$

so that the global model response is given by

$$f = \sum_i \xi_i f_i \qquad (4.32)$$

The optimization algorithm adjusts the parameter of each rule (4.29) to minimize the squared error Δf^2 between the model output and the training data set.

ANFIS can be accessed in two ways: either through a graphical user interface (GUI) or directly through MATLAB commands. The GUI approach is recommended for the beginner, whereas the experienced user may prefer to write the MATLAB script directly. Both methods will now be illustrated by reworking the two previous examples: the function approximation presented in Section 4.3.3 and the tank dynamical model of Section 4.4.2.

4.4.3.1   Introduction to the ANFIS Editor

The four basic operations that can be performed in the ANFIS editor are now briefly described. More information can be found in the MATLAB help. The screenshots showing these operations are illustrated in Figure 4.29 and they should be performed sequentially, as described below.

1. *Data loading*: The ANFIS editor requires the data to be organized in a matrix with each column representing one input and the last column containing the output data. Each row represents an observation. The data should be saved in ASCII format and not as a .mat binary file. The .dat extension is not essential, but it helps.
2. *FIS generation*: An initial FIS is generated with the selected type and number of mfs specified by the user. There are several FIS generating methods available, but the default (grid partition) is recommended for the beginner.
3. *FIS training*: After having specified the FIS structure, training can now begin, and the 'hybrid' method is recommended, applying a mix of least squares and backpropagation. The graphic panel will show the error trend over the training epochs.

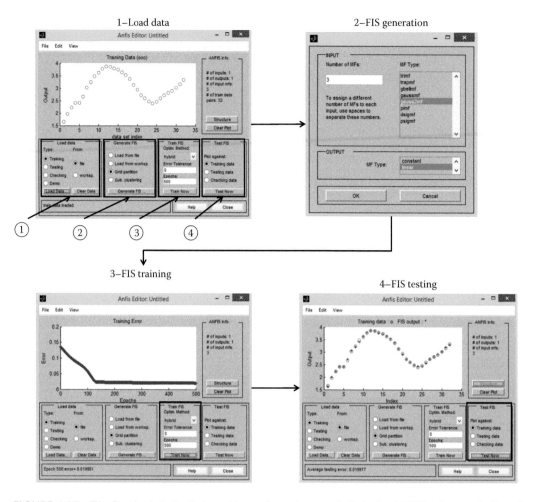

FIGURE 4.29   The four basic tasks that must be performed sequentially in the ANFIS editor are indicated by the circled numbers in (1), which is the ANFIS splash screen. Each active panel is then outlined in the subsequent screen shots.

4. *FIS testing*: When training is completed, the FIS is ready for testing. This can be done using the same training data or other data sets (checking and validation). If those data sets were not loaded during the initial phase (step 1), the user can go back to panel 1 and load the additional data, clicking on the pertinent radio button so that the editor can discriminate among training, testing and checking data sets.

The ANFIS editor is very handy, as the user can go back to any panel and redo the parts that require a second thought. The apparent downside is that of working in a closed environment from which it appears to be difficult to export any result. Luckily, this is not so, as the editor is capable of generating a.fis file that can be opened in the normal fuzzy editor or loaded directly onto the MATLAB workspace for further processing, as shown in Figure 4.30.

For those who prefer to do without the ANFIS editor, the same procedure can be programmed using the MATLAB commands `genfis1` and `anfis`. The first generates the initial FIS, while the second performs the adaptation. Figure 4.31 shows the two possible paths to produce an independent Simulink model containing an ANFIS-generated FIS. After splitting the data into training and validation subsets, the ANFIS can be set up either by writing a MATLAB script or by using the ANFIS editor.

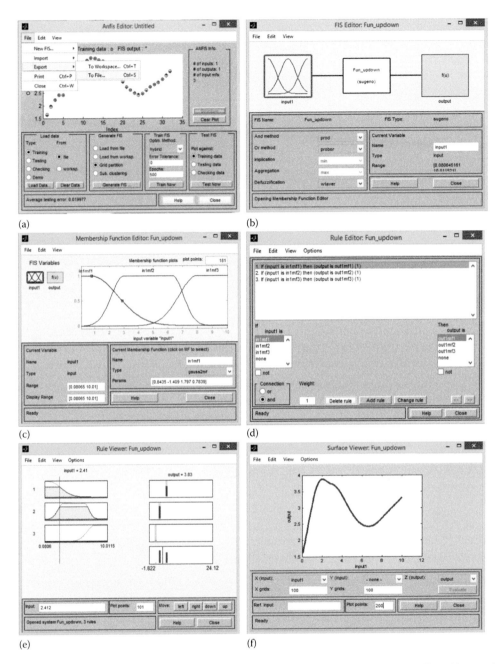

FIGURE 4.30 Exporting and managing the results of the ANFIS editor into the fuzzy editor. (a) exporting the FIS from the ANFIS editor, (b) importing the FIS into the fuzzy editor, (c) Viewing the mfs in the fuzzy editor, (d) viewing the rules in the fuzzy editor, (e) viewing the rules at work in the fuzzy editor and (f) viewing the control surface in the fuzzy editor.

In either case the resulting FIS can be saved onto the hard disk (or workspace) with the `writefis` command. The saved FIS can then be retrieved and imported into the MATLAB launch script with the `readfis` command that loads an FIS structure into the MATLAB Workspace, from where it can be incorporated into the fuzzy controller block of the Simulink model.

If we decide to do without the ANFIS editor, the steps required to create an ANFIS directly in a MATLAB script are illustrated in Table 4.7. After loading the data, coming either from a previous

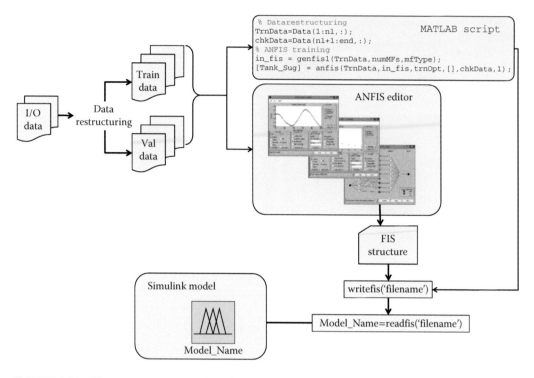

FIGURE 4.31 The two ways to export the FIS structure created in the ANFIS editor onto a Simulink model. After dividing the data into a training and a validation subset, the ANFIS can be created either by direct MATLAB programming (MATLAB script) or through the ANFIS editor. In either case, the resulting FIS can be saved to disk with the `writefis` command and imported into a Simulink model with the `readfis` command and fuzzy controller block.

TABLE 4.7

Detailed Command Sequence to Create and Export an ANFIS

Task	Command Lines in the MATLAB Script

Load simulation data and create the data structure	`load Sim_Data`
	 ....
	 `for k=4:n`
	 `Data(k-3,:) = [F(k-1) h(k-1) h(k)];`
	 `End`

Split the data into a training and a validation data set	`TrnData=Data(1:n1,:);` `chkData=Data(n1 + 1:end,:);`

Set ANFIS parameters	`numMFs = 3;`
	 `mfType = 'gauss2mf';`
	 `epoch_n = 10;`
	 `trnOpt=[epoch_n 0 0 0.8 1.2];`

Generate initial FIS	`in_fis = genfis1(TrnData, numMFs, mfType);`
Train ANFIS	`[`**`Tank_Sug`**`, error1,ss, CSTR_Sug2,error2] = anfis(TrnData,` `in_fis, trnOpt, [], chkData);`
Save FIS to file	`writefis(FIS_name,'`**`Tank_Sug`**`);`

simulation or from the field, they are split into two subsets (training and checking or validation). Then, the ANFIS structure is defined in terms of number and shape of the mfs. The number of training epochs should also be specified, together with other optimization parameters. An initial FIS is then generated from which the whole neuro-fuzzy structure is tuned with the training data and validated with the check data. The tuned ANFIS can then be saved to disk for independent use within a Simulink model.

4.4.3.2 Function Approximation with ANFIS

Let us now use ANFIS to revisit the function approximation problem already encountered in Section 4.3.3. The main difference from the previous attempt is that now both the antecedent mfs and the coefficients of the linear consequents are tuned, as opposed to the previous case when only the consequents were adjusted. A sample MATLAB script to obtain the function approximation of the previous example is listed in Box 4.5, whereas the influence of the number of mfs on the approximation accuracy is shown Figure 4.32.

It can be seen that if too few mfs are used (a) the approximation is poor, but using too many mfs (c) may result in 'overfitting', with the approximation running too close to the data, to the extent that even the noise or all sorts of numerical artefacts are included in the FIS. The right choice is shown in (c) where using five mfs provides a good functional approximation and yet leaves the noise out. The adjustment of the antecedent mfs operated by ANFIS is shown in Figure 4.33 comparing the initial and final mfs.

4.4.3.3 An ANFIS Tank Dynamical Model

The tank exercise of Section 4.4.2 is now revisited by generating an ANFIS model. First of all, it should be pointed out that in the original formulation (Jang, 1993) ANFIS has no internal feedback path, like that used in Figure 4.20 for the Mamdani dynamical model and in Figure 4.21 for its Sugeno counterpart. On the other hand, there is no provision for modifying the MATLAB ANFIS

BOX 4.5   MATLAB SCRIPT PERFORMING THE ANFIS TRAINING TO THE FUNCTIONAL DATA FUN_UP_DOWN

```
% Load the data representing the function
load Fun_up_down
% Assign the data to x and y vectors
x=Mis_exp(:,1);
y=Mis_exp(:,2);
% Build the training data set
trnData = [x y];
% define the number and type of mfs
numMFs = 15;
mfType = 'gbellmf';
% Set the number of training iterations
epoch_n = 20;
% Generate the initial FIS
in_fis = genfis1(trnData,numMFs,mfType);
% Train the ANFIS with the data
out_fis = anfis(trnData,in_fis,epoch_n);
% Compare the results with the data
plot(x,y,'o') % Plot the data
hold on
plot(x,evalfis(x,out_fis)); % Plot the ANFIS output
```

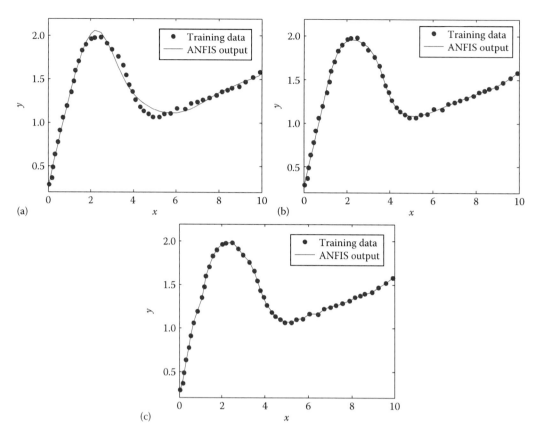

FIGURE 4.32   The number of mfs influences the accuracy of the FIS. In (a) using too few mfs results in a poor accuracy, whereas in (c) the high number of mfs produces an 'overfitting', with the approximation too close to the data, so that even the noise is modelled. The middle choice (b) uses five mfs and yields the best approximation. These graphs were obtained with the MATLAB script ANFIS _ Numeric _ Func _ approx.m in the ESA _ Matlab\Exercises\Chapter _ 4\ANFIS\ANFIS _ Funct _ Approx folder.

structure. So the only way of dealing with dynamical systems in the ANFIS context is to reshape the data in a format compatible with ANFIS. If we maintain the previous choice of limiting the model structure to the two last inputs and outputs, each row of the ANFIS input data matrix should be structured as

$$\text{Data}(k,:) = \left[\underbrace{F(k-1)\,F(k-2)\,h(k-1)\,h(k-2)}_{\text{input data}} \quad \underbrace{h(k)}_{\text{output data}} \right] \tag{4.33}$$

where k represents the generic sampling instant. Each line of the matrix (4.33) can be constructed by concatenating time-shifted copies of the recorded data, with the last column containing the output data. The data organization to produce the training data set is illustrated in Figure 4.34.

With the training data structured as in Equation 4.33, the generic Sugeno ANFIS rule looks like this:

$$R_i : \text{if } \left(F_{k-1} \text{ is } F_1^{(i)} \right) \text{ and } \left(F_{k-2} \text{ is } F_2^{(i)} \right) \text{ and } \left(h_{k-1} \text{ is } H_1^{(i)} \right) \text{ and } \left(h_{k-2} \text{ is } H_2^{(i)} \right)$$

$$\text{then } h_k^{(i)} = a_1^{(i)} \cdot h_{k-1} + a_2^{(i)} \cdot h_{k-2} + b_1^{(i)} \cdot F_{k-1} + b_2^{(i)} \cdot F_{k-2} + c^{(i)} \tag{4.34}$$

where $\left(F_1^{(i)}, F_2^{(i)}, H_1^{(i)}, H_2^{(i)} \right)$ are the antecedent mfs used in the ith rule. It is important to notice that in the consequent expression the past samples of the full output $\left(h_{k-1}, h_{k-2} \right)$ are used instead of the

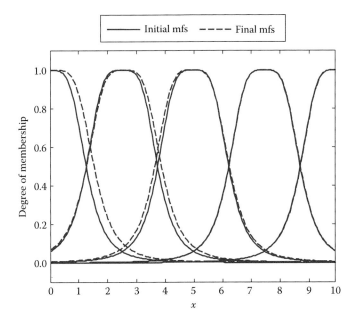

FIGURE 4.33 Comparison of initial (solid line) and adjusted (dashed line) mfs after adjusting the ANFIS to the function of Figure 4.32 with five mfs.

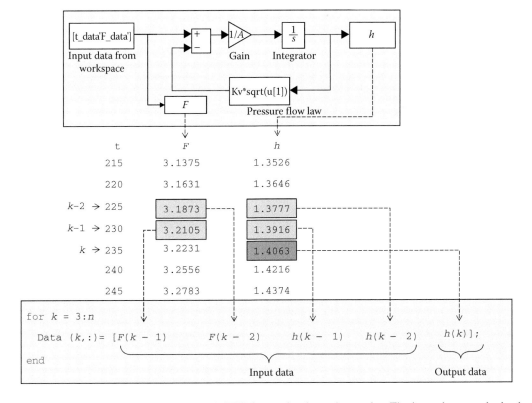

FIGURE 4.34 Data organization in the ANFIS format for the tank exercise. The input data are obtained from the simulation of the deterministic model (upper box). The `for` loop (lower box) organizes the data input matrix in which the first four columns are the delayed samples of flow (F) and height (h). The last column is the output (current height).

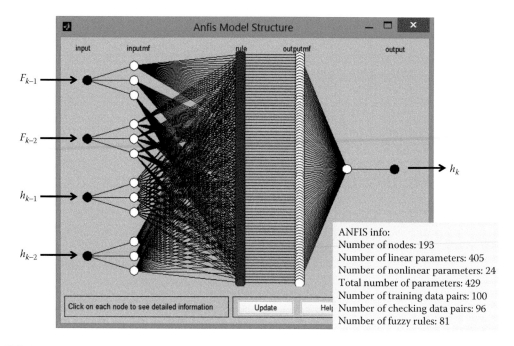

FIGURE 4.35 Structure of the Sugeno tank second-order model produced by ANFIS. The info box shows that the model contains the full paradigm of fuzzy rules (81), and that the number of model parameters is staggering (429!).

local model outputs $\left(h_{k-1}^{(i)}, h_{k-2}^{(i)} \right)$. ANFIS implements the whole rule paradigm, which according to Equation 4.24 consists of $3^4 = 81$ different rules if three mfs are defined for each antecedent. The overall FIS complexity is shown in Figure 4.35.

Though the result is satisfactory (see Figure 4.38), especially when compared to the previous Mamdani example of Section 4.4.2, the price to pay is a staggering model complexity in terms of number of rules (81) and tuning parameters (429!). Thus, it makes sense to ask whether a simpler, easier-to-tune, model would perform just as well. To this end, a reduced model structure is now considered, which uses only one-step-back information. The generic fuzzy rule of this first-order model has the form

$$R_i : \text{if } \left(F_{k-1} \text{ is } F_1^{(i)} \right) \text{ and } \left(h_{k-1} \text{ is } H_1^{(i)} \right) \text{ then } h_k^{(i)} = a_1^{(i)} \cdot h_{k-1} + b_1^{(i)} \cdot F_{k-1} + c^{(i)} \tag{4.35}$$

and the related data structure of Equation 4.33 simplifies into

$$\texttt{Data}(k,:) = \left[\underbrace{F(k-1) \, h(k-1)}_{\text{input data}} \quad \underbrace{h(k)}_{\text{output data}} \right] \tag{4.36}$$

After performing the usual ANFIS training, the resulting FIS can be exported to disk. It should be remarked once more that the FIS produced by ANFIS has exactly the same structure and attributes of an FIS created in the fuzzy editor described in Section 4.3.1. However, because this represents a dynamical model, its simulation requires the set up of a Simulink model, which includes the fuzzy controller block using the FIS just created, as already encountered in the previous exercise of Section 4.3.1.

The communication between the MATLAB script and the Simulink model is shown in Figure 4.36 for the first-order model and in Figure 4.37 for the second-order model.

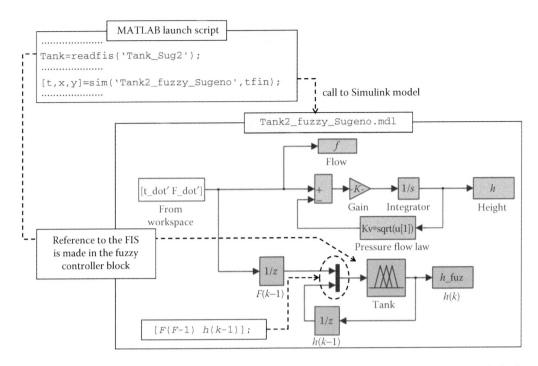

FIGURE 4.36 Incorporating the previously developed FIS in the first-order Simulink tank model via the MATLAB launch script. The `readfis` command loads the FIS into the Workspace where it is made available to the fuzzy controller Simulink block. The name of the FIS inside the latter must be the same created by `readfis` ('Tank' in this example). The mechanistic model (in grey) is just for output comparison. The pertinent MATLAB/Simulink codes can be retrieved from the `ESA _ Matlab\Exercises\Chapter _ 4\ANFIS\ANFIS _ Tank` folder.

Now, the tank exercise is completed, but before assessing the results, let us recap the various actions taken so far:

1. The data were generated by simulating the tank model (4.27). They were then divided into two equal portions for training and validation. The data were then reorganized as in Equation 4.33 or 4.36, as required by the selected data structure, as shown in Figure 4.34 for the second-order model.
2. An ANFIS neuro-fuzzy network was trained using the training data portion, and when a satisfactory performance was reached, the tuned FIS was exported.
3. A Simulink model was set up to include a fuzzy controller whose input is the data previously generated, reshaped as already mentioned. In any case the Mux connections are critical, and care must be place in connecting its inputs in exactly the same sequence (top to bottom) as defined by Equation 4.33 or 4.36, and by providing the right amount of unit delays.
4. A MATLAB launch script was set up to load the data, the FIS, perform the simulation and plot the results. The dialogue between the launch script and the Simulink model is shown in detail in Figure 4.36 for the first-order model, and in Figure 4.37 for the second-order model.

Now we can conclude the tank exercise by comparing the performance of the first- and second-order models, as shown in Figure 4.38 and Table 4.8. It can be seen that the simpler first-order model produces a performance comparable to that of the second-order model, though having a

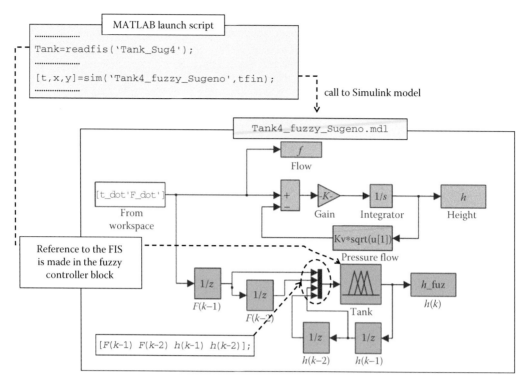

FIGURE 4.37 Communication between the MATLAB launch script and the Simulink model for the second-order Sugeno FIS. The Mux input sequence must follow the data organization scheme. The pertinent MATLAB/Simulink codes can be retrieved from the `ESA _ Matlab\Exercises\Chapter _ 4\ANFIS\ANFIS _ Tank` folder.

much simpler structure that makes its use much more attractive from the computational viewpoint. In fact, while the execution time for the second-order model was 0.285977 s, the first-order model took only 0.084738 s.

The fundamental lesson to be learned from this example is that with data-driven models an accurate calibration can more than make up for a limited model complexity. This largely holds for mechanistic models too, the general rule being that a simple well-calibrated model is almost always superior to a complex, poorly calibrated one.

4.5 FUZZY CLUSTERING: AGGREGATING DATA BY SIMILARITY

The term *Cluster Analysis* was first introduced by Tryon (1939) to indicate the grouping of observations according to some notion of similarity. Data in the same cluster are supposed to share a common trait to some extent. Cluster Analysis was originally applied to numerical taxonomy (Michener and Sokal, 1957; Sokal and Michener, 1958; Sokal, 1961; Sokal and Rohlf, 1962), and later was extensively applied to virtually every branch of science and technology, and particularly in the analysis of ecological data (Dillon and Goldstein, 1984; Pielou, 1984; Gotelli, 2001; Gotelli and Ellison, 2004; Legendre and Legendre, 2012). The objects considered in Cluster Analysis are observations described by a set of variables (features), in much the same way as the observations considered in principal component analysis, Section 3.5. Thus, the data will be represented by the same notation as a matrix $X \in \mathbb{R}^{N \times p}$, where each of the N rows represents one observation x, each possessing p features. The similarity measure is related to the relative 'distance' among data, and

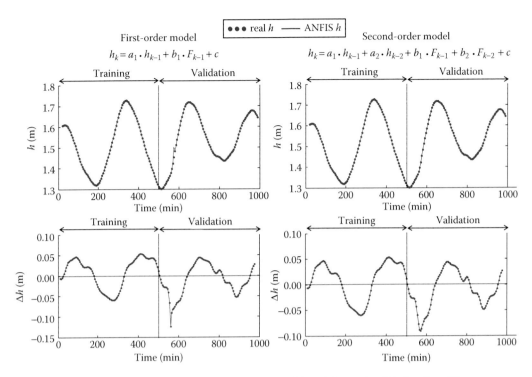

FIGURE 4.38 Performance of the ANFIS tank models with training and validation data. The upper row shows the original data and the output of the ANFIS models, while the lower row shows a magnified view of the modelling errors. The slight accuracy loss of the first-order model (notice the small spike around $t = 580$) is more than offset by its reduced complexity. These graphs were obtained with the MATLAB scripts Go _ Tank2 _ fuzzy.m and Go _ Tank4 _ fuzzy.m in the ESA _ Matlab\Exercises\Chapter _ 4\ ANFIS\ANFIS _ CSTR folder.

TABLE 4.8

Comparison of ANFIS Models for the Tank Exercise, on the Basis of the Data Structure, Network Complexity and Sum of Squared Errors (SSE)

First-Order Model	Second-Order Model
`[F(k-1)h(k-1) h(k)];`	`[F(k-1) F(k-2) h(k-1) h(k-2) h(k)];`
Number of nodes: 35	Number of nodes: 193
Number of linear parameters: 27	Number of linear parameters: 405
Number of nonlinear parameters: 24	Number of nonlinear parameters: 24
Total number of parameters: 51	Total number of parameters: 429
Number of training data pairs: 100	Number of training data pairs: 100
Number of checking data pairs: 96	Number of checking data pairs: 96
Number of fuzzy rules: 9	Number of fuzzy rules: 81
SSE = 0.2772	SSE = 0.2671
Execution time = 0.084738 s.	Execution time = 0.285977 s.

the many clustering algorithms differ in the way such distance is computed, according to the linkage criterion, the most popular of which, shown in Figure 4.39, are:

- *Average linkage*: considers the average of the distances between all pairs of data.
- *Single linkage* (nearest neighbour): aggregates data according to the minimum distance.
- *Complete linkage* (farthest neighbour): considers the maximum distance among data.

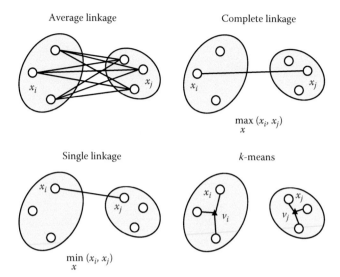

FIGURE 4.39 Grouping criteria based on the distance among points in a data set. The K-means differs from the other methods in that it considers distances from a non-member of the data set (the centroid v).

- *Centroid linkage* (K-means): unlike the previous methods, it minimizes the within-cluster distance between the data and a *centroid* (v). It is an iterative method in which each observation is assigned to the cluster with the nearest centroid. The number of clusters (c) is selected a priori and placed initially at random in the data set. The distance between each point and the centroids are then computed by assigning each point to the cluster with the nearest centroid. At each iteration the centroids are re-computed and the process ends when the centroids are stable.

The choice of the kind of linkage is highly discretionary and depends on the nature of the problem as well as on the pursuit of the analysis. Figure 4.40 shows how the single linkage or the complete linkage criteria perform with two data sets, one with an elongated shape, while the other has

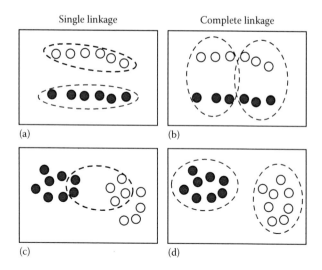

FIGURE 4.40 The single linkage criterion tends to produce elongated clusters, so it is well suited for the data in (a) but not for those in (c). Likewise, the complete linkage criterion forms compact spherical clusters, so it is appropriate for the data in (d) but not for those in (b).

a roughly spherical appearance. In the first instance, the single linkage is more appropriate (a), whereas the complete linkage yields a better separation (d) with the spherical data.

The main difference between the first three methods and the K-means is that the former are hierarchical methods in the sense that data are assigned into clusters with an increasing (agglomerative) or decreasing (divisive) criterion. The K-means method instead starts with an initial centroid guess and progressively associates data with clusters to minimize a function of the sum of squared distances. In the latter case, the number of clusters is arbitrarily assigned prior to the partition. All the algorithms discussed in the following are derived from the basic K-means iterative method.

From this brief introduction, it is clear that the definition of distance plays a crucial role in clustering. Among the many possibilities, those that will be used in what follows are:

1. *Euclidean distance*: squared sum of the differences between x_i and x_j

$$d_{ij}^2 = \left(x_i - x_j\right)^T \left(x_i - x_j\right) = \left\|x_i - x_j\right\|_2 \tag{4.37}$$

2. *Mahalanobis distance*: squared sum of the differences between x_i and x_j weighted by a norm-inducing matrix W

$$d_{ij,W}^2\left(x_i, x_j\right) = \left(x_i - x_j\right)^T W\left(x_i - x_j\right) = \left\|x_i - x_j\right\|_W \tag{4.38}$$

4.5.1 Deterministic *K*-Means

Let us start by considering a *hard* partition, meaning that if a point x_k is assigned to a cluster v_i $\left(\mu_{ij} = 1\right)$, its membership to all the other clusters must be zero. If we group all the membership in a matrix U with as many rows as clusters and as many columns as observations $\left(U \in \mathbb{R}^{c \times N}\right)$, this will consist of zeros and ones only. The K-means iterative method just described seeks to minimize the following partition criterion J_{hc}, where the subscript hc stands for hard cluster

$$J_{hc} = \min_{\mu_{ik}} \sum_{i=1}^{c} \sum_{k=1}^{N} d_{ik}^2 \quad \text{with} \quad d_{ik}^2 = \left\|x_k - v_i\right\|_2 \tag{4.39}$$

In Equation 4.39, c is the number of clusters and μ_{ik} is the (0/1) membership of the kth point x_k to the ith cluster C_i represented by its centroid v_i. Thus, the membership matrix will have the following aspect:

$$U = \begin{bmatrix} \mu_{11} & \mu_{12} & \cdots & \mu_{1N} \\ \vdots & \vdots & \vdots & \vdots \\ \mu_{c1} & \mu_{c2} & \cdots & \mu_{cN} \end{bmatrix} \in \mathbb{R}^{c \times N} \quad \text{with} \begin{cases} \mu_{ik} = 1 & \text{if } x_k \subset C_i \\ \mu_{ik} = 0 & \text{otherwise} \end{cases} \tag{4.40}$$

The K-means algorithm starts by choosing c centroids at random and repeatedly evaluates Equation 4.39 until two subsequent matrices differ in norm by less than a prescribed quantity $\varepsilon > 0$

$$\left\|U_t - U_{t-1}\right\| \leq \varepsilon \tag{4.41}$$

The limit of *hard* (or *crisp*) clustering is that it is unable to deal with *borderline* cases, where assigning a point to one cluster rather than another is problematical and a full membership $\left(\mu = 1\right)$ or a total exclusion $\left(\mu = 0\right)$ would be equally unjustified.

4.5.1.1 The Limits of Hard Clustering: The Butterfly Data

The so-called butterfly data are a well-known benchmark for clustering methods (Bezdek, 1981). The data, shown in Figure 4.41, are symmetrical about the central grey point and any hard partition with $c = 2$ would inconclusively assign it to either cluster. In fact, given the data coordinates

$$X = \begin{bmatrix} 0 & 2 & 4 & 1 & 2 & 3 & 2 & 2 & 2 & 1 & 2 & 3 & 0 & 2 & 4 \\ 0 & 0 & 0 & 1 & 1 & 1 & 2 & 3 & 4 & 5 & 5 & 5 & 6 & 6 & 6 \end{bmatrix}^T \qquad (4.42)$$

any of the two hard partitions U_1 and U_2

$$U_1 = \begin{bmatrix} 1 & 1 & 1 & 1 & 1 & 1 & 1 & 0 & 0 & 0 & 0 & 0 & 0 & 0 & 0 \\ 0 & 0 & 0 & 0 & 0 & 0 & 0 & 1 & 1 & 1 & 1 & 1 & 1 & 1 & 1 \end{bmatrix}$$

$$\uparrow$$
$$\text{central point} \qquad\qquad\qquad (4.43)$$
$$\downarrow$$

$$U_2 = \begin{bmatrix} 1 & 1 & 1 & 1 & 1 & 1 & 1 & 1 & 0 & 0 & 0 & 0 & 0 & 0 & 0 \\ 0 & 0 & 0 & 0 & 0 & 0 & 0 & 0 & 1 & 1 & 1 & 1 & 1 & 1 & 1 \end{bmatrix}$$

that differ by only the central point membership would prove equally unfair. In fact, the central point (8th column) is the one 'recalcitrant' element, which cannot be satisfactorily assigned to either cluster, meaning that its 'ambiguity' cannot be accommodated by deterministic clustering. We must conclude that this data set has no 'hard' satisfactory solution.

4.5.2 Fuzzy C-Means Clustering

To overcome the limitations of hard K-means clustering, Bezdek (1981) proposed its fuzzy counterpart (Fuzzy C-means or FCM for short), thus paving the way to a whole series of fuzzy clustering algorithms, which will be described in the following. Bezdek's idea was to 'fuzzify' the partition functional (4.39) by weighting the Euclidean distance (4.37) with the fuzzy membership, namely

$$J_{fc}(c,m) = \sum_{i=1}^{c} \sum_{k=1}^{N} \left(\mu_{ik}\right)^m d_{ik}^2 \quad \text{with } d_{ik}^2 = \left(x_k - v_i\right)^T \left(x_k - v_i\right), \qquad (4.44)$$

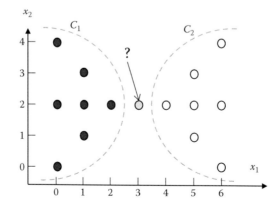

FIGURE 4.41 Graphical representation of the butterfly data, which draw their name by looking like the wings of a butterfly. Assigning the blue dots to one cluster and the white ones to another is an easy task, but what about the central grey dot? The dashed lines delimit the core of the clusters C_1 and C_2.

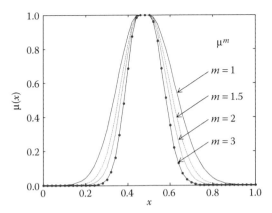

FIGURE 4.42 Effect of the fuzzy exponent m on the shape of a mf. As m increases the mf becomes narrower and the fuzziness decreases.

where v_i are the centroids and $m \in [1,\infty)$ is the *fuzzy exponent*. This coefficient plays a very important role in adjusting the fuzziness of the partition, in the sense that the larger is m, the less 'blurred' is the partition. To visualize its effect on a fuzzy mf, consider the Figure 4.42, in which a generic mf μ^m is plotted for increasing values of m. The shape of the original mf becomes narrower as m increases, making the distinction sharper. In the subsequent examples, a value $m = 2$ will be normally adopted, as with FCM this choice will suit most of the situations. However, with other clustering algorithms, we shall see that the choice of m has a major influence on the algorithm behaviour, and its choice may become critical.

In fuzzy clustering, the membership can smoothly vary between 0 (no membership) to 1 (full membership). Therefore each element can (partially) belong to more than one cluster, hence the ability to deal with ambiguous or borderline cases. However, the overall sum of the memberships must be 1 (complete assignment), and in a data set of N elements, the sum of the memberships to each cluster cannot be greater than N. Given these assumptions, the functional (4.44) has only a trivial solution unless we introduce the complete assignment constraint

$$\sum_{i=1}^{c} \mu_{ik} = 1 \quad k = 1,\ldots,N \tag{4.45}$$

Thus, the FCM problem can be reformulated as a constrained optimization problem by combining the fuzzy membership functional (4.44) with the constraint (4.45),

$$\min_{U,V}\left(J_{fc}(c,m)\right) = \sum_{i=1}^{c}\sum_{k=1}^{N}\left(\mu_{ik}\right)^m d_{ik}^2 \quad \text{subject to} \quad \sum_{i=1}^{c}\mu_{ik} = 1 \quad k = 1,\ldots,N \tag{4.46}$$

which may be turned into an unconstrained minimization problem via the Lagrange multiplier (λ),

$$\min_{(U,V)}\left(J_{fc}(c,m)\right) = \sum_{i=1}^{c}\sum_{k=1}^{N}\left(\mu_{ik}\right)^m d_{ik}^2 + \lambda\left(\sum_{i=1}^{c}\mu_{ik} - 1\right) \tag{4.47}$$

where the optimization involves both the membership matrix $\left(U \in \mathbb{R}^{c \times N}\right)$ and the centroid matrix $\left(V \in \mathbb{R}^{p \times c}\right)$ along the iterative scheme outlined in Box 4.6.

BOX 4.6   ITERATIVE STRUCTURE OF THE FCM ALGORITHM

Starting with a random $U^{(0)}$ repeat for $t = 1, 2, \ldots$.

1. Compute the centroids

$$v_i^{(t)} = \frac{\sum_{k=1}^{N} \left(\mu_{ik}^{(t-1)}\right)^m x_k}{\sum_{k=1}^{N} \left(\mu_{ik}^{(t-1)}\right)^m} \quad i = 1, \ldots, c$$

2. Compute the distances

$$d_{ik}^{2(t)} = \left\| x_k - v_i^{(t)} \right\|_2$$

3. Update the partition matrix

$$U^{(t)} : \mu_{ik}^{(t)} = \frac{1}{\sum_{\substack{j=1 \\ j \neq i}}^{c} \left(d_{ik}^{(t)} / d_{jk}^{(t)}\right)^{\frac{2}{m-1}}} \quad k = 1, \ldots, N$$

Until $\left\| U^{(t)} - U^{(t-1)} \right\| < \varepsilon$

The FCM algorithm (Bezdek, 1981; Bezdek and Pal, 1995) provides the optimal partition (U) and the centroids location (V) as

$$U : \mu_{ik} = \frac{1}{\sum_{\substack{j=1 \\ j \neq i}}^{c} \left(d_{ik} / d_{jk}\right)^{\frac{2}{m-1}}} \quad \begin{array}{l} k = 1, \ldots, N \\ i = 1, \ldots, c \end{array} \tag{4.48}$$

$$V : v_i = \frac{\sum_{k=1}^{N} \left(\mu_{ik}\right)^m x_k}{\sum_{k=1}^{N} \left(\mu_{ik}\right)^m} \quad i = 1, \ldots, c \tag{4.49}$$

where the distances are computed according to Equation 4.44. As in any K-means algorithm, the minimization of FCM functional (4.46) is solved iteratively as outlined in Box 4.6, and the two quantities $(\mu_{ik}$ and $v_i)$ are intertwined. Figure 4.43 shows a simple step-by-step example of computing the memberships and the centroids for a 2D data set. The procedure shown for the generic point x_k is to be repeated for the whole data set.

4.5.2.1   The Butterfly Data Revisited

Let us go back to the elusive butterfly data of Section 4.5.1.1 and apply the FCM algorithm with $m = 2$ to find a fuzzy cluster partition. Figure 4.44 shows the new partition with the central point correctly receiving a $\mu = 0.5$ membership to both clusters and the centroids being slightly displaced as a result of the 'push' exerted by the other cluster and the complete membership constraint of Equation 4.45.

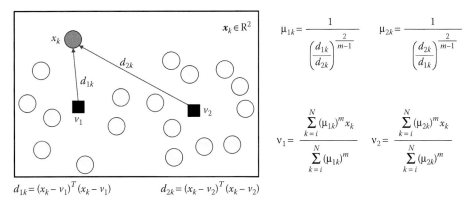

FIGURE 4.43   Practical computation of the FCM partitions with 2D data, showing how the distances and the memberships for the generic point x_k (blue dot) are computed with respect to the centroids v_1 and v_2, indicated as black squares.

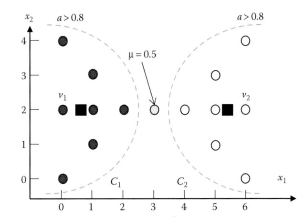

FIGURE 4.44   The butterfly data revisited with the FCM algorithm. Now the central point is (correctly) assigned a 0.5 membership to both clusters, acknowledging its borderline position. The two centroids (black squares) are slightly displaced inside each cluster because of the 'push' of the other cluster, and the dashed curves now have the meaning of α-cuts enclosing the portion of each cluster with membership greater than 0.8.

The centroids location is computed as

$$V = \begin{bmatrix} 0.85 & 5.14 \\ 2.00 & 2.00 \end{bmatrix}$$

$$\underset{v_1}{\uparrow} \quad \underset{v_2}{\uparrow}$$

(4.50)

whereas the new membership matrix is

$$U = \begin{bmatrix} 0.86 & 0.97 & 0.86 & 0.94 & 0.99 & 0.94 & 0.88 & 0.5 & 0.12 & 0.06 & 0.01 & 0.06 & 0.14 & 0.03 & 0.14 \\ 0.14 & 0.03 & 0.14 & 0.06 & 0.01 & 0.06 & 0.12 & 0.5 & 0.88 & 0.94 & 0.99 & 0.94 & 0.86 & 0.97 & 0.86 \end{bmatrix}$$

$$\uparrow$$
central point

(4.51)

where each column correctly sums to 1.

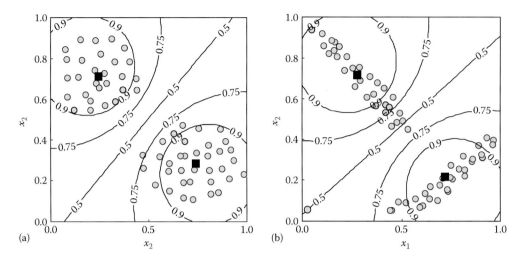

FIGURE 4.45 Suitable (a) and unsuitable (b) partitions produced by the FCM algorithm. In the first case (a), the data are distributed in roughly spherical patterns so that they are well explained by the FCM partition, whereas in (b) they follow an elongated pattern, which the Euclidean distance of FCM cannot accommodate. These graphs were obtained with the MATLAB script Go _ FCM.m in the ESA _ Matlab\Exercises\ Chapter _ 4\Fuzzy _ Clustering\FCM folder.

4.5.2.2   Limits of the FCM Clustering

The Euclidean distance used in Equation 4.44 tends to produce spherical clusters, which may not be appropriate if the data have an elongated shape. Figure 4.45 shows two examples of 2D data FCM clustering. In case (a), the data naturally have a roughly spherical distribution, which fits well with the FCM. In contrast, in case (b) the data are located in elongated patterns, which conflict with the propensity of the FCM to produce spherical clusters. The result is that improbable clusters are produced, with data clearly not associated with any cluster are assigned the 'no-man's-land' membership of 0.5. Notice also that the isolated point in the lower-left corner, which is undeservedly assigned a 0.5 membership as well, whereas its grade should be much lower, being alien to the overall data distribution. We shall come back to this oddity later, when we introduce the possibilistic clustering in Section 4.5.10.

4.5.3   Assessment of the Partition Efficiency

Before moving on to more flexible fuzzy clustering algorithms, let us pause to reconsider the impact of arbitrarily choosing the number of clusters (c). We have seen in Section 4.5.1 that K-means algorithms are not capable of determining the level of association of the data and therefore cannot suggest the best number of clusters, which has to the specified—arbitrarily for now—as an external parameter. But how can we be sure, apart from some naive visual check, that we have selected the 'best' number of clusters? Is there a way to assess the efficiency of the partition? To answer this question, let us remember that the purpose of clustering a data set is to 'bring order' in the data, in the sense that the information contained therein will hopefully come to the fore thanks to clustering. Now, we introduce some objective notion to quantify the improvement brought about by clustering in terms of data separation.

4.5.3.1   Partition Entropy

The first and foremost measure of partition efficiency is the *information entropy* (IE), which can be viewed as an indicator of how much order is brought into the data, or rather how much uncertainty is removed, by clustering. Derived from the second law of thermodynamics, where entropy is a measure of disorder, IE (Shannon, 1948) is largely used in communication to indicate the informative content of a coded message. In fuzzy sets, the normalized IE for a set of N data partitioned into c clusters is defined as (Bezdek, 1981)

$$H_n = -\frac{1}{1-(c/N)}\sum_{i=1}^{c}\sum_{k=1}^{N}\mu_{ik}\cdot\log(\mu_{ik})/N \qquad (4.52)$$

It measures the residual uncertainty after the data are partitioned into c clusters, so that the smaller it is the more 'certain' is the partition. H_n may vary between 0 ('hard' partition with no uncertainty) and a maximum of $\log(c)$ corresponding to all the memberships equal to $1/c$ (maximal uncertainty).

4.5.3.2 Separation Coefficient

This coefficient measures the clusters 'compactness' by comparing their 'radius', defined as the maximum fuzzy-weighted intra-cluster distance, with the distances among centroids,

$$r_i = \max_{1\leq k\leq N}\left(\mu_{ik}d_{ik}\right) \quad i=1,\dots,c \qquad (4.53)$$

The separation coefficient g is the maximum ratio between the sum of pairs of radii (r_i, r_j) and the inter-cluster distance d_{ij},

$$g = \max_{i,j\leq c}\left(\frac{r_i+r_j}{d_{ij}}\right) \qquad (4.54)$$

The notion of cluster radius and of separation coefficient are shown in Figure 4.46, together with the three possibilities of well-separated (a), tangent (b) and intersecting clusters (c), depending on the relative magnitude of g.

Figure 4.47 shows how the variations of the normalized entropy with respect to its maximum value $\log(c)$ can be used as an indicator of the partition efficiency. In fact, the relative partition entropy defined as

$$\frac{H_n}{H_{n_{max}}} = \frac{H_n}{\log(c)} \qquad (4.55)$$

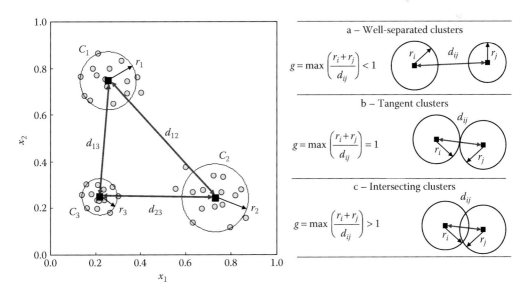

FIGURE 4.46 The separation coefficient g, comparing the clusters radii to their mutual distance (shown in the left plot), is a measure of the clusters compactness. The three possible situations (well-separated, tangent and intersecting clusters) are shown on the right.

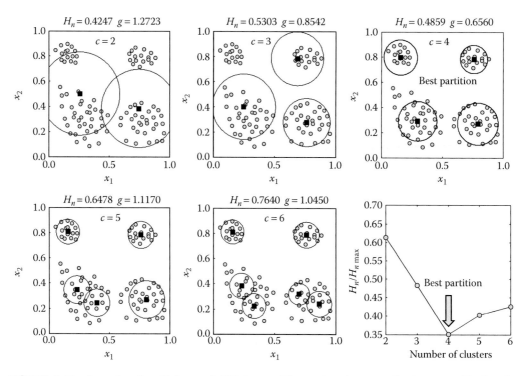

FIGURE 4.47 Assessing the efficiency of differing partitions using the normalized entropy. Plotting the ratio of the normalized entropy with respect to its maximum value shows that $c = 4$ (bold box) is the most efficient partition. The black squares indicate the location of the centroids.

shows where H_n stands with respect to its range. In this example, Equation 4.55 indicates that $c = 4$ is the most efficient partition.

4.5.4 Fuzzy Clusters as Antecedents

We have already seen that when a set of fuzzy rules involves many antecedents, each with several mfs, the number of rules may soon become overwhelming, as shown by Equation 4.11. Fuzzy clustering offers an alternative to this 'curse of dimensionality' because the degree of truth of a composite antecedent can be computed as the degree of membership to a cluster, so there will be as many rules as clusters, irrespective of the number of features of the data or the number of antecedent mfs. Figure 4.48 compares the two alternatives to obtain the overall degree of truth of the antecedent.

The conventional fuzzification procedure (a), where each of the n input variables is fuzzified with q mfs, results in the number of possible rules being q^n, which may quickly become explosive, even for moderate values of q and n. Figure 4.48b shows instead the clustered antecedents approach, where the number of rules depends only on the number of clusters. The rule syntax is obviously different, and surely more compact in case (b), as the following example shows. Let us consider two Sugeno fuzzy rules, one with conventional antecedents and the other with cluster antecedents:

conventional antecedents

$$R_i : \text{if } \left(x_1 \text{ is } X_1 \right) \text{ and } \left(x_2 \text{ is } X_2 \right) \dots \text{ and} \left(x_n \text{ is } X_n \right) \text{ then } y_i = f_i \left(x_1, x_2, \dots, x_n \right) \quad (4.56)$$

cluster antecedents

$$R_i : \text{if } \left(x_1, x_2, \dots, x_n \right) \subset C_i \text{ then } y_i = f_i \left(x_1, x_2, \dots, x_n \right) \quad (4.57)$$

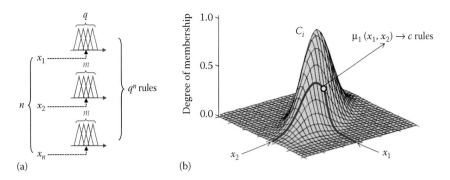

FIGURE 4.48 The antecedent degree of truth can be obtained in two ways. (a) shows the conventional fuzzification procedure, which may lead to an explosive proliferation of rules. (b) shows the clustered antecedents approach, where the number of rules depends only on the number of clusters.

where in the second rule the notation $(x \subset C_i)$ denotes the membership of the point x to the cluster C_i. In the conventional case, each antecedent variable is individually fuzzified using a bank of q mfs and then *and*-ed with all the others to produce the overall antecedent degree of truth. So there are q mfs for each of the n input variables, and consequently $m = q^n$ rules (see Figure 4.48a). Conversely, in the clustered antecedent approach, all the n input variables contribute *simultaneously* to determine the degree of truth with respect to each cluster. As a result, the number of rules is equal to the number of clusters, *irrespective of the number of inputs*. This new approach, though, has some repercussions on the way the rules are written and requires a previous 'training' to determine the clusters, which are completely specified by their centroids and the fuzzy exponent m. After that, any new point z can be classified by again applying Equation 4.48. However, in this case there is no iteration involved, because the centroids are already specified and the memberships of the new point z can be directly computed as

$$\mu_i(z) = \frac{1}{\sum_{\substack{j=1 \\ j \neq i}}^{c} \left(d_{iz}/d_{jz}\right)^{\frac{2}{m-1}}} \quad \Rightarrow \quad z \to \left[\mu_1(z),\ldots,\mu_c(z)\right] \tag{4.58}$$

The new point z receives its degree of membership with respect to the c clusters of the partition irrespective of its dimensions p. If $p \gg c$ this results in a considerable dimensional saving. Figure 4.49 shows the location of the new point z (triangle) and the computation of the distances producing the degree of membership to each cluster according to Equation 4.58.

Equation 4.58 can be used to classify any new entry (z) with respect to a set of pre-existing clusters $(C_1,\ldots,C_c)$. Following the example of Figure 4.49 the following FIS can be set up:

$$\text{if } (z \subset C_1) \text{ then } y_1 = f_1(z)$$

$$\text{else}$$

$$\text{if } (z \subset C_2) \text{ then } y_2 = f_2(z)$$

$$\text{else} \tag{4.59}$$

$$\text{if } (z \subset C_3) \text{ then } y_3 = f_3(z)$$

$$\text{else}$$

$$\text{if } (z \subset C_4) \text{ then } y_4 = f_4(z)$$

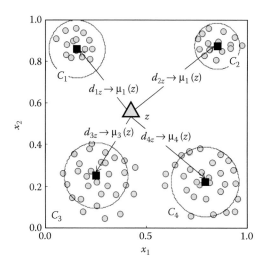

FIGURE 4.49 Classification of a new point z (triangle) with respect to the previously determined clusters $(C_1,\ldots,C_4)$ through Equation 4.58. The black squares indicate the position of the centroids.

where the notation $(z \subset C_i)$ produces $\mu_i(z)$. The actual output y is then obtained by defuzzification, according to Equation 4.16,

$$y = \frac{\sum_{i=1}^{4} y_i \cdot \mu_i(z)}{\sum_{i=1}^{4} \mu_i(z)} \tag{4.60}$$

It is important to emphasize the computational advantage of Equation 4.59, whose dimension is equal to the number of clusters (c) independently of the dimension of $z \in \mathbb{R}^p$. This approach can be profitably used in fuzzy modelling where the generic Sugeno modelling rule (4.25) could simply be written as

$$R_i : \text{if } \left(y_{k-1},\ldots,y_{k-n_y},u_{k-\delta},\ldots,u_{k-\delta-n_u} \right) \subset C_i \text{ then } y_k^{(i)} = f_i \left(y_{k-1},\ldots,y_{k-n_y},u_{k-\delta},\ldots,u_{k-\delta-n_u} \right) \tag{4.61}$$
$$i = 1,\ldots,c$$

The cluster-Sugeno model can then be described by the structure of Figure 4.50. The degrees of membership produced by the previously trained clusters $(C_1,\ldots,C_c)$ activate the consequents of Equation 4.61 which can be viewed as local models, each best describing the system behaviour when its variables are in the neighbourhood of the pertinent cluster. After defuzzification the past outputs are fed back into the model, together with the vector of past inputs, according to the model definition. This approach is amply described in Babuska (1998) while public domain MATLAB toolboxes with interesting exercises by the same author can be downloaded from the following web site http://www.dcsc.tudelft.nl/~rbabuska/.

4.5.5 Fuzzy Clustering as a Diagnostic Tool

Diagnosis is another possible application of fuzzy clustering. Suppose that a training data set has produced a reference partition, in terms of centroids location, which together with the value of the fuzzy exponent m form a *knowledge base* $\text{KB} = \{(v_1,\ldots,v_c),m\}$. Any new data can be classified with respect to that KB using Equation 4.58. If the clusters can be associated with any significant system condition, then the degree of membership to each cluster can be used as a diagnostic tool. Though the number of clusters c may be predetermined by other means in view of a specific diagnostic

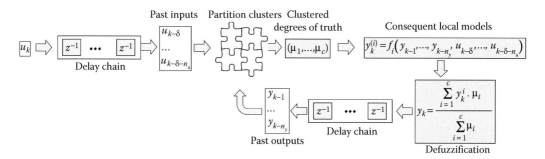

FIGURE 4.50 General structure of a Sugeno dynamical model with clustered antecedents. The delay chains are required for storing the past inputs and outputs.

purpose, nonetheless it is still a good idea to check the efficiency of the partition with the entropy defined by Equation 4.55.

4.5.5.1 Fuzzy Cluster Diagnosis of an Anaerobic Digester

The use of fuzzy cluster diagnostics can be illustrated by the following example (Marsili-Libelli and Beni, 1996; Marsili-Libelli and Müller, 1996; Müller et al., 1997). The state of an anaerobic digester, whose microbiology is briefly described in Section 7.2.11, can be inferred by the characteristics of the off-gas, which is a mixture of about 60% methane (CH_4) and 40% carbon dioxide (CO_2). An additional small part of hydrogen (H_2) is normally present, but a concentration exceeding a threshold of about 200–300 ppm, depending on the digester, may indicate a malfunction, for example an organic over-load (Marsili-Libelli and Beni, 1996). To obtain sufficient data relative to normal and pathological process operations, several organic overload shocks were administered to a pilot anaerobic digester and it was found that such data could be grouped into four typical behaviours in terms of hydrogen and gas flow: normal, overload, toxic, inhibition. Clustering these data with $c = 4$ and $m = 2$ produced the set of centroids indicated by the black squares in Figure 4.51. The typical response of the digester to an organic overload shock is indicated by the blue arrows in Figure 4.51.

4.5.6 ADAPTIVE FCM CLUSTERING

In the diagnostic use of FCM clustering the $KB = \{(v_1,\ldots,v_c), m\}$ is the yardstick against which each new information item is assessed by computing its membership via Equation 4.58. The validity of this procedure hinges on the assumption that the KB is constant in time, which seldom occurs in practice, where a shift in the reference conditions against which new data are compared is often observed. This leads to the conclusion that in many instances some form of adaptation of the KB must be introduced to keep the diagnostics up to date with the changing conditions. It is therefore desirable to include some trend-following capability and make the clustering procedure adaptive.

Now it remains to decide which data are allowed to update the KB by changing the centroid position, and consequently the diagnostic capability of the algorithm. This poses the problem of discriminating between 'good' and 'bad' data, depending on their information content with respect to the existing KB. Of the many possibilities (Pal and Bezdek, 1994), we describe here one method presented in Marsili-Libelli (1998), where the updating value of each new entry is assessed on the basis of the normalized partition entropy of Equation 4.52. The new point joins the cluster and alters the centroid position only if this results in a decrease (or a very moderate increase $\varepsilon > 0$) of the incremental partition entropy ΔH_n, namely

$$\Delta H_n = \frac{H_n(U_{\text{new}}) - H_n(U_{\text{old}})}{H_n(U_{\text{old}})} \leq \varepsilon \qquad (4.62)$$

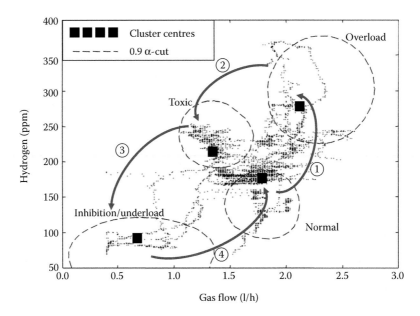

FIGURE 4.51 Mapping of gas flow and hydrogen concentration data obtained by administering an organic overload shock to a pilot digester. Four clusters with a clear process meaning were identified from the data and labelled to serve as a diagnostic tool. Typically, the response of the digester to the shock follows the sequel indicated by the arrows. (Reproduced with permission from Marsili-Libelli, S. and Müller, A., *Pattern Recogn. Lett.*, 17, 651–659, 1996.)

where U_{old} and U_{new} are the membership matrices before and after the inclusion of the new point. In this way, only the data which improve the order in the partition contribute to the KB updating. In practice, the updating algorithm can be based on a moving window of length N, equal to the dimension of the initial training set. At each new entry z, the operations shown in Box 4.7 are performed, with the oldest data being discarded and replaced by the new 'good' data, if they qualify as such. Though updating is certainly desirable when $\Delta H_n < 0$, the reason for allowing some small positive increase in ΔH_n can be justified with the need to follow the data evolution, even if this does not immediately result in a crisper partition.

BOX 4.7 ENTROPY-BASED UPDATING OF CENTROIDS IN FCM CLUSTERING

1. Classify z via Equation 4.58 $z \rightarrow \mu_i(z)$.
2. Add the new membership vector $\mu_i(z)$ as the rightmost (newest) column of U and discard the leftmost (oldest) column, corresponding to the first entry of the training set. There are now two matrices, representing the old partition (U_{old}) and the updated one (U_{new}).
3. Compute the partition entropy of both matrices via Equation 4.52 and evaluate the incremental entropy ΔH_n

$$\Delta H_n = \frac{H_n(U_{new}) - H_n(U_{old})}{H_n(U_{old})}$$

4. If $\Delta H_n < \varepsilon$ (with $\varepsilon > 0$ small), then allow z to update the partition by re-computing the centroids.

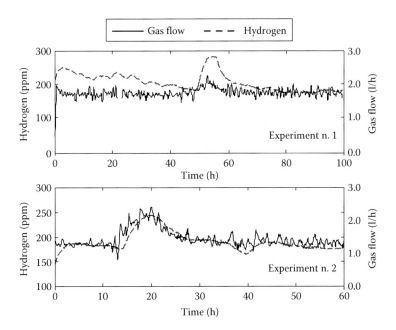

FIGURE 4.52 Two additional shock load experiments demonstrate the adaptive diagnostic method. Experiment 1 was used for training the initial KB, whereas experiment 2 was used to demonstrate the adaptation based on partition entropy.

The adaptive diagnostic just described is now applied to the anaerobic digester shock experiments already described in Section 4.5.5.1 and illustrated in Figure 4.51. We now consider two new shock experiments from the study by Müller et al. (1997), where the differing rates of hydrogen production relative to gas flow makes the diagnostics difficult, and yet the two experiments should be recognized as belonging to the same class of organic shocks. Of the two experiments, shown in Figure 4.52 the first was used for training and the second for updating the online detection.

In Figure 4.53, the experimental data are plotted in the detection variables space, together with the shock trend lines and detection results. The evolution of the test shock experiment 2 is depicted in the cluster space, showing that the centroids, trained with experiment 1, are 'pulled' towards this new trajectory by the adaptation mechanism.

Because the trend line in experiment 2 has a lower slope than in experiment 1, some adaptive capability is needed to adjust the trained clusters to the new situation. Adopting the updating mechanism of Equation 4.62 with $\varepsilon = 0.05$ only a fraction of the data in experiment 2 influences the clusters and produces an update of the centroids. Figure 4.54 shows the threshold memberships μ_{max} and μ_{min} relative to cluster 1 (overload) and cluster 2 (normal). The selection of the maximum permissible update rate ε is highly application-dependent and should be tuned on a case-by-case basis. A large ε would produce an excessive updating, with the danger of destroying the initial KB, whereas a too small value would hamper the adaptation, thus impairing the learning capability of the algorithm.

4.5.7 AUTOMATIC GENERATION OF MFS

The KB obtained by clustering a training data set can be used to generate mfs following the approach described in Section 4.5.4, which resulted in Equation 4.58. Given a scalar variable $x_{min} \leq x \leq x_{max}$ for which $KB = \{(v_1,\ldots,v_c), m\}$ has been trained, let us define an evenly spaced vector of values as $x = [\texttt{xmin}:\texttt{dx}:\texttt{xmax}]$ spanning the interval of interest (x_{min}, x_{max}) with resolution dx. This vector of values can be classified through Equation 4.58 with respect to the previous KB.

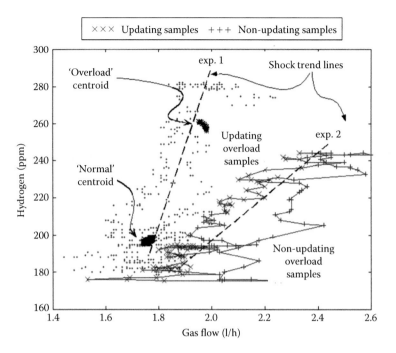

FIGURE 4.53 Diagnostics of the two shock load experiments shown in Figure 4.52, with the dots indicating the training data from experiment 1. The trend lines indicate that the two experiments are significantly different, and yet the adaptation mechanism 'pulls' the centroids towards the new trend, being influenced by the updating new data (x) only. (Reproduced with permission from Mueller, A. et al., *Water Res.*, 31, 3157–3167, 1997.)

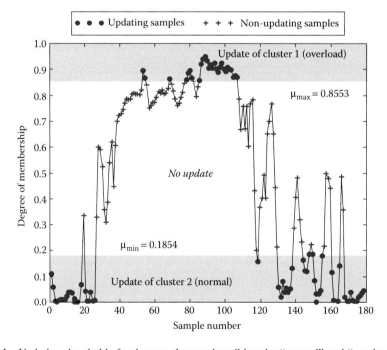

FIGURE 4.54 Updating thresholds for the two clusters describing the "normal" and "overload" conditions of the experiments of Figure 4.52. The dots in the grey bands indicate the samples that contribute to the centroid updating.

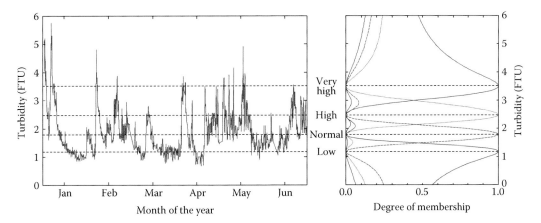

FIGURE 4.55 Building a KB of turbidity data, recorded at the potabilization plant of the city of Florence (Italy) during the first half of 2010, selecting $c = 4$ and $m = 2$. The mfs produced by the FCM are shown on the right. These graphs were obtained with the MATLAB script `Automatic_mfs_tuning.m` in the `ESA_Matlab\Exercises\Chapter_4\Membership_Estimation` folder.

These memberships can be visualized as a one-dimensional plot and can be used for diagnostics, as in the previous section, or they can represent the shape of antecedent mfs in a conventional Mamdani or Sugeno FIS.

As an example, let us consider a six-month record drawn from the potabilization plant of the City of Florence (Italy). It shows the output turbidity, expressed as formazine turbidity units (FTU), after the flocculation stage and before the sand filters. The water turbidity at this point in the potabilization process is critical because the sand filters yield an optimal performance only for a narrow range of turbidity, normally around 2 FTU. Whenever the turbidity departs from this range the flocculant dosing should be adjusted to bring the turbidity back at the desired value. Thus, turbidity diagnostics is the first step towards feedback control of the clarification–flocculation process. Figure 4.55 shows the KB obtained by clustering the turbidity data recorded during the first half of 2010 with a 30 min sampling interval. Clustering this record with $c = 4$ and $m = 2$ yields the following centroids $V = [1.1787 \ 2.4825 \ 1.7889 \ 3.5078]$ expressed as FTU. The mfs in the right graph were obtained by classifying a vector of turbidity values between 0 and 6 (FTU). It can be observed that each mf exhibits some 'humps' in the lower part (left part, considering the CW rotation). This is caused by the total membership constraint (4.45) embedded in the FCM algorithm. For the same reason, values of x much outside the range of interest are classified with the common value $\mu = 1/c$, which is far from being realistic.

Both these undesirable features, and the fact that the numerical mfs of Figure 4.55 cannot be transferred to any FIS where analytical mfs are required, can be removed by approximating them with one of the mathematically defined mfs in the MATLAB Fuzzy Toolbox library. We are again confronted with a parameter calibration problem, and while Box 4.8 outlines the general procedure for the mfs approximation, Box 4.9 shows the practical implementation for the calibration of a single mf. These operations must be repeated for all the FCM-generated numerical mfs, and applied to the data-generated mfs of Figure 4.55 to produce the approximated analytical mfs shown in Figure 4.56.

4.5.8 Gustafson–Kessel Fuzzy Clustering

We have seen in Section 4.5.2.2 that the main limit of the FCM clustering algorithm is its metric. In fact using the Euclidean norm forces the clusters to be spherical, regardless of the data shape. To overcome this limitation, the Gustafson–Kessel (GK) fuzzy clustering algorithm (Gustafson and

---

BOX 4.8   PROCEDURE FOR FITTING AN ANALYTICAL MF TO THE FCM-GENERATED CLUSTER DATA

1. *Select the mf type.* A suggestion is to use `gauss2mf` for the central memberships, `zmf` for the leftmost mf, and `smf` for the rightmost one. Once the mf type has been selected a *function handle* can be defined as `model = @gausmf` (a fancy name that has nothing to do with the MATLAB `gauss2mf`) meaning that in the optimization the objective function will be the `gauss2mf` mf. This is a handy way of specifying a function within another function, as illustrated in the next Box 4.9.
2. *Guess a set of initial parameters,* depending of the mf type. In the example of Box 4.9, the initial parameters of the `gauss2mf` are guessed from the mean and variance of the `xdata`. They are then grouped in the vector `start _ point`, to be later passed to the optimization.
3. *Define the body of the objective function.*
4. *Launch the optimization* with the built-in MATLAB command `estimates = fminsearch(model, start _ point, options)` where model points to the handle `@gausmf` previously defined.

BOX 4.9   MATLAB FUNCTION TO FIT AN ANALYTICAL MF TO THE FCM-GENERATED CLUSTER DATA

The script in the box refers to the middle mfs (`gauss2mf`), but the full MATLAB function (`fit _ mf.m` in the `ESA _ Matlab\ Exercises\Chapter _ 4\Membership _ Estimation` folder) fits also the first and last mfs using the `zmf` and `smf` fuzzy mfs.

```
function [estimates, model] = fit_mf(xdata,ydata,ord,c)
% Initial parameters estimates
    c1=0.9*mean(xdata);
    c2=1.1*mean(xdata);
    sig1=0.5*var(xdata);
    sig2=0.7*var(xdata);
% Form the initial parameters vector
start_point = [sig1 c1 sig2 c2];
% Define the function handle
model = @gausmf;
%-------------------------------------------------------------------
% Actual parameters optimization
options=optimset('MaxIter',1e8,'MaxFunEvals',1e8,
...'TolFun',1e-6,'TolX',1e-6);
estimates = fminsearch(model,start_point,options);
%-------------------------------------------------------------------
    function [sse, FittedCurve] = gausmf(par)
        % Objective function for parameter fitting
        % Fit the data to the selected analytical mf (gauss2mf)
        FittedCurve = gauss2mf(xdata,par);
        % Difference between numerical and analytical mfs
        ErrorVector = FittedCurve - ydata;
        sse = sum(ErrorVector.^2); % Sum of squared errors
```

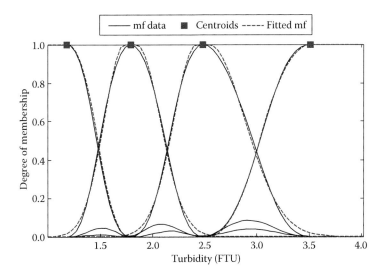

FIGURE 4.56 Fitting analytical mfs from the MATLAB Fuzzy Toolbox to the FCM-generated numerical membership data with the procedure outlined in Box 4.8. The leftmost mf was fitted to a zmf, the rightmost to an smf, while the central ones were adapted to a gauss2mf. Notice that the intermediate 'humps' of the numerical mfs (thin black lines) have disappeared, together with the asymptotic behaviour towards the $1/c$ common value. This graph was obtained with the MATLAB script Automatic_mfs_tuning.m in the ESA_Matlab\Exercises\Chapter_4\Membership_Estimation folder.

Kessel, 1979) used the Mahalanobis distance defined by Equation 4.38, which is now rewritten by introducing the norm-inducing matrices A_i,

$$d_{ik,A_i}^2 = \left(x_k - v_i \right)^T A_i \left(x_k - v_i \right) \quad \text{with } A_i \in \mathbb{R}^{p \times p} \tag{4.63}$$

The GK fuzzy partition functional becomes

$$J_{gk}(c, m, A_i) = \sum_{i=1}^{c} \sum_{k=1}^{N} \left(\mu_{ik} \right)^m d_{ik,A_i}^2 \tag{4.64}$$

Now, the optimization involves not only the membership matrix $\left(U \in \mathbb{R}^{c \times N} \right)$ and the centroid matrix $\left(V \in \mathbb{R}^{c \times p} \right)$, but also the matrices A_i, one for each cluster. It was suggested (Gustafson and Kessel, 1979; Babuska, 1998) that a feasible minimum of Equation 4.64 can be obtained by constraining the determinant of A_i to be positive,

$$\det \left(A_i \right) = \rho_i > 0. \tag{4.65}$$

Without specific information about the possible values of cluster volume, it can be assumed that $\rho = 1$. Minimizing Equation 4.64 with the constraint (4.65) yields the following expression for A_i (Babuska, 1998; Abonyi, 2003):

$$A_i = \left[\rho_i \det(F_i) \right]^{1/p} F_i^{-1} \tag{4.66}$$

where $F_i \in \mathbb{R}^{p \times p}$ is the fuzzy covariance matrix defined as

$$F_i = \frac{\sum_{k=1}^{N} \left(\mu_{ik} \right)^m \left(x_k - v_i \right)^T \left(x_k - v_i \right)}{\sum_{k=1}^{N} \left(\mu_{ik} \right)^m} \tag{4.67}$$

and the centroids $\left(v_i \in \mathbb{R}^p \mid i = 1, \ldots, c \right)$ are computed as

$$v_i = \frac{\sum_{k=1}^{N} (\mu_{ik})^m x_k}{\sum_{k=1}^{N} (\mu_{ik})^m} \quad i = 1,\ldots,c. \tag{4.68}$$

The point-to-centroid distance (4.63) can be rewritten by substituting Equation 4.66 for the norm-inducing matrix A_i as

$$d_{ik,F_i}^2 = (x_k - v_i)^T \left[\rho_i \det(F_i) \right]^{1/p} F_i^{-1} (x_k - v_i) \tag{4.69}$$

so that the mfs become

$$\mu_{ik} = \frac{1}{\sum_{\substack{j=1 \\ j \neq i}}^{c} \left(d_{ik,F_i} / d_{jk,F_i} \right)^{\frac{2}{m-1}}} \quad \begin{matrix} i = 1,\ldots,c \\ k = 1,\ldots,N \end{matrix} \tag{4.70}$$

where d_{ik,F_i} and d_{jk,F_i} are computed via Equation 4.69. Though the memberships (4.70) have the same formal expression of their FCM counterpart, the distances are computed by taking into account the fuzzy covariance matrix (4.67). The GK algorithm can adapt the shape of the cluster to the data by including the fuzzy covariance matrix in the distance computation (4.69), but this involves a greater computational complexity because the inverse and the determinant of the fuzzy covariance matrices F_i must be computed at each iteration, as shown in Box 4.10.

BOX 4.10   ITERATIVE STRUCTURE OF THE GK ALGORITHM

Starting with a random $U^{(0)}$ repeat for $t = 1, 2, \ldots$.

1. Compute the centroids

$$v_i^{(t)} = \frac{\sum_{k=1}^{N} \left(\mu_{ik}^{(t-1)} \right)^m x_k}{\sum_{k=1}^{N} \left(\mu_{ik}^{(t-1)} \right)^m} \quad i = 1,\ldots,c$$

2. Compute the fuzzy covariance matrix

$$F_i^{(t)} = \frac{\sum_{k=1}^{N} \left(\mu_{ik}^{(t-1)} \right)^m \left(x_k - v_i^{(t)} \right)^T \left(x_k - v_i^{(t)} \right)}{\sum_{k=1}^{N} \left(\mu_{ik}^{(t-1)} \right)^m}$$

3. Compute the distances

$$d_{ik,F_i}^{(t)2} = \left(x_k - v_i^{(t)} \right)^T \left[\rho_i \det \left(F_i^{(t)} \right) \right]^{1/p} \left(F_i^{(t)} \right)^{-1} \left(x_k - v_i^{(t)} \right)$$

4. Update the partition matrix

$$U^{(t)} : \mu_{ik}^{(t)} = \frac{1}{\sum_{\substack{j=1 \\ j \neq i}}^{c} \left(d_{ik,F_i}^{(t)} / d_{jk,F_i}^{(t)} \right)^{\frac{2}{m-1}}} \quad k = 1,\ldots,N$$

Until $\left\| U^{(t)} - U^{(t-1)} \right\| < \varepsilon$

4.5.8.1 Adaptive Volume GK Algorithm

Though the introduction of the fuzzy covariance into the norm-inducing matrix considerably improves the flexibility of the algorithm, the fixed volume assumption (4.65) does not fully exploit the potential of the adaptive norm. For this reason, several improvements to the basic GK algorithm were proposed (Krishnapuram and Kim, 1999; Babuska et al., 2002; Kaymak and Setnes, 2002) in which the cluster volumes ρ_i were estimated from the data. In particular, Krishnapuram and Kim (1999) proposed to estimate the cluster volume according to

$$\rho_i = \left[\frac{\min_j(\lambda_{ij})}{\det(\boldsymbol{F}_i)^{1/p}} \right]^p \tag{4.71}$$

where $\left(\lambda_{ij}, \ j = 1, \ldots, p \right)$ are the p eigenvalues of the fuzzy covariance matrix $\boldsymbol{F}_i$. The iterative scheme of the adaptive volume algorithm is shown in Box 4.11.

To demonstrate the better adaptability of the GK algorithm to the shape of the data and the additional improvement brought about by the volume adaptation, Figure 4.57 compares the two GK

BOX 4.11   ITERATIVE STRUCTURE OF THE GK ALGORITHM WITH ADAPTIVE CLUSTER VOLUMES

Starting with a random $\boldsymbol{U}^{(0)}$ repeat for $t = 1, 2, \ldots$.

1. Compute the centroids

$$\boldsymbol{v}_i^{(t)} = \frac{\sum_{k=1}^{N} \left(\mu_{ik}^{(t-1)} \right)^m \boldsymbol{x}_k}{\sum_{k=1}^{N} \left(\mu_{ik}^{(t-1)} \right)^m} \quad i = 1, \ldots, c$$

2. Compute the fuzzy covariance matrix

$$\boldsymbol{F}_i^{(t)} = \frac{\sum_{k=1}^{N} \left(\mu_{ik}^{(t-1)} \right)^m \left(\boldsymbol{x}_k - \boldsymbol{v}_i^{(t)} \right)^T \left(\boldsymbol{x}_k - \boldsymbol{v}_i^{(t)} \right)}{\sum_{k=1}^{N} \left(\mu_{ik}^{(t-1)} \right)^m}$$

3. Estimate the cluster volumes

$$\rho_i^{(t)} = \left[\frac{\min_k \left(\lambda_{ik}^{(t)} \right)}{\det \left(\boldsymbol{F}_i^{(t)} \right)^{1/p}} \right]^p$$

4. Compute the distances

$$d_{ik,F_i}^{(t)2} = \left(\boldsymbol{x}_k - \boldsymbol{v}_i^{(t)} \right)^T \left[\rho_i^{(t)} \det \left(\boldsymbol{F}_i^{(t)} \right) \right]^{1/p} \left(\boldsymbol{F}_i^{(t)} \right)^{-1} \left(\boldsymbol{x}_k - \boldsymbol{v}_i^{(t)} \right)$$

5. Update the partition matrix

$$\boldsymbol{U}^{(t)} : \mu_{ik}^{(t)} = \frac{1}{\sum_{\substack{j=1 \\ j \neq i}}^{c} \left(d_{ik,F_i}^{(t)} / d_{jk,F_i}^{(t)} \right)^{\frac{2}{m-1}}} \quad k = 1, \ldots, N$$

Until $\left\| \boldsymbol{U}^{(t)} - \boldsymbol{U}^{(t-1)} \right\| < \varepsilon$

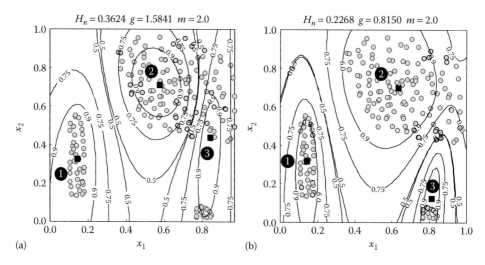

FIGURE 4.57 Comparison of the two versions of the GK clustering algorithm. In (a) the constant volume constraint produces intersecting clusters ($g > 1$) with the rightmost centroid (3) straddling two data clouds. In contrast, the adaptive volume version (b) produces a much more definite separation ($g < 1$), which is reflected by the lower H_n value, and a much more reasonable centroid location. These graphs were obtained with the MATLAB script `Go _ GK _ Vol.m` in the `ESA _ Matlab\Exercises\Chapter _ 4\Fuzzy _ Clustering\GK` folder.

versions with a data set in which the nature of the data would result in an awkward FCM clustering, given their non-spherical shape. Although the basic GK algorithm (a) performs better than the FCM, still the centroid n. 3 is placed in an unrealistic position, straddling part of the data in clusters 2 and 3. Conversely, the adaptive volume version (b) keeps all the data of cluster 2 together and places the third centroid where it is expected to be, near the small clump of data in the lower right corner. The better discrimination of the adaptive version is confirmed by the normalized entropy value, which is significantly lower, and of the separation coefficient, which indicates that the clusters are now well separated $(g < 1)$, whereas in the constant volume version they were intersecting.

4.5.9 Fuzzy Maximum Likelihood Estimates

Further progress in adapting the cluster shape to the data is provided by the fuzzy maximum likelihood estimates (FMLE) clustering, which uses a distance norm based on the fuzzy maximum likelihood estimates

$$d_{ik,\Phi_i} = \frac{\left(\det \Phi_i\right)^{\frac{1}{2}}}{\left(1/N\right)\sum_{k=1}^{N}\mu_{ik}} \exp\left(\frac{1}{2}\left(x_k - v_i\right)^T \Phi_i^{-1}\left(x_k - v_i\right)\right) \tag{4.72}$$

This differs from the GK distance (4.69) in two ways: the distance is weighted exponentially, and the matrix Φ_i

$$\Phi_i = \frac{\sum_{k=1}^{N}\mu_{ik}\left(x_k - v_i\right)^T\left(x_k - v_i\right)}{\sum_{k=1}^{N}\mu_{ik}} \tag{4.73}$$

is similar to $\boldsymbol{F}_i$ of Equation 4.67, save for the fact that the membership μ_{ik} is not raised to the exponent m. The other major difference with the GK algorithm is the lack of any constraint in the cluster volume, given the 'natural' ability of this algorithm to conform to the data through the distance computation (4.72). The iterative scheme, outlined in Box 4.12, is the same as that of FCM and GK, but the exponential distance makes the convergence more problematical, so that FMLE requires a careful initialization, possibly provided by a previous GK partition. Figure 4.58 compares the performance of the FMLE clustering algorithm to the adaptive volume GK over the same data set. It can be seen that FMLE achieved an even greater separation, delineates the cluster boundary better, and is less sensitive to the value of the fuzzy exponent m.

4.5.10   POSSIBILISTIC C-MEANS CLUSTERING

We now examine a clustering method that strives to eliminate, or at least relax, the total membership constraint (4.45) embedded in all of the previous methods. This new algorithm, termed Possibilistic C-means (PCM), follows the same approach of the previous algorithms, in the

---

BOX 4.12   ITERATIVE STRUCTURE OF THE FMLE ALGORITHM

Starting with a random $\boldsymbol{U}^{(0)}$ repeat for $t = 1, 2, \dots$.

1. Compute the centroids

$$v_i^{(t)} = \frac{\sum_{k=1}^{N} \left(\mu_{ik}^{(t-1)}\right)^m x_k}{\sum_{k=1}^{N} \left(\mu_{ik}^{(t-1)}\right)^m} \quad i = 1, \dots, c$$

2. Compute the matrices

$$\Phi_i^{(t)} = \frac{\sum_{k=1}^{N} \mu_{ik}^{(t-1)} \left(x_k - v_i^{(t)}\right)^T \left(x_k - v_i^{(t)}\right)}{\sum_{k=1}^{N} \mu_{ik}^{(t-1)}}$$

3. Compute the distances

$$d_{ik,\Phi_i}^{(t)} = \frac{\left(\det \Phi_i^{(t)}\right)^{\frac{1}{2}}}{(1/N)\sum_{k=1}^{N} \mu_{ik}^{(t-1)}} \exp\left(\frac{1}{2}\left(x_k - v_i^{(t)}\right)^T \left(\Phi_i^{(t)}\right)^{-1} \left(x_k - v_i^{(t)}\right)\right)$$

4. Update the partition matrix

$$\boldsymbol{U}^{(t)} : \mu_{ik}^{(t)} = \frac{1}{\sum_{\substack{j=1 \\ j \neq i}}^{c} \left(d_{ik,\Phi_i}^{(t)} \big/ d_{jk,\Phi_i}^{(t)}\right)^{\frac{2}{m-1}}} \quad k = 1, \dots, N$$

Until $\left\| \boldsymbol{U}^{(t)} - \boldsymbol{U}^{(t-1)} \right\| < \varepsilon$

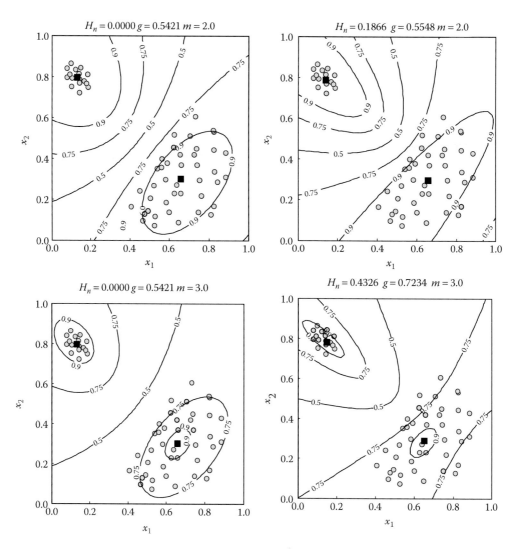

FIGURE 4.58 Comparison of FLME (left) and adaptive volume GK (right) clustering algorithms of the same data set. The separation achieved by the FMLE is superior to that provided by the GK algorithm, even in the adaptive volume version. Note how the FMLE is less sensitive to the value of the fuzzy exponent m and its contour lines better conform to the data. These graphs were obtained with the MATLAB script `Go_FMLE.m` in the `ESA_Matlab\Exercises\Chapter_4\Fuzzy_Clustering\FMLE` folder.

sense that the constraint is embedded in the objective functional via a Lagrange multiplier, but in this case the term $(\Sigma\mu-1)$ is weighted by a multiplier that decreases its relative importance (Krishnapuram and Keller, 1993; Babuska, 1998; Pal et al., 2005). The objective functional thus becomes

$$\min_{\mu,\eta} J(X,\eta) = \sum_{i=1}^{c}\sum_{k=1}^{N}(\mu_{ik})^{m}\cdot\|x_k-v_i\|_{A_i} + \sum_{i=1}^{c}\eta_i\cdot\sum_{k=1}^{N}(1-\mu_{ik})^{m} \qquad (4.74)$$

where A_i are norm-inducing matrices that define the point-to-centroid distance and $(\eta_i>0,\ i=1,\ldots,c)$ are positive constants, hopefully small enough to effectively relax the total membership constraint.

If we restrict to the case $A_i = I$ in the functional (4.74), we are back at the FCM algorithm, but with a relaxed membership constraint. The introduction of the new multipliers η_i greatly complicates the minimization of Equation 4.74 and a suggested choice (Krishnapuram and Keller, 1993) is to relate the new weights to the distances, that is,

$$\eta_i = K \frac{\sum_{k=1}^{N} (\mu_{ik})^m \cdot d_{ik}^2}{\sum_{k=1}^{N} (\mu_{ik})^m} \quad \text{with } K > 0 \tag{4.75}$$

where the distance d_{ik} are the Euclidean norm of Equation 4.37. The memberships are then computed as a function of the constraint coefficients η_i,

$$\mu_{ik} = \frac{1}{1 + (d_{ik}/\eta_i)^{\frac{2}{m-1}}} \tag{4.76}$$

and the centroids are computed as in the FCM algorithm according to the iterative scheme of Box 4.13. Because the minimization of Equation 4.74 is fairly critical, it is advisable to abandon the random initialization and start with a feasible partition such as that produced by a preliminary FCM clustering.

Figure 4.59 compares the results of PCM and FCM clustering on the same set of data with two very problematical data, which could be regarded as outliers and do not deserve to be assigned to

---

BOX 4.13   ITERATIVE STRUCTURE OF THE PCM ALGORITHM

Starting with a good $U^{(0)}$ (e.g., pre-computed with the FCM algorithm) repeat for $t = 1, 2, \ldots$

1. Compute the centroids

$$v_i^{(t)} = \frac{\sum_{k=1}^{N} \left(\mu_{ik}^{(t-1)}\right)^m x_k}{\sum_{k=1}^{N} \left(\mu_{ik}^{(t-1)}\right)^m} \quad i = 1, \ldots, c$$

2. Compute the distances

$$d_{ik}^{(t)^2} = \left(x_k - v_i^{(t)}\right)^T \left(x_k - v_i^{(t)}\right)$$

3. Compute the possibilistic constraint

$$\eta_i^{(t)} = K \frac{\sum_{k=1}^{N} \left(\mu_{ik}^{(t-1)}\right)^m \cdot d_{ik}^{(t)^2}}{\sum_{k=1}^{N} \left(\mu_{ik}^{(t-1)}\right)^m} \quad \text{with } K > 0$$

4. Update the partition matrix

$$U^{(t)} : \mu_{ik}^{(t)} = \frac{1}{\sum_{\substack{j=1 \\ j \neq i}}^{c} \left(d_{ik}^{(t)}/\eta_i\right)^{\frac{2}{m-1}}} \quad k = 1, \ldots, N$$

Until $\left\| U^{(t)} - U^{(t-1)} \right\| < \varepsilon$

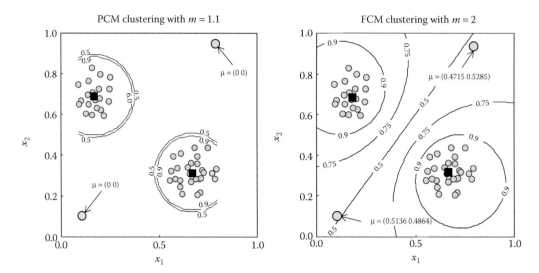

FIGURE 4.59 Comparison of PCM and FCM clustering of a data set with two outliers. The possibilistic clustering (left) is able to exclude these points from both clusters, whereas the FCM algorithm (right) yields an ambiguous classification. These graphs were obtained with the MATLAB script `Go _ PCM.m` in the `ESA _ Matlab\Exercises\Chapter _ 4\Fuzzy _ Clustering\PCM` folder.

either cluster. The PCM correctly classifies them with a zero membership, whereas the FCM assigns a very unrealistic ~0.5 membership as a consequence of the total memberships constraint. Notice also the very low value of m, very near the admissible minimum ($m = 1$), which makes the PCM algorithm very discriminative.

5 Population Dynamics Modelling

Either we limit our population growth or the natural world will do it for us, and the natural world is doing it for us right now.

David Attenborough

The evolution of populations has always been a central theme in mathematical modelling and in Chapter 1 we have already mentioned that after the pioneering efforts by Maynard Smith (1974) mathematical modelling of populations, viewed as the central players in the ecosystem, has become a mainstay in ecology (Odum, 1983; Ricklefs and Miller, 1999; Gotelli, 2001; Odum and Barrett, 2004; Rockwood, 2006; Vandermeer and Goldberg, 2013). In this chapter, we shall examine the basics of population dynamical models, starting with single-species modelling, and then expanding the analysis to include the interactions between competing species, eventually culminating in the modelling of the entire food web.

The goal of population dynamics is to provide a quantitative description of the evolution of a group of individuals (the population), belonging to one or more species, which share the resources of the same environment. When working with quantitative models, it is important to define the units in which the population is expressed. The basic unit will always be *population density*, understood as the number of individuals per unit of habitat (surface or volume), and if we deal with single-species models, then population density alone (e.g., individuals/unit surface or individuals/unit volume) might suffice. However, when we consider several species, which might range from the flea to the elephant, simply counting individuals could become impossible, particularly when we describe the interactions among them and with the environment. Furthermore, when it comes to considering their impact on the ecosystem, surely a flea does not exert the same environmental pressure as an elephant; nonetheless it is a necessary component of the ecosystem. Thus, we must agree on a common unit, such as the energy content of each organism, which will be expressed in kcal/unit surface or kcal/unit volume. Using this common unit, energy transfer across differing trophic levels can be established regardless of the species in each level, as we have already considered in Chapter 1. This approach has been fully exploited in constructing thermodynamical models (Jorgensen and Svirezhev, 2004), as briefly mentioned in Chapter 1. As to the scale of modelling, several options are available (Odum and Odum, 2000). We will start with simple single-species models (Maynard Smith, 1974; Pielou, 1977; Begon and Mortimer, 1986; Renshaw, 1991; Vandermeer and Goldberg, 2013), and then move on to multispecies models (Turchin, 2003; Mittelbach, 2012).

5.1 SINGLE-SPECIES CONTINUOUS-TIME POPULATION MODELS

The growth of a single-species population is amply treated in many ecology textbooks, so only the main results will be recalled here. A single-species growth model is based on the following simplifying assumptions, which will be progressively relaxed as we move towards more realistic models. For now, we assume that the individuals in the population are all constantly reproductive, that each new-born is immediately fertile, and that the growth rate is proportional to the size of the population. Under these assumptions (that will be relaxed later), the growth of a population x can be described by the following ordinary differential equation:

$$\frac{dx}{dt} = x \cdot F(x) \tag{5.1}$$

where $F(x)$ is the population growth function, which should satisfy the following hypotheses to have a stable and limited model:

$$
\begin{aligned}
&1. \quad F(x) > 0 \quad \forall x > 0 \\
&2. \quad F(\bar{x}) = 0 \\
&3. \quad \frac{dF(x)}{dx} < 0 \quad \forall x > 0
\end{aligned}
\tag{5.2}
$$

The first condition states that $F(x)$ must be positive for all x. The second states that at the equilibrium $\bar{x}$, the function $F(x)$ must vanish, whereas the third requires its derivative to be a decreasing function, implying that the growth must relent as the population approaches its equilibrium. Figure 5.1a shows some possible shapes for $F(x)$, from which some departure is possible due to the Allee effect (b), which states that a population cannot grow if its density is below a 'critical' value so small that mating and reproduction is very difficult. For this reason, at very low density the growth function may become negative, indicating that the impaired reproductive ability will make the population decrease even more. This departure from the theoretical shape of the growth function is indicated by the blue shading in Figure 5.1b.

5.1.1 Stability of the Equilibrium of a Single-Species Growth Model

Apart from the Malthusian model, which considers an unlimited growth, ecological and biological factors always impose a limit to growth. This translates in the mathematical property of *equilibrium* and its *stability*, as already seen in Chapter 1. Because stability is a property of the equilibrium, before investigating the model stability we must compute its equilibrium and then decide whether this is stable or not. A necessary condition for the equilibrium of the growth model (Equation 5.1) is the vanishing of its derivative, so we seek the solutions to the stationary equation

$$
0 = x \cdot F(x)
\tag{5.3}
$$

Because we assume a non-zero population $(x > 0)$, Equation 5.3 reduces to solving for $F(x) = 0$. The equilibrium solutions of Equation 5.3 are therefore

$$
\bar{x} : F(\bar{x}) = 0
\tag{5.4}
$$

To analyse the stability of the, generally multiple, equilibria $\bar{x}$ in the sense already treated in Section 1.9.2, we consider small perturbations about the equilibrium $\bar{x}$, that is, $x = \bar{x} + \tilde{x}$, so that $\dot{x} = \dot{\tilde{x}}$. Linearizing the model (Equation 5.1) in this neighbourhood according to Equation 1.84 yields

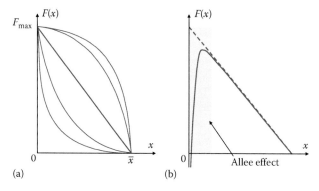

FIGURE 5.1   General aspect of the growth function (a) and possible departure due to the Allee effect (b).

$$\frac{d\tilde{x}}{dt} = \left(\left.\frac{\partial F(x)}{\partial x}\right|_{\bar{x}} \right) \cdot \tilde{x} \tag{5.5}$$

where the term in parentheses can be regarded as the one-dimensional Jacobian of the system. The (local) stability of the linearized model (Equation 5.5) depends on the sign of the derivative of the growth function at the equilibria. Figure 5.2 shows a growth function exhibiting *depensation*, that is, negative growth rate for low population densities (the Allee effect). This growth function intersects the *x*-axis at three points, two of which (1 and 3) yield a negative derivative of the Jacobian function in Equation 5.5, and are therefore stable, whereas in point 2 the derivative is positive, hence this point is unstable. To dispel the intuitive notion that stability is 'good' and instability 'bad', let us consider that a stable equilibrium will attract the trajectories: in the case of point 1 that will mean extinction. Likewise the instability of point 2 has a positive meaning, because it fosters population growth away from extinction and towards the other stable equilibrium (3). We shall soon get back to the ecological implications of stability, but for now let us recall that Equation 5.5 is the general tool for assessing the stability of a growth model.

5.1.2 THE LOGISTIC GROWTH FUNCTION

Undoubtedly, the most popular single-species growth model is the logistic equation, which has enjoyed great popularity for some time as a significant improvement to the unrealistic Malthusian model, which predicted unlimited growth. Because limitation is observed in any population growing in a finite environment and depending on its resources for its development, this equation introduces self-limitation in the form

$$\frac{dx}{dt} = r \cdot x \cdot \left(1 - \frac{x}{K}\right) \tag{5.6}$$

where:

r is the growth rate
K is the *carrying capacity*

The latter plays a major role in population growth as it represents the maximum density that the population can reach in *that* environment. Consequently, it represents the role of the environment in fostering or limiting the development of the population, in terms of food, shelter, nesting sites and so on. However, the environment does not have an influence on the growth rate *r*, which is an intrinsic characteristic of each species. According to Equation 5.6, a population with an initial density x_o grows until it reaches the carrying capacity *K*, which represents a stable equilibrium, as depicted in

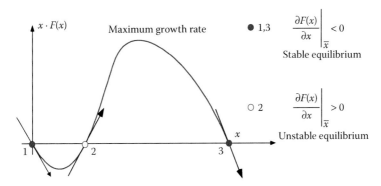

FIGURE 5.2 Graphical and analytical assessment of the equilibria of a single-species growth model. Equilibria 1 and 3 are stable, whereas the equilibrium 2 is unstable.

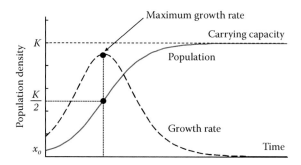

FIGURE 5.3 Time evolution of a population according to the logistic equation (solid line) and its growth rate (dashed line). The maximum growth coincides with the inflexion point.

Figure 5.3, in which the growth rate is also shown. The growth curve has an inflexion point when the population density is half the carrying capacity, at which level the growth rate is maximal. The population density producing the maximum growth rate v_{max} can be determined by solving for the value x_{max} at which the derivative of the growth function with respect to x vanishes, namely

$$x_{max} \Leftrightarrow \frac{d}{dx}\left(rx\left(1-\frac{x}{k}\right)\right) = r\left(1-\frac{2x}{K}\right)\frac{dx}{dt} = 0$$

but   (5.7)

$$\frac{dx}{dt} \neq 0 \quad \text{for } x < K \rightarrow 1 - \frac{2x}{K} = 0 \rightarrow x_{max} = \frac{K}{2}$$

This justifies the previous observation that the logistic growth has an inflexion point at half the maximum (carrying) capacity, irrespective of the initial population density x_o. Substituting x_{max} into the growth function yields the maximum growth rate as

$$v_{max} = r \cdot x_{max}\left(1 - \frac{x_{max}}{K}\right) = r \cdot \frac{K}{2}\left(1 - \frac{K}{2}\cdot\frac{1}{K}\right) = \frac{rK}{4}$$ (5.8)

The relationship among the various quantities in Equations 5.7 and 5.8 are illustrated in Figure 5.4.

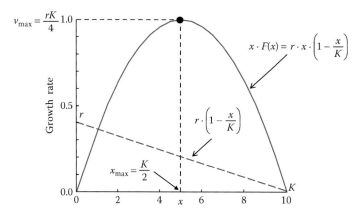

FIGURE 5.4 Analysis of the logistic growth function. Its curve (solid line) is symmetrical with respect to the middle growth point where the maximum growth occurs. The growth function (dashed line) is linear and terminates at the equilibrium $x = K$.

The carrying capacity represents a *globally* stable equilibrium, and the population in that state is able to cope with perturbations that tend to shift it away from the equilibrium. Linearizing the growth function around $x = K$ yields

$$\left. \frac{\partial}{\partial x}\left(r \cdot x \cdot \left(1 - \frac{x}{K}\right)\right)\right|_{x=K} = -r \tag{5.9}$$

so after a perturbation displacing the population from K, the population will return to the equilibrium with an approximate dynamics

$$\frac{d\tilde{x}}{dt} = -r \cdot \tilde{x} \rightarrow \tilde{x} = \tilde{x}_o \cdot e^{-r \cdot t} \tag{5.10}$$

where x_o represents the initial displacement from equilibrium. The time constant $T_r = 1/r$ is sometimes termed the 'return time'.

The importance of the logistic model is primarily pedagogical to introduce the subject of population dynamics. However, there are cases where it can actually be fitted to laboratory or field observation, without the need to use more complex models, which will be considered in the following. For example, in Figure 5.5 shows a logistic growth model fitted to the data from two populations of the same species of snails growing in differing environments. The interesting fact about this exercise is that in both cases the growth rate (r) is the same, whereas the carrying capacity (K) varies according to the ability of the environment to provide favourable growth conditions.

It is worthwhile to mention that the generality of the logistic equation has made it applicable in scientific areas outside ecology, such as in the neurological sciences, where it has been used to robustly model the human intracranial pressure–volume relationship. Lakin and Gross (1992) developed a logistic fit for the experimentally determined pressure–volume relationship to replace previous fits used for this relationship, which were usually exponential in nature. While exponential

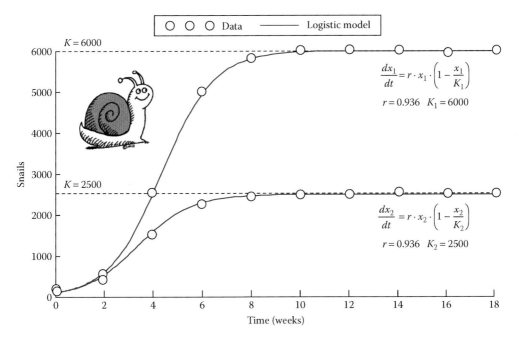

FIGURE 5.5 Fitting the logistic model to snail populations grown in differing environments. Calibrating the model to the two sets of data yields the same growth rate but differing carrying capacities.

models can provide reasonably good fits to experimental data on the lower reaches of the pressure–volume curve, close to the resting pressure where compliances are relatively large because volume changes can be easily accommodated by the intracranial venous system, the exponential fits increasingly deviate from the measured data on the upper reaches of the pressure–volume curve (beyond the inflexion point). By contrast, the logistic fit provides excellent agreement with the experimental data over the entire (non-fatal) range of interest of the pressure–volume curve. Use of the logistic equation also facilitated a clear explanation of the physiological mechanisms that lead to the pressure–volume relationship. In the context of a compartmental mathematical model for the human systemic circulatory system, Stevens and Lakin (2005) used logistic functions as mathematical representations that provide a smooth connection between a maximum effect and a minimum effect for the autonomic and central nervous system reflexes that maintain arterial pressure, cardiac output and cerebral blood flow. Indeed, these logistic representations form the heart of the mathematical model. The power of the logistic representations is such that, even though the compartment model itself is quite simple, it was capable of accurately describing the pressure, volume and flow dynamics of the human circulatory system over the full physiological range of human pressures and volumes.

5.1.2.1 Delayed Recruitment Logistic Growth

Let us now begin to relax some of the conditions under which the logistic model was derived. First of all, we may challenge the hypothesis that an individual is reproductive right after birth. If we introduce a delay between the birth and the onset of fertility, this implies that the population actually contributing to growth is not the entire population at time t, but only that portion born at an earlier time $t - \delta$, where δ is the time delay required by the newborn individuals to reach the reproductive maturity. For example, only adult insects lay eggs, whereas the larval stage accounts for the bulk of the population. In general, delayed growth occurs when the environment affects mostly the young age groups and influences their sexual maturity. To account for the presence of the delay, the logistic equation (5.6) can be rewritten as

$$\frac{dx}{dt} = r \cdot x \cdot \left(1 - \frac{x(t - \delta)}{K} \right) \tag{5.11}$$

which has the same equilibrium $\bar{x} = K$, but a very different dynamic behaviour. Figure 5.6 shows that the delay has a destabilizing effect and may eventually lead to limit cycles, examined in Section 1.9.4. The stability of the equilibrium around the carrying capacity can again be investigated by considering the linearized model around K, where $x = K + \tilde{x}$

$$\frac{d\tilde{x}}{dt} = r \cdot \left(K + \tilde{x}(t) \right) \left(1 - \frac{K + \tilde{x}(t - \delta)}{K} \right)$$

$$= -r \cdot \left(K + \tilde{x}(t) \right) \frac{\tilde{x}(t - \delta)}{K} = -r \cdot \tilde{x}(t - \delta) - r \frac{\tilde{x}(t) \cdot \tilde{x}(t - \delta)}{K} \tag{5.12}$$

Neglecting the second-order term $\tilde{x}(t) \cdot \tilde{x}(t - \delta)$, the linearized dynamics around the carrying capacity is

$$\frac{d\tilde{x}}{dt} + r \cdot \tilde{x}(t - \delta) = 0 \tag{5.13}$$

It might appear that r is the only factor influencing the dynamics, but this is not the case given the presence of the delayed population $\tilde{x}(t - \delta)$. In fact, Equation 5.13 has infinite roots, introduced by the delay. This becomes evident when studying the associated characteristic equation obtained via the Laplace transform

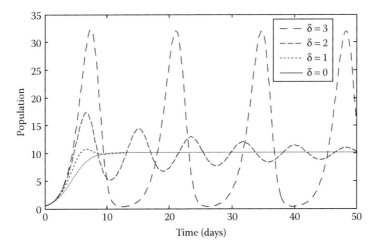

FIGURE 5.6 Effect of reproductive delay on the dynamic behaviour of the logistic model with delay equation (5.11).

$$s \cdot \tilde{x}(s) + r \cdot \tilde{x}(s) \cdot e^{-\delta \cdot s} = 0 \Rightarrow s + r \cdot e^{-s \cdot \delta} = 0 \tag{5.14}$$

where s is the complex Laplace variable. Equation 5.14 is a transcendental equation and as such it has infinite solutions. However, its exponential term can be approximated by a second-order Taylor expansion

$$e^{-\delta \cdot s} \cong 1 - \delta \cdot s + \frac{\delta^2 \cdot s^2}{2!} \tag{5.15}$$

Substituting this term into Equation 5.14 yields

$$s^2 + 2\frac{1 - r\delta}{r\delta^2}s + \frac{2}{\delta^2} = 0 \tag{5.16}$$

which describes the behaviour of the delay equation (5.13) in the neighbourhood of the equilibrium $\bar{x} = K$. The root locus of Figure 5.7 describes the location of the two roots as a function of the delay δ: for $\delta < \delta_1$ the two roots are both negative real, but as the delay increases beyond δ_1, they bifurcate becoming complex conjugate, corresponding to an oscillatory behaviour, which is less and less damped as δ increases because their real part decreases, until it vanishes for $\delta = \delta_2$, where the locus crosses the imaginary axis. From there on the real part becomes positive as $\delta \to \infty$, resulting in an unstable behaviour. This is a case in which the linearized approximation fails (see Table 1.7) and only numerical simulation can give the real picture of the model behaviour, which is shown in Figure 5.6, where the initially (for small δ) damped oscillation turns into sustained oscillations and eventually produces a limit cycle.

A thorough review on predator–prey models with discrete delay can be found in Ruan (2009). What might appear as a purely mathematical trick is in fact observed in nature, where there is ample evidence of reproductive delay inducing periodic oscillations, as in the well-studied case of the sheep blowfly (*Lucilia cuprina*). Periodic fluctuations in this population were observed by Nicholson (1954) and were later analysed in the context of complex behaviours of population models (May, 1974, 1976b; Mueller and Joshi, 2000). More recently, a fuzzy model was also proposed (Rashkovsky and Margaliot, 2007), and this example is still being analysed in the advanced

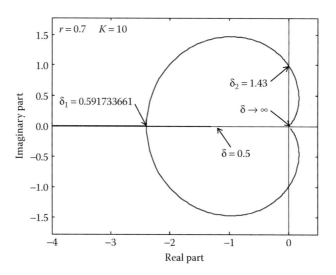

FIGURE 5.7 Stability analysis of the linearized equation (5.16). For small values of the delay, the roots are both negative real, implying exponential stability, until at δ_1 they bifurcate in a couple of complex conjugate roots, implying an oscillatory behaviour. The other interesting value is δ_2 where the conjugate roots are purely imaginary, and the model produces a limit cycle. For $\delta > \delta_2$, the roots have a positive real part, so that the model becomes unstable. This graph was obtained with the MATLAB script `Go_root_locus_delay.m` in the `ESA_Matlab\Exercises\Chapter_5\Single\Single_continuous_time` folder.

statistical framework (Brillinger, 2013). Other examples of fluctuating populations, for example, of Rotifers, are available in the literature (Snell and Serra, 1998; Gonzalez and Descamp-Julien, 2004).

5.1.2.2   Analytical Solution of the Logistic Equation

Equation 5.6 with an initial condition x_o can be solved by calculus to yield

$$x(t) = \frac{K}{1 + ((K - x_o)/x_o)e^{-r \cdot t}} = \frac{K}{1 + e^{(\beta - r \cdot t)}} \quad \text{with } \beta = \ln\left(\frac{K - x_o}{x_o}\right) \tag{5.17}$$

We have seen that the main limitation of the logistic equation is that the resulting sigmoid response has a fixed inflexion at $x = K/2$ as a consequence of the linear exponential term $(\beta - r \cdot t)$. Many alternate expressions for modelling growth have been proposed, each of them derived in different contexts and based on differing biological assumptions. Their common endeavour was to add flexibility to the logistic response, in particular making the sigmoid non-symmetrical. These models introduce a non-linear exponential term in place of the linear term $(\beta - r \cdot t)$ so that Equation 5.17 can be generalized into

$$x(t) = \frac{K}{1 + e^{\Phi(t)}} \tag{5.18}$$

with the exponential term satisfying the usual conditions for a growth function given by Equation 5.2. In particular,

$$\begin{aligned}
&\text{1.} \quad \frac{d\Phi(t)}{dt} < 0 \\
&\text{2.} \quad \lim_{t \to \infty} e^{\Phi(t)} = 0
\end{aligned} \tag{5.19}$$

5.1.3   THE GOMPERTEZ GROWTH EQUATION

In the search of a more flexible growth model, the Gompertez model is surely one of the oldest, having been conceived as early as 1825(!) to explain certain census data. Today, it is largely used in plant growth dynamics for its ability to incorporate the ageing mechanism typical of plant leaves (Berger, 1981; Causton and Venus, 1982). Its differential form is

$$\frac{dx}{dt} = r_g \cdot x \cdot \left(\ln K - \ln x \right) \tag{5.20}$$

which can be solved analytically to yield

$$x(t) = K \cdot e^{\left(e^{-(\beta - r \cdot t)} \right)} \tag{5.21}$$

It can be seen that the growth rate varies in time as

$$R(t) = r_g \left(\ln K - \ln x(t) \right) = r_g \left(\ln K - \ln \left(K e^{-e^{(\beta - r \cdot t)}} \right) \right)$$

$$= r_g \left(\ln K - \ln K + e^{(\beta - r \cdot t)} \right) = r_g \cdot e^{(\beta - r \cdot t)} \tag{5.22}$$

The maximum growth rate can be found by deriving Equation 5.20 with respect to x and setting the derivative to zero.

$$\frac{d}{dx} \left(\frac{dx}{dt} \right) = r \left(\ln K - \ln x \right) + \left(rx \left(-\frac{1}{x} \right) \right) = r \cdot \ln K - r \cdot \ln x - r \cdot x \cdot \frac{1}{x} = 0 \tag{5.23}$$

Solving for x yields

$$\dot{x}_{max} = e^{\left(\ln K - 1 \right)} \tag{5.24}$$

Figure 5.8 shows some examples of population growth described by the Gompertez equation (5.20) with a varying growth coefficient r, whereas the composite growth rate, given by Equation 5.22, is

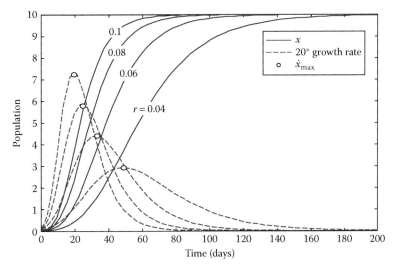

FIGURE 5.8   Population growth according to the Gompertez equation (5.20) for several values of the growth rate r. As required, the maximum growth rate (dots) is now flexible.

shown by the dashed lines. The Gompertez dynamics is based on the assumption that growth relents as time progresses as a result of ageing. In other words, the exponential term in Equation 5.22 expresses the fact that the reproductive capability decreases as the average age of the population increases.

5.1.4 THE RICHARDS GROWTH EQUATION

Another flexible sigmoid model used to describe the vegetation growth from a single leaf to a whole plant (Causton and Venus, 1982) is due to Richards, which can be viewed as a further generalization of the logistic growth function. Its differential form is

$$\frac{dx}{dt} = x \cdot \frac{r_r}{n_r}\left(1 - \frac{x^{n_r}}{K^{n_r}}\right) \tag{5.25}$$

With respect to the logistic equation, it introduces a third parameter (n_r) that controls the point of inflexion $(\dot{x}_{max})$, where the maximum growth rate occurs. Applying the same procedure used for the Gompertez model, the point of inflexion is determined as

$$\dot{x}_{max} = \frac{K}{\left(1 + n_r\right)^{1/n_r}} \tag{5.26}$$

Of course, Equation 5.25 reduces to the logistic for $n = 1$, but otherwise it is able to anticipate or delay the population development, because the point of inflexion depends on K and n_r. Figure 5.9 shows the influence of the growth coefficient r_r on the growth, with the inflexion point remaining at the same population level determined by Equation 5.26, whereas Figure 5.10 shows the effect of the coefficient n_r in delaying the population development. Thus, the Richards and the Gompertez models have opposite effects, in the sense that while growth decreases with ageing in the Gompertez model, in the Richards model mature populations have a higher growth rate.

An application of the Richards model to the growth of the algae *Selenastrum capricornutum* is described later in Section 8.6. This microalga, now renamed *Raphidocelis subcapitata*, has a moon-shaped body of length between 8 and 14 μm, and a width between 2 and 3 μm. It is considered as a

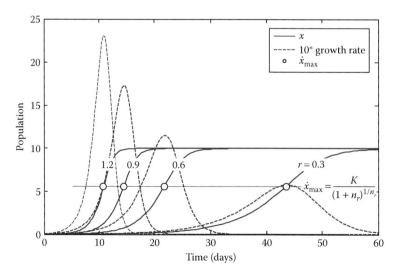

FIGURE 5.9 Population growth according to the Richards equation (5.25) for several values of the growth rate r_r, while K and n_r, which define the inflexion point (grey dots) through Equation 5.26, are kept constant.

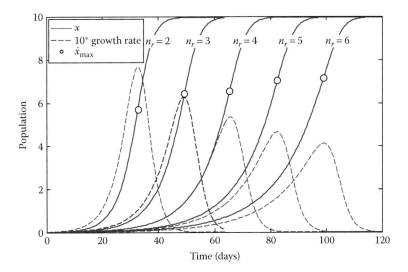

FIGURE 5.10 Population growth according to the Richards equation (5.25) for several values of the coefficient n_r, which has the effect of delaying the growth and of changing the inflexion point (grey dots).

primary bio-indicator species to assess the levels of nutrients or toxic substances in freshwater environments. The calibrated Richards model is compared to *S. capricornutum* growth data in Figure 8.43.

5.1.5 THE BERTALANFFY GROWTH EQUATION

This model (Von Bertalanffy, 1969) is yet another extension to a non-symmetric sigmoid and has the differential form

$$\frac{dx}{dt} = r_b \cdot x^{n_b} - k_d \cdot x \tag{5.27}$$

The hypothesis underlying this model equation is that the population dynamics is the net result of growth $\left(r_b \cdot x^{n_b} \right)$ and decay $\left(k_d \cdot x \right)$, both related to the biomass (x). This model was originally derived for bacterial populations in which nutrient uptake, and therefore growth, is assumed to be proportional to bacterial surface. For this reason, the exponent should theoretically be in the interval $2/3 \leq n_b \leq 1$. The decay term is assumed to be density-dependent, that is, proportional to the biomass itself. The carrying capacity does not appear explicitly in the Bertalanffy model, but it can be computed from the equilibrium by setting $x = K$, to yield

$$0 = r_b \cdot K^{n_b} - k_d \cdot K \tag{5.28}$$

which can be solved for the decay coefficient k_d

$$k_d = r_b \cdot K^{(n_b - 1)} \tag{5.29}$$

or for the growth rate factor r_b

$$r_b = \frac{k_d}{K^{(n_b - 1)}} \tag{5.30}$$

As a concluding remark on these single-species growth models, their evolution towards the equilibrium, represented by the carrying capacity, and their growth rate along this trajectory are shown in Figure 5.11.

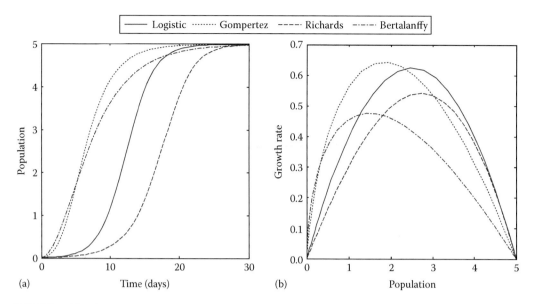

FIGURE 5.11 Comparison of single-species growth functions. (a) shows the evolution of the population towards its equilibrium (carrying capacity), whereas (b) compares the growth rates as the population evolves along the trajectories shown in (a). These graphs were obtained with the MATLAB script `Growth_Compare.m` in the `ESA_Matlab\Exercises\Chapter_5\Single\Single_continuous_time` folder.

In this figure, parameters were adjusted to reach the same steady state for all models $\left(\bar{x} = K \right)$, and the increased flexibility with respect to the logistic equation introduced by the additional parameters in the other models is evident. The maximum growth rate of the logistic is symmetric with respect to the population, occurring at $\dot{x}_{max} = K/2$, whereas for the Gompertez equation the maximum is reached for lower densities, that is, for younger populations. In fact, this model accounts for the diminished reproductive capacity of ageing populations. The Bertalanffy model shares the characteristic of anticipating the maximum growth rate, whereas the Richards model has an opposite behaviour, shifting the maximum growth rate towards a higher (more mature) population. The Simulink® model used to produce the graphs of Figure 5.11 is shown in Figure 5.12, and the MATLAB® launch script is listed in Box 5.1.

5.1.6 A FINAL APPRAISAL OF THE LOGISTIC EQUATION

The Bertalanffy model is appealing because it separates growth and decay in the population dynamics of simple organisms. In fact, that model was originally developed to describe microbial growth. This consideration might lead us to criticize the logistic model for not having a similar birth-and-death structure. In fact, however, this is not so, as can be easily demonstrated as follows. Let us consider the birth and death processes as population-dependent, with the birth rate decreasing as the population grows, remembering that the carrying capacity can be viewed as a limit to growth. The death rate, on the other hand, is increased by food scarcity or, more generally, by a crowding effect. Thus

$$\begin{array}{ll} \text{Birth process:} & b - p_1 \cdot x \\ \text{Death process:} & d + p_2 \cdot x \end{array} \tag{5.31}$$

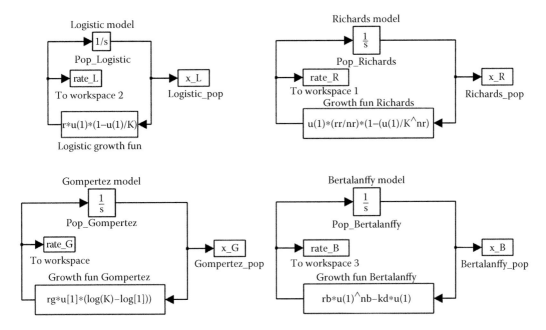

FIGURE 5.12 Simulink model to simulate the four single-species growth models described in Figure 5.11. The launch MATLAB script is listed in Box 5.1, while the Simulink model is `Growth _ Models.mdl` in the `ESA _ Matlab\Exercises\Chapter _ 5\Single\Single _ continuous _ time` folder.

Because the model must (dynamically) balance these two processes, the growth equation will be

$$\frac{dx}{dt} = x \cdot \left[\left(b - p_1 \cdot x \right) - \left(d + p_2 \cdot x \right) \right]$$

$$= x \cdot \left[b - d - x(p_1 + p_2) \right] = x \left(b - d \right) \left[1 - \frac{x(p_1 + p_2)}{b - d} \right]$$

(5.32)

Now, defining the new coefficients

$$r = b - d \quad \text{and} \quad K = \frac{b - d}{p_1 + p_2}$$

(5.33)

we recover the logistic equation

$$\frac{dx}{dt} = r \cdot x \cdot \left(1 - \frac{x}{K} \right)$$

(5.34)

This allows us to view the logistic model in a new light, particularly by considering that the two parameters r and K are not independent from one another, but can be viewed as balancing the opposite forces of birth and death. In this sense, $r = b - d$ represents the *net* growth rate at the very early stage of population development $\left(x \ll K \right)$, and of course, for a population to grow, it must be $b > d$. The structuring of K is more complex as it includes the density-dependent parameters p_1 and p_2, which determine the sensitivity of the population to the crowding effect.

An example of a usable birth-and-death model is that employed by the International Whaling Commission (Morales-Zárate et al., 2004; Basille et al., 2008; Freeman, 2008) to estimate the whale population around the globe

---

**BOX 5.1   LAUNCH MATLAB SCRIPT FOR THE COMPARISON
OF THE FOUR SINGLE-SPECIES CONTINUOUS-TIME
POPULATION MODELS DESCRIBED IN SECTION 5.1**

```
xo=0.01; % Initial population
K= 5; % Common Carrying capacity
nr=1.5; % Richards exponent
nb=0.667; % Bertalanffy exponent
kd=0.6436; % Bertalanffy decay rate
r=0.5; % Logistic growth rate
rg=0.35; % Gompertez growth rate
rr=0.5; % Richards growth rate
rb=1.1; % Bertalanffy growth rate
tfin=30; % Simulation horizon
figure(1)
set(1,'Position',[683 493 784 464])

% Simulate models
[t,x,y]=sim('Growth_Models',tfin);
% Plot results
subplot(121)
plot(t,x_L,'b',t,x_G,':b',t,x_R,'--b',t,x_B,'-.b')
set(gca,'FontName','Arial','FontSize',14)
xlabel('time (days)')
ylabel('population')
subplot(122)
plot(x_L,rate_L,'b',x_G,rate_G,':b',x_R,rate_R,'--b',x_B,rate_B,'-.b')
set(gca,'FontName','Arial','FontSize',14)
xlabel('population')
ylabel('Growth rate')
legend('Logistic','Gompertez','Richards','Bertalanffy','location','north',
'orientation','horizontal')
```

$$\frac{dx}{dt} = x(t) \cdot \left[p + q \left(1 - \frac{x(t-\delta)}{K} \right)^z \right] - k_d \cdot x(t) \tag{5.35}$$

where the parameters have a clear biological meaning:

p is the pregnancy rate

q is the probability of survival to the adult stage

z is the sensitivity to the depensation effect

δ is the time required to reach the reproductive maturity (6 years for the blue whale, 25 years for the sperm whale)

5.2   SINGLE-SPECIES DISCRETE-TIME POPULATION MODELS

We are now prepared to challenge two of the basic assumptions under which the continuous-time models were developed, that is, the continuous reproduction hypothesis and (save for the logistic model with delay) the lack of time lag between birth and reproductive maturity. These assumptions may become crucial if the population is sparse, or the developmental lag is a considerable fraction of the mean lifespan. In many cases, a discrete-time representation of growth may be more realistic, especially with non-overlapping generations. Insects are an example here, as their adult life is relatively short and mainly devoted to egg laying. When the next generation comes to life, the previous one would have already died, so they are non-overlapping. On the other hand, in many species, generations do overlap because their biology is such that individuals remain fertile for a considerable fraction of their lifespan. Discrete-time growth models should not be regarded only as approximation of continuous-time processes. In many ways, they portray more closely what actually happens in nature, because they easily allow for the inclusion of the time lag required by newborn individuals to reach the reproductive stage. In addition to explaining some naturally observed behaviours, discrete-time growth equations have also been studied for their complex behaviour, which is not paralleled by their continuous-time counterparts (May, 1974, 1976, 2001; Turchin, 2003; Solé and Bascompte, 2006).

5.2.1 THE DISCRETE-TIME LOGISTIC EQUATION

To develop a discrete-time growth model, consider again the logistic growth Equation 5.6 assuming that the population is observed only at discrete-time sampling intervals h so that the time derivative can be approximated with a finite difference

$$\frac{dx}{dt} \cong \frac{x(t+h) - x(t)}{h} = r \cdot x(t) \cdot \left(1 - \frac{x(t)}{K}\right) \tag{5.36}$$

Considering the index $k \in \mathbb{Z}^+$ in place of the physical sampling times $t, t+h, t+2h, \ldots$ as in Equation 1.26, yields the discrete-time model

$$x_{k+1} = \alpha \cdot x_k \left(1 - \frac{\beta}{\alpha} x_k\right) \quad \text{with} \quad \begin{cases} \alpha = 1 + h \cdot r \\ \beta = \dfrac{h \cdot r}{K} \end{cases} \tag{5.37}$$

This discrete representation can be further simplified by introducing the new variable $z = (\beta/\alpha)x$ to yield

$$z_{k+1} = \alpha \cdot z_k \cdot (1 - z_k) \tag{5.38}$$

In spite of its innocent look, Equation 5.38 has received considerable attention for its varied and surprising behaviour, as thoroughly investigated by May (1974, 1976a) who studied its stability with the graphical tools of Section 1.9.6. As already discussed in Chapter 1, the stability of Equation 5.38 depends on the magnitude of the Jacobian at the equilibrium, which can be computed by setting $z_{k+1} = z_k = \overline{z}$

$$\overline{z} = a\overline{z} \cdot (1 - \overline{z}) \Rightarrow \overline{z} = \frac{\alpha - 1}{\alpha} \tag{5.39}$$

producing the Jacobian at $\overline{z}$ as

$$\frac{d}{dz}\left[\alpha \cdot z \cdot (1-z)\right]\Big|_{\overline{z}} = \alpha\left(1 - 2\frac{\alpha-1}{\alpha}\right) = 2 - \alpha \tag{5.40}$$

The linearized equation around $\overline{z}$ is therefore

$$\tilde{z}_{k+1} = (2 - \alpha)\tilde{z}_k \tag{5.41}$$

The equilibrium requires that

$$-1 < 2 - \alpha < 1 \Rightarrow 1 < \alpha < 3 \Rightarrow \begin{cases} 1 < \alpha < 2 & \text{monotonic response} \\ 2 < \alpha < 3 & \text{oscillatory response} \end{cases} \tag{5.42}$$

For α in the stability region, the linearized analysis is confirmed by the first two plots of Figure 5.13 showing that, depending on the value of α, the response changes from monotonic (a) to a damped oscillation (b), as predicted by Equation 5.42. For $\alpha > 3$, the linearized analysis fails to explain the model behaviour, which in fact exhibits a limit cycle as soon as α exceeds 3, eventually moving towards chaos when α is further increased.

Another aspect of Equation 5.38 studied by May (1976a) is the so-called period-doubling, which can be studied by rewriting this equation in two successive steps

$$z_{k+1} = \alpha \cdot z_k \left(1 - z_k\right)$$

$$z_{k+2} = \alpha \cdot z_{k+1} \left(1 - z_{k+1}\right) \tag{5.43}$$

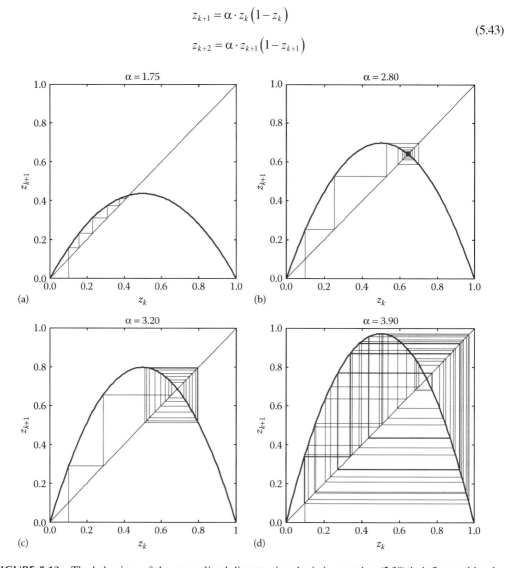

FIGURE 5.13 The behaviour of the normalized discrete-time logistic equation (5.38) is influenced by the value of α, ranging from a monotonic response (a), to damped oscillations (b), to a limit cycle (c), and eventually to chaos (d). These graphs were obtained with the MATLAB script Go _ Logistics _ dt.m in the ESA _ Matlab\Exercises\Chapter _ 5\Single\Single _ discrete _ time folder.

Substituting the first equation into the second and eliminating the intermediate variable z_{k+1} yields the two-step-ahead discrete-time equation

$$z_{k+2} = \alpha^2 \left(1 - \alpha\left(1 - z_k\right)z_k\right)\left(1 - z_k\right)z_k \tag{5.44}$$

whose equilibrium can be found by solving the steady-state equation

$$\alpha z^3 - 2\alpha z^2 + \left(1 + \alpha\right)z - 1 + \frac{1}{\alpha^2} = 0 \tag{5.45}$$

Equation 5.45 has three roots, one of them being the same as for the one-step-ahead equation, that is, $\bar{z}_1 = (\alpha - 1)/\alpha$, whereas the other two roots are the solutions of the second-order algebraic equation

$$\alpha z^2 - \left(1 + \alpha\right)z + \left(1 + \frac{1}{\alpha}\right) = 0 \tag{5.46}$$

For $\alpha > 3$, a series of bifurcations of the steady state appear, first by generating a limit cycle that oscillates between two differing steady states. As α increases, more bifurcations occur, and the period doubling becomes more and more frequent until it turns into chaos, as shown in Figure 5.14.

The progression of steady states as α increases, better known as bifurcation diagram or the 'Feigenbaum bubbles', is shown in Figure 5.15.

Each dot in the graph represents one steady state, recorded after simulating a sufficiently long growth period. In the first part, there is only one equilibrium, and hence a single dot appears. The first bifurcation occurs when the system oscillates between two values (limit cycle), and then at the

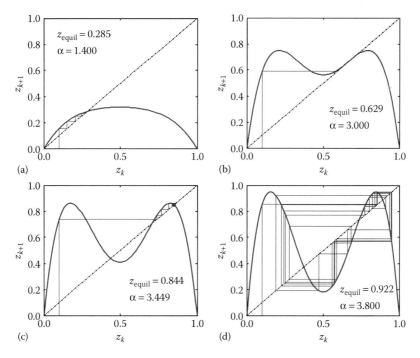

FIGURE 5.14 The behaviour of the two-step-ahead normalized discrete-time logistic equation (5.44) is influenced by the value of α, ranging from a monotonic response (a), to damped oscillations (b), to a limit cycle (c), eventually to chaos (d). These graphs were obtained with the MATLAB script Go_Logist_dt_period2.m in the ESA_Matlab\Exercises\Chapter_5\Single\Single_discrete_time folder.

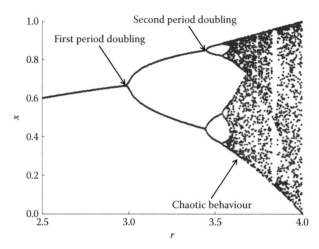

FIGURE 5.15 Bifurcation of equilibria of the discretized logistic equation (5.38) as a function of the growth rate r. The proliferation of period doublings eventually leads to a chaotic behaviour. This graph was obtained with the MATLAB script `Feigenbaum_Logistic.m` in the `ESA_Matlab\Exercises\ Chapter_5\Single\Single_discrete_time` folder.

second bifurcation the stationary oscillation involves four points, and so on until chaos ensues: there is no proper steady state, and the population fluctuates all over the place'. May (1976a) showed that for $3 < \alpha < 1 + \sqrt{6}$ a limit cycle develops, then for α between 3.44949 and 3.54409 there is a sustained oscillation among four values. As α increases, each period of oscillation bifurcates into two more periods, so that the population jumps among 8, 16, 32,..., 2^n equilibria, until for $\alpha > 3.56995$ the behaviour become chaotic. The ratio between the lengths of two successive such bifurcation intervals approaches the Feigenbaum constant $\delta = 4.66920$ (Feigenbaum, 1978).

5.2.2 OTHER DISCRETE-TIME POPULATION GROWTH EQUATIONS

Other discrete-time population models have been proposed in the context of natural resources management (Ginzburg and Golemberg, 1985; Begon and Mortimer, 1986; Levin et al., 1989; Akçakaya et al., 1999). The general form of these model can be written as

$$x_{k+1} = x_k \cdot R\left(x_k\right) \tag{5.47}$$

where the growth function R, itself a function of x_k, can include several sources of population fluctuations, such as environmental variability due to climate or habitat changes, as well as intrinsic variability such as variations in the reproductive rate. The most popular of such models are the Beverton–Holt model (also known as Contest), which has been used for fishery management,

$$x_{k+1} = \frac{R_{max} \cdot K}{R_{max} \cdot x_k - x_k + K} \cdot x_k \tag{5.48}$$

and the Ricker model (also known as Scramble), which has been applied to model wild populations,

$$x_{k+1} = x_k \cdot R_{max}^{\left(1 - (x_k/K)\right)} \tag{5.49}$$

These two models exhibit radically different behaviours, and their growth rates $R\left(x_k\right)$, with deeply differing meanings and values, are plotted in Figure 5.16.

Both models have their equilibrium at the carrying capacity K, but whereas the Contest model (Equation 5.48) reaches its steady state monotonically, the Ricker model (Equation 5.49) can

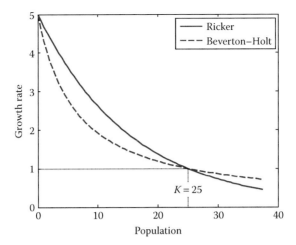

FIGURE 5.16 Comparison of population-dependent growth rates of the Ricker and Beverton–Holt discrete-time models. In both models the growth rate approaches unity as the population reaches the carrying capacity, in this case $K = 25$.

produce several strikingly different behaviours, ranging from monotonic, oscillatory, limit cycle to eventually chaotic, in much the same way as the discrete-time logistic equation.

The common feature of these two models is that the growth factor $R(x_k)$ in Equation 5.47 tends to unity as the population tends to the equilibrium. This is always true for the Beverton–Holt model, whereas for the Ricker model this holds only if the growth factor is such as to produce a monotonic or damped oscillatory response. The behaviour of the Ricker model (Equation 5.49) can be studied with the same graphical approach used for the discrete-time logistic. Figure 5.17 shows the

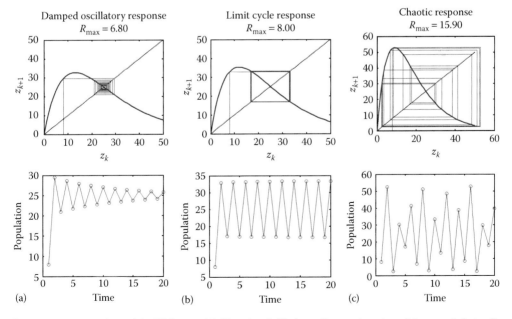

FIGURE 5.17 Behaviour of the Ricker model (Equation 5.49) depending on the value of the growth factor R_{max}. The upper graphs show the (z_k, z_{k+1}) plane for the equilibrium analysis, whereas the lower graphs show the corresponding time portrait. These graphs were obtained with the MATLAB script `Go_Ricker.m` in the `ESA_Matlab\Exercises\Chapter_5\Single\Single_discrete_time` folder.

graphical equilibrium analysis (upper row) and the corresponding time response of this model as R_{max} varies, producing a damped oscillatory response (a), a limit cycle (b), and a chaotic behaviour as R_{max} increases.

Like the discrete-time logistic model, the Ricker model has the period-doubling property, as shown in Figure 5.18, that was generated in the same way as Figure 5.15 and presents a similar aspect, with period doubling becoming more and more frequent oscillations until eventually giving rise to chaos.

5.2.3 INTRODUCING UNCERTAINTY AS DEMOGRAPHIC VARIABILITY

So far, we have assumed that the model parameters remain constant throughout the simulation horizon, but in nature several factors, such as climate and habitat changes, human intervention, influence the reproduction. This affects the population growth, which can be decomposed into the sum of a 'structural' species-dependent base growth factor $\bar{R}$ plus a climatic variation ΔR, as shown in Figure 5.19.

As an example, the growth rate of certain Shrews (*Crocidura russula*) inhabiting Swiss gardens was found to be influenced by several external factors (Akçakaya et al., 1999)

$$\Delta R = 0.73 \cdot P - 0.78 \cdot S + 0.50 \cdot T_s - 0.83 \cdot T_w \tag{5.50}$$

where:
P is the spring monthly precipitation
S is the winter snowfall
T_s is the average monthly temperature in summer
T_w is its winter counterpart

It is not easy to incorporate such fluctuations into the growth mechanism, unless the environment–species interactions are analysed in detail on a case-by-case basis. Here, it suffices to introduce the uncertainty on population development as a generic demographic variability, leaving it to specific studies to establish the precise relationship between the cause (environmental influence) and

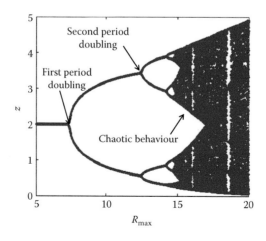

FIGURE 5.18 Period doubling bifurcations, eventually leading to chaos, in the Ricker model (Equation 5.49). This graph was obtained with the MATLAB script Feigenbaum _ Ricker.m in the ESA _ Matlab\ Exercises\Chapter _ 5\Single\Single _ discrete _ time folder.

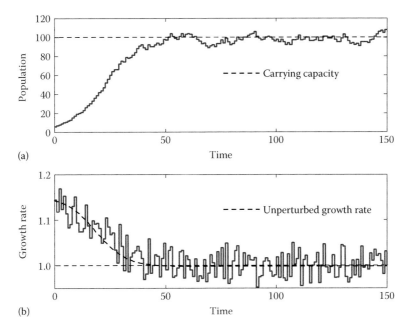

FIGURE 5.19 The effect of demographic variability. In (a), the population fluctuates around its carrying capacity. At the same time, the growth rate fluctuates around the steady-state value (b). The dashed lines represent the noise-free ideal situation.

the effect (growth rate variations). Let us consider the Beverton–Holt model, which is certainly 'tame' enough to withstand this structural extension, and let us consider that, according to Equation 5.47, the growth rate can be viewed as the ratio between the population at two successive sampling times, that is,

$$R(t) = \frac{x_{k+1}}{x_k} \tag{5.51}$$

Figure 5.16 has already shown that the growth rates of both the Ricker and the Beverton–Holt models tends to unity as the population approaches its carrying capacity, but now we want to structure the growth rate R a bit further, assuming that, in addition to its deterministic value given by Equation 5.48 (Beverton–Holt) or by Equation 5.49 (Ricker), the growth rate of Equation 5.51 also has a random part to model the demographic variability, namely

$$R(t) = \bar{R}(t) + v(t) \quad \text{with } v(t) = N(0, \sigma_p) \tag{5.52}$$

In this representation, $v(t)$ is the random component, modelled as a Gaussian white noise with zero mean and σ_p standard deviation. Figure 5.19 shows the effect of demographic variability in causing fluctuations around the steady-state population level (carrying capacity) and the corresponding variations in the growth rate around the unit value reached at steady state. The characteristics of a fluctuating population can be analysed from a statistical viewpoint, considering the distribution of the mean values of the population after the carrying capacity has been nominally reached. Furthermore, if we know which factor has caused the demographic variability, we could attempt to establish a relationship between the statistical distribution of the driving factor (cause) and the statistic of the steady-state population (effect). This study involves a Monte Carlo approach, where a large number of simulations are performed, each with a perturbed parameter drawn from a random process with known characteristics, as shown in Figure 5.20. An example of this Monte Carlo analysis is given in Figure 5.21, in which (a) shows the histogram of the fluctuating populations between time 80 and 160, whereas (b) shows the cumulative frequency distribution with the 5th and 95th percentiles, that can be related to the probability of extinction or explosion.

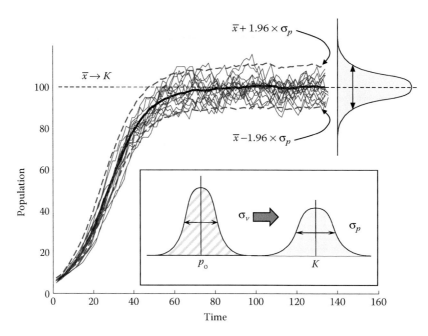

FIGURE 5.20   A possible link between the observed demographic variability around the steady state and the distribution of the perturbing factor around its nominal value p_o. After performing a large number of simulations (10,000) with perturbed growth factor, the population confidence bands are traced for $\bar{x} \pm 1.96 \cdot \sigma_p$, with the thick black line representing the average population. The inset hints at the possibility of establishing a relationship between the variability of p and that of the steady-state population.

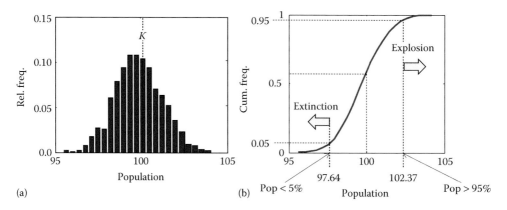

FIGURE 5.21   Statistical analysis of the previous Monte Carlo simulations. (a) shows the histogram of the fluctuating populations between time 80 and 130, whereas (b) shows the cumulative frequency distribution with the 5th and 95th percentiles, which can be related to the probability of extinction or explosion.

5.2.4   AGE-STRUCTURED POPULATIONS

We are now going to challenge another initial hypothesis on growth modelling, the one claiming that the all the individuals of a population have the same age, or rather that age is not a distinctive feature. Practical observations contradict this assumption and, in particular, it is clear that fecundity is unevenly distributed across the individual's lifespan, with an infertile period from birth to sexual maturity, and a declining reproductive capacity caused by ageing. The initial idea of dividing the individuals of a population into age groups was initially proposed by Leslie (1945), and age-structured population models are now widely used in ecological studies (Begon and Mortimer, 1986; Pastor, 2008). By necessity, age-structured models have a discrete-time representation because they consider a finite number of age classes, and therefore the transition from one class to the next is timed accordingly. In the Leslie model, the population at time k (X_k) is divided into n age groups

$$X_k = \begin{bmatrix} x_{k,1} & x_{k,2} & \cdots & x_{k,n} \end{bmatrix}^T \tag{5.53}$$

where the first class includes all the newborns that survive to enter the second class. The transition from one age group to another is governed by the graph of Figure 5.22: the fraction of individuals from the generic age-group $x_{k,i}$ at time k that survive until the next time (σ_i) contribute to the generation of newborns (first class) with their fertility rate (f_i). Of course the model considers the female population.

So the model describing the transition from one age class to the next and the generation of newborns is

$$\begin{cases} x_{k+1,1} = \sigma_0 f_1 x_{k,1} + \sigma_0 f_2 x_{k,2} + \sigma_0 f_3 x_{k,3} + \cdots \sigma_0 f_n x_{k,n} \\ x_{k+1,i} = \sigma_{i-1} x_{k,i-1} \end{cases} \tag{5.54}$$

where:

σ_0 is the survival rate of the newborns
f_i is the fertility of the ith class

Equation 5.54 can be written in compact matrix notation as

$$X_{k+1} = M \cdot X_k \tag{5.55}$$

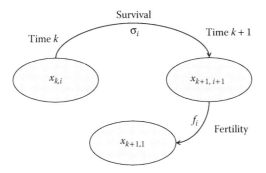

FIGURE 5.22   Evolution of the generic ith age group in the Leslie model. The survivors of the ith class at time k contribute to the newborn in the next time step $k + 1$.

where the system matrix is structured as

$$
M = \begin{bmatrix}
\sigma_0 f_1 & \sigma_0 f_2 & \sigma_0 f_3 & \dots & \sigma_0 f_n \\
\sigma_1 & 0 & 0 & \dots & 0 \\
0 & \sigma_2 & 0 & \dots & 0 \\
\dots & \dots & \dots & \dots & 0 \\
0 & 0 & 0 & \sigma_{n-1} & \sigma_n
\end{bmatrix}
\tag{5.56}
$$

The survival rates can be estimated as the ratio between the individuals in the i-th class at the current time (k) and the individuals in the younger class (i-1) at the previous time (k-1), that is,

$$
\sigma_i = \frac{x_{k,i}}{x_{k-1,i-1}} \quad i = 1, \dots, n-1
\tag{5.57}
$$

The elements in the first row of the matrix in Equation 5.56 are the product of the fraction of newborns produced by each age group (f_i) multiplied by the survival rate (σ_0) of the first age group. The last row also needs some explanation, being different from the previous ones. By matrix multiplication, the last model equation would read

$$
x_{k+1,n} = \sigma_{n-1} \cdot x_{k,n-1} + \sigma_n \cdot x_{k,n}
\tag{5.58}
$$

which makes sense if we consider the last age group as an open-ended group in the sense that it includes the individuals of age *n or more*, up to an indefinite upper limit. Therefore, in Equation 5.58 the first term represents the survival rate of the individuals previously in the $n - 1$ age group, whereas the second term accounts for the individuals who were already in the nth class at the previous time (k) and have survived (in the same class of course) until the next time ($k + 1$). In this way, σ_n may be regarded as a true survival rate within the nth age group. An alternative option is to consider the nth age group just as a *sink* that receives the surviving individuals from the $(n - 1)$th group, in which case $\sigma_n = 0$. This assumption will be used throughout Section 5.2.4.

The practical implementation of Equation 5.57 is not as straightforward as it may appear, because the population varies, and therefore the ratios may fluctuate. For this reason, it is important to perform a census survey at a time when the population can be assumed to be fairly stationary. However, the problem of reliably sampling a population, especially in the wild, poses major problems to field ecologists, and has given rise to many elaborate techniques that are used to estimate populations in the field (Krebs, 1998, 2009; Morris and Doak, 2002; Magurran, 2004).

5.2.4.1 Stable Age Distributions

In age-structured models, two concepts have a special importance: the stability of the population as a whole, and the stability of the age distribution. In particular, one may ask whether there are stable age distributions that guarantee the global stability (meaning that the total number of individuals of the population is stable). These two concepts are interrelated, as will be shown next. The age-structured model (Equation 5.55) is linear, so we can use the methods illustrated in Chapter 1 to study its stability, which depends on the eigenvalues of the matrix M of Equation 5.56, that can be computed by solving the matrix equation

$$
\det(M - \lambda I) = \det \begin{bmatrix}
\sigma_0 f_1 - \lambda & \sigma_0 f_2 & \sigma_0 f_3 & \dots & \sigma_0 f_n \\
\sigma_1 & -\lambda & 0 & \dots & 0 \\
0 & \sigma_2 & -\lambda & \dots & 0 \\
\dots & \dots & \dots & \dots & \dots \\
0 & 0 & 0 & \sigma_{n-1} & -\lambda
\end{bmatrix} = 0
\tag{5.59}
$$

which yields the nth-order polynomial equation

$$\lambda^n - \lambda^{n-1}\sigma_0 f_1 - \lambda^{n-2}\sigma_0\sigma_1 f_2 - \cdots - \sigma_0\sigma_1\sigma_2 \cdots \sigma_{n-1} f_n = 0 \qquad (5.60)$$

Defining $p_i = \sigma_0\sigma_1 \ldots \sigma_{i-1}$, Equation 5.60 can be written as

$$\lambda^n - \lambda^{n-1}p_1 f_1 - \lambda^{n-2}p_2 f_2 - \cdots \lambda^{n-i}p_i f_i - \cdots - p_n f_n = 0 \qquad (5.61)$$

so that its eigenvalues are functions of the survival and fertility rates. The Perron–Frobenius theorem (see, e.g., Gantmacher, 2000) states that a non-negative irreducible matrix has a dominant, real and positive, eigenvalue (λ_1) that is a simple root of the characteristic equation, and its module is larger than that of any other eigenvalue. Furthermore, all of the components of the corresponding eigenvector are positive. This theorem provides us with the important result that the asymptotic evolution of the population as a whole depends only on λ_1, that is,

$$\lim_{k \to \infty} X_k = M^k \cdot X_o \to c\lambda_1^k v \qquad (5.62)$$

where the constant c depends on the initial population (X_0). Asymptotically, the evolution of the population and its age-class distribution separate into the following facts:

1. The whole population grows with a rate given by λ_1

$$X_{k+1} = \lambda_1 \cdot X_k \Rightarrow \lim_{k \to \infty} \frac{x_{k+1,i}}{x_{k,1}} = \lambda_1 \qquad (5.63)$$

2. The asymptotic age-class distribution is determined by the eigenvector v associated with λ_1

$$\lim_{k \to \infty} \frac{x_{k,1}}{x_{k,i+1}} = \frac{v_i}{v_{i+1}} = \frac{\lambda_1}{\sigma_i} \qquad (5.64)$$

A simple, yet instructive, application of the Perron–Frobenius theorem to the Leslie matrix is the so-called Fibonacci's rabbits example. Consider a rabbit population with the following characteristics:

- The population is divided into three age groups
- Each rabbit survives exactly three classes
- Each female generates one female while in the first age-class and another one while in the second class
- The initial population includes one female

According to these definitions, the Leslie matrix for the rabbit example is

$$M = \begin{bmatrix} 1 & 1 & 0 \\ 1 & 0 & 0 \\ 0 & 1 & 0 \end{bmatrix} \qquad (5.65)$$

whose characteristic equation is

$$\lambda(\lambda^2 - \lambda - 1) = 0 \qquad (5.66)$$

The dominant eigenvalue is $\lambda_1 = (1 + \sqrt{5})/2 = 1.618$, which is precisely the Fibonacci ratio

$$F_r = \frac{F_{k+1}}{F_k} \quad \text{with } F_k = F_{k-1} + F_{k-2} \tag{5.67}$$

and in fact the population grows with rate

$$\lim_{k \to \infty} \frac{x_{k+1,i}}{x_{k,1}} = \lambda_1 = 1.618 \tag{5.68}$$

as shown in Figure 5.23.

Now let us reconsider Equation 5.63 to better understand the asymptotic behaviour of the Leslie model. Because the evolution of the population is ultimately governed by the dominant eigenvalue, that is, $Mv = \lambda_1 v$, we can write

$$\begin{bmatrix} \sigma_0 f_1 & \sigma_0 f_2 & \sigma_0 f_3 & \dots & \sigma_0 f_n \\ \sigma_1 & 0 & 0 & \dots & 0 \\ 0 & \sigma_2 & 0 & \dots & 0 \\ \dots & \dots & \dots & \dots & \dots \\ 0 & 0 & 0 & \sigma_{n-1} & 0 \end{bmatrix} \times \begin{bmatrix} v_1 \\ v_2 \\ v_3 \\ \dots \\ v_n \end{bmatrix} = \lambda_1 \begin{bmatrix} v_1 \\ v_2 \\ v_3 \\ \dots \\ v_n \end{bmatrix} \tag{5.69}$$

where $v = \begin{bmatrix} v_1 \dots v_n \end{bmatrix}^T$ is the stationary age distribution. Expanding Equation 5.69 row-wise yields

$$\begin{aligned} \sigma_0 f_1 v_1 + \sigma_0 f_2 v_2 + \dots + \sigma_0 f_n v_n &= \lambda_1 v_1 \quad &\text{for } i = 1 \\ \sigma_{i-1} v_{i-1} &= \lambda_1 v_i \quad &\text{for } i = 2, \dots, n \end{aligned} \tag{5.70}$$

The components of the second equation can be expressed in terms of v_1

$$\frac{\sigma_{i-1} v_{i-1}}{\lambda_1} = v_i \Rightarrow v_2 = \frac{\sigma_1 v_1}{\lambda_1}; v_3 = \frac{\sigma_2 v_2}{\lambda_1} = \frac{\sigma_1 \sigma_2 v_1}{\lambda_1^2}; \dots, v_i = \frac{\sigma_1 \sigma_2 \cdots \sigma_{i-1}}{\lambda_1^{i-1}} v_1 \quad i = 2, \dots, n \tag{5.71}$$

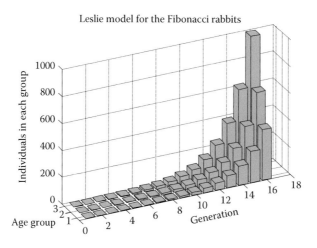

FIGURE 5.23   Evolution of the Fibonacci rabbit population.

Defining the quantities

$$p_i = \sigma_1 \ldots \sigma_{i-1} \quad i = 2, \ldots, n$$

$$p_1 = 1$$

(5.72)

so that

$$v_i = \frac{p_i}{\lambda_1^{i-1}} v_1$$

(5.73)

and substituting Equation 5.71 into Equation 5.70 yields

$$\sigma_0 f_1 + \frac{p_2 f_2}{\lambda_1} + \cdots + \frac{p_n f_n}{\lambda_1^{n-1}} = \lambda_1 \Rightarrow \sum_{i=1}^{n} p_i f_i \lambda_1^{-i} = 1$$

(5.74)

The stationary distribution in the age groups can be found by again using Equation 5.73 as

$$\pi_i = \frac{v_i}{v_1 + v_2 + \cdots + v_n} = \frac{(\sigma_1 \cdots \sigma_{i-1}/\lambda_1^i) v_1}{v_1 + (\sigma_1/\lambda_1) v_1 + (\sigma_1 \sigma_2/\lambda_1^2) v_1 + \cdots + (\sigma_1 \sigma_2 \cdots \sigma_{n-1}/\lambda_1^{n-1}) v_1}$$

$$= \frac{p_i \cdot \lambda_1^{-i}}{\sum_{j=1}^{n} p_j \cdot \lambda_1^{j-1}}$$

(5.75)

which shows that the stationary age distribution depends on the dominant eigenvalue λ_1 as well as on the survival coefficients $\sigma_1, \ldots, \sigma_{n-1}$. Therefore, Equations 5.74 and 5.75 define the characteristics of the asymptotic population. If, for example, we are interested in finding the conditions for zero growth, then we substitute $\lambda_1 = 1$ in Equation 5.74, which becomes

$$\sum_{i=1}^{n} p_i f_i = 1$$

$$\pi_i^0 = \frac{p_i}{\sum_{j=1}^{n} p_j}$$

(5.76)

involving both the fertility and survival coefficients.

5.2.4.2 A Simple Example of Leslie Model: The Great Tit (*Parus major*)

As an example, consider the following Leslie matrix used to model a population of Great tit (*Parus major*). The time-step in this case is 1 year and six classes are enough to cover the average lifespan of this avian species. The survival rates were obtained from census data, and we ask what number of eggs laid by each age group could stabilize the population. In this case, the fertility coefficients are the number of eggs, and $\sigma_0 = 0.09$ means that only 9% of the eggs give rise to a young chick surviving until the next year

$$M_{gt} = \begin{bmatrix} 0 & \sigma_0 f_2 & \sigma_0 f_2 & \sigma_0 f_2 & \sigma_0 f_2 & \sigma_0 f_2 \\ 0.6667 & 0 & 0 & 0 & 0 & 0 \\ 0 & 0.3333 & 0 & 0 & 0 & 0 \\ 0 & 0 & 0.45 & 0 & 0 & 0 \\ 0 & 0 & 0 & 0.6667 & 0 & 0 \\ 0 & 0 & 0 & 0 & 0.1667 & 0 \end{bmatrix}$$

(5.77)

Egg laying is largely influenced by the habitat (food availability, nesting sites, etc.). If, in a given habitat, each female lays three eggs, the matrix of Equation 5.77 becomes

$$\boldsymbol{M}_{gt} = \begin{bmatrix} 0 & 0.27 & 0.27 & 0.27 & 0.27 & 0.27 \\ 0.6667 & 0 & 0 & 0 & 0 & 0 \\ 0 & 0.3333 & 0 & 0 & 0 & 0 \\ 0 & 0 & 0.45 & 0 & 0 & 0 \\ 0 & 0 & 0 & 0.6667 & 0 & 0 \\ 0 & 0 & 0 & 0 & 0.1667 & 0 \end{bmatrix} \tag{5.78}$$

and the population is doomed because the dominant eigenvalues is a meagre $\lambda_1 = 0.6481$. To have a stable population, each female should lay on average 10.416 eggs, which yields the following matrix

$$\boldsymbol{M}_{gt} = \begin{bmatrix} 0 & 0.9374 & 0.9374 & 0.9374 & 0.9374 & 0.9374 \\ 0.6667 & 0 & 0 & 0 & 0 & 0 \\ 0 & 0.3333 & 0 & 0 & 0 & 0 \\ 0 & 0 & 0.45 & 0 & 0 & 0 \\ 0 & 0 & 0 & 0.6667 & 0 & 0 \\ 0 & 0 & 0 & 0 & 0.1667 & 0 \end{bmatrix} \tag{5.79}$$

This new matrix has indeed $\lambda_1 = 1$. The resulting stable population is shown in Figure 5.24.

5.2.4.3 A More Structured Leslie Model: The *Poa annua* Weed

So far, we have supposed that the Leslie survival and fertility coefficients are constant throughout the simulation, but there are cases in which these factors may vary as a function of the population density. This self-regulatory feedback mechanism controls the density variations by limiting the

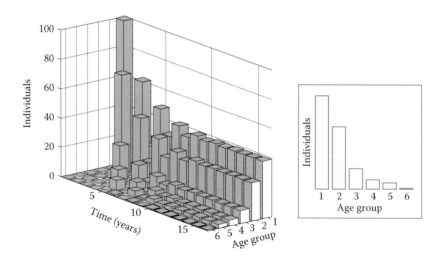

FIGURE 5.24 Evolution of the Great tit (*Parus major*) population with the Leslie matrix of Equation 5.79. The stable distribution is shown in the inset. These graphs were obtained with the MATLAB script `Go_Leslie_Great_tit.m` in the `ESA_Matlab\Exercises\Chapter_5\Single\Leslie_Models` folder.

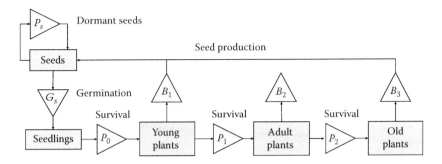

FIGURE 5.25 Model structure for the *P. annua* weed. The triangles represent the survival and fertility coefficients, whereas the boxes indicate the age groups.

new generations. Such is the case for the *Poa annua* weed (Begon and Mortimer, 1986), for which a five-class Leslie model was developed with the structure of Figure 5.25.

The population is divided in to five age groups: seeds, seedlings, young plant, adult plant, old plants, for which the Leslie matrix can be written as follows:

$$\boldsymbol{M}_{pa} = \begin{bmatrix} P_s & 0 & B_1 & B_2 & B_3 \\ G_s & 0 & 0 & 0 & 0 \\ 0 & P_0 & 0 & 0 & 0 \\ 0 & 0 & P_1 & 0 & 0 \\ 0 & 0 & 0 & P_2 & 0 \end{bmatrix} \tag{5.80}$$

Field studies have suggested that the matrix coefficients can be structured as a function of the total *P. annua* population as follows:

$$P_0 = \begin{cases} 0.75 - 0.25e^{(0.00005N)} & \text{for } N < 27726 \\ 0 & \text{for } N > 27726 \end{cases}$$

$$B_1 = B_3 = 100e^{-0.0001N}$$

$$B_2 = 200e^{-0.0001N} \tag{5.81}$$

$$P_s = 0.2$$

$$G_s = 0.05$$

where N is the total population density. The zero in the second place of the first row of $\boldsymbol{M}_{pa}$ means that the seedlings do not generate seeds, whereas their survival (P_0) strongly depends on the total weed density, as do the seed generation coefficients (B_1, B_2, B_3) of the fully developed plants. Simulating the model (Equation 5.80) with the suggested parameters (Equation 5.81) yields the weed evolution of Figure 5.26, showing an almost stable age distribution (white bars at the right end).

The eigenvalues change during the simulation as a result of the density functions of Equation 5.81, as shown in Figure 5.27, until they reach their final location, indicated by the grey dots. The population is almost stationary because the final matrix and its eigenvalues are complex conjugate, save for the dominant λ_1 of course.

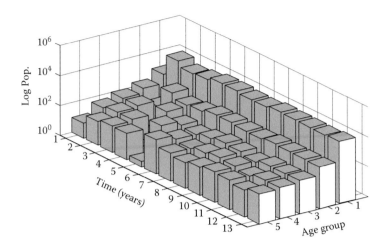

FIGURE 5.26 Evolution of the age-structured model (Equations 5.80 and 5.81) until a stable zero-growth distribution is reached. The white bars on the right indicate the stable age distribution. The spread of the population values suggested to plot the logarithm of the results. This graph was obtained with the MATLAB script `Go _ Leslie _ Poa.m` in the `ESA _ Matlab\Exercises\Chapter _ 5\Single\Leslie _ Models` folder.

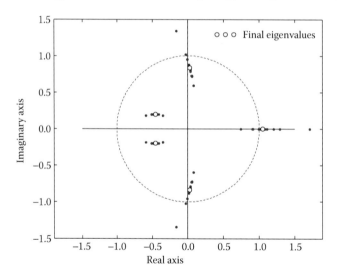

FIGURE 5.27 Evolution of the eigenvalues of the matrix (5.80) during the simulation of Figure 5.26. The major grey dots indicate their final location. This graph was obtained with the MATLAB script `Go _ Leslie _ Poa.m` in the `ESA _ Matlab\Exercises\Chapter _ 5\Single\Leslie _ Models` folder.

$$
\boldsymbol{M}_{pa} = \begin{bmatrix}
0.2000 & 0.0000 & 25.3172 & 50.6344 & 25.3172 \\
0.0500 & 0.0000 & 0.0000 & 0.0000 & 0.0000 \\
0.0000 & 0.2531 & 0.0000 & 0.0000 & 0.0000 \\
0.0000 & 0.0000 & 0.7500 & 0.0000 & 0.0000 \\
0.0000 & 0.0000 & 0.0000 & 0.7500 & 0.0000
\end{bmatrix}
\tag{5.82}
$$

$$\lambda_1 = 1.0513$$

$$\lambda_{2,3} = 0.0274 \pm j0.8359$$

$$\lambda_{4,5} = -0.4530 \pm j0.1997$$

That λ_1 is slightly larger than one is due to the fact that the population has not yet reached its steady state, or its 'climax' in ecological terms. Lengthening the simulation horizon, for example, to 100 years results in $\lambda_1 = 1.0000$, thus confirming that the system indeed tends to a stable population.

5.3   HARVESTED POPULATIONS

Populations may have an economic value, and their exploitation has been the subject of many bio-economic studies. Models for determining the amount of harvest that would not impair the population development have been proposed, with a special emphasis to fisheries (Clark, 1990) and wildlife management (Starfeld and Bleloch, 1986). Setting up a single-species exploitation model requires adding a harvesting term to the previous growth models, to yield

$$\frac{dx}{dt} = x \cdot F(x) - h(x) \tag{5.83}$$

where in addition to the growth function $F(x)$, we now have a subtractive term $h(x)$ representing the rate at which the resource (i.e., part of the population) is withdrawn from its habitat. We shall now proceed to assess the stability of Equation 5.83, which in this case means *sustainability of the exploitation*, as a function of the harvesting term $h(x)$. The aim of the analysis will be the determination of a limit to the withdrawal that will reconcile the economic return with the population survival.

5.3.1   CONSTANT HARVESTING

One of the major problems in implementing a sustainable harvesting policy is the knowledge, or at least a good estimate, of the actual wild population stock (Young et al., 1994; Krebs, 1998, 2009; Morris and Doak 2002; Gonzalez and Descamp-Julien, 2004; Freeman, 2008; Mittelbach, 2012). Let us begin our analysis by considering a constant withdrawal policy that does not require any stock estimate. This means that the harvesting term $h(x) = h$ is constant. Adopting a logistic model for the population, Equation 5.83 then becomes

$$\frac{dx}{dt} = r \cdot x \cdot \left(1 - \frac{x}{K}\right) - h \tag{5.84}$$

The question now is how much *sustainable* harvesting effort can be applied without bringing the population to extinction. The steady-state population can be computed by setting the derivative of Equation 5.84 to zero and solving for x. Figure 5.28 shows the three possible settings, depending on the level at which the constant harvesting line crosses the logistic curve.

   For the lower line, there are two intersection points, with differing stability characteristics. Point A corresponds to a positive derivative of the growth curve, and therefore is unstable, as already discussed in Section 5.1.1, whereas point B is stable, having a negative derivative. Consequently, this latter point is the stable equilibrium under harvesting, and it will attract the system trajectories provided they start from the right of point A, so that the population will settle at the stable value $\bar{x} < K$, instead of K if it were not harvested. However, this policy is not entirely safe, because if the harvesting begins *before* point A has been surpassed, that is, for populations $x < x^*$, the population will be pushed back towards extinction. Figure 5.29 shows several trajectories for differing harvesting start times. If harvesting begins before the population has grown beyond x^*, the unstable equilibrium will push it back towards extinction. Conversely, any harvesting starting at $x > x^*$ will terminate at the stable population $\bar{x}$.

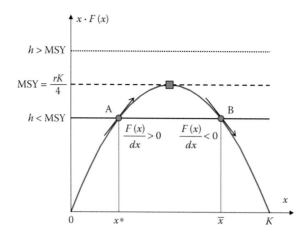

FIGURE 5.28 Three possible situations in the constant harvesting policy. Only the lower one (solid line) is sustainable, whereas the others either are metastable, or directly lead to extinction. MSY stands for maximum sustainable yield. The harvested population settles at the stable $\bar{x} < K$ value, whereas x^* represents the unstable equilibrium.

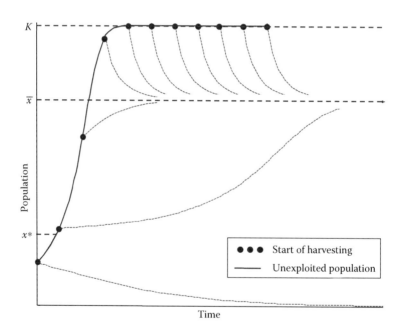

FIGURE 5.29 Population trajectories depending on the harvesting start time. If the exploitation begins when the population has not yet exceeded the unstable equilibrium (x^*), the trajectory will be pushed towards the extinction. On the contrary, any harvesting beginning with a population $x_{start} > x^*$ will reach the stable harvested level $\bar{x}$. This graph was obtained with the MATLAB script Go _ const _ harvest.m in the ESA _ Matlab\Exercises\Chapter _ 5\Harvest folder.

A relationship between the harvesting effort h and the steady population $\bar{x}$ can be obtained by solving Equation 5.84 for x after setting the derivative to zero and considering only the stable solution, which will necessarily be for $x > K/2$

$$0 = r \cdot x \cdot \left(1 - \frac{x}{K}\right) - h = x^2 - Kx + \frac{Kh}{r} \Rightarrow \bar{x} = \frac{K + \sqrt{K^2 - 4(Kh/r)}}{2} > \frac{K}{2} \qquad (5.85)$$

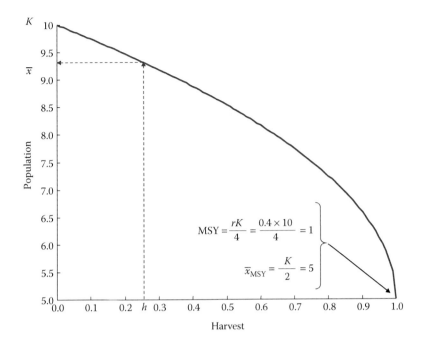

FIGURE 5.30   Amount of harvest as a function of the exploitation level h, for the logistic parameter values indicated in the graph. The possible harvest values range from a maximum corresponding to the minimum sustainable population $\bar{x}_{MSY}$ to nil in the case of no harvesting.

Figure 5.30 shows the amount of possible return obtained by applying a certain harvesting effort, ranging from the MSY corresponding to the minimum sustainable population x_{MSY}, to nil in the case of no harvesting.

The constant harvesting policy seems safe enough, but there is one last caveat regarding the kind of population being exploited. Figure 5.31 shows the impact that the same harvesting variation Δh may have on two populations with differing growth rates. Species with a low growth rate, usually referred to as *K-strategists* by ecologists, rely on the stability of the population and have

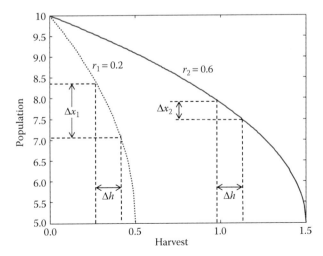

FIGURE 5.31   Differing effects of harvesting depending on the population growth rate. If the growth rate is relatively low, even a small harvesting variation may have a major impact on the population.

a low growth rate. If the population level is altered, returning to the previous level may be difficult, and hence the perturbation can be considerable. Conversely, populations with a high growth rate, known as *r-strategists*, are quick to replace the missing individuals thanks to their robust fertility, which makes them much more resilient to harvesting variations.

5.3.2 PROPORTIONAL HARVESTING

In the previous section, we have examined the risk of implementing a constant harvesting policy, which may result in an irreversible depletion of the resource. This drawback has motivated a safer exploitation policy based on *proportional* harvesting. We shall see that this approach always produces a stable harvesting policy, but it has the disadvantage of requiring a knowledge of the exploited stock which, as previously pointed out, is not an easy task. The proportionally harvested population dynamics, still assuming an intrinsic logistic growth, is therefore

$$\frac{dx}{dt} = r \cdot x \cdot \left(1 - \frac{x}{K}\right) - g \cdot x \tag{5.86}$$

where g represents the fraction of the stock x that is withdrawn per unit of time.

Again, we now analyse the equilibrium of Equation 5.86 as a function of the harvesting factor g by setting the derivative to zero. The equilibrium population can be computed as

$$0 = r \cdot x \cdot \left(1 - \frac{x}{K}\right) - g \cdot x \Rightarrow \bar{x} = K\left(1 - \frac{g}{r}\right) \tag{5.87}$$

and the yield is defined as

$$Y = g \cdot \bar{x} \tag{5.88}$$

The stability condition requires that the *total* derivative with respect to x to be negative, thus

$$\left.\frac{dF}{dx}\right|_{\bar{x}} = \frac{d}{dx}\left[rx\left(1 - \frac{x}{K}\right) - gx\right]\Bigg|_{\bar{x}} = r\left(1 - \frac{2\bar{x}}{K}\right) - g$$

$$= r\left(1 - \frac{2K\left(1 - (g/r)\right)}{K}\right) - g = r\left(1 - 2\left(1 - \frac{g}{r}\right)\right) - g \tag{5.89}$$

$$= r - 2r + 2g - g = -r + g < 0 \Rightarrow g < r$$

By inspecting Figure 5.32 it can be seen that this stability constraint $(g < r)$ is satisfied so long as the harvesting line $(g \cdot x)$ intersects the growth curve, whose maximum derivative (r) representing its upper limit for $x = 0$. It is easy to show that the growth curve derivative is always less than g

$$\frac{d\left(x \cdot F(x)\right)}{dx} = \frac{d}{dx}\left[rx\left(1 - \frac{x}{K}\right)\right]\Bigg|_{\bar{x}} = r - 2\frac{\bar{x}}{K} < r \quad \text{for } 0 < \bar{x} < K \tag{5.90}$$

The only exception, as will be shown later, is the case in which the population has *depensation* in its initial part. Equation 5.87 yields a harvest/population relationship similar to Equation 5.85, which shows that the MSY still occurs for

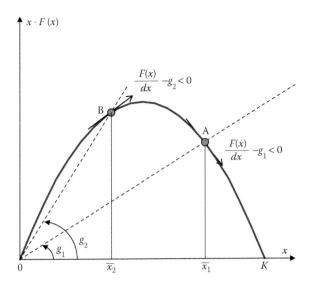

FIGURE 5.32 Two possible proportional harvesting policies, both resulting in a stable equilibrium.

$$\overline{x}_{MSY} = \frac{K}{2} \Rightarrow g_{MSY} = r\left(1 - \frac{\overline{x}_{MSY}}{K}\right) = r\left(1 - \frac{K}{2K}\right) = \frac{r}{2} \Rightarrow Y_{max} = g_{MSY} \cdot \overline{x}_{MSY} \quad (5.91)$$

However, unlike with constant harvesting, a withdrawal $g > r/2$ does not destabilize the population, though it does result in a smaller yield. The proportional harvesting situation is shown in Figure 5.33, in which (a) shows the intersection between the growth curve and the harvest line with slope g produces the equilibrium population $\overline{x}$.

The resulting yield $Y = g \cdot \overline{x}$ is then plotted in (b) versus the proportional harvest g. It can be seen that the (b) curve has the same maximum yield as the constant policy for $\overline{x} = K/2$, but a further increase of the exploitation does not destabilize the population, though resulting in a smaller yield.

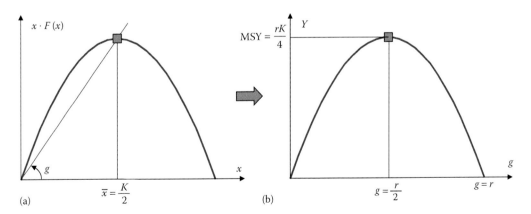

FIGURE 5.33 Computation of the yield curve for proportional harvesting. The intersection between the growth curve and the harvest line (a) produces the yield that is plotted as a function of g in (b). The MSY is the same as that of the constant harvest, but a large g does not destabilize the population, though it results in a lower yield. Proportional harvest can be simulated with the MATLAB script Go_prop_harvest.m in the ESA _ Matlab\Exercises\Chapter _ 5\Harvest folder.

5.3.2.1 Harvesting Depensated Populations

The only case in which proportional harvesting can produce a destabilizing effect is when the growth curve exhibits a depensation, consisting of a reduced growth for small population values. This representation is borrowed from the fishery biological literature to indicate growth curves with an initial upward convexity, resulting in lesser growth. Two possible depensation curves exist, depending on whether the growth is just reduced (simple depensation) or becomes negative (critical depensation), as shown in Figure 5.34. Proportional harvesting of a depensated population may lead to a violation of the stability condition (Equation 5.89), as shown in Figure 5.35, in which B is unstable because the derivative of the growth curve is greater than g, whereas A is conditionally stable (metastable), unlike C which has no unstable counterpart. The maximum sustainable, though metastable, equilibrium is when g is the tangent to the growth curve.

Regarding the stability of the harvesting policy, if we repeat the procedure leading to the construction of the yield curve of Figure 5.33, the situation becomes problematic because, as a consequence of the inflexion in the growth curve, we have seen that the intersection with the g line may produce two solutions, one of which is unstable.

In this case, the yield curve will have two branches, a stable one and an unstable one, with the former being subject to being destabilized by the unstable counterpart, depending on the level of

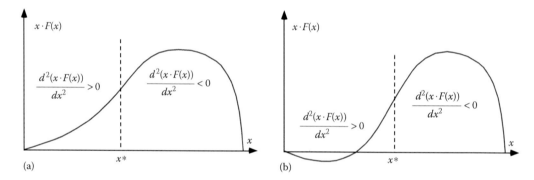

FIGURE 5.34 Depensated growth curves. Simple depensation (a) still provides positive growth for all x, whereas critical depensation (b) may involve a negative growth rate for small populations. In both cases, depensation presents an initially upward convexity, for $0 < x < x^*$.

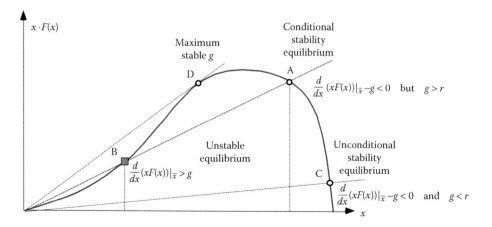

FIGURE 5.35 Harvesting a depensated population may result in two equilibria (A and B) if the harvest line crosses the growth curve in two points. In point B, the derivative condition is violated, and hence B is unstable, whereas A is stable. Point C is unconditionally stable because it has no counterpart in the ascending branch of the growth curve. D represents the metastable point where the harvest line is tangent to the growth curve.

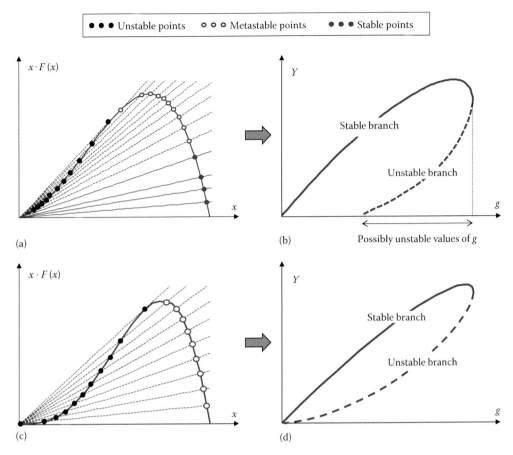

FIGURE 5.36 Proportional harvesting with a depensated growth curve. In (a), the dashed harvest lines produce two equilibria, the leftmost of which is unstable (black dots) and a (meta)stable one (grey dots), whereas the solid harvest lines produce only one stable equilibrium (blue dots). The corresponding effort/yield curve (b) has an upper stable branch (solid line) and an underlying unstable branch (dashed line) for the values of g corresponding to the dashed lines. The lower graphs (c) and (d) show the same computation for a critically depensated growth curve. In this case all the effort values produce an unstable equilibrium. These graph were obtained with the MATLAB scripts `Yield_depensated.m` and `Yield_critical_depens.m` in the ESA_Matlab\ Exercises\Chapter_5\Harvest folder.

the population, which may fall in the domain of attraction of the unstable point and thus be pushed back to the origin (extinction). Figure 5.36 shows how these solution branches are formed in the case of simply or critically depensated populations.

5.4 COMPETITION

So far, we have considered a single-species population, a highly improbable case in practice, because the same habitat is normally shared by several species. Sharing the same resource inevitably creates a competitive situation, which we are now going to consider by extending the logistic model to n species as follows:

$$\frac{dx_i}{dt} = x_i \left[r_i - \sum_{j=1}^{n} \gamma_{ij} x_j \right] \quad i = 1,\ldots,n \tag{5.92}$$

where

r_i is the growth rate of each species,

γ_{ii} are the intraspecific competition coefficients,

γ_{ij} with $i \neq j$ are the interspecific competition coefficients.

Let us begin by considering the competition between two species, so that Equation 5.92 becomes

$$\begin{cases} \dfrac{dx_1}{dt} = x_1 \left(r_1 - \gamma_{11}x_1 - \gamma_{12}x_2 \right) \\ \dfrac{dx_2}{dt} = x_2 \left(r_2 - \gamma_{22}x_2 - \gamma_{21}x_1 \right) \end{cases} \tag{5.93}$$

The equilibrium can be determined by setting the derivative to zero. Solving for x_1 and x_2 yields

$$\bar{x}_1 = \frac{r_1 - \gamma_{12}\bar{x}_2}{\gamma_{11}} \quad \bar{x}_2 = \frac{r_2 - \gamma_{21}\bar{x}_1}{\gamma_{22}} \tag{5.94}$$

It can be seen that Equation 5.94 represents two lines in the (x_1, x_2) plane. The equilibrium will be *admissible* if they cross in the first quadrant, that is, $(\bar{x}_1, \bar{x}_2) > 0$, while its *feasibility* depends on the Jacobian of Equation 5.93. This matrix can be obtained by deriving with respect to x_1 and x_2 at the equilibrium of Equation 5.94

$$J = \begin{bmatrix} \dfrac{d}{dx_1}\left(r_1 x_1 - \gamma_{11}x_1^2 - \gamma_{12}x_1 x_2 \right) & \dfrac{d}{dx_2}\left(r_1 x_1 - \gamma_{11}x_1^2 - \gamma_{12}x_1 x_2 \right) \\ \dfrac{d}{dx_1}\left(r_2 x_2 - \gamma_{22}x_2^2 - \gamma_{21}x_1 x_2 \right) & \dfrac{d}{dx_2}\left(r_2 x_2 - \gamma_{22}x_2^2 - \gamma_{21}x_1 x_2 \right) \end{bmatrix}$$

$$= \begin{bmatrix} r_1 - 2\gamma_{11}x_1 - \gamma_{12}x_2 & -\gamma_{12}x_1 \\ -\gamma_{21}x_2 & r_2 - 2\gamma_{22}x_2 - \gamma_{21}x_1 \end{bmatrix}_{(\bar{x}_1, \bar{x}_2)} \tag{5.95}$$

Substituting the equilibrium values of Equation 5.94 into Equation 5.95 produces the Jacobian at the equilibrium

$$J = \begin{bmatrix} r_1 - 2\gamma_{11}\bar{x}_1 - \gamma_{12}\bar{x}_2 & -\gamma_{12}\bar{x}_1 \\ -\gamma_{21}\bar{x}_2 & r_2 - 2\gamma_{22}\bar{x}_2 - \gamma_{21}\bar{x}_1 \end{bmatrix} \tag{5.96}$$

which yields information about the stability of the equilibrium through its eigenvalues, as discussed in Chapter 1 for continuous-time systems. However, there is a more intuitive approach to the analysis of the equilibrium using the isocline approach (Section 1.9.5). Recalling their definition,

$$\begin{matrix} \dfrac{dx_1}{dt} = f_1(x_1, x_2) \\ \dfrac{dx_2}{dt} = f_2(x_1, x_2) \end{matrix} \Rightarrow \text{slope}: \dfrac{dx_2}{dx_1} = \dfrac{dx_2/dt}{dx_1/dt} \Rightarrow \begin{cases} \text{horizontal isocline} & \dfrac{dx_2}{dt} = 0 \\ \\ \text{vertical isocline} & \dfrac{dx_1}{dt} = 0 \end{cases} \tag{5.97}$$

whereby the $\dot{x}_1 = 0$ isocline (equilibrium of the first species x_1) are crossed by vertical trajectories, and the $\dot{x}_2 = 0$ (equilibrium of the second species x_2) are traversed horizontally. Equation 5.94 is in fact our equilibrium isocline, and its portrait is shown in Figure 5.37 in which, from Equation 5.94, the points of intersection with the axes are given by

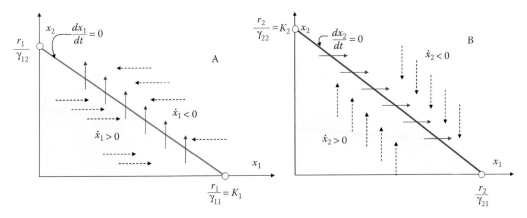

FIGURE 5.37 Graphical analysis of the two-species competitive equilibrium with the method of isoclines. A shows the equilibrium isocline for x_1, and B depicts the equilibrium isoclines for x_2. The pertinent variable has a positive time derivative in the light blue areas and negative one in the grey area. The dashed arrows indicate the component of the trajectories in approaching the equilibrium.

$$x_1 = \frac{r_1 - \gamma_{12}x_2}{\gamma_{11}} \rightarrow \begin{cases} x_1 = 0 \Leftrightarrow r_1 = \gamma_{12}x_2 \Rightarrow x_2 = \dfrac{r_1}{\gamma_{12}} \\ \\ x_2 = 0 \Leftrightarrow x_1 = \dfrac{r_1}{\gamma_{11}} = K_1 \end{cases}$$

$$x_2 = \frac{r_2 - \gamma_{21}x_1}{\gamma_{22}} \rightarrow \begin{cases} x_1 = 0 \Leftrightarrow r_1 = x_2 = \dfrac{r_2}{\gamma_{22}} = K_2 \\ \\ x_2 = 0 \Leftrightarrow x_1 = \dfrac{r_2}{\gamma_{21}} \end{cases}$$

(5.98)

Notice that in Equation 5.98, the carrying capacities for the two species can be determined as the ratio between the growth rate and the intraspecific competition coefficient. Because each isocline is an equilibrium locus, the trajectory will tend to it in the sense indicated by the dashed arrows. In the light-blue areas, the species is below its carrying capacity and hence the time derivative is positive, whereas in the grey areas, the species exceeds its carrying capacity, and therefore its time derivative is negative. The four possible combinations of the relative isoclines positions, and hence of the kind of equilibrium, are shown in Figure 5.38.

In the first two cases (A and B), the intersection of the isoclines is outside the first quadrant, which means that no feasible $(x_1, x_2) > 0$ equilibrium is possible, and only the species whose isocline is external survives (species 1 in A and species 2 in B). In the other two cases (C and D), the isoclines do cross in the first quadrant, but only C results in a stable coexistence, whereas in D which species prevails depends on the initial conditions. Now let us consider the four cases in detail, examining the trajectory from an arbitrary initial condition to the equilibrium. Figure 5.39 shows the survival of species 1, and Figure 5.40 the survival of species 2. In general, when the intersection is outside the first quadrant, the species with the lower isocline is doomed.

Figure 5.41 instead shows the competitive coexistence where the intersection of the isoclines is a stable point, whereas Figure 5.42 shows the case in which the intersection is unstable and repels the trajectories that end in either species' carrying capacity while the other becomes extinct.

The dashed curve passing through the intersection of the two isoclines is called a *separatrix* because it separates the domains of attraction of the two stable equilibria. Being very difficult to determine analytically, it is normally estimated by numerical means by producing a large number of trajectories near the expected boundary and approximating the divide by observing their direction.

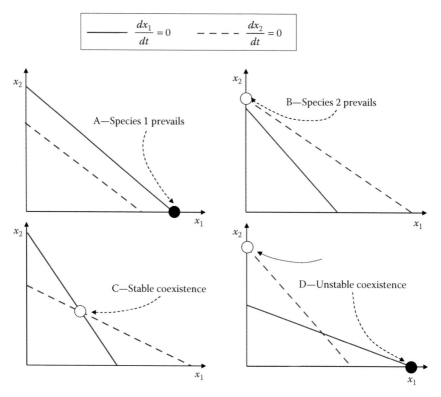

FIGURE 5.38 Possible mutual positions of the isoclines giving rise to four differing equilibria. Only C results in a stable coexistence, whereas in all other cases only one species survives while the other becomes extinct.

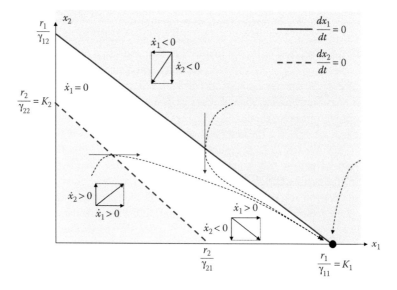

FIGURE 5.39 Trajectories in the case of the survival of species 1, whose isocline is above that of species 2. The coloured areas show the differing sign of the time derivative, and all the trajectories point to the carrying capacity of species 1, irrespective of the initial conditions, while species 2 becomes extinct. This graph was obtained with the MATLAB script Go _ Extinction _ 2.m in the ESA _ Matlab\Exercises\ Chapter _ 5\Competition folder.

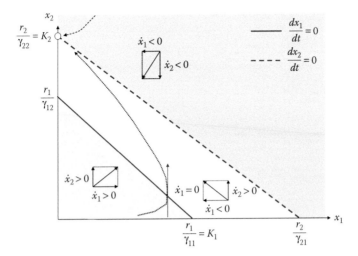

FIGURE 5.40 Trajectories in the case of the survival of species 2, whose isocline is above that of species 1. The coloured areas show the differing sign of the time derivative, and all the trajectories point to the carrying capacity of species 2, irrespective of the initial conditions, while species 1 becomes extinct. This graph was obtained with the MATLAB script `Go _ Extinction _ 1.m` in the `ESA _ Matlab\Exercises\ Chapter _ 5\Competition` folder.

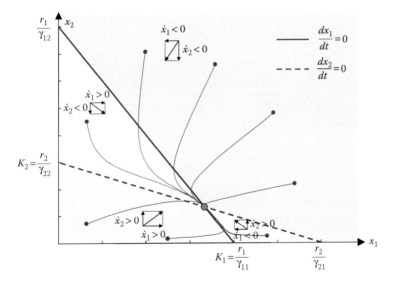

FIGURE 5.41 Stable coexistence of both species. The equilibrium point is globally stable because it is reached from any initial condition. This graph was obtained with the MATLAB script `Go _ Comp _ Equil.m` in the `ESA _ Matlab\Exercises\Chapter _ 5\Competition` folder.

We are now going to work out a set of analytical conditions for the competitive equilibrium. Let us consider again the equilibrium equations

$$
\begin{cases}
\dfrac{dx_1}{dt} = x_1\left(r_1 - \gamma_{11}x_1 - \gamma_{12}x_2\right) \\
\dfrac{dx_2}{dt} = x_2\left(r_2 - \gamma_{22}x_2 - \gamma_{21}x_2\right)
\end{cases}
\xRightarrow{\frac{d}{dt}=0}
\begin{cases}
r_1 = \gamma_{11}x_1 + \gamma_{12}x_2 \\
r_2 = \gamma_{22}x_2 + \gamma_{21}x_2
\end{cases}
\tag{5.99}
$$

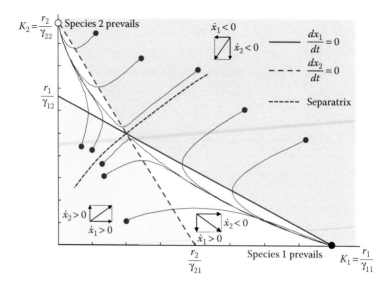

FIGURE 5.42 Competitive exclusion. The prevailing species depends on the initial conditions. Notice that the intersection point is unstable and repels the approaching trajectories. The separatrix curve divides the domains of attraction of the two equilibria and obviously passes through the unstable equilibrium at the intersection. This graph was obtained with the MATLAB script Go _ Comp _ Exclusion.m in the ESA _ Matlab\Exercises\Chapter _ 5\Competition folder.

from which the linear system on the right-hand side can be solved for x_1 and x_2

$$\begin{bmatrix} \gamma_{11} & \gamma_{12} \\ \gamma_{21} & \gamma_{22} \end{bmatrix}\begin{bmatrix} x_1 \\ x_2 \end{bmatrix} = \begin{bmatrix} r_1 \\ r_2 \end{bmatrix} \Rightarrow \begin{aligned} x_1 &= \dfrac{r_1\gamma_{22} - r_2\gamma_{12}}{\gamma_{11}\gamma_{22} - \gamma_{12}\gamma_{21}} \\ x_1 &= \dfrac{r_2\gamma_{11} - r_1\gamma_{21}}{\gamma_{11}\gamma_{22} - \gamma_{12}\gamma_{21}} \end{aligned} \tag{5.100}$$

Requiring the solution to be in the first quadrant $(x_1, x_2) > 0$ yields

$$\begin{aligned} r_1\gamma_{22} - r_2\gamma_{12} > 0 \\ r_2\gamma_{11} - r_1\gamma_{21} > 0 \end{aligned} \Rightarrow \begin{cases} \dfrac{r_1}{\gamma_{12}} > \dfrac{r_2}{\gamma_{22}} = K_2 \\ \dfrac{r_2}{\gamma_{21}} > \dfrac{r_1}{\gamma_{11}} = K_1 \end{cases} \tag{5.101}$$

This result confirms the relative position in which the isoclines in Figure 5.41 cross the x_1 and x_2 axes. The biological interpretation of this conclusion is that stable coexistence is possible only if the carrying capacity of each species is less than the ratio between the other species' growth rate divided by the intraspecific competition coefficient. In other words, coexistence requires the interspecific competition to be moderate with respect to the carrying capacity. Conversely, when interspecific competition is strong the exclusion principle holds, with the prevailing species being determined by the initial conditions.

5.4.1 MATLAB ANALYSIS OF COMPETITION

Before proceeding further, let us consider the MATLAB implementation of the competitive interactions just introduced. The Simulink model implementing the two-species competition is shown in Figure 5.43 and consists of two function blocks containing the interaction expressions, followed by two integrators. The simulation output is exported to the workspace via the matrix simout.

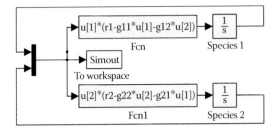

FIGURE 5.43 Simulink two-species competition model. The simulation results are exported to the work-space as the simout matrix. This Simulink model (Coexist.mdl) can be retrieved from the ESA _ Matlab\Exercises\Chapter _ 5\Competition folder.

This model has been used to simulate each of the four possible equilibria, with the launch MATLAB script setting the conditions to obtain any of them. The main body of the script is listed in Box 5.2 for the stable coexistence arrangement of Figure 5.41. To simulate the other possibilities (survival of one species or competitive exclusion) the parameters should be selected as indicated in the pertinent column of Box 5.3.

BOX 5.2   MATLAB LAUNCH SCRIPT FOR THE COMPETITIVE EQUILIBRIUM MODEL, MAKING USE OF THE COEXIST.MDL SIMULINK MODEL

```
% Simulation of competitive equilibrium
clear
% Set the model coefficients
r1=0.2;
r2=0.15;
g11=0.01;
g12=0.015;
g21=0.003;
g22=0.03;
% Set the final simulation time
tfin=300;
figure(1)
hold off;clf
% Extremal isocline points
% species 1
K1=r1/g11;
P1=r1/g12;
%species 2
K2=r2/g22;
P2=r2/g21;
figure(1)
hold off;clf
plot([K1,0],[0,P1],'b']) % Species 1 isocline
hold on
plot([P2,0],[0,K2],':b') % Species 2 isocline
% Enter the initial conditions
butt=1;
while butt==1
[x1o,x2o,butt]=ginput(1);
plot(x1o,x2o,'o')
```

Box 5.2 continued

Box 5.2 continued

```
[t,x,y]=sim('coexist',tfin);
plot(simout(:,1),simout(:,2))
drawnow
xlabel('x1')
ylabel('x2')
end
% Form the competition coefficients matrix
G=[g11 g12;g21 g22];
R=[r1 r2]';
% Compute the equilibrium
Xo=G\R
% Compute the Jacobian
J=[r1-2*g11*Xo(1)-g12*Xo(2) -g12*Xo(1);-g21*Xo(2) r2-g21*Xo(1)
-2*g22*Xo(2)];
% Compute the eigenvalues
L=eig(J)
```

BOX 5.3　SUGGESTED MODEL COEFFICIENT TO OBTAIN EACH OF THE THREE REMAINING EQUILIBRIA

Competitive exclusion	Species 1 survival	Species 2 survival

r1=0.4;
r2=0.2;
g11=0.01;
g22=0.02;
g12=0.06;
g21=0.01;
K1=r1/g11;
K2=r2/g22;
	r1=0.2;
	r2=0.15;
	g12=0.0067;
	g21=0.0107;
	g11=0.0100;
	g22=0.0060;
	K1=20;
	K2=25;
		r1=0.2;
		r2=0.15;
		g12=0.0114;
		g21=0.0062;
		g11=0.0100;
		g22=0.0060;
		K1=20;
		K2=25;

All the MATLAB software to simulate the four possible conditions of the competitive equilibrium is available from the `ESA _ Matlab\Exercises\Chapter _ 5\Competition` folder.

5.4.2　An Example of Competition: Gause's Protozoa

We are now going to revisit a historical experiment that the Russian ecologist Georgy Gause (1934) carried out by growing two species of protozoa (*Paramecium aurelia* and *Paramecium caudatum*), first separately, and then in the same environment. He observed that while each species alone developed up to the carrying capacity, when grown together one species prevailed over the other, which was brought near extinction, though there was ample food for both. The data from his experiments were later analysed by Leslie (1957), who was the first to apply mathematical modelling to these data to explain the observed behaviour. The analysis can be made in two steps: first a logistic model is fit to the data of each species grown alone (Figure 5.44), then a competition model is fit to the combined growth data (Figure 5.45).

It can be seen that the *P. aurelia* becomes dominant at the expense of the declining population of *P. caudatum*, although food is plentiful for both species. The parameters of the three models are grouped in Table 5.1, from which it can be seen that the competition has a major impact on both species parameters.

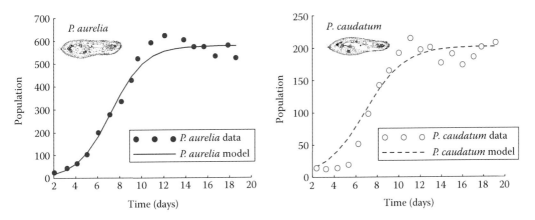

FIGURE 5.44 Fitting the Gause Paramecia data (Gause, 1934) when each species is grown separately.

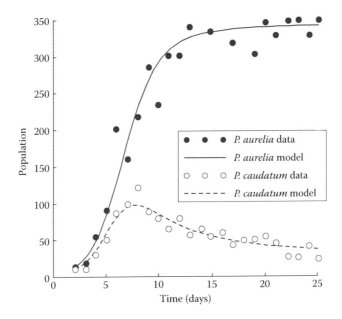

FIGURE 5.45 Fitting the Gause Paramecia data (Gause, 1934) when both species share the same environment, giving rise to competition. The *P. caudatum* is clearly the succumbing species.

TABLE 5.1

Comparison of Protozoa Logistic Parameters in Separate and Common Growth Conditions

Parameter	Separate Growth (Logistic Model)		Common Growth (Competition Model)	
	P. aurelia	*P. caudatum*	*P. aurelia*	*P. caudatum*
r	0.6555	0.5472	0.7045	0.7901
K	576.7464	202.1490	354.4325	218.3363
γ_{11}	0.00113655	–	0.0019876	–
γ_{22}	–	0.00270691	–	0.0036185
γ_{12}	–	–	0.000629	–
γ_{21}	–	–	–	0.001977

Consider that while in the separate growth experiments the carrying capacity K was directly estimated, when fitting the competitive model it is computed as the ratio between the growth factor (r) and the intraspecific competition coefficients, that is, $K_1 = r_1/\gamma_{11}$ and $K_2 = r_2/\gamma_{22}$. Figure 5.46 presents the same data in the now familiar species phase plane. The competition equilibrium is located at the intersection of the two isoclines with a very low value for the *P. caudatum*, confirming that this species is greatly affected by the competition and forced near to extinction by the pressure of the competing species *P. aurelia*. As a result of these experiments, Gause formulated the competitive exclusion principle, stating that *two species competing for the same resource cannot coexist at constant population values if other ecological factors remain constant*. The species with the largest ratio $r_i/\gamma_{ii} = K_i$ will prevail in the long term, either leading to the extinction of its competitor, or forcing it to shift towards a different ecological niche. This long-term behaviour is obviously not apparent from the short time experiments considered here, so the equilibrium shown in Figure 5.46 may be temporary, and eventually lead to the extinction of the *P. caudatum*.

5.4.3 Competition for a Limited Resource

So far, we have assumed an unlimited common resource, but if the two species compete for a limited resource, how will their coexistence be affected? To investigate this new aspect of competition, consider a model where, in addition to growth, each species is influenced by the availability of a resource T and its exploitation by the two species, that is, $f(T, x_1, x_2)$. The competitive model thus becomes

$$\begin{cases} \dfrac{dx_1}{dt} = x_1\left(r_1 - \alpha_1 f\left(T, x_1, x_2\right)\right) \\ \dfrac{dx_2}{dt} = x_2\left(r_2 - \alpha_2 f\left(T, x_1, x_2\right)\right) \end{cases} \tag{5.102}$$

If we assume a linear exploitation model for the resource T

$$f\left(T, x_1 x_2\right) = T - \beta_1 x_1 - \beta_2 x_2 \tag{5.103}$$

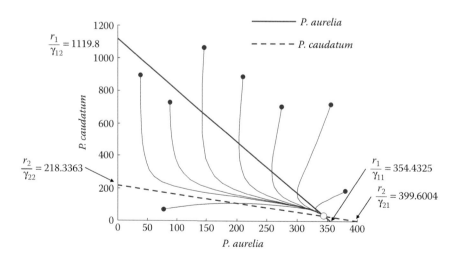

FIGURE 5.46 Phase-plane portrait for the Paramecia competition, confirming Gause's conclusions. The stable coexistence is largely imbalanced in favour of the *P. aurelia*, which is able to maintain a much larger population than the *P. caudatum*, which is driven close to extinction as a consequence of competition.

meaning that the resource T is linearly depleted by the exploitation, Equation 5.102 can be written as

$$
\begin{cases}
\dfrac{dx_1}{dt} = x_1\left(r_1 - T + \alpha_1\beta_1 x_1 + \alpha_1\beta_2 x_2\right) \\[3mm]
\dfrac{dx_2}{dt} = x_2\left(r_2 - T + \alpha_2\beta_1 x_1 + \alpha_2\beta_2 x_2\right)
\end{cases}
\tag{5.104}
$$

The question is whether this new interaction may lead to stable coexistence, and the answer is, alas, No. Computing the equilibrium isoclines from Equation 5.104 yields

$$
\begin{cases}
0 = r_1 - T + \alpha_1\beta_1 x_1 + \alpha_1\beta_2 x_2 \\[2mm]
0 = r_2 - T + \alpha_2\beta_1 x_1 + \alpha_2\beta_2 x_2
\end{cases}
\Rightarrow
\begin{cases}
x_1 = -\dfrac{\beta_2}{\beta_1} x_2 + \dfrac{T - r_1}{\alpha_1\beta_1} \\[3mm]
x_1 = -\dfrac{\beta_2}{\beta_1} x_2 + \dfrac{T - r_2}{\alpha_2\beta_1}
\end{cases}
\tag{5.105}
$$

Clearly, the two isoclines have the same slope $\left(-\beta_2/\beta_1\right)$, and the intercepts with the axes are

$$
\begin{cases}
\text{Isocline species 1}
\begin{cases}
x_1 = \dfrac{T - r_1}{\alpha_1\beta_1} \\[3mm]
x_2 = 0
\end{cases}
;
\begin{cases}
x_1 = 0 \\[3mm]
x_2 = \dfrac{T - r_1}{\alpha_1\beta_2}
\end{cases} \\[10mm]
\text{Isocline species 2}
\begin{cases}
x_1 = \dfrac{T - r_2}{\alpha_2\beta_2} \\[3mm]
x_2 = 0
\end{cases}
;
\begin{cases}
x_1 = 0 \\[3mm]
x_2 = \dfrac{T - r_2}{\alpha_2\beta_1}
\end{cases}
\end{cases}
\tag{5.106}
$$

as shown in Figure 5.47. Because there is no intersection between the two isoclines, no coexistence is possible, and the prevailing species will be that with the 'outer' isocline, for which the following inequality holds

$$
\frac{T - r_1}{\alpha_1\beta_1} > \frac{T - r_2}{\alpha_2\beta_2}
\tag{5.107}
$$

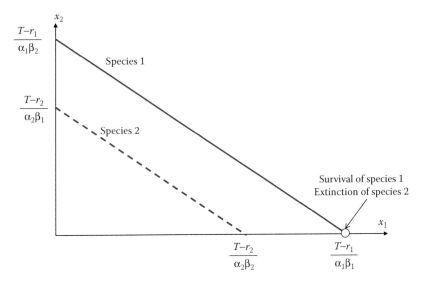

FIGURE 5.47 Competitive exclusion in the case of a limited resource. The two isoclines have the same slope, therefore no coexistence is possible, and the species with the outward isocline prevails.

If the inequality is reversed, then the species 2 will prevail.

Possible extensions to the above linear resource model could come from including secondary food sources, which could force the isoclines to cross somewhere in the first quadrant and produce an equilibrium, whose stability would have to be investigated on a case-by-case basis. In practice, the exploitation terms could be non-linear, limiting the exploitation of the resource (this issue will be considered when dealing with the functional response of a predatory population) and creating some type of saturation. Furthermore, the environmental factors may induce a fluctuating pressure of the dominating population, giving some relief to the other species, which may occasionally prevail. As a result, several equilibria (either stable or unstable) may develop.

5.4.4 Extension to Multispecies Competition

Now let us return to the general n-species competition model (Equation 5.92) and consider the linear term, which can be written in matrix form

$$\left(r_i - \sum_{j=1}^{n} \gamma_{ij} x_j\right) \Rightarrow r - AX = 0 \quad \text{with } r = \begin{bmatrix} r_1 \\ r_2 \\ \cdots \\ r_m \end{bmatrix} \quad A = \begin{bmatrix} \gamma_{11} & \gamma_{12} & \cdots & \gamma_{1m} \\ \gamma_{21} & \gamma_{22} & \cdots & \gamma_{2m} \\ \cdots & \cdots & \cdots & \cdots \\ \gamma_{m1} & \gamma_{m2} & \cdots & \gamma_{mn} \end{bmatrix} \tag{5.108}$$

so that the equilibrium can be computed by solving a linear system of equations

$$0 = r - AX \Rightarrow \bar{X} = A^{-1} r \quad \bar{X} > 0 \tag{5.109}$$

The stability of the multiple species equilibrium can be studied with the Lyapunov method treated in Section 1.9.7, according to which the system (5.92) is *globally stable* if and only if the following conditions hold (Goh, 1976, 1977)

1. $\bar{X} > 0$ (admissible equilibrium)
2. $\exists C = \text{diag}(c_1, c_2, \ldots, c_m) > 0 \Rightarrow CA + A^T C \leq 0$
3. Having defined

$$V = \sum_{i=1}^{n} c_i \left(x_i - \bar{x}_i - \bar{x}_i \ln \frac{x_i}{\bar{x}_i} \right) \quad \text{with } c_i > 0 \quad i = 1, \ldots, n \tag{5.110}$$

4. Its time derivative $\dot{V} = (1/2)(X - \bar{X})^T (CA + A^T C)(X - \bar{X})$ vanishes only in $\bar{X}$.

5.4.4.1 Vulnerability

Suppose that the environment induces a perturbation through some piecewise continuous functions $u_i(t)$, which might represent a fractional population withdrawal

$$\frac{dx_i}{dt} = x_i \left(r_i - \sum_{j=1}^{n} \gamma_{ij} x_j \right) + u_i(t) \cdot x_i \tag{5.111}$$

and suppose that these perturbation functions are bounded, that is,

$$-\xi \leq u_i(t) \leq \xi \tag{5.112}$$

then the derivative of the previous Lyapunov function (5.110), computed along the solutions of the model (5.111), becomes

$$\dot{V} = \frac{1}{2}\left(X - \bar{X}\right)^T \left(CA + A^T C\right)\left(X - \bar{X}\right) + \left(X - \bar{X}\right)^T Cu \quad \text{with } u = \left[u_1 \ldots u_n\right]^T \quad (5.113)$$

where the matrix $C = \mathrm{diag}\left(c_1, \ldots, c_n\right)$ is selected to have its smallest element equal to 1. Furthermore, because the matrix $-\frac{1}{2}\left(CA + A^T C\right)$ is symmetric and negative definite, its eigenvalues are positive real. An *invulnerability* condition can be derived as follows (Goh, 1976, 1979, 1980). Given a sphere of radius R around the equilibrium point $\bar{X}$ if the following inequality holds

$$\lambda_0 > \frac{\bar{c}\xi\sqrt{n}}{R} \quad (5.114)$$

where:

$$\lambda_0 = \text{minimum eigenvalue of} -\tfrac{1}{2}\left(CA + A^T C\right)$$

$$C = \mathrm{diag}\left(c_1, \ldots, c_n\right)$$

$$\bar{c} = \max\left(c_1, \ldots, c_n\right) \quad (5.115)$$

$$1 = \min\left(c_1, \ldots, c_n\right)$$

then $\dot{V} \leq 0$ for every X inside that sphere of radius R

$$\left(X - \bar{X}\right)^T \left(X - \bar{X}\right) = R^2 \quad (5.116)$$

Thus, for every population X the Lyapunov function (5.113) and its derivative have the following properties:

$$V\left(X\right) \leq \mu \quad \text{and} \quad \dot{V}\left(X\right) \leq 0 \quad (5.117)$$

where μ is the maximum value of the Lyapunov function on the boundary of the sphere of radius R. Conditions (5.117) imply that every trajectory originating in this region will remain inside the sphere indefinitely, at least as long as the assumptions of Equation 5.112 on the input function hold. Therefore, the radius R and the upper value of the Lyapunov function μ define the region of invulnerability with respect to the class of disturbances u bounded by Equation 5.112.

Consider next the following example (Goh, 1976) with two competing populations with dynamics

$$\frac{dx_1}{dt} = x_1 \cdot \left(22 - 0.1 \cdot x_1 - 2.1 \cdot x_2\right)$$
$$\frac{dx_2}{dt} = x_2 \cdot \left(-5 + 0.7 \cdot x_1 - 0.2 \cdot x_2\right) \quad (5.118)$$

which has a non-trivial equilibrium at $x_1 = 10, x_2 = 10$. Defining $C = \mathrm{diag}\left(1,3\right)$ yields

$$-\frac{1}{2}\left(CA + A^T C\right) = -\frac{1}{2}\left(\begin{bmatrix} 1 & 0 \\ 0 & 3 \end{bmatrix}\begin{bmatrix} -0.1 & -2.1 \\ 0.7 & -0.2 \end{bmatrix} + \begin{bmatrix} -0.1 & 0.7 \\ -2.1 & -0.2 \end{bmatrix}\begin{bmatrix} 1 & 0 \\ 0 & 3 \end{bmatrix}\right) = \begin{bmatrix} 0.1 & 0 \\ 0 & 0.6 \end{bmatrix} \quad (5.119)$$

and thus $\lambda_0 = 0.1$. From Equation 5.114, the perturbation upper bound can be computed as

$$\xi < \frac{\lambda_0 R}{\bar{c}\sqrt{n}} = \frac{0.1 \cdot 5}{3\sqrt{2}} = 0.1179 \quad (5.120)$$

The populations inside the invulnerability region must jointly satisfy the Lyapunov function (5.110) and the spherical condition constraint (5.116), that is,

$$\mu = \max_{x_i} \left(V = \sum_{i=1}^{n} c_i \left(x_i - \overline{x}_i - \overline{x}_i \ln \frac{x_i}{\overline{x}_i} \right) \right) \text{ subject to } \left(X - \overline{X} \right)^T \left(X - \overline{X} \right) = R^2 \quad (5.121)$$

Therefore, for any initial condition in R, the Lyapunov function will be such that

$$V(x_1, x_2) \leq \mu \quad \text{and} \quad \dot{V} \leq 0 \quad (5.122)$$

which means that the trajectory will be confined in the bounded region $V(x_1, x_2) \leq \mu$ around the equilibrium, even in the presence of a bounded disturbance. This example was simulated with the Simulink model of Figure 5.48, and the phase-plane portrait of the two populations is shown in Figure 5.49, in which the common point between the sphere of radius R and the Lyapunov function is indicated, yielding the maximum of the latter $V = 5.7935$.

5.5 PREY–PREDATOR MODELS

In the previous section, we considered two or more species competing for a common resource, but what happens when one species becomes the food resource of the other? In this case, we have a *prey–predator* relationship between the two species, and its extension to many species ecosystem becomes a *food chain* or a *food web*, as will be discussed later. Now let us begin by setting out some basic hypotheses on which the prey–predator interaction hinges:

1. The resource is not unlimited, being represented by the exploited prey population
2. Without predation, the prey behaves like a single-species population, reaching its carrying capacity
3. Without prey the predator cannot survive, assuming that no alternative food source is available
4. The (lossy) energy transfer is unidirectional from the prey to the predator

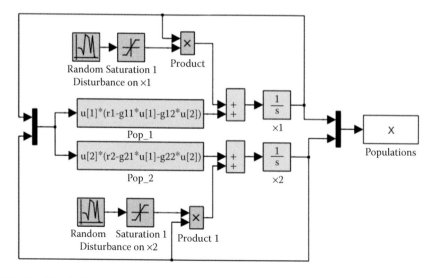

FIGURE 5.48 Simulink model to demonstrate the vulnerability condition (5.121). Two separate noise generators were used, one for each population, to obtain independent disturbances. This Simulink model (Lyap _ Vulnerabity.mdl) can be retrieved from the ESA _ Matlab\Exercises\Chapter _ 5\ Competition folder.

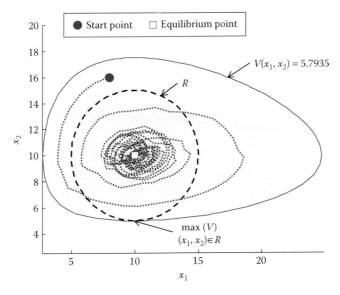

FIGURE 5.49 Phase-plane portrait of the two populations, whose trajectories are inside the shaded area, where the Lyapunov function satisfies the conditions (5.122). This graph was produced with the MATLAB script `Go _ Lyap _ Vulnerability.m` in the `ESA _ Matlab\Exercises\Chapter _ 5\Competition` folder.

Under these assumptions, the dynamical prey–predator model can be written as follows:

$$\text{Prey} \qquad \frac{dx}{dt} = x \cdot F(x) - y \cdot p(x)$$

$$\text{Predator} \qquad \frac{dy}{dy} = -m \cdot y + c \cdot y.p(x) \tag{5.123}$$

where:

$F(x)$ is the prey growth function

$p(x)$ is the *functional response* of the predator, to be examined later

m is the extinction rate of the predator

c is the energy conversion efficiency of the predation

The many prey–predator models that have been proposed over the years basically differ in the forms of the two functions $F(x)$ and $p(x)$. Some options are now examined.

5.5.1 LOTKA–VOLTERRA PREY–PREDATOR MODEL

This is undoubtedly the first and still the most popular model of this kind. It considers a prey logistic growth with growth rate r and carrying capacity K and a simple, multiplicative interaction for the predation term, which is equivalent to a linear unbounded predation rate. Under these assumptions the Lotka–Volterra model can be written as

$$\text{Prey} \qquad \frac{dx}{dt} = r \cdot x \cdot \left(1 - \frac{x}{K}\right) - b_{12} \cdot x \cdot y$$

$$\text{Predator} \qquad \frac{dy}{dy} = -m \cdot y + b_{21} \cdot x \cdot y \tag{5.124}$$

The lossy energy flow requires that $b_{21} < b_{12}$, meaning that the energy subtracted from the prey is not fully transferred to the predator, for reasons that will be examined in detail later. The equilibrium of the system (5.124) can be found by computing the isoclines of both species as

$$
\begin{aligned}
0 &= r \cdot x \cdot \left(1 - \frac{x}{K}\right) - b_{12} \cdot x \cdot y \\
0 &= -m \cdot y + b_{21} \cdot x \cdot y
\end{aligned}
\Rightarrow
\quad
\begin{aligned}
r \cdot \left(1 - \frac{x}{K}\right) &= b_{12} \cdot y \\
x &= \frac{m}{b_{21}}
\end{aligned}
\tag{5.125}
$$

It can be seen that the prey isocline has a negative-slope, whereas the predator isocline is a vertical line. Substituting the second equation into the first, the following equilibrium is obtained as

$$
\begin{cases}
\bar{y} = \dfrac{r}{b_{12}}\left(1 - \dfrac{m}{b_{21}K}\right) \\[2ex]
\bar{x} = \dfrac{m}{b_{21}}
\end{cases}
\tag{5.126}
$$

which represents a feasible $(x, y) > 0$ equilibrium only if

$$
K > \frac{m}{b_{21}}
\tag{5.127}
$$

In fact, if this condition is not satisfied, the two isoclines do not cross in the first quadrant. The two possible equilibria are shown in Figure 5.50, which presents the feasible equilibrium (a) given by the intersection of the two isoclines, and the unfeasible solution (b) when the condition (5.127) is not met, leading to the extinction of the predator and the full development of the prey, which reaches its carrying capacity.

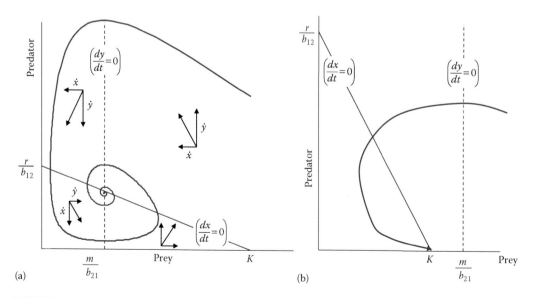

FIGURE 5.50 Graphical computation of the Lotka–Volterra prey–predator system. In (a), the feasible equilibrium is determined by the intersection of the equilibrium isoclines, whereas in (b), no feasible equilibrium is possible, leading to the extinction of the predator, while the prey reaches its carrying capacity K.

Figure 5.50 also shows the phase-plane trajectories from an arbitrary initial condition to the equilibrium, whichever this may be. In the feasible equilibrium case (a), the isoclines divide the first quadrant in four sectors with differing derivative signs that cause the trajectory to spiral in a counterclockwise direction. Other possible trajectories will be considered soon, but first let us discuss the stability of the equilibrium (5.126) under the condition (5.127). From the model (Equation 5.124), the derivatives with respect to x and y can be computed as

$$
\begin{cases}
\dfrac{\partial f_1}{\partial x} = r\left(1 - 2\dfrac{\bar{x}}{K}\right) - b_{12}\bar{y} = r\left(1 - 2\dfrac{m}{b_{21}K}\right) - b_{12}\dfrac{r}{b_{12}}\left(1 - \dfrac{m}{b_{21}K}\right) = -\dfrac{mr}{b_{21}K} \\[3mm]
\dfrac{\partial f_1}{\partial y} = -b_{12}\bar{x} = -m\dfrac{b_{12}}{b_{21}} \\[3mm]
\dfrac{\partial f_2}{\partial x} = b_{21}\bar{y} = b_{21}\dfrac{r}{b_{12}}\left(1 - \dfrac{m}{b_{21}K}\right) \\[3mm]
\dfrac{\partial f_2}{\partial y} = -m + b_{21}\bar{x} = -m + b_{21}\dfrac{m}{b_{21}} = 0
\end{cases}
\tag{5.128}
$$

from which the Jacobian can be determined as

$$
\boldsymbol{J} = \begin{bmatrix}
-\dfrac{m \cdot r}{b_{21} \cdot K} & -m\dfrac{b_{12}}{b_{21}} \\[4mm]
b_{21}\dfrac{r}{b_{12}}\left(1 - \dfrac{m}{b_{21} \cdot K}\right) & 0
\end{bmatrix}
\tag{5.129}
$$

Its eigenvalues are

$$
\det(\lambda \boldsymbol{I} - \boldsymbol{J}) = \det\begin{bmatrix}
\lambda + \dfrac{m \cdot r}{b_{21} \cdot K} & m\dfrac{b_{12}}{b_{21}} \\[4mm]
-b_{21}\dfrac{r}{b_{12}}\left(1 - \dfrac{m}{b_{21} \cdot K}\right) & \lambda
\end{bmatrix}
$$

$$
= \lambda \cdot \left(\lambda + \dfrac{m \cdot r}{b_{21} \cdot K}\right) + m \cdot r \cdot \left(1 - \dfrac{m}{b_{21} \cdot K}\right)
\tag{5.130}
$$

$$
= \lambda^2 + \lambda\dfrac{m \cdot r}{b_{21} \cdot K} + m \cdot r \cdot \left(1 - \dfrac{m}{b_{21} \cdot K}\right) = 0
$$

$$
\lambda = \dfrac{-((m \cdot r)/(b_{21} \cdot K)) \pm \sqrt{((m \cdot r)/(b_{21} \cdot K))^2 - 4 \cdot m \cdot r\left(1 - (m/(b_{21} \cdot K))\right)}}{2}
\tag{5.131}
$$

From Equation 5.131, we find that the eigenvalues have a negative real part if

$$
1 - \dfrac{m}{b_{21} \cdot K} > 0 \Rightarrow K > \dfrac{m}{b_{21}}
\tag{5.132}
$$

so the previous feasibility condition (Equation 5.127) also implies stability of the equilibrium. Real negative eigenvalues are obtained for

$$
\dfrac{m \cdot r}{b_{21} \cdot K} > 4\left(b_{21} \cdot K - m\right)
\tag{5.133}
$$

TABLE 5.2

Possible Behaviours of the Lotka–Volterra Prey–Predator Model as a Function of Its Parameters

Parameter	Feasible Equilibrium (Oscillatory)	Feasible Equilibrium (Monotonic)	Unfeasible Equilibrium (Predator Extinction)
r	0.4	30	1.4
K	20	18	10
m	0.5	5	4.6
b_{12}	0.33	1.5	0.3
b_{21}	0.1	0.5	0.4
Eigenvalues	$\lambda_{1,2} = -0.0500$	$\lambda_1 = -10.0000$	$\lambda_1 = -2.0754$
	$\pm j0.3841$	$\lambda_2 = -6.6667$	$\lambda_2 = 0.4654$

and in this case the model response is monotonic. Table 5.2 shows some parameter sets that produce the three possible behaviours of the Lotka–Volterra model: stable equilibria (oscillatory or monotonic) or unstable equilibrium with predator extinction.

Notice that, in the latter case, one eigenvalue is positive real, indicating instability. The model responses for the parameter values of Table 5.2 are shown in Figure 5.51 for the stable oscillatory case, in Figure 5.52 for the stable monotonic case, and in Figure 5.53 for the predator extinction case.

5.5.1.1 A Helping Hand

In the latter case, one may wonder whether some corrective action could be taken to stabilize the system and avoid the predator extinction. One possibility is to provide an external food supply to make up for the insufficient prey. This is often done to support starving herbivore populations in natural parks during the winter months. To model this additional input, Equation 5.124 can be rewritten as

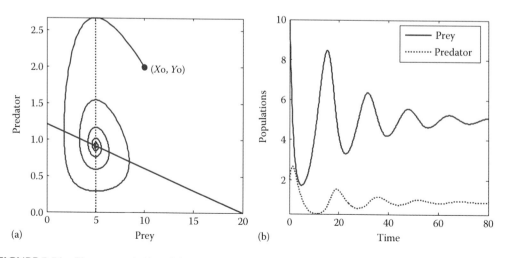

FIGURE 5.51 Phase portrait (a) and time trend (b) of the Lotka–Volterra model in the case of a stable oscillatory equilibrium, obtained with the parameters values taken from the pertinent column of Table 5.2. These graphs were produced with the MATLAB script Go _ Lotka _ Volterra.m in the ESA _ Matlab\ Exercises\Chapter _ 5\Prey _ Predator\Lotka _ Volterra folder.

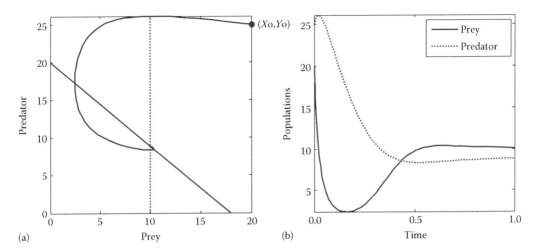

FIGURE 5.52   Phase portrait (a) and time trend (b) of the Lotka–Volterra model in the case of a stable mono-tonic equilibrium, obtained with the parameters values taken from the pertinent column of Table 5.2. These graphs were produced with the MATLAB script `Go_Lotka_Volterra.m` in the `ESA_Matlab\Exercises\Chapter_5\Prey_Predator\Lotka_Volterra` folder.

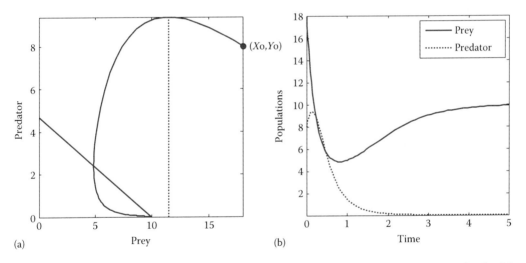

FIGURE 5.53   Phase portrait (a) and time trend (b) of the Lotka–Volterra model in the case of unfeasible equilibrium leading to the predator extinction, obtained with the parameters values taken from the pertinent column of Table 5.2. These graphs were produced with the MATLAB script `Go_Lotka_Volterra.m` in the `ESA_Matlab\Exercises\Chapter_5\Prey_Predator\Lotka_Volterra` folder.

$$\text{Prey} \qquad \frac{dx}{dt} = r \cdot x \cdot \left(1 - \frac{x}{K}\right) - b_{12} \cdot x \cdot y + F$$

$$\text{Predator} \qquad \frac{dy}{dy} = -m \cdot y + b_{21} \cdot x \cdot y$$

(5.134)

where F can be viewed as the additional food supply rate, which of course has the dimension of energy over time. Plotting the predator isocline with F as a parameter yields the family of curves generated by the new the prey isocline which, contrary to the previous linear case of Equation 5.125, is non-linear.

$$F = b_{12} \cdot x \cdot y - r \cdot x \cdot \left(1 - \frac{x}{K}\right) \tag{5.135}$$

With this additional manipulated variable F, we can decide the level of the predator that we want to sustain by recomputing the equilibrium, while keeping the prey level of the original steady state, that is, $\bar{x} = m/b_{21}$. Substituting this value in Equation 5.135, we obtain the required additional food supply

$$\bar{F} = b_{12} \cdot \frac{m}{b_{21}} \cdot \bar{\bar{y}} - r \cdot \frac{m}{b_{21}} \cdot \left(1 - \frac{m}{b_{21} \cdot K}\right) \tag{5.136}$$

where $\bar{\bar{y}}$ is the desired predator level. Figure 5.54 shows how the prey isoclines change as a function of the additional energy supply (a) and how the new equilibrium is reached (b).

5.5.2 Predator Functional Response

The Lotka–Volterra model supposes that there is an unbounded predation rate, whereas in nature it is well known that the predator is limited in its voracity by several factors. They represent an upper bound on the predation by introducing some form of saturation that is globally referred to as the predator *functional response*. Generally, three kinds of functional responses are known, whose shapes are shown in Figure 5.55.

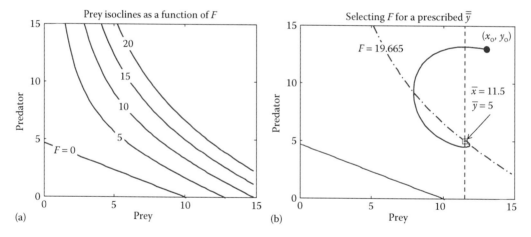

(a) (b)

FIGURE 5.54 Changing the prey isoclines by adding an external foodsource (a) and the new equilibrium, with the amount of food selected to sustain a prescribed predator population (b). The right-hand graph was produced with the MATLAB script Go _ Food _ Supply _ LV.m in the ESA _ Matlab\Exercises\ Chapter _ 5\Prey _ Predator\Lotka _ Volterra folder.

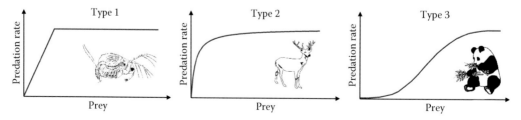

FIGURE 5.55 The three possible shapes of the predator functional response. The insets show typical animals exhibiting this kind of response: the crustacean *Daphnia*, the Deer and the Panda.

The type 1 functional response is typical of a filtering organism, such as the crustacean *Daphnia*, which feeds by filtering water and retaining the plankton. Its feeding capacity is limited by the maximum water flow through the filtering apparatus. The type 2 response is typical of herbivores that gradually reach satiety, and its analytical form can be derived as follows. Define the following feeding parameters:

- T = total feeding time
- T_r = prey search time
- T_m = total prey feeding time
- t_m = single prey feeding time = search area per time unit
- p_a = prey capture probability
- x = prey density

Then, the amount of prey captured during the search time T_r is

$$D = p_a \cdot q \cdot x \cdot T_r \tag{5.137}$$

and the total feeding time T_m amounts to

$$T = T_m + T_r = T_r \cdot \left(1 + t_m \cdot p_a \cdot q \cdot x\right) \tag{5.138}$$

Thus, the total feeding time is

$$T = T_m + T_r = T_r \cdot \left(1 + t_m \cdot p_a \cdot q \cdot x\right) \tag{5.139}$$

Thus, the quantity of prey consumed in the time unit, that is, the *functional response*, can be expressed as

$$p(x) = \frac{D}{T} = \frac{p_a \cdot q \cdot x \cdot T_r}{\left(1 + t_m \cdot p_a \cdot q \cdot x\right) T_r} = \frac{\alpha \cdot x}{1 + t_m \cdot \alpha \cdot x} \quad \text{with} \quad \alpha = p_a \cdot q \tag{5.140}$$

which tends to the saturation value $\left(1/t_m\right)$ as the prey density grows indefinitely. This response is typical of animals with a good mobility, for which feeding is primarily limited by the time required for prey manipulation. The type 3 response is found in animals with lesser mobility, which have problems in finding food when this is scarce. The type 3 functional response can be described either by an exponential sigmoid, providing a better control for low prey density, or by a polynomial sigmoid, which is more appropriate when good control is required for high prey density

$$\text{Exponential sigmoid} \quad p(x) = \frac{a}{1 + b \cdot e^{-c \cdot x}}$$
$$\text{Polynomial sigmoid} \quad p(x) = \frac{a}{1 + b \cdot x^{-n}} \tag{5.141}$$

5.5.3 GENERALIZING THE PREY–PREDATOR MODEL

Now that we have better defined the predator functional response, the prey–predator model can be generalized into what is known as the Kolmogorov generalized prey–predator model (Freedman, 1980; Brauer and Castillo-Chaves, 2014)

$$\begin{cases} \text{Prey} & \dfrac{dx}{dt} = x \cdot g(x) - y \cdot p(x) \\[3mm] \text{Predator} & \dfrac{dy}{dt} = -m \cdot y + c \cdot y \cdot p(x) \end{cases} \tag{5.142}$$

with the following conditions for the growth $g(x)$ and the predator response $p(x)$ functions:

$$
\begin{array}{cc}
\text{conditions on } g(x) & \text{conditions on } p(x) \\
g(0) > 0 & p(0) = 0 \\
g(K) = 0 & \lim_{x \to \infty} p(x) = \text{const.} \\
\dfrac{dg}{dx} < 0 & \dfrac{dp}{dx} > 0
\end{array}
\tag{5.143}
$$

The conditions on $g(x)$ are basically the same that were defined for a single-species growth model, whereas those on $f(x)$ are connected to the predator functional response. Two broad classes of functions satisfying the previous conditions (5.143) are

$$
g(x) = r\left[1 - \left(\frac{x}{K} \right)^p \right] \quad p > 0 \quad p(x) = \frac{ax^n}{X_s^n + x^n} \quad n > 0
\tag{5.144}
$$

which encompass every kind of growth, including depensation, and all functional responses of type 2 and 3. Figure 5.56 shows the possible shapes of the two functions as their parameters p and n vary. It can be seen that for $p = 1$, the linear logistic growth function is obtained. The exponent n controls the shape of the functional response of the predator: $n \le 1$ produces a type 2 response, whereas for $n > 1$ a type 3 response is obtained.

Now we substitute the functions (5.144) in the generalized model (Equation 5.142)

$$
\begin{cases}
\dfrac{dx}{dt} = x \cdot r\left[1 - \left(\frac{x}{K} \right)^p \right] - y \cdot \dfrac{ax^n}{X_s^n + x^n} \\
\dfrac{dy}{dt} = y \cdot \left[-m + c \cdot \dfrac{ax^n}{X_s^n + x^n} \right]
\end{cases}
\tag{5.145}
$$

and set out to compute the system equilibrium by setting the derivatives to zero

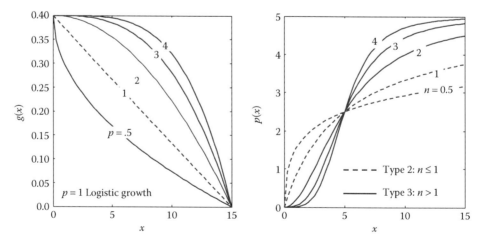

FIGURE 5.56 Possible shapes of the generalized prey–predator model functions $g(x)$ and $p(x)$ defined by Equation 5.144.

$$\begin{cases} 0 = x \cdot r \left[1 - \left(\frac{x}{K} \right)^p \right] - y \cdot \frac{ax^n}{X_s^n + x^n} \\ 0 = y \cdot \left[-m + c \cdot \frac{ax^n}{X_s^n + x^n} \right] \end{cases} \tag{5.146}$$

from which the two equilibrium isoclines are obtained

$$\text{Prey isocline} \qquad y = \frac{r}{a} \left[1 - \left(\frac{x}{K} \right)^p \right] \cdot \frac{X_s^n + x^n}{x^{n-1}}$$

$$\text{Predator isocline} \qquad \bar{x} = X_s \left(\frac{m}{c \cdot a - m} \right)^{1/n} \tag{5.147}$$

Substituting $\bar{x}$ from the second equation (5.147) into the first yields the equilibrium $\bar{y}$ from which the Jacobian can be computed as

$$J = \begin{bmatrix} r \left(1 - (p+1) \left(\frac{\bar{x}}{K} \right)^p \right) - \bar{y} \cdot a \cdot X_s^n \frac{n \cdot \bar{x}^{n-1}}{\left(X_s^n + \bar{x}^n \right)^2} & -\frac{a \cdot x^n}{X_s^n + x^n} \\ \bar{y} \cdot c \cdot a \cdot X_s^n \frac{n \cdot \bar{x}^{n-1}}{\left(X_s^n + \bar{x}^n \right)^2} & 0 \end{bmatrix} \tag{5.148}$$

To show the flexibility of the generalized prey–predator model (5.145), Figure 5.57 shows its equilibrium isoclines obtained with differing values of the predation coefficient m while all the other parameters were kept constant at the values shown in Table 5.3.

The m parameter influences only the predator isocline (vertical line), while the prey isocline is the same for all m. The equilibria, resulting from the intersection of the two isoclines, are all stable in the large and represent widely differing behaviours, ranging from monotonic to a limit cycle, as listed in Table 5.3. Some of these results may come as a surprise from a biological viewpoint. In particular, the case of equilibrium 1 in Figure 5.57 and Table 5.3 shows that the predator population

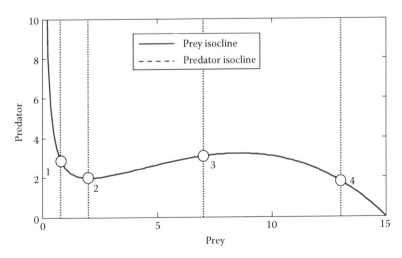

FIGURE 5.57 Possible equilibria over the isoclines of the generalized model (5.147) as a function of the predation coefficient m. See Table 5.3 for the values of the parameters.

TABLE 5.3

Possible Behaviours of the Generalized Prey–Predator Model (Equation 5.145) While m Is Varied and the Other Parameters Have the Following Constant Values: $r = 1$; $K = 15$; $p = 2.2$; $n = 2$; $X_s = 2$; $a = 2$; $c = 0.08$

Point	m	Equilibrium	Eigenvalues	Behaviour
1	0.0221	$\bar{x} = 0.8000$	$\lambda_1 = -0.6698$	Stable monotonic
		$\bar{y} = 2.8954$	$\lambda_2 = -0.0567$	
2	0.0800	$\bar{x} = 2.000$	$\lambda_1 = -0.0131 + j0.2809$	Stable oscillatory
		$\bar{y} = 1.9762$	$\lambda_2 = -0.0131 - j0.2809$	
3	0.1479	$\bar{x} = 7.0000$	$\lambda_1 = 0.1755$	Stable limit cycle
		$\bar{y} = 3.0778$	$\lambda_2 = 0.1035$	
4	0.1563	$\bar{x} = 13.0000$	$\lambda_1 = -1.34688$	Stable monotonic
		$\bar{y} = 1.7971$	$\lambda_2 = -0.0014$	(slow-fast mixed modes)

is greater than the prey, as though the predator could sustain itself with a very small predation factor, in fact $m = 0.0221$ in this case. As m increases the pressure on the prey increases too, and for $m > 0.08$ the predator population is less than the prey.

The phase-plane portraits (Figure 5.58) and time behaviour (Figure 5.59) of the model (5.145) deserve more comments: at equilibrium point 1 in Figure 5.57, the eigenvalues are real negative

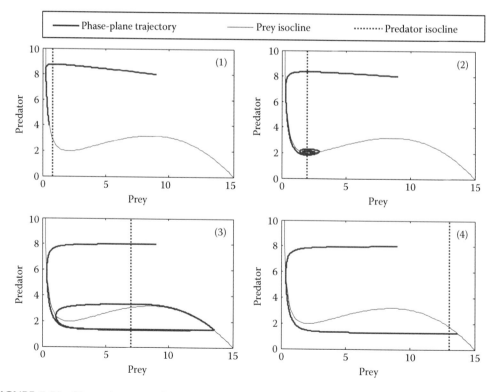

FIGURE 5.58 Phase-plane portraits of the generalized prey–predator model (5.147) as a function of the predation coefficient m. See Table 5.3 for the values of the parameters. The numbers refer to the equilibria of Figure 5.57 and of Table 5.3. These graphs were produced with the MATLAB script Go _ Kolmogorov.m in the ESA _ Matlab\Exercises\Chapter _ 5\Prey _ Predator\Kolmogorov folder.

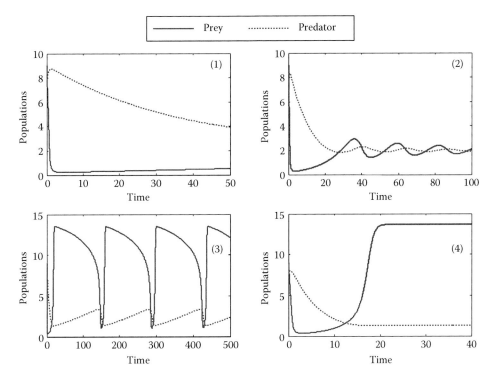

FIGURE 5.59 Time behaviour of the generalized prey–predator model (5.147) as a function of the predation coefficient m. See Table 5.3 for the values of the parameters. The numbers refer to the equilibria of Figure 5.57 and of Table 5.3.

(see Table 5.3), so the system is expected to converge monotonically to the steady state, but what is surprising is that the speed of the prey in reaching the equilibrium (Figure 5.59) is much faster than that of the predator at the beginning of the simulation, and then it slows down considerably. In fact, the phase-plane trajectory (Figure 5.58) is almost superimposed to the prey isocline, which is the locus where its time derivative tends to zero. Thus, the speed of the trajectory in the neighbourhood of the isocline is very slow. This explains the apparently strange behaviour of the prey, reflected in the time plot of Figure 5.59 (1). The general rule is that trajectories are moving fast when they are far from the isoclines and slow when they are near. This also explains the behaviours in the other cases. By comparing the phase-plane portraits of Figure 5.58 to the time plots of Figure 5.59, it also appears that the limit cycle has fast and slow branches, but the extreme case is that of point 4, where the descending branch of the prey isocline is reached very fast. However, once the trajectory hugs this locus, it takes ages to reach the equilibrium. The fast/slow behaviour around equilibria 1 and, even more, 4 is reflected in the differing order of magnitude of the eigenvalues (see Table 5.3) controlling the slow and fast modes of the system.

5.5.3.1 A Pasture–Herbivore Model

As an application of the previous generalized prey–predator model, let us next consider a pasture grazed by a population of herbivores. In this case, the prey (vegetation)—predator (herbivores) model can be set up under the following assumptions:

- The vegetation (V) is modelled with a birth-and-death model where the growth is a function of the rainfall $r(t)$, and the decay (p) is proportional to the vegetation itself
- Additionally, the vegetation is consumed by the herbivores with a grazing coefficient q

- The herbivores behave like a real predator in the sense that without pasture the population will die with extinction rate m
- The herbivores normalized functional response is of type 2
- The grazing conversion efficiency s is less than q

Under these assumptions, the model can be written as

$$\begin{cases} \text{Vegetation} & \dfrac{dV}{dt} = r(t) - p \cdot V - q \cdot f(V) \cdot H \\[2mm] \text{Herbivores} & \dfrac{dH}{dt} = -m \cdot H + s \cdot f(V) \cdot H \end{cases} \tag{5.149}$$

where

$$f(V) = \frac{V}{K_v + V} \tag{5.150}$$

From Equations 5.149 and 5.150, the equilibrium isoclines can be computed as

$$\begin{cases} \dfrac{dV}{dt} = 0 \Rightarrow & \bar{r} - p \cdot V = q \cdot f(V) \cdot H \Rightarrow H = \dfrac{(\bar{r} - p \cdot \bar{V})(K_v + \bar{V})}{q \cdot \bar{V}} \\[4mm] \dfrac{dH}{dt} = 0 \Rightarrow & m \cdot \bar{H} = s \dfrac{\bar{V}}{K_v + \bar{V}} \bar{H} \Rightarrow \bar{V} = \dfrac{K_v}{((s/m) - 1)} \end{cases} \tag{5.151}$$

where the average rain-induced growth $\bar{r} = \text{mean}(r(t))$ was considered instead of the time-varying rainfall samples. Figure 5.60 shows the equilibrium point at the intersection of the isoclines, with the arrows in each quadrant indicating the direction of the trajectories, whereas Figure 5.61 shows the Simulink model implementing Equation 5.149. Simulating the system with a fluctuating rainfall yields the results of Figure 5.62, where in the phase-plane portrait (a), the trajectory meanders about the equilibrium because of the fluctuating rainfall. The steady state could eventually be reached

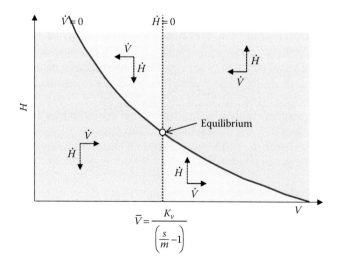

FIGURE 5.60 Isoclines and equilibrium of the vegetation–herbivores model (Equation 5.151). In each quadrant, the sign of the derivatives is indicated by a vector (solid arrow for vegetation, dashed arrow for herbivores).

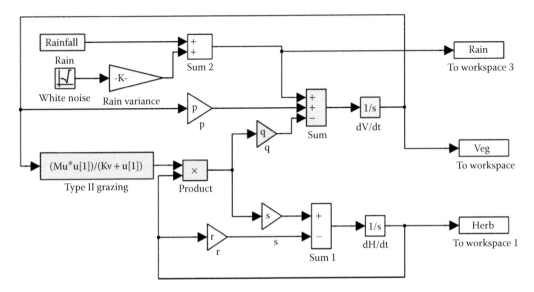

FIGURE 5.61 Simulink diagram implementing the vegetation–herbivores model (Equation 5.151). This Simulink model is available as `veg_herb.mdl` in the `ESA_Matlab\Exercises\Chapter_5\ Pasture` folder.

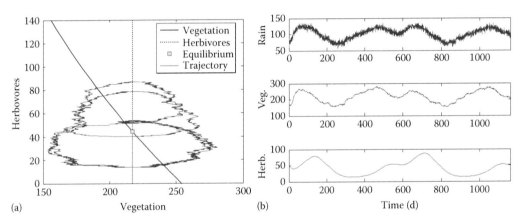

FIGURE 5.62 Simulation of the vegetation–herbivores model (Equation 5.151) with a fluctuating rainfall. In the phase-plane portrait (a), the trajectory meander around the equilibrium, which it would reach only for a constant rainfall. In the time portrait (b), the herbivores lag the vegetation as a result of its internal dynamics. These graphs were obtained with the MATLAB script `Go_Vegetation_Herbivores.m` in the `ESA_Matlab\Exercises\Chapter_5\Pasture` folder.

only if the rainfall would remain constant at its average value $\bar{r}$, which was used to compute the equilibrium point. In the time portrait (b), the herbivore population clearly lags the vegetation fluctuations, because of its internal dynamics.

5.5.3.2 The Strange Case of the Spruce Budworm

An interesting example of vegetation–insect interaction, with a major ecological impact, is undoubtedly that of the spruce budworm (*Choristoneura fumiferana*, Clemens) that causes defoliation in vast areas of Canada and North America during its periodical outbreaks. This insect, and its adverse effects on the spruce and fir forests, have been studied for more than half a century

(Morris and Miller, 1954). Mathematical models have been produced to describe its outbreaks, which causes extensive forest defoliation (Ludwig et al., 1978, 1979; Royama, 1984; Royama et al., 2005). The last major outbreak occurred in the 1970s and damaged more than 50 million ha of Canadian forest. Minor infestations were reported between 1999 and 2003, and after a lull, an increase of forest damage has been observed since 2006, which is suspected to be a warning sign of a new possible massive outbreak. More information on this major pest can be found in the National Resources Canada web site (http://www.nrcan.gc.ca/forests/insects-diseases/13383) from where this information was obtained. The spruce budworm life cycle is outlined in Figure 5.63, in which the larvae feeding period is highlighted. This period represents the critical phase when defoliation occurs, causing extensive tree mortality, because the fallen leaves (or needles) cannot regrow fast enough to sustain the plant life.

In building the spruce budworm model (Ludwig et al., 1978), the following assumptions were made:

- The budworm population follows a logistic growth in the absence of predation
- The carrying capacity K is related to the foliage density, and it is possible to estimate its value
- The budworm is subject to predation by insectivorous birds, whose functional response is of type 3, meaning that when the budworm density is low, the predation is very low, but when the density increases the predation increases at a faster rate, as described by Equation 5.141. In particular, in this case, the following form has been selected for the functional response

$$p(B) = \beta \cdot \frac{B^2}{\alpha^2 + B^2} \tag{5.152}$$

where:
B represents the budworm density
α, β are parameters shaping the functional response

Under these assumptions, the spruce budworm model is the following:

$$\frac{dB}{dt} = r_B \cdot B \cdot \left(1 - \frac{B}{K_B}\right) - \beta \frac{B^2}{\alpha^2 + B^2} \tag{5.153}$$

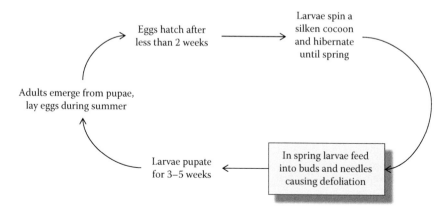

FIGURE 5.63 Life cycle of the spruce budworm. The dangerous phase, when the larvae start feeding, has a coloured background.

The behaviour of this model is best studied by defining the normalized budworm density $x = B/\alpha$, so that Equation 5.153, at steady state $(dB/dt = 0)$, can be written as

$$r_B \cdot \left(1 - \frac{\alpha x}{K_B}\right) - \beta \cdot \frac{\alpha x}{\alpha^2 (1 + x^2)} = 0 \qquad (5.154)$$

which after multiplication by α/β and defining the quantities

$$R = \frac{\alpha \cdot r_B}{\beta} \quad Q = \frac{K_B}{\alpha} \qquad (5.155)$$

yields the dimensionless steady-state equation

$$R \cdot \left(1 - \frac{x}{Q}\right) - \frac{x}{1 + x^2} = 0 \qquad (5.156)$$

whereas the normalized dynamical model is

$$\frac{dx}{dt} = R \cdot \left(1 - \frac{x}{Q}\right) - \frac{x}{1 + x^2} \qquad (5.157)$$

Equation 5.156 has three roots, but they are all real at the intersections between the logistic growth rate (left) and the type 3 functional response. These solutions are controlled by the parameter R, which plays the role of a growth rate, and Q, which is a normalized carrying capacity. Figure 5.64 shows the possible relative positions of the two terms of Equation 5.156. In (a), it can be seen that the number and location of the equilibria depends on the value of R. In general, there is only one intersection, but for a critical range of R values a third (unstable) solution appears in between the two stable equilibria, repelling the trajectories towards the other equilibria. Figure 5.64b plots the time derivative, starting from differing initial conditions, and invariably ending in a stable equilibrium point. The two trajectories starting very near to points 1 and 2, close to the unstable point (white square), terminate in either the square or dot stable equilibria, depending on whose domain of attraction the initial condition was placed.

By varying both R and Q, the domain of these parameters is divided as shown in Figure 5.65, in which a 'cusp' shaped area exists in which there are three real roots, that is, three equilibria occur, one of which is unstable.

The time behaviour of the model (5.157) in Figure 5.66a shows a non-linear oscillation between the two stable equilibria, while the slow sine trend of R is shown in (c).

Because the R variations span the critical range in which a third unstable equilibrium develops, a hysteresis results from the variable x being unable to settle at the unstable point and instead being attracted to either stable equilibrium, depending on its previous values. The (b) graph shows the values of R as a function of the equilibria. The range of instability is indicated by the valley (dotted curve) and is relative to the multi-valued branch of the equilibrium curve $R(x)$. When x decreases from a high value, it travels through the upper branch of the hysteresis loop, but when it comes from the low values it goes through the lower branch. The two branches reunite when the $R(x)$ function again becomes single valued, asymptotically tending to Q for large R values. Here we have considered the limited case of a single dynamical equation (5.157), but in Ludwig et al. (1978) the study is carried much further by removing the hypothesis of constant parameters and including two more variables related to the forest stand and its energy content, eventually explaining the conditions leading to the pest outbreaks.

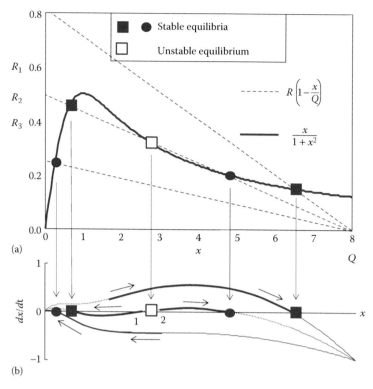

(a)

(b)

FIGURE 5.64 Possible equilibria of the steady-state equation (5.156) (a) when the logistic normalized rate R varies and $Q = 8$. Given the shape of the normalized functional response, one or three intersections may occur. In the latter case, the middle point (white square) is unstable, whereas all the others are stable (blue circles or squares). In (b), the dynamical trajectories of Equation 5.157 (thick lines) show that the system converges to a stable point (blue dots/squares), even if the initial condition (1 or 2) is very near the unstable equilibrium point (white square), eventually settling in either stable equilibria. These graphs were obtained with the MATLAB script Go _ spruce _ budworm.m in the ESA _ Matlab\Exercises\Chapter _ 5\Spruce _ budworm folder.

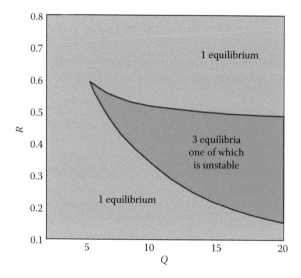

FIGURE 5.65 Exploring the distribution of the equilibria in the (R,Q) parameter space. In the grey area, only one stable equilibrium is possible, but in the 'cusp' shaped blue area, three equilibria are generated by the mechanism illustrated in Figure 5.64, one of which is unstable.

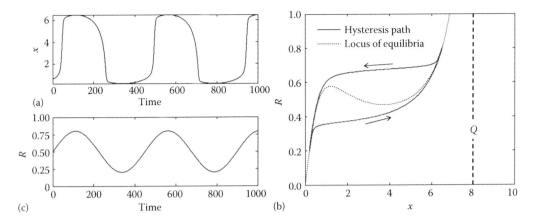

FIGURE 5.66 (a) Time evolution of the normalized budworm model (Equation 5.156) with a slow-varying growth rate R spanning the 'cusp' region of Figure 5.65 (c). In this range of R values, a hysteresis loop is developed (b), depending on whether R is increasing or decreasing. Q is the asymptotic value of the extreme branch of the R/x curve.

5.6 FOOD CHAINS

When more species share resources in the same habitat, they define a whole set of trophic relations that can be viewed as a *food chain* if there is a cascaded path from one species to the next, or rather as a *food web* if there are several paths connecting the trophic levels. Specifying the details of a trophic web is a very complex task and requires a deep knowledge of the habitat and the inhabiting species. A notable example of a fishery-oriented ecosystem modelling is Ecopath (Christensen and Pauly, 1992; Christensen and Walters, 2004; Christensen et al., 2008), which can now be considered as the leading software for modelling aquatic trophic networks, particularly for assessing the viability of a marine ecosystem and for fisheries management. Ecopath, in the present version 6, is coupled with a dynamical simulation platform (Ecosim) and is freely downloadable from its web site (www.ecopath.org). As with many software packages dealing with environmental systems, Ecopath is considerably data hungry, requiring a vast amount of information to produce reliable results, including a detailed species inventory, their diet and other habitat data. It's output is equally complex and yields operational directions that can be used for fishery management.

On a much more modest level, and for pedagogical purposes, a simple food chain will be illustrated here and implemented in MATLAB to gain some hands-on experience. Figure 5.67 illustrates the main energy and matter cycling paths in a food chain consisting of three main trophic levels (Primary Producers, Herbivores and Carnivores), each transferring—with considerable losses—energy to the next trophic level. The Decomposers is a fourth trophic level, whose task is to recycle the dead matter from each level into the nutrients pool, to be used by the Primary Producers, whereas the Respiration compartment accounts for all non-recoverable energy losses, using the oxygen produced by photosynthesis.

The energy balance for each trophic level is given by the algebraic sum of incoming and outgoing energy fluxes. Consequently, the generic energy balance will look like

$$\frac{dx_i}{dt} = F_{\text{in}} - F_{\text{out}} - F_{\text{resp}} - F_{\text{decomp}} \tag{5.158}$$

where x_i represents the energy budget of the ith trophic level. Using the notations of Figure 5.67, that trophic chain translates into the following energy balances:

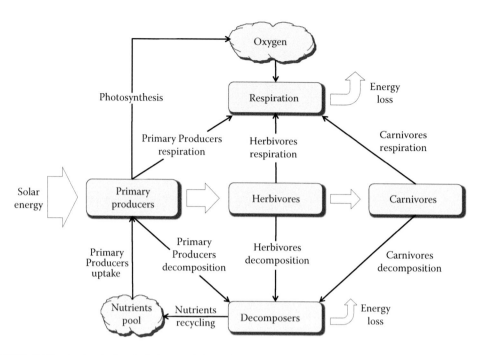

FIGURE 5.67 A simple food chain consisting of three trophic levels: Primary Producers, Herbivores, Carnivores, connected by narrowing arrows representing the diminishing energy transfer from one trophic level to the next. The Decomposer provides a nutrients recycle path, whereas the Respiration compartment represents an unrecoverable energy loss.

$$
\begin{cases}
\text{Primary Producers} & \dfrac{dx_1}{dt} = F_{01} - F_{12} - F_{14} - F_{10} \\[2mm]
\text{Herbivores} & \dfrac{dx_2}{dt} = F_{12} - F_{23} - F_{24} - F_{20} \\[2mm]
\text{Carnivores} & \dfrac{dx_3}{dt} = F_{23} - F_{34} - F_{30} \\[2mm]
\text{Decomposers} & \dfrac{dx_4}{dt} = F_{14} + F_{24} + F_{34}
\end{cases}
\tag{5.159}
$$

Because the solar energy is the ultimate energy source $\left(F_{01}\right)$ for the entire ecosystem, this source must be accounted for by the energy budget of each compartment, with the net loss due to respiration, so we have a 'global' balance in the form

$$
\sum_i \frac{dx_i}{dt} = F_{01} - \sum_i F_{i0}
\tag{5.160}
$$

where:
  i is extended to all the trophic levels in the ecosystem
  the F_{i0} represent the energy losses due to respiration (including that of the Decomposers)

Now, the form of the energy fluxes remains to be specified. There are several options, assuming that the energy flows from the donor compartment (x_i) to the acceptor compartment (x_j):

$$
\begin{cases}
\text{Donor dependent} & F_{ij} = k_{ij} \times x_i \\
\text{Receptor dependent} & F_{ij} = k_{ij} \times x_j \\
\text{Mutual dependency} & F_{ij} = k_{ij} \cdot x_i \cdot x_j \\
\text{Lotka} - \text{Volterra} & F_{ij} = k_{ij} \cdot x_j \cdot \left(1 - \beta_{ij} \cdot x_j\right) \\
\text{Competitive} & F_{ij} = k_{ij} \cdot x_j \cdot \left(1 - \alpha_{ij} \cdot x_i - \beta_{ij} \cdot x_j\right) \\
\text{Functional response} & F_{ij} = k_{ij} \dfrac{x_i \cdot x_j}{\gamma + x_i}
\end{cases}
\tag{5.161}
$$

A simple model for the trophic chain of Figure 5.67 will now be set up assuming that the solar energy varies with a yearly cycle that can be approximated with a sinusoid, and that the energy transfer between compartments can be modelled with a functional response of type 2. Then, the generic model (Equation 5.159) can be specified as follows, supposing that we are modelling a pasture where herbivores graze and are, in turn, predated by carnivores

$$
\begin{cases}
\text{Pasture} & \dfrac{dx_1}{dt} = r_g \cdot \underbrace{\dfrac{s(t)}{K_g + s(t)} \cdot x_1}_{\text{Grass growth}} - \dfrac{1}{Y_h} \cdot g_h \cdot \underbrace{\dfrac{x_1}{P_g + x_1} \cdot x_2}_{\text{Grazing}} - \underbrace{k_{d_g} \cdot x_1}_{\text{Decay}} \\[4ex]
\text{Herbivores} & \dfrac{dx_2}{dt} = g_h \cdot \underbrace{\dfrac{x_1}{P_g + x_1} \cdot x_2}_{\text{Herbivores growth}} - \dfrac{1}{Y_c} \cdot g_c \cdot \underbrace{\dfrac{x_2}{P_c + x_2} \cdot x_3}_{\text{Carnivores predation}} - \underbrace{k_{d_h} \cdot x_2}_{\text{Decay}} \\[4ex]
\text{Carnivores} & \dfrac{dx_3}{dt} = g_c \cdot \underbrace{\dfrac{x_2}{P_c + x_2} \cdot x_3}_{\text{Carnivores growth}} - \underbrace{k_{d_c} \cdot x_3}_{\text{Decay}}
\end{cases}
\tag{5.162}
$$

where the solar energy function $s(t)$ can be synthesized from the data as explained in Chapter 3. The Decomposer compartment was left out of this simplified model as it only becomes important when the recycled material is considered, as amply treated in Agren and Bosatta (1996) for the growth/predation coupled dynamics between primary production (grass) and herbivores, and between herbivores and carnivores. The decay terms include both decomposition and respiration. If we reintroduce a Decomposer compartment, then the two components of this term should be separated again. In Equation 5.162, the predation/growth terms are coupled by the yield coefficients that account for the lossy energy transfer. They should be interpreted as the amount of predator growth per unit of up taken prey, so $Y < 1$. The numerical values of the parameters of the model (Equation 5.162) are listed in Table 5.4. Figure 5.68 shows the Simulink model implementing the three equations (5.162).

For a clean representation, the blocks implementing each equation were grouped into a subsystem, which can be easily expanded to reveal its internal structure just by double-clicking on it, as shown in Figure 5.69 for the Pasture subsystem, in Figure 5.70 for the Herbivores subsystem, and in Figure 5.71 for the Carnivores subsystem. A period of 3.8 years was simulated with the parameter values of Table 5.4, and the model output are shown in Figure 5.72, together with the fluctuating solar radiation. It can be seen that all of the variables stabilize into a steady-state yearly fluctuations.

TABLE 5.4

Parameter Values and Initial Conditions for the Food Chain Model (Equation 5.162)

Parameter Value	Meaning and Units
$r_g = 0.8721$	Grass growth rate (1/d)
$K_g = 83$	Solar radiation half velocity constant (kcal/m²·d)
$P_g = 668$	Grass half velocity constant (kg/m²)
$Y_h = 0.141$	Herbivores grazing yield (—)
$k_{d_g} = 0.23$	Grass decay rate (1/d)
$g_h = 0.02224$	Herbivores growth rate (1/d)
$k_{d_c} = 0.02103$	Grass decay rate (1/d)
$g_c = 0.00673$	Carnivores growth rate (1/d)
$P_c = 854$	Carnivores half velocity constant (kg/m²)
$Y_c = 0.025$	Carnivores yield constant (—)
$k_{d_c} = 0.00122$	Carnivores decay rate (1/d)
$G_o = 12,500$	Initial pasture energy (kcal/m²)
$H_o = 360$	Initial herbivores energy (kcal/m²)
$C_o = 17$	Initial carnivores energy (kcal/m²)

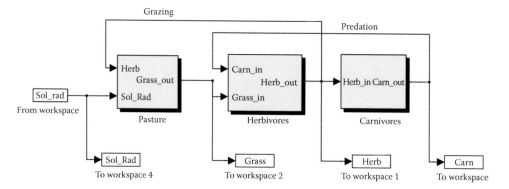

FIGURE 5.68 Simulink diagram implementing the food chain model (Equation 5.162). Each compartment has been grouped as a subsystem. Double-clicking on each blue block shows the inner structure of each subsystem, as illustrated in the following figures.

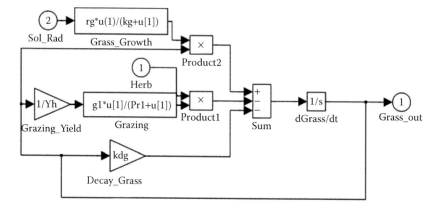

FIGURE 5.69 Inner structure of the pasture subsystem. The circled numbers indicate the input/output ports.

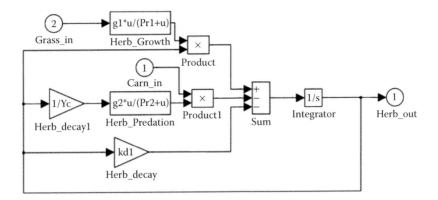

FIGURE 5.70 Inner structure of the herbivores subsystem. The circled numbers indicate the input/output ports.

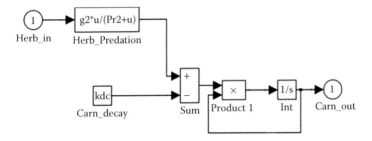

FIGURE 5.71 Inner structure of the carnivores subsystem. The circled numbers indicate the input/output ports.

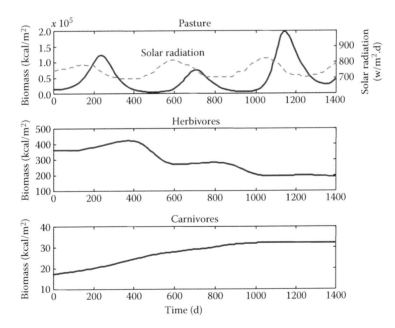

FIGURE 5.72 Simulation of the food chain model (Equation 5.162) with the parameter values and initial conditions of Table 5.4. It can be seen that the fluctuations induced by a varying solar radiation dampen down along the food chain where the carnivores eventually stabilize. The considerable loss of energy along the chain is typical, though widely variable, of all kinds of ecosystems. These graphs were obtained with the MATLAB script `Go_Food_Chain_Solar_Rad.m` in the `ESA_Matlab\Exercises\Chapter_5\Food_Chain` folder.

6 Flow Reactor Modelling

In rivers, the water that you touch is the last of what has passed and the first of that which comes.

Leonardo da Vinci

We are now going to take a closer look at the inner workings of the environmental processes, broadly outlined in Chapter 1. In particular, we have already encountered the problem of setting the ecosystem boundaries and of modelling the inner dynamics. Since we shall be considering aquatic ecosystems, either natural or man-made, the hydraulics of the waterbody will be of primary importance to describe the overall system dynamics. This chapter will be devoted to the modelling of the reaction vessel, be it an artificial reactor used for wastewater treatment, or a natural waterbody, such as a river or a lake.

6.1 CONTINUOUS-FLOW REACTOR MODELLING

A liquid takes the shape of its containment vessel, which defines its shape but cannot control its internal movements. Since these water movements are as important as the (bio)chemical reactions which develop therein in defining the overall behaviour of the aquatic environmental system, equal attention will be devoted to modelling both these aspects. This chapter will be devoted to the hydraulics of the reacting environment, while the chemical and biochemical aspects will be considered in Chapter 7. Figure 6.1 shows the possible schemes of the reaction volume.

The batch reactor (a) is a closed vessel where no input/output flows are present during the reaction, which is determined by the initial conditions only. Batch reactors are very common in the pharmaceutical or food industry, where substrate and microbial inoculum are initially mixed in the vessel, and the reaction proceeds with a given time sequence (the 'recipe') in terms of temperature profile, aeration, mixing, etc. and, at the end of the process, the products are extracted and separated (downstream processing). In the natural environment, river impoundments can be approximated with a batch reactor, at least during the periods in which this volume is isolated from the rest of the waterbody, for example, during low flow. There are no flux transport terms in a batch reactor, and the reaction is determined only on the basis of the kinetics terms. We have already seen such an example in Section 1.10.4.1, introducing the Monod kinetics in a closed vessel.

The single continuously stirred tank reactor (CSTR) (b) is the next possibility, when a continuous flow is constantly routed through the vessel, which is continuously stirred to ensure a complete mixing, hence the acronym CSTR. In this case, both the hydraulic and the kinetic terms contribute to the overall reaction rate.

The cascaded CSTR (c) is the next choice, used when a single CSTR does not suffice to reach the required degree of reaction.

The plug-flow (PF) reactor (d) will be shown to be the limit of the previous case when an infinite number of infinitely small CSTR is cascaded. It critically rests on the assumption that each liquid element does not mix with the adjacent elements, but travels through the tubular reactor as an isolated 'plug' (hence its name). Its optimality with respect to the previous case (c) will be demonstrated, as well as its usefulness in approximating certain wastewater treatment plants or river reaches.

The diffusive reactor (e) is the real-world version of the PF when longitudinal diffusion is significant and cannot be overlooked. This scheme is the most versatile in approximating real-world river reaches.

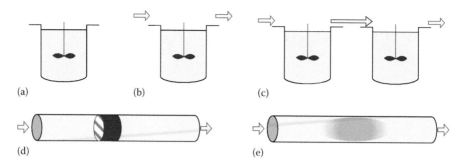

(a)   (b)   (c)

(d)   (e)

FIGURE 6.1 Possible reactor configurations from the batch (a) to the single CSTR (b), cascaded CSTRs (c), plug-flow (d), and diffusive reactor (e), each resulting in differing hydraulic regimes, which greatly influences the system behaviour, as will be illustrated in this chapter.

6.1.1 SINGLE CSTR

The batch reactor of Figure 6.1a will be considered later when dealing with the reaction kinetics, but for now, let us consider the single CSTR of Figure 6.1b, which we have already briefly encountered in Section 1.3.1, where a cylindrical vessel was used to demonstrate basic mass conservation principles. In that example, we have already examined the general case of time-varying volume. Now, we consider instead the most frequent case of constant volume. This case automatically makes the total mass balance unnecessary, because the input and output flows balance at any one time, that is, $F_{in} = F_{out} = F$. Under this assumption, the component mass balance Equation 1.13 can now be written as

$$\frac{dC}{dt} = \frac{F}{V}\left(C_{in} - C\right) \pm r\left(C\right)$$ (6.1)

where the first term accounts for the CSTR hydraulics, and the reaction rate $r\left(C\right)$ will be specified later.

6.1.1.1 Hydraulic Retention Time

Let us consider the *dilution rate* $q = F/V$ and its reciprocal $\vartheta = 1/q = V/F$. This term, often referred to as *hydraulic retention time* (HRT), is an important parameter in the CSTR dynamic behaviour. To clarify the meaning and role of the HRT, consider injecting a known mass of inert tracer M_o into the system. If the injection time is short enough to be approximated by a Dirac pulse of strength M_o, the injection immediately sets the concentration in the reactor, previously zero, to $C_o = M_o/V$. Since the tracer is inert, $r\left(C\right) = 0$, and if no further tracer is injected, that is, $C_{in} = 0$, for any $t > 0$ Equation 6.1 becomes

$$\frac{dC}{dt} = -\frac{C}{\vartheta} \quad \text{with initial condition } C\left(0\right) = C_o$$ (6.2)

Equation 6.2 can be integrated by separation of the variables to yield

$$C\left(t\right) = C_o \cdot e^{-t/\vartheta}$$ (6.3)

The plot of Equation 6.3 is shown to the right of the CSTR in Figure 6.2. The shaded area represents the amount of tracer still inside the reactor for $t \geq t^*$, which can be computed as

$$M\left(t^*\right) = \int_{t^*}^{\infty} F \cdot C\left(t\right) \cdot dt$$ (6.4)

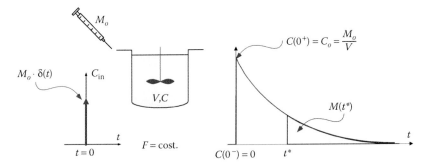

FIGURE 6.2 Defining the HRT through a tracer pulse injection. The impulsive nature of the injection causes the concentration to rise instantly from 0 to $C_o = M_o/V$.

If we consider $C(t)$ given by Equation 6.3 as a distribution of concentrations inside the reactor, then the average residence time (t_m) can be computed as

$$t_m = \frac{\int_0^\infty t \cdot C_o \cdot e^{-t/\vartheta} \cdot dt}{\int_0^\infty C_o \cdot e^{-t/\vartheta} \cdot dt} = \frac{-t \cdot \vartheta \cdot e^{-t/\vartheta}\Big|_0^\infty + \vartheta \int_0^\infty e^{-t/\vartheta} \cdot dt}{\int_0^\infty e^{-t/\vartheta} \cdot dt} = \vartheta \quad \text{because } t \cdot \vartheta \cdot e^{-t/\vartheta}\Big|_0^\infty = 0 \quad (6.5)$$

Equation 6.5 shows that ϑ can indeed be regarded as the average residence time of the fluid particles inside the reactor. Further, if the tracer is non-reactive, the full amount of injected mass M_o must be recovered at the output. Thus,

$$M_o = VC_o = \int_0^\infty FC(t)\,dt = FC_o \int_0^\infty e^{-t/\vartheta}\,dt \quad (6.6)$$

but

$$\int_0^\infty e^{-t/\vartheta}\,dt = -\vartheta e^{-qt}\Big|_0^\infty = \vartheta \quad (6.7)$$

Substituting in Equation 6.6 yields

$$M_o = FC_o \int_0^\infty e^{-t/\vartheta}\,dt = FC_o \cdot \vartheta = V \cdot C_o \quad (6.8)$$

which shows that indeed all of the tracer is recovered at the output.

6.1.1.2   Which Input for the CSTR?

There are two possible inputs to the CSTR: flow F and input concentration C_{in}. Though, in general, both may be assumed to be time varying, let us first consider the two special cases in which one is kept constant while the other is allowed to vary. Assuming a first-order kinetic term $(-k \cdot C)$, in the two cases the resulting models are fundamentally different

$$\begin{aligned} \text{Constant feed concentration} \quad & \frac{dC}{dt} = \frac{F}{V} \cdot C_{in} - \frac{F}{V} \cdot C - k \cdot C \\[2mm] \text{Constant flow} \quad & \frac{dC}{dt} = -\left(\frac{F}{V} + k\right)C + \frac{F}{V}C_{in} \end{aligned} \qquad (6.9)$$

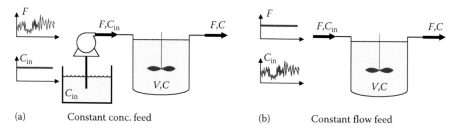

(a)   Constant conc. feed   (b)   Constant flow feed

FIGURE 6.3 Possible input choices for the CSTR. In (a), the input concentration is kept constant, while the flow varies. In (b), instead, the flow is kept constant, while the input concentration varies.

Figure 6.3 shows the two input situations described by the Equation 6.9. In the constant input concentration case, the flow is allowed to vary, with its variations influencing the HRT. The resulting model is nonlinear because the middle term is the product between the input F and the state variable C. In the constant-flow case, the model is instead linear in the concentration, because the input F is assumed to be constant. Hence, the dynamic coefficient $(1/\vartheta+k)$ is composed of two terms, the first being the contribution of the hydraulics, through the HRT, and the second being the kinetic contribution.

Let us now consider the constant-flow case, which is described by a linear system, still assuming a first-order kinetics. The steady-state relationship between the input and output concentrations can be determined as

$$qC_{in} = (q+k)C \quad \Rightarrow \quad \frac{C_{out}}{C_{in}} = \frac{q}{q+k} = \frac{1}{1+\vartheta \cdot k} = \frac{1}{1+N_{Da}} \tag{6.10}$$

where the product of the kinetic constant and the HRT $(\vartheta = 1/q)$ is referred to as the Damköhler number (N_{Da}) in the chemical engineering literature (see, e.g., Levenspiel, 1972). Equation 6.10 can also be used in reverse to compute the HRT (ϑ^*) required to obtain a prescribed extent of reaction $\beta = C_{out}/C_{in}$ given the kinetic rate k

$$\vartheta^* = \frac{1}{k}\left[\left(\frac{C_{in}}{C_{out}}\right)-1\right] = \frac{1}{k}\left[\left(\frac{1}{\beta}\right)-1\right] \tag{6.11}$$

6.1.1.3 Linearization of the CSTR Dynamics

If the flow is assumed to be the system input, then we have already seen that the system is nonlinear. Suppose that the flow switches between two values F_1 and F_2 with $F_1 > F_2$. The CSTR response is asymmetrical because the flow is part of the time constant

$$-\left(\frac{F}{V}+k\right) \rightarrow \begin{cases} -\left(\dfrac{F_1}{V}+k\right) & \text{Fast time constant} \\[3mm] -\left(\dfrac{F_2}{V}+k\right) & \text{Slow time constant} \end{cases} \tag{6.12}$$

and the asymmetrical output concentration is shown in Figure 6.4.

This system can be linearized by selecting an equilibrium pair (q_o, C_{in}^o) from which the steady-state output concentration can be computed as

$$C_{out}^o = \frac{q_o}{q_o+k}C_{in}^o \tag{6.13}$$

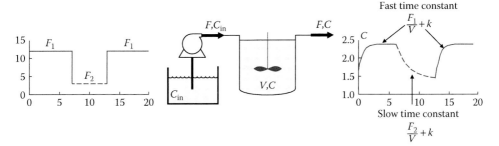

FIGURE 6.4 A square-wave input flow produces an asymmetrical response of the CSTR with first-order kinetics, because the varying flow changes the reactor time constant. This simulation was obtained with the MATLAB script Go _ CSTR _ square _ flow.m in the ESA _ MATLAB\Exercises\Chapter _ 6\CSTR folder.

The incremental linear model now has the form

$$\frac{d\tilde{C}}{dt} = \frac{\partial f}{\partial C}\bigg|_{C_o} \tilde{C} + \frac{\partial f}{\partial q}\bigg|_{q_o} \tilde{q} + \frac{\partial f}{\partial C_{in}}\bigg|_{C_{in}^o} \tilde{C}_i \tag{6.14}$$

which by computing the partial derivatives, yields

$$\left.\begin{aligned}
\frac{\partial f}{\partial C}\bigg|_{C_o} &= -(q_o + k) \\
\frac{\partial f}{\partial q}\bigg|_{q_o} &= C_{in}^o - C_o \\
\frac{\partial f}{\partial C_i}\bigg|_{C_{io}} &= q_o
\end{aligned}\right\} \Rightarrow \frac{d\tilde{C}}{dt} = -(q_o + k)\tilde{C} + (C_{in}^o - C_o)\tilde{q} + q_o\tilde{C}_i \tag{6.15}$$

If instead, the kinetic is of second order, that is, $-k \cdot C^2$, the reactor dynamics becomes

$$\frac{dC}{dt} = q(C_{in} - C) - k \cdot C^2 \tag{6.16}$$

and its equilibrium value can be computed by solving the second-order algebraic equation

$$k \cdot C_o^2 + q_o \cdot C_o - q_o \cdot C_{in}^o = 0 \tag{6.17}$$

whose only feasible solution is

$$C_o = \frac{-q_o + \sqrt{q_o^2 + 4kq_oC_{in}^o}}{2k} \tag{6.18}$$

The linearized model in this case is

$$\left.\begin{aligned}
\frac{\partial f}{\partial C}\bigg|_{C_o} &= -(q_o + 2kC_o) \\
\frac{\partial f}{\partial q}\bigg|_{q_o} &= C_{in}^o - C_o \\
\frac{\partial f}{\partial C_i}\bigg|_{C_{io}} &= q_o
\end{aligned}\right\} \Rightarrow \frac{d\tilde{C}}{dt} = -(q_o + 2kC_o)\tilde{C} + (C_{in}^o - C_o)\tilde{q} + q_o\tilde{C}_i \tag{6.19}$$

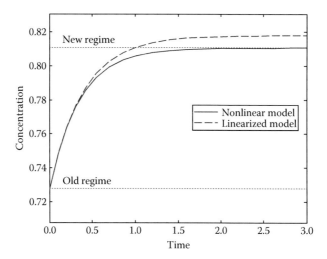

FIGURE 6.5 Comparison between the step response of a CSTR with second-order kinetics and its linearized approximation, which is accurate in the neighbourhood of the old regime, but grossly wrong near the new regime. This simulation was obtained with the MATLAB script `Go_CSTR_secons_ord.m` in the `ESA_MATLAB\Exercises\Chapter_6\CSTR` folder.

The comparison between the step response of a CSTR with second-order kinetics and its linearized approximation according to Equation 6.19 is shown in Figure 6.5.

6.1.2 CHAIN OF CASCADED CSTR

The amount of reaction provided by a single CSTR may not provide the required extent of reaction. For this reason, CSTRs are often cascaded, as we have already seen in the case of the Nash chain of reservoirs, Section 1.8.3.1. Let us first consider a pair of cascaded CSTRs with a first-order kinetics, as depicted in Figure 6.6, in which the output of the first CSTR (C_1) is the input to the second. Assuming a constant flow throughout the system, the combined model is

$$\frac{dC_1}{dt} = \frac{F}{V_1}C_{in} - \frac{F}{V_1}C_1 - kC_1$$

$$\frac{dC_2}{dt} = \frac{F}{V_2}C_1 - \frac{F}{V_2}C_2 - kC_2$$

(6.20)

which can be written in matrix form as

$$\begin{bmatrix} \dot{C}_1 \\ \dot{C}_2 \end{bmatrix} = \begin{bmatrix} -\dfrac{F}{V_1} - k & 0 \\ \dfrac{F}{V_2} & -\dfrac{F}{V_2} - k \end{bmatrix} \times \begin{bmatrix} C_1 \\ C_2 \end{bmatrix} + \begin{bmatrix} \dfrac{F}{V_1} \\ 0 \end{bmatrix} C_{in}$$

(6.21)

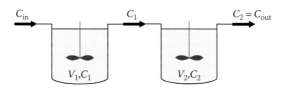

FIGURE 6.6 A cascade of two CSTRs in series.

By inspection, it can be seen that the system matrix is lower–triangular, and its eigenvalues are

$$\lambda_1 = -\frac{F}{V_1} - k \quad \lambda_2 = -\frac{F}{V_2} - k \quad \lambda_1, \lambda_2 < 0 \tag{6.22}$$

So we can conclude that the system is always stable because its eigenvalues are real negative.

The cascaded arrangement can be extended to n CSTRs, and the overall model then becomes

$$\frac{dC_1}{dt} = \frac{F}{V_1} C_{in} - \frac{F}{V_1} C_1 - kC_1$$

$$\frac{dC_2}{dt} = \frac{F}{V_2} C_1 - \frac{F}{V_2} C_2 - kC_2 \tag{6.23}$$

$$\vdots$$

$$\frac{dC_n}{dt} = \frac{F}{V_n} C_{n-1} - \frac{F}{V_n} C_n - kC_n$$

or, in matrix form,

$$\begin{bmatrix} \dot{C}_1 \\ \dot{C}_2 \\ \dots \\ \dot{C}_n \end{bmatrix} = \begin{bmatrix} -\frac{F}{V_1} - k & 0 & \dots & 0 \\ \frac{F}{V_2} & -\frac{F}{V_2} - k & \dots & 0 \\ \dots & \dots & \dots & 0 \\ 0 & 0 & \frac{F}{V_n} & -\frac{F}{V_n} - k \end{bmatrix} \times \begin{bmatrix} C_1 \\ C_2 \\ \dots \\ C_n \end{bmatrix} + \begin{bmatrix} \frac{F}{V_1} \\ 0 \\ \dots \\ 0 \end{bmatrix} C_{in} \tag{6.24}$$

The system matrix is still lower–triangular, with eigenvalues

$$\lambda_1 = -\frac{F}{V_1} - k \qquad \lambda_2 = -\frac{F}{V_2} - k \qquad \lambda_n = -\frac{F}{V_n} - k \qquad \lambda_1, \lambda_2, \dots, \lambda_n < 0 \tag{6.25}$$

Therefore, the cascade of n CSTRs is unconditionally stable. Regarding the time response of the CSTR chain, Figure 6.7 shows that the impulse injection changes the initial condition of the first tank from zero to M_o/V while all the others remain at zero. Hence, the impulse response of the first tank is radically different from that of the following tanks, which start from their naturally zero initial conditions. The concentration peak is progressively delayed along the reactor chain, but the area of each response is still equal to the injected mass M_o.

In analysing the impulse response already treated in Section 1.8.3.1 we now take a step further. Let us suppose that a unit pulse of inert tracer is injected into the tank. We have already computed the response as

$$1 = \int_0^\infty C(t) \, dt = A \int_0^\infty t^{n-1} e^{-qt} \, dt \quad = A \times \frac{(n-1)!}{q^n} \Rightarrow \quad A = \frac{q^n}{(n-1)!} \tag{6.26}$$

where:
  q is the dilution rate constant and
  A is a normalization factor, so that the output concentration can be written as

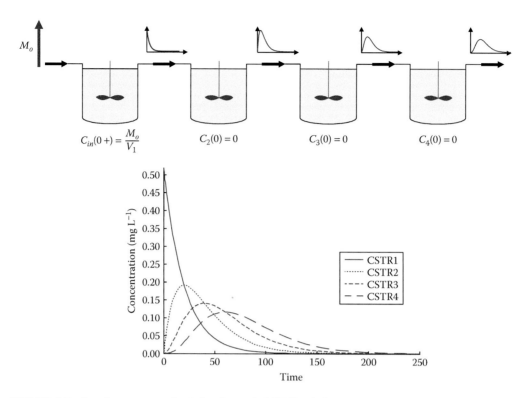

FIGURE 6.7 Impulse response of a chain of cascaded CSTRs. A finite amount of tracer (M_o) is injected into the first reactor and propagates through the chain as shown in the lower graph.

$$C(t) = \frac{q^n}{(n-1)!} t^{n-1} e^{-qt}$$

$$(6.27)$$

If the amount of injected tracer is M_o, then

$$M_o = VC_o = F\left(\frac{C_0}{q}\right) \quad \Rightarrow \quad C(t) = C_o \frac{q^{n-1}}{(n-1)!} t^{n-1} e^{-qt} \qquad (6.28)$$

6.1.2.1 MATLAB Modelling of the Cascaded CSTRs

Before examining more reactor structures, let us pause and consider the MATLAB® implementation of the reactor chain that produced the graphs in Figure 6.7. First, a Simulink® model is set up for the elementary CSTR, and then this structure is replicated for the whole chain by connecting identical subsystems, as shown in Figure 6.8.

The launch MATLAB script is conceived to simulate both the impulse and the step response. We have already seen that the former is obtained by changing the initial condition of the first reactor to $C_o = M_o/V$ while leaving all the other initial conditions and the input concentration C_{in} to zero. The step response, instead, is obtained by leaving all the initial conditions to zero and setting the input concentration to the desired value C_{in}. If the same script is to simulate both regimes, it is convenient to use a `switch` command with labels 'step' and 'impulse' to set up the appropriate simulation conditions, as listed in Box 6.1. Figure 6.9 shows the simulated response obtained with the Simulink model of Figure 6.8 and the MATLAB launch script of Box 6.1.

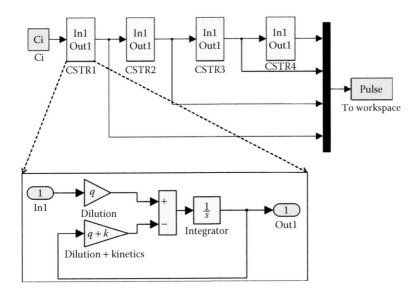

FIGURE 6.8 Simulink diagram of the CSTR _ cascade.mdl model, showing the system modularity: a cascade of CSTRs can be formed by modelling the single CSTR (lower inset) and connecting them in series. Of course, the initial conditions of each subsystem must be set up individually (see launch MATLAB script in Box 6.1).

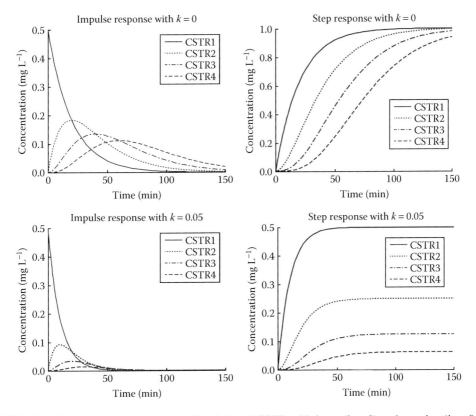

FIGURE 6.9 Impulse and step response of a chain of CSTR with inert ($k = 0$) and reactive ($k = 0.05$) tracer obtained with the combination of the Simulink model of Figure 6.8 and of the MATLAB script of Box 6.1. The full launch MATLAB script is Go _ CSTR _ chain.m in the ESA _ MATLAB\Exercises\ Chapter _ 6\Cascade _ CSTR folder.

BOX 6.1   MATLAB LAUNCH SCRIPT TO SIMULATE THE IMPULSE AND STEP RESPONSE OF THE CSTR CHAIN

```
% Select the response
% Select the input (uncomment the desired one)
% input_mode='pulse';
input_mode='step';
V=20;     % Volume of each CSTR (L)
F=1;      % Flow (L/min)
q=F/V;    % Dilution rate
k=0.05;   % Reaction first-order kinetics
switch input_mode
  case 'pulse'
      Mo=10;    % Amount of injected tracer (mg)
      Co=Mo/V;  % Initial concentration of first CSTR
      Ci=0;     % No input
  case 'step'
      Ci=1;     % Input concentration for t>0
      Co=0;     % Zero initial conditions for all CSTRs
end
tfin=250;
[t,x,y]=sim('CSTR_cascade',tfin);
figure(1)
plot(t,pulse(:,4),'b',t,pulse(:,3),':b',t,pulse(:,2),'-.b',
t,pulse(:,1),'--b','LineWidth',2)
set(gca,'FontName','Arial','FontSize',12)
switch input_mode
  case 'pulse'
    legend('CSTR1','CSTR2','CSTR3','CSTR4','location','northeast')
  case 'step'
    legend('CSTR1','CSTR2','CSTR3','CSTR4','location','east')
end
xlabel('time (min)')
ylabel('Concentration (mg/L)')
```

6.1.3   STEADY-STATE ANALYSIS OF THE CASCADED CSTRs

Though the kinetic constant is the same for all tanks, Figure 6.9d shows more clearly than the pulse response of Figure 6.9c that the tracer abatement is not uniform along the reactor chain, with the first tank producing the largest reduction, and the subsequent tanks apparently contributing by progressively smaller amounts. How can this be explained? Let us recall that, according to the first-order kinetics, the reaction rate is proportional to the concentration, that is,

$$\frac{dC}{dt} = -k \cdot C \tag{6.29}$$

Therefore, the higher the concentration C, the faster will be its decay, and hence the decreasing concentration along the chain produces a progressively slower reaction rate, reducing the contribution of each reactor along the cascade. Now, we set out to investigate the efficiency of the cascade under two differing viewpoints, by posing the following two optimization problems:

1. *Given a prescribed extent of reaction* $\beta = C_{out}/C_{in}$, determine the minimum total retention time (Θ_{min}), supposing that all the tanks have equal HRT $\vartheta = \Theta/n$.
2. *Given the total retention time* Θ, find the largest achievable extent of reaction β_{max}.

Let us now start by considering the first problem. Suppose that the reaction follows a first-order kinetics, so that the extent of reaction is related to the kinetic rate and the retention time of each CSTR (ϑ) by the relation

$$\beta = \frac{C_{out}}{C_{in}} = \left(\frac{1}{1+k \cdot \vartheta}\right)^{n} \qquad (6.30)$$

First, let us show that the total HRT required to achieve the same extent of reaction with two cascaded reactors (ϑ_2) is less than that of a single CSTR (ϑ_1). If both systems are to provide the same extent of reaction

$$\beta = \underbrace{\left(\frac{1}{1+k \cdot \vartheta_2}\right)^{2}}_{\substack{\text{Two cascaded} \\ \text{CSTR}}} = \underbrace{\frac{1}{1+k \cdot \vartheta_1}}_{\substack{\text{One single} \\ \text{CSTR}}} \qquad (6.31)$$

solving for ϑ_1 yields

$$\vartheta_1 = \vartheta_2 (2 + k \cdot \vartheta_2) > 2 \cdot \vartheta_2 \qquad (6.32)$$

Thus, the total HRT of the single CSTR (ϑ_1) is more than twice the HRT of each CSTR (ϑ_2) of the chain, demonstrating that the same extent of reaction can be achieved with a lower HRT, that is, a smaller total volume in the cascade arrangement. This result can be generalized to a chain of n cascaded elements by solving Equation 6.30 for the HRT of the single reactor in the chain

$$\vartheta_n = \frac{1}{k}\left[\left(\frac{1}{\beta}\right)^{\frac{1}{n}} - 1\right] \qquad (6.33)$$

Figure 6.10a shows the reduction of the HRT of each reactor of the cascade (ϑ_n) with the increasing number n of elements in the chain. The dashed line is the no-improvement case, where the total volume is simply divided by the number of elements. Since the cascaded CSTRs (dotted line) are always below the no-improvement locus (dashed line), there is indeed an improvement in splitting a given reaction volume in a chain of equal CSTRs. Figure 6.10a also shows that the improvement provided by the chain is higher at the beginning, and then tapers out, so that the reactors added after the first few elements bring little, if any, improvement. Consequently, it makes sense to look for the minimum overall HRT (Θ_{min}) that can be achieved, regardless of the number of reactors.

This can be computed as the limit of $\Theta = n \cdot \vartheta_n$, with ϑ_n given by Equation 6.33, for $n \to \infty$.

$$\Theta_{min} = \lim_{n \to \infty}(n \cdot \vartheta_n) = \lim_{n \to \infty} n \times \frac{1}{k}\left[\left(\frac{1}{\beta}\right)^{1/n} - 1\right] \quad \text{defining} \quad \begin{cases} c = \dfrac{1}{\beta} \\ x = \dfrac{1}{n} \end{cases} \quad \text{then} \lim_{n \to \infty} n\left[\left(\frac{1}{\beta}\right)^{1/n} - 1\right]$$

$$= \lim_{x \to 0}\frac{c^x - 1}{x} = \lim_{x \to 0}\frac{e^{x\ln(c)} - 1}{x} = \frac{e^{x\ln(c)} - 1}{x \cdot \ln(c)}\ln(c) \quad \text{but} \quad \lim_{x \to 0}\frac{e^{x\ln(c)} - 1}{x \cdot \ln(c)} = 1 \quad \text{thus,} \qquad (6.34)$$

$$\Theta_{min} = \frac{1}{k}\ln\left(\frac{1}{\beta}\right)$$

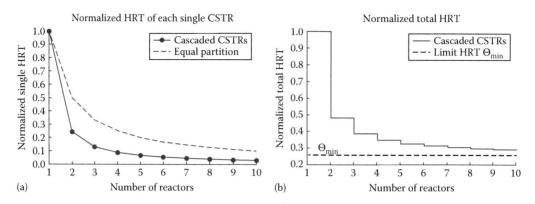

FIGURE 6.10 (a) Improvement in the HRT of each cascaded reactor $\left(\vartheta_n\right)$ to achieve a prescribed extent of reaction β. By comparison, the dashed curve shows the case of no improvement. (b) Total HRT as a function of the number of chain elements to achieve the same β. The dashed line represents the minimum HRT $\left(\Theta_{\min}\right)$ of Equation 6.34 achievable with an infinite chain of CSTRs.

Equation 6.34 solves the first of the two problems that we have just posed. It yields the minimum overall HRT $\left(\Theta_{\min}\right)$ for a prescribed extent of reaction β. The advantage of splitting decreases for large values of *n*, and Figure 6.10b shows that the total HRT of the chain can never be less than $\Theta_{\min}$ given by Equation 6.34.

Now, we turn our attention to problem 2: find the largest achievable extent of reaction $\beta_{\max}$, for a prescribed total retention time Θ. Suppose that the chain is composed of equal reactors, each with HRT ϑ_n equal to a fraction of the total HRT, that is, $\vartheta_n = \Theta/n$. Then, the extent of reaction can be computed by rearranging Equation 6.30 in terms of the overall HRT

$$\beta = \frac{C_{\text{out}}}{C_{\text{in}}} = \left(\frac{1}{1+k\cdot\dfrac{\Theta}{n}}\right)^n \tag{6.35}$$

Considering the logarithm of β and taking the limit for $n \to \infty$ yield

$$\lim_{n\to\infty}\left(\ln\frac{C_{\text{out}}}{C_{\text{in}}}\right) = \lim_{n\to\infty}\left(-\ln\frac{1+(k\Theta/n)}{1/n}\right) = \lim_{n\to\infty}-\frac{\left[1/\left(1+(k\Theta/n)\right)\right]\left[-\left(k\Theta/n^2\right)\right]}{-1/n^2} = -k\Theta \tag{6.36}$$

Thus, the maximum achievable extent of reaction $\beta_{\max}$ for a prescribed total HRT Θ is

$$\beta_{\max} = \frac{C_{\text{out}}}{C_{\text{in}}} = e^{-k\Theta} \tag{6.37}$$

Figure 6.11 shows the decreasing normalized output concentration C_{out} at the output of each CSTR in the chain for a varying number of HRT partitioning, computed as

$$C_{\text{out}}(i) = \frac{C_{\text{out}}(i-1)}{1+k\cdot(\Theta/i)} \quad \text{with} \quad 2 \le i \le n \quad \text{and} \quad C_{\text{out}}(1) = C_{\text{in}} = 1 \tag{6.38}$$

By comparison, the output obtained by Equation 6.37 as the normalized HRT varies from 0 to 1 is also shown in the figure.

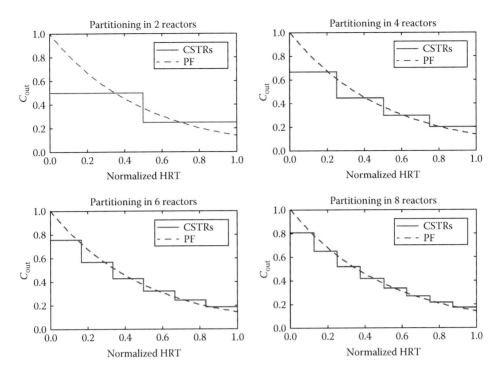

FIGURE 6.11 Normalized concentration at the output of each CSTR of the chain for differing numbers of reactors as a function of the normalized HRT. The dashed curve indicates the minimum output concentration achieved by the PF reactor (Section 6.2). These graphs were obtained with the MATLAB script Go _ Fractionation.m in the ESA _ MATLAB\Exercises\Chapter _ 6\Cascade _ CSTR folder.

$$C_{out} = C_{in} \cdot e^{-k \cdot \vartheta} \quad \text{with} \quad 0 \leq \vartheta \leq 1 \quad \text{and} \quad C_{in} = 1 \tag{6.39}$$

Note that the normalized concentration at the end of the system is always above the value of the dashed curve, indicating that the abatement achieved by Equation 6.39 is the limit extent of reaction. We shall see in Section 6.2 that a physical reactor, the PF, is indeed capable of achieving this extreme extent of reaction β_{max} for a given volume. Equation 6.37 solves the second optimization problem of determining the maximum achievable extent of reaction for a given HRT Θ.

6.1.4 Feedback CSTR

In some cases, the flow along the CSTR chain is not unidirectional, as there may exist a counter-current returning part of the flow to the upstream reactor. There are several practical instances of this bidirectional flow, as exemplified in Figure 6.12. In (a), two communicating waterbodies of largely differing dimensions, for example, a lake and a embayment, can exchange mass through diffusion; in (b) two counter-rotating impellers in a wastewater oxidation tank create opposing currents with a bidirectional mass exchange; (c) shows the situation of a stratified lake during the summer months. The lake is practically divided into two horizontal layers, the upper Epilimnion and the lower Hypolimnion, which exchange mass through vertical mixing; finally, in (d) the mass exchange takes place between the waterbody of a lake or river and the bottom sediment through the opposite processes of settling and resuspension. The difficulty in modelling any of these situations rests in the fact that the internal exchange flows are not easily measurable, especially when they are not *bulk* fluxes, but are driven by diffusion, as will be explained shortly.

Let us model the feedback two-tank system according to the scheme of Figure 6.13, with F_{21} representing the feedback flow from tank 2 back into tank 1, while F_{12} is the forward flow. Supposing

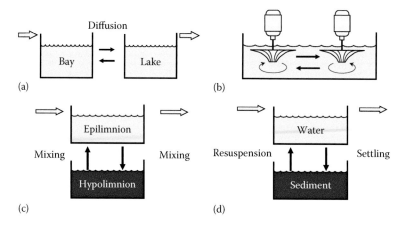

FIGURE 6.12 Examples of feedback CSTRs. (a) A lake and a smaller embayment can exchange mass through diffusion. (b) Two counter-rotating impellers in a wastewater treatment tank create counter currents. (c) During the summer stratification, the two horizontal layers of a lake have a limited mass exchange through vertical mixing. (d) Mass exchanges between the waterbody and the sediment in a lake or river occur through the opposite processes of settling and resuspension.

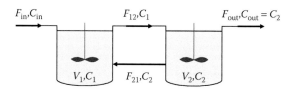

FIGURE 6.13 Ideal scheme of a two-tank feedback system, with the counter-current flow F_{21} representing the feedback path.

that a chemical reaction takes place in the system with a first-order kinetic, and that the kinetic rate constant k is the same for both tanks, the model of the feedback system can be written as

tank 1 $\dfrac{dC_1}{dt} = \dfrac{F_{in}}{V_1} C_{in} + \dfrac{F_{21}}{V_1} C_2 - \dfrac{F_{12}}{V_1} C_1 - kC_1$

tank 2 $\dfrac{dC_2}{dt} = \dfrac{F_{12}}{V_2} C_1 - \dfrac{F_{21}}{V_2} C_2 - \dfrac{F_{out}}{V_2} C_2 - kC_2$

(6.40)

which, in compact matrix notation, becomes

$$\begin{bmatrix} \dot{C}_1 \\ \dot{C}_2 \end{bmatrix} = \begin{bmatrix} -\dfrac{F_{12}}{V_1} - k & \dfrac{F_{21}}{V_1} \\ \dfrac{F_{12}}{V_2} & -\dfrac{F_{21}}{V_2} - \dfrac{F_{out}}{V_2} - k \end{bmatrix} \times \begin{bmatrix} C_1 \\ C_2 \end{bmatrix} + \begin{bmatrix} \dfrac{F_{in}}{V_1} \\ 0 \end{bmatrix} C_{in}$$

(6.41)

The steady-state solution of Equation 6.41 can be computed as

$$\bar{C} = -A^{-1} \cdot b \cdot C_{in} = -\begin{bmatrix} -\dfrac{F_{12}}{V_1} - k & \dfrac{F_{21}}{V_1} \\ \dfrac{F_{12}}{V_2} & -\dfrac{F_{21}}{V_2} - \dfrac{F_{out}}{V_2} - k \end{bmatrix}^{-1} \times \begin{bmatrix} \dfrac{F_{in}}{V_1} \\ 0 \end{bmatrix} \times C_{in}$$

(6.42)

and its stability depends on the eigenvalues of the system matrix A. Their real part is negative if

$$tr(A) = (\lambda_1 + \lambda_2) < 0 \quad \text{and} \quad \det(A) = \lambda_1 \cdot \lambda_2 > 0 \tag{6.43}$$

which, in this case, yields

$$tr(A) = \left(-\frac{F_{12}}{V_1} - \frac{F_{21}}{V_2} - \frac{F_{out}}{V_2} - 2k\right) < 0$$

$$\det(A) = \left[\left(\frac{F_{12}}{V_1} + k\right) \cdot \left(\frac{F_{21}}{V_2} + \frac{F_{out}}{V_2} + k\right) - \frac{F_{21}}{V_1} \cdot \frac{F_{12}}{V_2}\right] > 0 \tag{6.44}$$

The eigenvalues are mainly influenced by the size of the two volumes. If V_1 and V_2 differ by one or more orders of magnitude, then the eigenvalues, which are always negative real as shown by Equation 6.44, also differ by one or more orders of magnitude, thus generating a *stiff* system (Quinney, 1987; Dormand, 1996; Butcher, 2003), with a *fast* and a *slow* mode. This is reflected in the variable behaviour of C_1 in Figure 6.14, where the initial fast mode is gradually replaced by the slow mode. Figure 6.15 confirms, on a logarithmic scale, the dependence of the eigenvalues λ_1 and λ_2 on the volume ratio. While the fast mode (λ_2) remains almost constant for a volume ratio above five, the slow mode progressively approaches zero, thus making the slow mode more and more dominant.

6.1.5 Generalized CSTR

We are now in a position to generalize the CSTR model by writing a balance across the boundaries of the control volume, indicated by the dashed marquee in Figure 6.16, considering the following inputs and outputs:

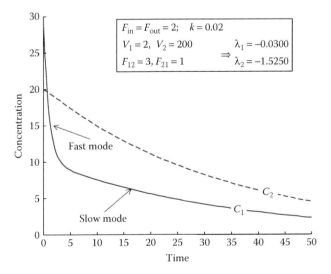

FIGURE 6.14 The feedback system of Figure 6.13 may become stiff if the volumes are significantly different. In this case, the system response includes both a fast and a slow mode.

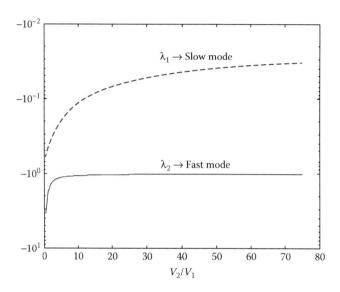

FIGURE 6.15 Eigenvalues of the feedback system of Figure 6.14 as a function of the volume ratio.

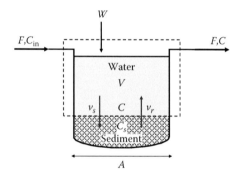

FIGURE 6.16 Conceptual scheme of a generalized CSTR, which includes external mass loading (W), resuspension (v_r) and sedimentation (v_s). The control volume for the mass balance is indicated by the dashed box.

- Dilution input $\left(F \cdot C_{\text{in}}\right)$
- Mass loading input $\left(W\right)$
- Sediment resuspension flux from the sediment to the water compartment through the cross section A and upward velocity $v_r \left(v_r \, A \, C_s\right)$
- Sedimentation flux from the water compartment to the sediment through the cross section A and downward velocity $v_s \left(v_s \, A \, C\right)$

Combining the above features yields the dynamics of the water compartment as

$$V \frac{dC}{dt} = W + F \cdot C_{\text{in}} + J_r - J_s - F \cdot C - k \cdot C \cdot V \tag{6.45}$$

The sedimentation and resuspension fluxes J_r and J_s can be regarded as mass transfer terms involving the particulate matter and can be parametrized as first-order kinetics, defining an average equivalent depth $h = V/A$

$$h = \frac{V}{A} \Rightarrow J = v \, AC = v \frac{V}{h} C = V \, k \, C \quad \text{with} \quad k = \frac{v}{h} \tag{6.46}$$

so the two mass fluxes across the water-sediment boundary can be written as

$$J_r = V\,k_r\,C_s \quad \text{with} \quad k_r = \frac{v_r}{h}$$

(6.47)

$$J_s = V\,k_s\,C \quad \text{with} \quad k_s = \frac{v_s}{h}$$

Substituting Equation 6.47 into the dynamic balance Equation 6.45 and dividing by the volume V yield the generalized CSTR model

$$\frac{dC}{dt} = \frac{W}{V} + q \cdot C_{\text{in}} + k_r \cdot C_s - k_s \cdot C - q \cdot C - k \cdot C$$

(6.48)

6.2 PLUG-FLOW REACTOR

Extending to the limit, the chain of CSTRs of Section 6.1.2, Equation 6.37 has demonstrated that in principle a reactor exists which can maximize the extent of reaction β_{max} for a given HRT. Such a reactor is not a mathematical abstraction, though, but an actual tubular reactor, provided that the hydraulic regime is such that each element of fluid (the 'plug') does not mix with the adjacent ones, implementing in fact a cascade of tiny CSTRs in each of which the reaction goes on without any interference from the neighbouring elements, that is, without any cross-mixing. Many physical examples of tubular reactors exist that approximate the PF condition well, such as a slow-flowing river or, in a wastewater treatment plant, a reaction tank whose length is much greater than its cross section.

The PF is best modelled as a steady-state system in which the hydraulic regime is constant and the flow time t_s is related to the longitudinal dimension (x) of the reactor by the relation

$$t_s = \frac{x}{u(x)}$$

(6.49)

where:
  x is the distance from the upstream end of the reactor and
  $u(x)$ is the flow velocity, averaged across the reactor section

Without chemical kinetics, the PF introduces a mere input–output delay so that its output is a delayed version of the input

$$C_{\text{out}}(t) = C_{\text{in}}(t - \delta)$$

(6.50)

The delay δ coincides with the PF HRT ϑ_{PF}. In fact, if V is the reactor volume, from Equation 6.49 it follows that

$$\left.\begin{array}{c} V = S \cdot x \\ F = S \cdot u \end{array}\right\} \Rightarrow \vartheta_{\text{PF}} = \frac{V}{F} = \frac{S \cdot x}{S \cdot u} = \frac{x}{u} = \delta$$

(6.51)

where S is the reactor cross section. If a chemical reaction takes place inside the tubular reactor, then its development is the same as would occur in a batch reactor travelling through the tubular reactor with velocity u defined by Equation 6.49. Combining the travel time with the reaction rate, the concentration along the reactor can be expressed either as a function of the flow time t_s or as a function of the position along the reactor length (x) as

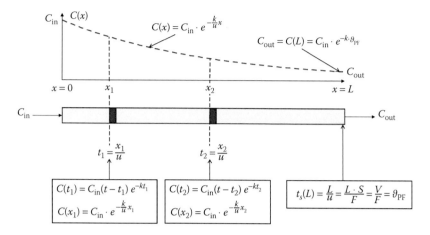

FIGURE 6.17 Concentration along a tubular reactor operated with a PF regime, expressed either as a function of flow time or as a function of position. The travel time along the reactor is equal to its HRT ϑ_{PF}.

$$C\left(t_s\right) = C_{in}\left(t - t_s\right) \cdot e^{-k \cdot t_s}$$

$$C\left(x\right) = C_{in} \cdot e^{-k\frac{x}{u}}$$

(6.52)

Figure 6.17 shows the concentration along the PF, computed either as a function of flow time t_s or as a function of position along the reactor length, according to Equation 6.52.

If the PF is considered to be in a hydraulic steady state, the delay due to travel time becomes irrelevant and the extent of reaction can be computed by considering the spatial dimension only. Assuming again a first-order kinetic, the chemical transformation along the PF can be modelled as

$$\frac{dC}{dt_s} = -k \cdot C \quad \Rightarrow \quad \frac{dC}{dx/u} = -k \cdot C \quad \Rightarrow \quad \frac{dC}{dx} = -\frac{k}{u}C$$

(6.53)

Separating the variables and integrating along the reactor length yields

$$\int_{C_{in}}^{C_{out}} \frac{dC}{C} = -k \int_0^L \frac{dx}{u}$$

$$\ln\left(\frac{C_{out}}{C_{in}}\right) = -\frac{k}{u}L = -k\frac{S}{F}L = -k \cdot \vartheta$$

(6.54)

$$C_{out} = C_{in} \cdot e^{-k \cdot \vartheta}$$

which coincides with the limit value of the CSTR chain Equation 6.39, apart from the normalization factor $\left(C_{in} = 1\right)$. Consider that in the PF model, the input term does not appear explicitly, but is incorporated in the initial condition C_{in}.

So far, we have considered a first-order kinetic, which is the most widespread, but there are other possibilities, and it is interesting to compare the performance of the PF to that of the CSTR using a few other types of kinetics.

Zero-order kinetics. In this case the reaction rate is constant, irrespective of the concentration, that is,

$$\text{CSTR} \quad \frac{dC}{dt} = q \cdot \left(C_{in} - C \right) - k$$

$$\text{PF} \quad \frac{dC}{dx} = -\frac{k}{u} \tag{6.55}$$

from which the HRTs can be computed as

$$\text{CSTR} \quad \vartheta_{CSTR}^{(0)} = \frac{1}{k} \left(C_{in} - C_{out} \right)$$

$$\text{PF} \quad \int_{C_{in}}^{C_{out}} dC = -\frac{k}{u} \int_0^L dx = -\frac{k}{u} L = -k \cdot \vartheta_{PF} \quad \Rightarrow \quad \vartheta_{PF}^{(0)} = \frac{1}{k} \left(C_{in} - C_{out} \right) \tag{6.56}$$

Equation 6.56 shows that in the case of a zero-order reaction, the PF does not offer any advantage over the CSTR.

Second-order kinetics. In this case, the CSTR dynamics is given by Equation 6.16, from which the steady-state equivalent HRT $\left(\vartheta_{CSTR}^{(2)} \right)$ can be computed as

$$\vartheta_{CSTR}^{(2)} = \frac{1}{k \cdot C_{out}} \left(\frac{C_{in}}{C_{out}} - 1 \right) \tag{6.57}$$

where as in the PF case

$$\frac{dC}{dx} = -\frac{k}{u} C^2 \tag{6.58}$$

from which the HRT can be computed by integrating along the reactor length L

$$\int_{C_{in}}^{C_{out}} \frac{dC}{C^2} = -\frac{k}{u} \int_0^L dx \quad \Rightarrow \quad = \vartheta_{PF}^{(2)} = \frac{1}{k \cdot C_{out}} \left(1 - \frac{C_{out}}{C_{in}} \right) \tag{6.59}$$

Comparing the two HRTs given by Equations 6.57 and 6.59, it can be seen that $\vartheta_{PF}^{(2)} < \vartheta_{CSTR}^{(2)}$, so the PF does yield a better performance, in terms of a smaller volume, in the case of a second-order kinetics.

Monod kinetics reaction. If the reaction is enzyme controlled and follows a Monod kinetics, as introduced in Section 1.10.4.1 and further explored in Chapter 7, then the CSTR model is

$$\frac{dC}{dt} = q \left(C_{in} - C \right) - \frac{\mu_{max}}{K_s + C} C \tag{6.60}$$

from which the HRT can be computed as

$$\vartheta_{CSTR}^{(M)} = \frac{\left(C_{in} - C_{out} \right) \left(K_s + C_{out} \right)}{\mu_{max} \cdot C_{out}} \tag{6.61}$$

while the corresponding PF model is

$$\frac{dC}{dx} = -\frac{1}{u} \frac{\mu_{max}}{K_s + C} C \tag{6.62}$$

Separating the variables and integrating along the reactor length L yield

$$\int_{C_{in}}^{C_{out}} \frac{K_s + C}{C} dC = -\frac{\mu_{max}}{u} \int_0^L dx = -\mu_{max} \cdot \vartheta_{PF}^{(M)} \tag{6.63}$$

from which the HRT of the PF can be obtained as

$$\vartheta_{PF}^{(M)} = \frac{1}{\mu_{max}} \left[K_s \cdot \ln\left(\frac{C_{in}}{C_{out}}\right) - \left(C_{in} - C_{out}\right) \right] \tag{6.64}$$

Comparing Equations 6.61 and 6.64, we can see that $\vartheta_{PF}^{(M)} < \vartheta_{CSTR}^{(M)}$. So even in the Monod case, the PF represents an improvement with respect to the CSTR, because it provides the same extent of the reaction with a smaller volume.

6.3   DIFFUSIVE REACTOR

We have previously examined the two extreme cases with respect to diffusion. At one end we had the CSTR with its perfect mixing, that is, infinite diffusion. At the other end, we considered the PF, where there is no diffusion at all. In between these two extremes we had diffusion creeping in, when we considered the case of the feedback CSTRs (Section 6.1.4), where the return flow is often represented by diffusion, or the generalized CSTR (Section 6.1.5), where resuspension and sedimentation are essentially diffusive processes. All of these examples have anticipated the widespread natural process of diffusion, whereby a certain amount of a substance dissolved in the water tends to expand in the opposite direction of its gradient, that is, from high concentration to low concentration zones. This principle is expressed by the Fick law

$$J = -D\frac{dC}{dx} \tag{6.65}$$

which states that the diffusive flux J is inversely proportional to the gradient of the concentration C along the direction of diffusion x, with D being the diffusion coefficient. In Equation 6.65, we have assumed for simplicity that diffusion is unidirectional along the longitudinal dimension (x), but diffusion could well take place in two or all three of the spatial coordinates. Diffusion, as opposed to advection, does not transfer mass through bulk transport, but only by its concentration gradient and therefore can take place in a still liquid. Figure 6.18 shows the difference between advection, based on bulk transport (a), diffusion in still water (b), and the combination of both (c), which makes the initial 'blob' gradually expand as it diffuses outward.

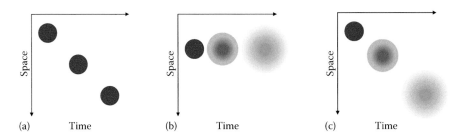

FIGURE 6.18   In pure advection (a), particles move along the bulk flux and maintain their initial shape (PF), whereas in (b), diffusion in still water makes the initial mass expand due to concentration gradient. In (c), the two processes are combined so that the initial mass both expands and is transported downstream.

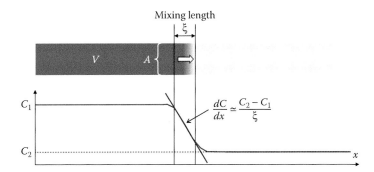

FIGURE 6.19 The mixing length ξ is defined as the thickness of the intermediate layer through which diffusion occurs.

A practical problem with gradients is to delimit their spatial extent, which in this case could be referred to the mixing length ξ illustrated in Figure 6.19.

The gradient is limited to the boundary layer of surface A and thickness ξ between the two areas with constant concentrations C_1 and C_2, and the diffusive flux can be approximated as

$$V\frac{dC_1}{dt} = -JA = DA\frac{dC}{dx}\Big|_{x\in\xi} \cong -DA\frac{C_1-C_2}{\xi} \tag{6.66}$$

A practical problem with diffusion is the accurate determination of the diffusive coefficient D, whose dimensions are $[L^2 T^{-1}]$. Also, the diffusive surface A and the mixing length ξ are elusive quantities, and generally they are all grouped into a single parameter D' defined as the diffusion mass transfer coefficient

$$D' = \frac{DA}{\xi} \quad \left[L^3\,T^{-1}\right] \tag{6.67}$$

With this substitution, the diffusion Equation 6.66 takes the simple form

$$V\frac{dC_1}{dt} = D'\left(C_2 - C_1\right) \quad \text{with } C_1 > C_2 \tag{6.68}$$

6.3.1 An Example of Feedback CSTRs with Diffusion: The Lake Huron–Saginaw Bay System

The Lake Huron–Saginaw Bay system, a part of the Great Lakes across the US–Canadian border, provides an example of waterbodies that communicate through diffusion. This system, shown in Figure 6.20, has been studied by Chapra (1979, 1997) to illustrate the role of diffusion, as opposed to bulk transport, and to estimate the diffusion mass transfer coefficient D' of Equation 6.67.

The Lake–Bay system can be modelled as two CSTRs coupled by diffusion, for which the feedback arrangement of Section 6.1.4 can be used, with the loading situation shown in Figure 6.21. If we consider the dynamics of a conservative substance, such as chlorides, the Equation 6.40 can be adapted by making the following assumptions:

- The chlorides enter the system by mass loading in the bay only, that is, $W_1 = 0$; $C_{1_{in}} = C_{2_{in}} = 0$.
- Being a conservative substance, there are no kinetic terms, that is, $k_1 = k_2 = 0$.
- The mass of chlorides is transferred by diffusion between the lake and the bay.

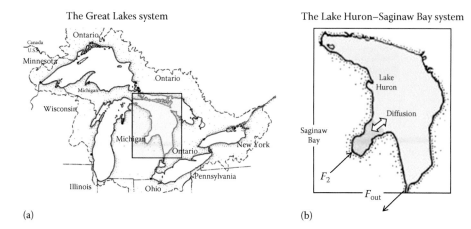

FIGURE 6.20   In the Great Lakes system (a), the Lake Huron–Saginaw Bay (b) system is an example of two waterbodies (shown in differing colour shades in the inset) linked by diffusion.

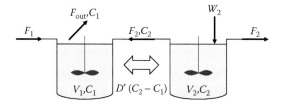

FIGURE 6.21   Modelling of the Lake Huron–Saginaw Bay system as two feedback CSTRs connected by diffusion.

According to the scheme of Figure 6.21, the dynamical equations of the system can be written as follows:

$$
\text{Lake}\quad V_1\frac{dC_1}{dt}=F_2\cdot C_2+D'\left(C_2-C_1\right)-F_{\text{out}}\cdot C_1
$$
$$
\text{Bay}\quad V_2\frac{dC_2}{dt}=W_2-F_2\cdot C_2-D'\left(C_2-C_1\right)
$$

(6.69)

which can be written in matrix form as

$$
\begin{bmatrix}\dot C_1\\ \dot C_2\end{bmatrix}=\begin{bmatrix}-\dfrac{F_{\text{out}}+D'}{V_1} & \dfrac{F_2+D'}{V_1}\\[3mm] \dfrac{D'}{V_2} & -\dfrac{F_2+D'}{V_2}\end{bmatrix}\times\begin{bmatrix}C_1\\ C_2\end{bmatrix}+\begin{bmatrix}0\\ \dfrac{1}{V_2}\end{bmatrix}\cdot W_2
$$

(6.70)

To estimate the diffusion mass transfer coefficient D', Chapra (1997) suggested setting the derivative of the Bay Equation 6.69 to zero and solving for D'

$$
D'=\frac{W_2+F_2C_2}{C_2-C_1}
$$

(6.71)

Using the numerical values in Table 6.1 provided by Chapra (1997) yields

$$
D'=\frac{0.353\times10^{12}-7\times10^{9}\cdot15.2}{15.2-5.4}=2.5163\times10^{10}\ \left(\text{m}^3\,\text{year}^{-1}\right)
$$

(6.72)

TABLE 6.1

Parameter Values for the Lake Huron–Saginaw Bay Example

Water Body	Parameter	Symbol	Value	Units
Lake Huron	Volume	V_1	3507×10^9	m^3
	Outflow	F_{out}	161×10^9	$m^3\,yr^{-1}$
	Chloride conc.	C_1	5.4	$g\,m^{-3}$
Saginaw Bay	Volume	V_2	8×10^9	m^3
	Outflow	F_2	7×10^9	$m^3\,yr^{-1}$
	Chloride conc.	C_2	15.2	$g\,m^{-3}$

Source: Chapra, S.C., *Surface Water-Quality Modeling*, McGraw-Hill, New York, 844, 1997.

In that study, Chapra (1997) did not specify whether the chloride concentrations in Table 6.1 used to compute D' through Equation 6.72 could be considered as steady-state values, but from a check of the system Equation 6.69, it appears that they are not. In fact, summing the two Equations 6.69 at steady state yields

$$\bar{C}_1 = \frac{W_2}{F_{out}} = \frac{353 \times 10^9}{161 \times 10^9} = 2.1925 \left(g\,m^{-3}\right) \tag{6.73}$$

which is independent of the diffusion coefficient D'. Further, solving the matrix Equation 6.70 with zero derivatives yields the steady-state concentrations

$$\bar{C}_1 = 2.1925 \left(g\,m^{-3}\right) \quad \bar{C}_2 = 12.6906 \left(g\,m^{-3}\right) \tag{6.74}$$

Obviously the values of $\bar{C}_1$ computed by the two methods coincide, but they differ from the value used by Chapra (1997). Simulating the dynamical system (Equation 6.69) with the values of Table 6.1 and the diffusion D' obtained from Equation 6.72 yields the results of Figure 6.22a, where the concentrations converge to the steady-state values predicted by Equation 6.74. Given the stiff nature of the system, with the eigenvalues of the system matrix of Equation 6.70 being

$$\lambda_1 = -0.0458 \quad \lambda_2 = -4.0277 \tag{6.75}$$

initializing the system with the value $\bar{C}_1$ given by Equations 6.73 or 6.74 results in a constant value of C_1, whereas initializing C_2 with a high value shows the fast/slow mode typical of stiff systems, with the short-lived fast mode shown in the inset of Figure 6.22a. Finally, to investigate the influence of the diffusion coefficient D' on the steady-state regime, the embayment model (6.69) was simulated for a wide range of D' and the steady-state concentrations were plotted in Figure 6.22b, from which it can be seen that the concentration gradient decreases while D' increases, because a greater diffusion tends to equalize the concentrations.

6.3.2 DIFFUSION AND DISPERSION

Dispersion is often used as a comprehensive term indicating the dispersal of substances in natural waters due to many coexisting processes, such as molecular diffusion, turbulent mixing and mixing due to shear. Given the high complexity of dispersion and mixing, the interested reader is referred to specialized texts (Fisher et al., 1979; Benedini and Tsakiris, 2013), and only a brief review is included here. We have seen that the molecular diffusion considered in the previous section is due

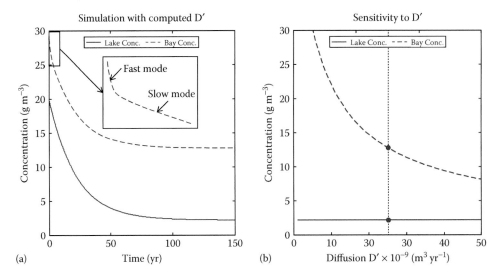

(a) Time (yr)   (b) Diffusion $D' \times 10^{-9}$ (m^3 yr^{-1})

FIGURE 6.22 (a) The Lake Huron—Saginaw Bay embayment model was simulated with D' computed by Equation 6.72. The inset shows a magnified view of the short-lived fast mode controlled by λ_2 of Equation 6.75, which dies out in about 6 years, then making way for the slow mode controlled by λ_1. In (b), the sensitivity of the steady-state concentrations to D' was computed. The dashed vertical line refers to the steady state obtained with the computed D' value used in (a). These graphs were obtained with the MATLAB script Go _ Bay _ Lake _ Chlorides.m in the ESA _ MATLAB\Exercises\Chapter _ 6\Bay _ Lake folder.

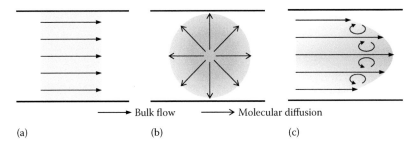

Bulk flow   Molecular diffusion

(a)   (b)   (c)

FIGURE 6.23 Possible mechanisms of particle mixing in a channel flow. (a) No mixing (plug-flow); (b) molecular diffusion due to concentration gradient in a still fluid; (c) dispersion due to turbulent mixing caused by eddies originated by axial velocity gradient.

to a concentration gradient across the mixing zone, but this is not the only possible mixing process, as shown in Figure 6.23, where the difference between diffusion (b) and turbulent mixing (c) is illustrated and compared with the no-mixing PF regime.

This is a consequence of the lack of axial velocity differences across the plug front, as in Figure 6.23a. Conversely, the concentration gradient is the only force inducing molecular diffusion, acting even in a still fluid, as in Figure 6.23b, whereas (c) refers to a third possibility, involving turbulent diffusion produced by axial velocity gradient, which results in eddies that mix particles with differing speeds. Whatever the cause, dispersion, with units of (cm^2 s^{-1}), according to Chapra (1997) is related to the dimension of the waterbody, with turbulent mixing being the main dispersal force in large lakes and streams with values ranging from 10^0 to 10^8 cm^2 s^{-1}, while molecular diffusion is in the order of 10^{-6}–10^{-2} cm^2 s^{-1}. The latter, therefore, becomes negligible with respect to turbulent mixing in large waterbodies. Chapra (1997) also reported a linear relation with a 4/3 slope between the length scale and the turbulent diffusion for large water masses, such as the Great Lakes and oceans.

For slow-flowing rivers having Froude numbers $\left(\text{Fr} = u/\sqrt{g \cdot h}\right)$ less than 0.5, the Fisher formula (Fisher, 1967; Fisher et al., 1979) yields an estimate of the diffusion as

$$D = 0.011 \frac{u^2 w^2}{h\sqrt{ghs}} \quad \left(\text{m}^2\text{s}^{-1}\right) \tag{6.76}$$

where:
u (m s^{-1}) is the average stream velocity
w is the river width (m)
s is the slope
h is the average depth (m)
g is the gravity acceleration (m s^{-2})

In river modelling, axial dispersion is usually greater than radial dispersion, which is often neglected unless the stream is exceptionally wide and slow flowing. Table 6.2 compares the computation of dispersion with the Fisher formula (6.76) for a slow-flowing river and a mountain stream.

6.3.3 DIFFUSIVE REACTOR MODELLING

We can now summarize the various factors governing the movements of a fluid in a channel, including advection and diffusion, by setting up a model for the diffusive reactor, as depicted in Figure 6.1e. Figure 6.24 considers a cylindrical section of thickness dx in a diffusive reactor of section S. Assuming that the flow variations are much slower than those of the concentration, we can consider the flow velocity a function of the abscissa x alone, that is, $u = u(x)$, while C is a function of both abscissa and time, that is, $C = C(x,t)$. The following mass balance can be written around the differential cylinder of volume $dV = S \cdot dx$ of Figure 6.24 considering the following components and, assuming for simplicity, that the diffusion coefficient D is constant across dx:

- Bulk transport input through the upstream section $u(x) \cdot S \cdot C(x,t)$.
- Dispersion input through the upstream section $-D \cdot S \left.\dfrac{\partial C}{\partial x}\right|_{x,t}$.
- Bulk transport output through the downstream section $-u(x+dx) \cdot S \cdot C(x+dx,t)$.
- Dispersion output through the downstream section $-D \cdot S \left.\dfrac{\partial C}{\partial x}\right|_{x+dx,t}$.
- Internal chemical transformation $S \cdot dx \cdot f(C)$.

The dynamical concentration balance inside the differential cylinder dV can then be written by combining the above terms to yield

---

TABLE 6.2

Examples of Dispersion Computation with the Fisher Formula (6.76)

Component	Symbol	Slow-Flowing River	Mountain Stream
Velocity (m s^{-1})	u	0.45	0.7
River width (m)	w	10	4
Slope	s	0.0005	0.005
Average depth (m)	h	2	0.2
Froude number	Fr	0.1016	0.4997
Fisher Dispersion (m^2 s^{-1})	D	1.1245	4.3536

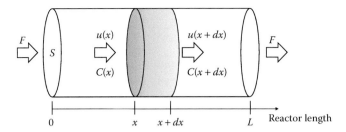

FIGURE 6.24 Schematization of a diffusive reactor with length L and cross section S. The mass balance is computed around the coloured cylinder of thickness dx.

$$S \cdot dx \frac{\partial C}{\partial t}\bigg|_{x,t} = \left\{ u \cdot S \cdot C - DS \frac{\partial C}{\partial x} \right\}\bigg|_{x,t} - \left\{ u \cdot S \cdot C - D \cdot S \frac{\partial C}{\partial x} \right\}\bigg|_{x+dx,t} + S \cdot dx \cdot f\left[C\left(x,t \right) \right] \quad (6.77)$$

Then, dividing by the differential volume $dV = S \cdot dx$ and letting $dx \to 0$ yield, after rearranging, the longitudinal dispersion dynamics

$$\frac{\partial C}{\partial t} = -u \frac{\partial C}{\partial x} + D \frac{\partial^2 C}{\partial x^2} + f\left(C \right) \quad (6.78)$$

This well-known partial differential equation (PDE) describes in a very concise way the monodimensional diffusive evolution in time and space of a substance dissolved in the moving water, but it also conceals some important difficulties. First of all, it contains two parameters (u and D) that must be specified by other means before solving Equation 6.78. The stream velocity u must be computed separately by means of specific numerical hydraulics software, such as the public domain HEC–RAS, kindly made available by the US Army Corps of Engineers (http://www.hec.usace.army.mil/software/hec-ras/). This well-diffused and reliable software will provide the velocity field u, given the river morphological characteristics and the selected flow regime. The practical determination of the diffusion is more complex, as it requires tracer experiments, some of which will now be described. Alternatively, diffusion may be approximated by the Fisher Equation 6.76. In any case, it is important to understand that until these two quantities are known, at least in approximate form, no attempt can be made at solving Equation 6.78 by the numerical approach that will be described later.

6.3.3.1 Impulse Response of the Diffusive Reactor

Though we have just said that the diffusive Equation 6.78 is not very easy to handle, there are, however, some instances for which an analytical solution can be obtained. One such case is an upstream tracer pulse of mass M injected in a very short time lapse δt, thus approximating an impulse input. This special loading condition translates into the following boundary conditions (Levenspiel, 1972; Chapra, 1997; Tchobanoglous and Schroeder, 1985; Palmeri et al., 2014):

$$\begin{cases} C(x,0) = 0 & x > 0 \\ \lim_{x \to \infty} C(x,t) = 0 & \forall t > \delta t \\ C(0,t) = 0 & t < 0 \\ C(0,t) = \dfrac{M}{F \cdot \delta t} & t \in (0, \delta t) \\ C(0,t) = 0 & t > \delta t \end{cases} \quad (6.79)$$

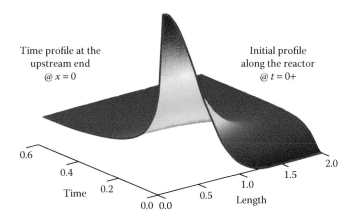

FIGURE 6.25 Diffusive reactor pulse response computed with Equation 6.80 with the following parameters: $u = 0.6$; $M = 0.1$; $S = 1$; $D = 9$. This graph can be obtained with the MATLAB script `Go _ Pulse.m` in the `ESA _ MATLAB\Exercises\Chapter _ 6` folder.

for which the diffusion Equation 6.78 has the following analytical solution:

$$C_{\delta t}(x,t) = \frac{M}{2 \cdot S \sqrt{\pi \cdot D \cdot t}} e^{\left(-(x - u \cdot t)^2 / 4 \cdot D \cdot t\right)} \tag{6.80}$$

reminiscent of a Gaussian function, if it were not for the time t at the denominator. Figure 6.25 shows the numerical representation of Equation 6.80 with suitable parameters. It can be seen that the pulse changes the initial conditions, as already examined in Section 6.1.2, and defines the initial concentration profile along the reactor.

Apart from its theoretical importance, Equation 6.80 can actually be used to estimate the diffusion coefficient D of real water streams. If the length x is fixed at the end of the reach being investigated, then Equation 6.80 can be adapted to describe the variation of tracer at the downstream end of the reach, by setting $x = L$, so that the impulse response is only a function of time

$$C_{\delta t}(L,t) = \frac{M}{2 \cdot S \sqrt{\pi \cdot D \cdot t}} e^{\left(\frac{(L - u \cdot t)^2}{4 \cdot D \cdot t}\right)} \tag{6.81}$$

To test the agreement between Equation 6.81 and real tracer data, the experimental laboratory equipment shown in Figure 6.26 was set up.

The experimental rig, which is thoroughly described elsewhere (Marsili-Libelli, 1997), consists of a 4-metre-long pipe equipped with a flow-control loop to ensure a constant flow, and a conductivity cell at the downstream end, measuring the conductivity variations induced by a pulse-like injection of saturated salt water at the upstream end of the reactor. Equation 6.81 was then fitted to the data, with D being the only unknown parameter. The result of this estimation exercise is shown in Figure 6.27, from which a rough value of the diffusion can be estimated, although the agreement between the data and the impulse response Equation 6.81 is far from satisfactory. In addition to missing the steep leading edge of the pulse, there is a considerable model/data disagreement in the trailing part (the 'tail') of the experiment, where the data seem to fluctuate randomly. This is due to possible dead zones in the conductivity flow-cell, where tracer particles may be held up for a considerable time lapse to be released later, giving rise to the disagreement between the data and the theoretical response in the trailing part of the experiment. Hence, we need a better model for pulse response of the diffusive reactor.

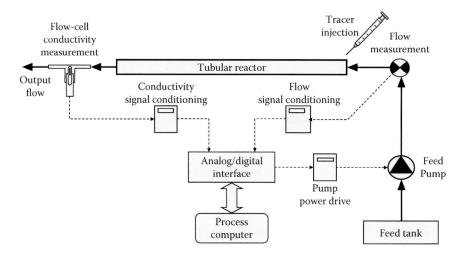

FIGURE 6.26 Experimental set-up to determine the diffusion in a tubular reactor. (Reproduced with permission from Marsili-Libelli, S. and Colzi, A., *Control Eng. Pract.*, 6, 707–713, 1998.)

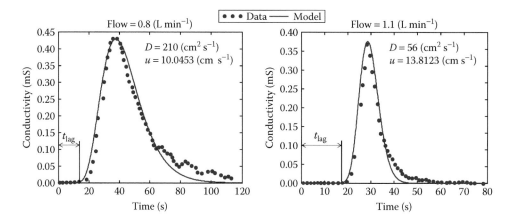

FIGURE 6.27 Fitting the theoretical impulse response Equation 6.81 with the data obtained from the experimental set-up of Figure 6.26, with t_{lag} representing the transport delay before the tracer pulse begins to show at the outlet. The poor data agreement in the descending branch of the response may be due to tracer particles trapped in the flow cell, but it also shows the limits of the theoretical Equation 6.81 in modelling real tracer data.

6.3.3.2 Transport Delay Modelling

We have seen that dispersion, due to a combination of diffusion and turbulent mixing, 'spreads' the initially narrow tracer pulse along the flow direction. However, the first concentration changes at the outlet are observed after a time lag due to the travel time of the pulse along the reactor. This transport delay $\left(t_{lag}\right)$, which is clearly a function of the flow rate, can be estimated in several ways. In principle, it can be related to flow by a hyperbolic law

$$t_{lag} = A + \frac{B}{F} \tag{6.82}$$

where:

F is the flow rate

A and B are suitable constants, depending on the reactor shape

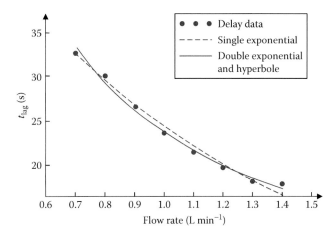

FIGURE 6.28 Experimental data relating the transport delay (t_{lag}) to the flow rate, obtained with the experimental set-up of Figure 6.26, and the function fitting of Table 6.3.

However, other approximations are possible, and to assess their validity an experiment was carried out using the same laboratory set-up of Figure 6.26 to find a relationship between the flow rate and the transport delay, as shown in Figure 6.28. Three approximating functions were tested: single exponential, double exponential and hyperbolic law.

The fitting results are reported in Table 6.3 and confirm the suitability of the hyperbolic function. As an alternative, the double exponential yields a slightly better fit, but this is at the expenses of a much more complex function. Further, the wide confidence intervals of the estimated parameters denote the scarce reliability of this model.

6.3.3.3 Lumped-Parameter Approximation of the Diffusive Reactor

Having separated the transport delay $\left(t_{lag}\right)$ from the internal dynamics opens the way to a lumped-parameter approximation of the diffusive reactor, which can be decomposed into the structure of Figure 6.29, where the initial PF introduces the transport delay t_{lag} and the CSTR chain models the internal dynamics.

If all the CSTRs have equal HRT $\left(\vartheta\right)$, the impulse response can be expressed by Equation 6.28, where the peak concentration C_{max} can be found by setting its time derivative to zero, to yield

TABLE 6.3

Possible Approximations of the Flow Rate–Transport Delay Data of Figure 6.28

Single Exponential	Double Exponential	Hyperbole
$t_{lag} = a\,e^{-bF}$	$t_{lag} = a_1\,e^{-b_1 F} + a_2\,e^{-b_2 F} + a_3$	$t_{lag} = A + \dfrac{B}{F}$
$a = 62.8532 \pm 2.96$	$a_1 = 75.218 \pm 70.04$	$A = 2.1124 \pm 0.81$
$b = 0.93133 \pm 0.098$	$a_2 = 30.218 \pm 28.76$	$B = 22.036 \pm 0.83$
	$a_3 = 0.442 \pm 0.38$	
	$b_1 = 2.723 \pm 0.89$	
	$b_2 = 0.473 \pm 2.87$	
$\chi^2 = 3.1943$	$\chi^2 = 1.4243$	$\chi^2 = 1.6152$

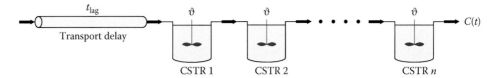

FIGURE 6.29 Diffusive reactor approximation consisting of a PF, to simulate the transport delay (t_{lag}), and a chain of CSTRs.

$$t_{max} = \vartheta \cdot (n-1)$$

$$C_{max} = C_o \frac{1}{\vartheta \cdot (n-1)!}(n-1)^{n-1} \cdot e^{-(n-1)} \tag{6.83}$$

where t_{max} is computed from the onset of the output pulse, obtained by subtracting t_{lag} from the experimental time. Figure 6.30a defines the impulse response parameters t_{max} and C_{max} of Equation 6.83, while Figure 6.30b shows its fitting to the tracer experiment of Figure 6.27b.

By comparing these two fittings, it can be concluded that the lumped-parameter model of Figure 6.29 represents a better description of the diffusive reactor than that obtained with the impulse response method of Equation 6.81. There are instances, however, where even this model appears to lack the flexibility required to fit experimental diffusion data. Particularly, in the model of Figure 6.29, the shape of the pulse response is a function of the number of CSTRs (n), and the equal CSTRs assumption may represent a limit to good fitting. More complex, and flexible, lumped-parameter structures to approximate a diffusive reactor have been proposed (Marsili-Libelli, 1997; Marsili-Libelli and Checchi, 2005), and these will be examined in more detail in the next chapters.

6.3.3.4 Diffusion Estimation from the Pulse Response

The pulse response Equation 6.81 is skewed by the presence of the time factor in the denominator, but as the measuring point is moved downstream, the response becomes more and more symmetrical, and it can be approximated with a normal curve

$$G(t,t_m,\sigma_m) = C_m \cdot e^{-\frac{u^2(t-t_m)^2}{2\cdot\sigma_m^2}} \tag{6.84}$$

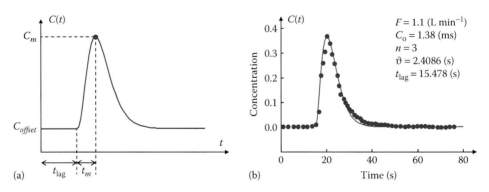

FIGURE 6.30 Output impulse response of the diffusive reactor approximation of Figure 6.29. (a) Theoretical response with relevant parameters. (b) Fitting the curve (a) to the tracer experiment of Figure 6.27b. The last graph was obtained with the MATLAB script `Diffusive_pulse_CSTR_chain.m` in the ESA_MATLAB\Exercises\Chapter_6\Cascade_CSTR folder.

with C_m and t_m already defined in Figure 6.30a. Thus, the diffusion coefficient D can be estimated by equating the exponential terms of Equations 6.81 and 6.84

$$\frac{\left(L-u\cdot t\right)^2}{4\cdot D\cdot t} = \frac{u^2\left(t-t_m\right)^2}{2\cdot\sigma_m^2} \tag{6.85}$$

Substituting $L \simeq t_m \cdot u$ in the left-hand expression yields

$$\frac{u^2\left(t_m-t\right)^2}{4\cdot D\cdot t} = \frac{u^2\left(t_m-t\right)^2}{2\cdot\sigma_m^2} \tag{6.86}$$

Equating the denominators and solving for D at the time of the output peak t_m yields

$$D \simeq \frac{\sigma_m^2}{2\cdot t_m} \tag{6.87}$$

Figure 6.31 shows that the pulse response approaches a normal distribution as the measuring point is moved downstream, making the approximation of Equation 6.86 more and more accurate. The smaller the D value, the better the normal approximation.

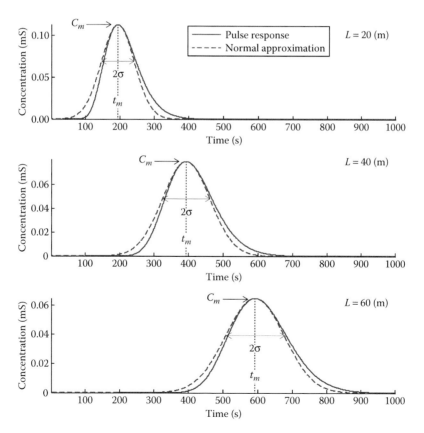

FIGURE 6.31 Extending the reactor length far from the injection point shows that the pulse response can be approximated with the normal curve, whose variance increases with the distance as a consequence of mixing. Using the normal approximation, the diffusion coefficient can be approximated with Equation 6.87. This graph was obtained with the MATLAB script `Diffusive_Length_Gauss.m` in the `ESA_MATLAB\Exercises\Chapter_6\Diffusive_River` folder.

6.3.4 The 'Follow-the-Plug' Model, a Lagrangian Approach

Let us go back to the basic diffusional PDE (6.78) and try to work out a simplified solution in the case of hydraulic constant regime and no diffusion $(D = 0)$. The first textbook which considered this aspect in details was Rinaldi et al. (1979). In this case the general Equation 6.78 simplifies into

$$\frac{\partial C}{\partial t} + u\frac{\partial C}{\partial x} = -f(C) \tag{6.88}$$

which can be solved in a simple way that will lead to the Streeter and Phelps model already encountered. Computing the differential of the concentration yields

$$dC = \frac{\partial C}{\partial t}dt + \frac{\partial C}{\partial x}dx \tag{6.89}$$

which, dividing by total time differential dt, produces

$$\frac{dC}{dt} = \frac{\partial C}{\partial t} + \frac{\partial C}{\partial x}\frac{dx}{dt}$$
$$= \frac{\partial C}{\partial t} + \frac{\partial C}{\partial x}u = -f(C) \tag{6.90}$$

Thus, the PDE (6.88) reduces to an ordinary differential equation (ODE) provided that the domain of integration is a *characteristic line*, along which the following relations hold

$$\begin{cases} \dfrac{dt}{dt_s} = 1 \\[2mm] \dfrac{dx}{dt_s} = u \end{cases} \tag{6.91}$$

Leonardo da Vinci's quotation at the beginning of this chapter, in modern terms, may be viewed as an Eulerian observer, looking from a fixed location at the fluid as it flows past. Conversely, the approach followed here is based upon a Lagrangian assumption, which requires that the observer follows an individual element of fluid as it moves through space and time (Batchelor, 2000). Figure 6.32 shows the practical consequences of taking samples along a characteristic line: if each sample is timed so that the distances along the river and the sampling times are in agreement with the characteristic line defined by Equation 6.91, then, according to Equation 6.90, Equation 6.88 simply reduces to

$$\frac{dC}{d\tau} = -f(C) \quad \Rightarrow \quad \frac{dC}{dx} = -\frac{1}{u}f(C) \tag{6.92}$$

If the Streeter and Phelps kinetics are substituted into Equation 6.92, the model can be expressed either in flow-time (t_s) or in the reach length (x)

$$\begin{cases} \dfrac{dB}{dt_s} = -k_b \cdot B \\[2mm] \dfrac{dC}{dt_s} = k_c(C_{sat} - C) - k_b \cdot B \end{cases} \qquad \begin{cases} \dfrac{dB}{dx} = -\dfrac{1}{u}k_b \cdot B \\[2mm] \dfrac{dC}{dx} = \dfrac{1}{u}k_c(C_{sat} - C) - \dfrac{1}{u}k_b \cdot B \end{cases} \tag{6.93}$$

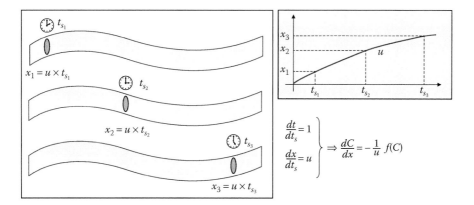

FIGURE 6.32   Implementation of the 'follow-the-plug' sampling strategy, which simplifies the original PDE (6.88) without diffusion into the ODE (6.92). The equivalence between the time kinetics and the space kinetics holds only if the PDE is solved along a characteristic line corresponding to a specific hydraulic regime u.

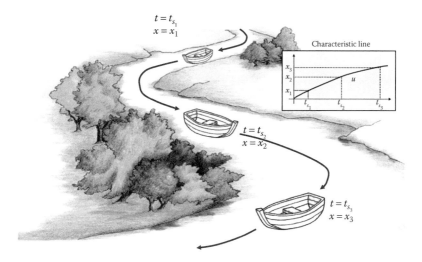

FIGURE 6.33   Pictorial representation of the 'follow-the-plug' concept. Imagine an observer taking samples from a boat without oars, which moves with the current. The samples taken at various instants $\left(t_{s_1}, t_{s_2}, t_{s_3}\right)$ in the water next to the boat are related to the positions $\left(x_1, x_2, x_3\right)$ by the characteristics line shown in the inset.

where the relation between the two independent variables is given by Equation 6.91. Figure 6.33 gives a pictorial view of a hypothetical observer taking samples from a boat which, being without oars, is forced to glide with the stream, that is, to 'follow-the-plug'.

6.3.5   Numerical Treatment of the Monodimensional Diffusive Equation

So far, we have considered a special solution to the PDE (6.88), by considering either a special class of inputs (the pulse) or by neglecting the diffusion and choosing a very particular domain of integration, such as a characteristic line. In the first instance, the purpose of that exercise was the estimation of the diffusion coefficient, either from the pulse response function Equation 6.81, or from the normal distribution curve Equation 6.86, using the lumped-parameter approximation of Figure 6.29. This approach is widely used in chemical engineering practice, where the reactor has a well-defined, regular shape and the loading conditions are precisely controlled. In the 'follow-the-plug' case, the aim was to obtain a very simple model, in which only the kinetics terms would be required to fully specify the model evolution, without much concern about boundary conditions, which became normal initial conditions.

We now turn our attention to the more general problem of approximating the solution of the diffusive equation when the goal is to represent the features of a natural water course whose shape may have all the irregularities typical of natural environments. This implies that its hydraulic regime may be variable, and that diffusion, be it molecularly or most likely turbulent, cannot be ignored anymore. If the diffusive Equation 6.78 is to describe the transport, diffusion and chemical transformation of a dissolved pollutant, the following boundary conditions should be introduced:

$$\text{Upstream boundary profile} \quad C(0,t) \text{ for } \forall t > 0 \tag{6.94}$$

$$\text{Initial concentration profile} \quad C(x,0) \text{ for } 0 \le x \le L \tag{6.95}$$

The spatio-temporal domain of the solutions to the diffusive model with these boundary conditions is illustrated in Figure 6.34, showing the feasible trajectories of pollutant particles travelling downstream.

The first boundary condition Equation 6.94 represents the upstream input pollutant concentration in time, while Equation 6.95 provides the initial concentration profile in space along the whole reach. Starting from a steady-state condition is not strictly required, but it helps, making the simulation start from a consistent profile and avoiding lengthy numerical transients in the simulation, which may even lead to numerical instability. According to Figure 6.34, a volume of water starting

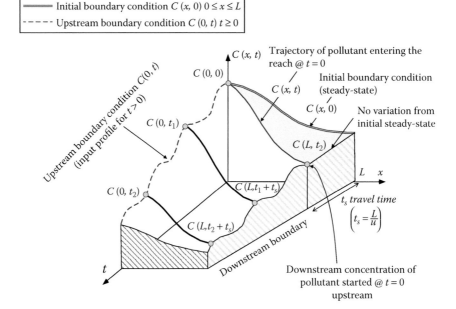

FIGURE 6.34 Graphic representation of the boundary conditions of Equations 6.94 and 6.95 in the unidirectional diffusive domain. The pollutant feasible trajectories originating from the upstream end follow the black curves (characteristic lines) and reach the downstream boundary after the travel time t_s. Only trajectories starting at $t < 0$ can exist in the blue triangle. The initial steady state is represented by the double blue line and represents the space boundary condition, while the upstream time boundary condition is the blue dashed line on the left.

at the initial time $(t = 0)$ from the upstream end $(x = 0)$ with pollutant concentration $C(0,0)$ travels downstream along a *characteristic line* described by the relationship $u = x/t_s$ where t_s is the flow time and u is the stream velocity. Water volumes, starting at later times t_1 and t_2, each travel along their own characteristic lines and reach the downstream end at $t_1 + t_s$ and $t_2 + t_s$, respectively, where $t_s = L/u$ is the travel time across the reach length (L). If the pollutant is reactive, internal transformations may occur in the water volumes while they travel along the reach.

6.3.5.1   Selecting the Proper Boundary Conditions

While the first boundary condition Equation 6.94 requires the knowledge of the input concentration $C(0,t)$ for all $t > 0$, together with the velocity profile $u(x,t)$ along the reach, the second condition is more elaborate, as it requires the computation of a steady-state solution to the diffusive Equation 6.78. As noted, starting from a steady-state condition is not strictly required, but it helps. Therefore, let us now consider the steady-state form of Equation 6.78 by setting the left-hand time derivative to zero so that the original PDE becomes an ODE in the spatial dimension (x) only

$$0 = -u\frac{dC}{dx} + D\frac{d^2C}{dx^2} + f(C) \tag{6.96}$$

Equation 6.88 is best handled in a normalized form by introducing the new variable $z = x/L$, to yield

$$\frac{d^2C}{dz^2} = \frac{uL}{D}\cdot\frac{dC}{dz} + \frac{kL^2}{D}\cdot C \tag{6.97}$$

which can be put in matrix form by defining the two new variables $C_1 = C$ and $C_2 = \dfrac{dC}{dz}$

$$\begin{bmatrix} \dfrac{dC_1}{dz} \\ \dfrac{dC_2}{dz} \end{bmatrix} = \begin{bmatrix} 0 & 1 \\ \dfrac{L^2 k}{D} & \dfrac{uL}{D} \end{bmatrix} \cdot \begin{bmatrix} C_1 \\ C_2 \end{bmatrix} \tag{6.98}$$

Its characteristic equation is

$$\lambda^2 - \frac{uL}{D}\lambda - \frac{kL^2}{D} = 0 \tag{6.99}$$

which can be solved for the eigenvalues

$$\lambda = \frac{\dfrac{uL}{D} \pm \sqrt{\left(\dfrac{uL}{D}\right)^2 + 4\dfrac{kL}{D}}}{2} \tag{6.100}$$

Equation 6.100 shows that the two eigenvalues are real with opposite signs, so we have the *saddle* situation illustrated in Figure 1.36, and we can conclude that the steady-state boundary Equation 6.96 is *intrinsically unstable* for any value of its parameters (k, D, u) whose influence on the eigenvalues is shown in Figure 6.35.

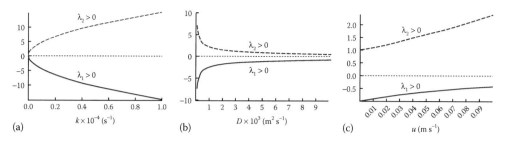

(a)

(b)

(c)

FIGURE 6.35 The eigenvalues of the steady-state boundary Equation 6.96 are influenced by the value of the parameters. In (a) increasing the kinetic constant makes the eigenvalue diverge, while in (b) diffusion D has the opposite effect. In (c) the effect of the velocity (u) is mixed, with the stable eigenvalue becoming more and more dominant. In any case, one eigenvalue is always positive real (dashed curves) and therefore the system is unstable.

Each factor has a differing influence on the eigenvalues. Figure 6.35a shows that an increasing kinetic constant increases the divergence of the eigenvalues, while Figure 6.35b shows that an increase of the diffusion coefficient D dampens the system response, with the eigenvalues approaching zero from opposite directions. Finally, Figure 6.35c demonstrates that the effect of the velocity has an opposite effect on the eigenvalues, increasing the unstable one (λ_2), while the stable λ_1 approaches the zero value, thus becoming dominant. In summary, any combination of the three parameters results in an unstable saddle configuration, as shown by the two sample trajectories in the phase plane portrayed in Figure 6.36. This, however, does not imply that Equation 6.98 cannot be used. In fact, if the simulation ends well before the influence of the unstable eigenvector dominates the trajectory, this does not prevent the use of this model for computing a feasible space boundary condition.

Figure 6.37 shows the two trajectories of Figure 6.36 in the spatial domain of the normalized reach length. In (a), a monotonic concentration decrease of the pollutant is shown, while (b) displays a very strange case in which the pollutant increases in the final part of the reach. This very

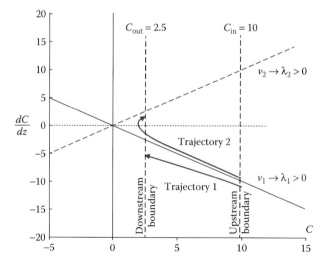

FIGURE 6.36 Phase plane portrait of two example solutions to the steady-state diffusive Equation 6.98. The straight blue lines indicate the direction of the eigenvectors, with v_1 (solid line) corresponding to the stable eigenvalue λ_1, while v_2 (dashed line) corresponds to the unstable eigenvalue λ_2. Both trajectories (thick curves) are eventually attracted by the unstable eigenvector v_2, but the simulations end well before it dominates the solution.

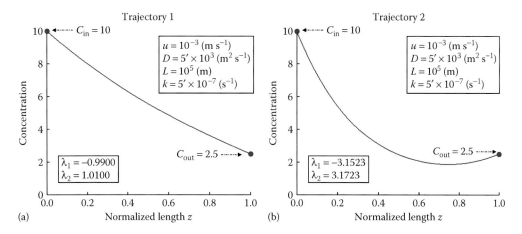

FIGURE 6.37 The two trajectories of Figure 6.36, with differing values of the kinetic constant, plotted versus the normalized length. In (a), trajectory 1 reaches the output concentration monotonically. In (b), trajectory 2 has an internal minimum, a situation with no chemical justification.

unnatural behaviour is discussed next and is due to the mathematical boundary constraints, rather than to a plausible chemical reason.

6.3.5.2 A Boundary Value Initialization Problem

The numerical simulation of the stationary concentration profile, either in the form of Equation 6.97 or in the state-space representation of Equation 6.98, being a second-order ODE, requires two initial conditions. On the other hand, the numerical methods illustrated in Section 1.10.2 use forward integration starting from initial conditions, which should be specified at the beginning of the integration domain. This is easily done with a first-order differential equation, for which the initial value can be estimated or obtained from experimental observation, while in the case of the second-order equation it also requires the knowledge of the initial derivate C'_{in}, for which it is difficult to obtain a good estimate. It would be easier, instead, to specify the pollutant concentration at both ends, that is, $C_{in} = C(0)$ and $C_{out} = C(1)$, which is referred to as a *boundary value problem* (BVP), while the Simulink integration algorithms are all *initial value problem*–oriented methods. Consequently, we are confronted with the problem of transforming the constraint on $C(1)$ into a condition for C'_{in}. There are several approaches to infer C'_{in} from $C(1)$, and the most popular is the so-called *shooting* method (Quinney, 1987). This method operates by successive approximation of the initial derivative C'_{in} as a function of the final 'shooting' error, that is, the difference between the target final concentration C_{out} and the actual end point of the simulation $C(1)$. Thus, the BVP translates into an initial derivative search that can be posed as

$$C'^{opt}_{in} = \arg\min_{C'_{in}}\left[C(1)-C_{out}\right]^2 \tag{6.101}$$

which ends when the squared difference is less than a prescribed quantity $\varepsilon > 0$. The optimal shooting search Equation 6.101 is implemented by linking the Simulink implementation of Equation 6.97 and the search for the optimal initial derivative, as shown in Figure 6.38. This method solves the BVP and determines the initial derivative C'_{in} so that the solution differs from the prescribed end point C_{out} by less than $\varepsilon > 0$. But how feasible is this solution? Although the solution in Figure 6.37a looks feasible (monotonic decrease of the pollutant concentration), the one in Figure 6.37b is certainly not. In fact, the final upward branch is very difficult to explain from a chemical viewpoint, though it makes sense mathematically, because it satisfies the given boundary conditions. It appears that a large chemical kinetic rate degrades the pollutant too quickly, with the result that the final

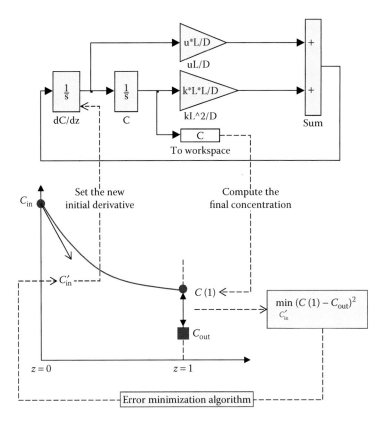

FIGURE 6.38 Implementation of the shooting method to find the initial concentration derivative C'_{in} to obtain the prescribed final value C_{out}. This computational scheme can be launched by the MATLAB script `Steady_Diffusive_BVP.m` in the `ESA_MATLAB\Exercises\Chapter_6\Steady_State_Diffusive_Reactor` folder.

concentration is reached well before the end of the reach, thus requiring a hypothetical additional injection to make up for the missing pollutant and meet the end boundary condition, which is awkward. Thus the BVP, which looks very neat from a mathematical point of view, may yield an environmentally inconsistent solution, and therefore it should be regarded with some suspicion.

6.3.5.3 A Least-Squares Initialization Problem

Now, we take a much more practical approach to the solution of the steady-state profile Equation 6.97. In practice, it is likely that pollutant concentration measurements C_i^{exp} are available at some sampling points along the reach, rather than at the end points only. In this case, it makes sense to initialize the model with an initial derivative C'_{in} such that the sum of squared differences between the experimental points C_i^{exp} and the corresponding model responses C_i^{exp} is minimized. Thus, the objective function (6.101) now becomes

$$C'^{opt}_{in} = \arg\min_{C'_{in}} \sum_i \varepsilon_i^2 = \arg\min_{C'_{in}} \sum_i \left(C_i - C_i^{exp}\right)^2 \tag{6.102}$$

which is very similar to the error functional encountered in Chapter 2 when dealing with the parameter estimation problem. In this case, however, the goal of Equation 6.102 is to find the best initial derivative C'_{in} so that the model solution approximates the experimental observations C_i^{exp} in the least square sense. The previous computational scheme of Figure 6.38 must now be modified to be consistent with Equation 6.102, as shown in Figure 6.39, where the shooting algorithm is replaced

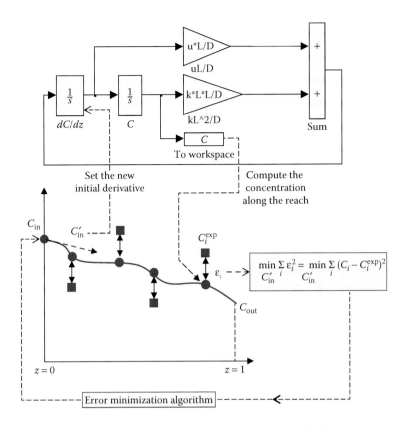

FIGURE 6.39 Implementation of the least-squares method to find the initial concentration derivative C'_{in} minimizing the sum of squared differences between the model output (blue dots) and the data (blue squares) without regard to the final value C_{out}. This computational scheme can be launched by the MATLAB script `Steady_Diffusive_LS.m` in the `ESA_MATLAB\Exercises\Chapter_6\Steady_State_Diffusive_Reactor` folder.

by an error minimization search. A least-squares solution is sought because, in general, the number of constraints, represented by the experimental observations, is greater than the two boundary conditions $\left(C_{in} \text{ and } C'_{in}\right)$. For this reason, this solution is much more robust than the previous BVP and rules out absurd results such as that of Figure 6.37b.

Figure 6.40 compares the two solutions, Least Squares (LS) and Boundary Value Problem (BVP), to the steady-state diffusive Equation 6.97, showing that the LS solution is influenced by all the intermediate data along the reach. Since the model parameters are well posed, the two solutions are very close, but the LS approach is by far more robust, being overdetermined. By contrast, the BVP solution considers only the extreme constraints, and anything can happen in between the end points, as demonstrated in Figure 6.37b.

6.3.6 SPATIAL DISCRETIZATION OF THE DIFFUSIVE REACH MODEL

So far, we have set the boundary conditions required for the integration of the unidirectional diffusive Equation 6.78. Now, while retaining the time continuity, we consider its actual numerical solution by spatial discretization. The reach is partitioned into n cells Δx long, connected by advection and diffusion, each with its internal chemical kinetics, according to the discretization scheme shown in Figure 6.41. To derive the discrete-space cell dynamics, consider the generic cell located at the abscissa x, while the upstream cell is located at $\left(x - \Delta x\right)$, and the downstream cell at $\left(x + \Delta x\right)$. Adjacent cells communicate via advection $\left(u_i\right)$ and diffusion $\left(D_i\right)$. The pollutant concentration

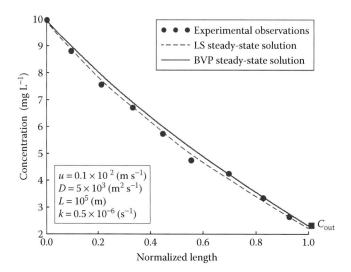

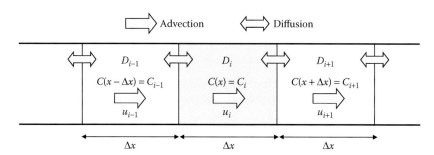

FIGURE 6.40 Comparison of LS and BVP solutions to the steady-state diffusion Equation 6.97. It can be seen that the LS solution is determined by the data along the normalized length, while the BVP solution considers only the final constraint. These graphs were obtained with the MATLAB scripts `Steady_Diffusive_BVP.m` and `Steady_Diffusive_LS.m` in the `ESA_MATLAB\Exercises\Chapter_6\Steady_State_Diffusive_Reactor` folder.

FIGURE 6.41 Spatial discretization scheme of the river reach to derive the time dynamics for each cell, which can be considered as a CSTR that communicates with the adjacent cells via advection (monodirectional) and diffusion (bidirectional).

inside the cell (C_i) is assumed to be homogeneous (CSTR hypothesis), and chemical transformations are governed by the internal kinetics $f(C_i)$. Now, approximating the first- and second-order spatial derivatives in Equation 6.78 with a second-order Taylor expansion based on central finite differences yields the following quantities:

$$C(x+\Delta x) \simeq C(x) + \left.\frac{\partial C}{\partial x}\right|_x \Delta x + \frac{1}{2}\left.\frac{\partial^2 C}{\partial x^2}\right|_x \Delta x^2 + O^2$$

$$C(x-\Delta x) \simeq C(x) - \left.\frac{\partial C}{\partial x}\right|_x \Delta x + \frac{1}{2}\left.\frac{\partial^2 C}{\partial x^2}\right|_x \Delta x^2 + O^2$$

(6.103)

To simplify the notation, assign the subscript (i) to the central cell, with the light blue shading in Figure 6.41. The upstream cell will have the index $(i-1)$, while the downstream cell will be indexed as $(i+1)$.

With this new notation, by summing or subtracting the Equation 6.103, the approximate first- and second-order derivatives are obtained as

$$\frac{\partial C_i}{\partial x} \simeq \frac{C_{i+1} - C_{i-1}}{2 \cdot \Delta x_i}$$

$$\frac{\partial^2 C_i}{\partial x^2} \simeq \frac{C_{i+1} - 2 \cdot C_i + C_{i-1}}{\Delta x_i^2} \qquad (6.104)$$

Substituting these approximations into the Equation 6.78 yields the discrete-space approximation of the unidimensional diffusion equation

$$\frac{dC_i}{dt} = \underbrace{-u_i \frac{C_{i+1} - C_{i-1}}{2\Delta x_i}}_{\text{advection}} + \underbrace{D_i \frac{C_{i+1} - 2C_i + C_{i-1}}{\Delta x_i^2}}_{\text{diffusion}} + \underbrace{f(C_i)}_{\substack{\text{chemical} \\ \text{transformations}}} \qquad i = 1,\dots,n \qquad (6.105)$$

The spatial discretization (6.105) has transformed the single PDE 6.78 into a set of n ODEs, whose initial conditions $\left[C_i(0), i = 1,\dots,n\right]$ can be obtained by the initial steady-state profile discussed in Section 6.3.5. Assuming for simplicity that the chemical transformation in each cell follows a first-order kinetic $-k_i \cdot C_i$, the generic Equation 6.105 may be written by grouping the terms common to C_{i-1}, C_i and C_{i+1}

$$\frac{dC_i}{dt} = \alpha_i C_{i-1} + \left(\beta_i - k_i\right)C_i + \gamma_i C_{i+1} \qquad (6.106)$$

with the combined parameters

$$\alpha_i = \frac{u_i}{2\Delta x_i} + \frac{D_i}{\Delta x_i^2} \qquad \beta_i = -\frac{2D_i}{\Delta x_i^2} \qquad \gamma_i = \frac{D_i}{\Delta x_i^2} - \frac{u_i}{2\Delta x_i} \qquad (6.107)$$

resulting in a set of linear dynamical system composed of n cells

$$\begin{cases} \dfrac{dC_1}{dt} = \alpha_1 C_{\text{in}} + \left(\beta_1 - k_1\right)C_1 + \gamma_1 C_2 \\[2mm] \dfrac{dC_2}{dt} = \alpha_2 C_1 + \left(\beta_2 - k_2\right)C_2 + \gamma_2 C_3 \\[2mm] \qquad\qquad\qquad \vdots \\[2mm] \dfrac{dC_n}{dt} = \alpha_n C_{n-1} + \left(\beta_n - k_n + \gamma_n\right)C_n \end{cases} \qquad (6.108)$$

or in matrix form

$$\begin{bmatrix} \dot{C}_1 \\ \dot{C}_2 \\ \dot{C}_3 \\ \dots \\ \dot{C}_n \end{bmatrix} = \begin{bmatrix} \beta_1 - k_1 & \gamma_1 & 0 & 0 & 0 \\ \alpha_2 & \beta_2 - k_2 & \gamma_2 & 0 & 0 \\ 0 & \alpha_3 & \beta_3 - k_3 & \gamma_3 & 0 \\ 0 & 0 & \dots & \dots & \dots \\ 0 & \dots & 0 & \alpha_n & \beta_n - k_n + \gamma_n \end{bmatrix} \begin{bmatrix} C_1 \\ C_2 \\ C_3 \\ \dots \\ C_n \end{bmatrix} + \begin{bmatrix} \alpha_1 \\ 0 \\ 0 \\ \dots \\ 0 \end{bmatrix} C_{\text{in}} \qquad (6.109)$$

By inspection, it can be seen that the system matrix has a tridiagonal structure with similar terms, save for the last row, which requires a simple explanation. The last, *n*th, cell has no downstream counterpart, and hence its diffusion term would be missing. But the model was developed under the 'open-end' assumption, meaning that the river condition would remain the same beyond the downstream boundary. It is therefore necessary to introduce a further, virtual cell—a 'ghost cell' if you like—to preserve the diffusional continuity at the downstream end. This ghost cell is assumed to have the same concentration as the previous nth cell and no kinetics, so that $C_{n+1} = C_n$. For this reason, the diagonal term $\beta_i - k_1$ is changed into $\beta_n - k_n + \gamma_n$ in the last row. The space-discretized continuous-time diffusive reach model is shown in Figure 6.42, where each cell is linked to the adjacent ones via advection and diffusion. The first cell receives the upstream input $C_{in}(t)$ that represents the time boundary condition, while the steady-state profile obtained by solving the steady-state spatial Equation 6.97 with either the BVP or the LS method provides the initial condition for each cell.

Here, we have a problem of redundancy, because the solution to Equation 6.97 represents a continuous concentration profile, whereas for the initialization of the n cells, only n discrete initial conditions are required. There are two possibilities to obtain a meaningful initial condition for each cell from the continuous profile, by taking either the average concentration or the middle cell concentration, as illustrated in Figure 6.43.

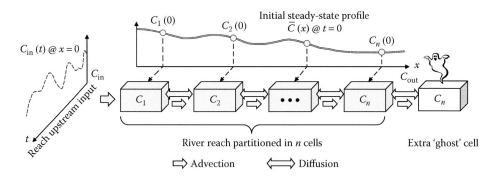

FIGURE 6.42 Spatial discretization of the mono-dimensional diffusion equation. The reach is partitioned into n completely mixed cells that communicate by means of advection and diffusion. The 'open-end' assumption requires the introduction of a 'ghost' cell after the last cell, with the same concentration. The previously computed steady-state profile provides the initial condition for each cell.

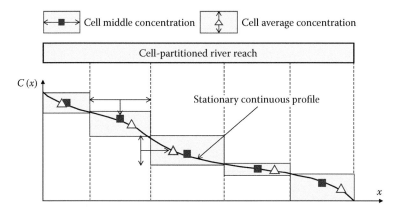

FIGURE 6.43 Differing methods for extracting the cell initial conditions from the steady-state concentration profile. The triangles represent the within-cell average concentration, while the squares indicate the mid-cell concentration.

6.3.6.1 Numerical Simulation of the Diffusive Reach Model in Time and Space

Let us now recap the work done so far: the system structure has been space-discretized and its boundary conditions have been specified using one of the procedures described in Section 6.3.5, including both the initial steady-state profile and the upstream input time-series $C_{in}(t)$. In the following example, we are going to simulate the river system with the linear model Equation 6.109, assuming a first-order chemical kinetics. Figure 6.44 shows how the eigenvalues of the system matrix Equation 6.109 are influenced by the velocity (a) and by the diffusion (b). In both cases, their negative real part is constrained by the kinetic constant (k), thus ensuring the system stability. The vertical line at $-k$ constraints the eigenvalues in the left-half-plane, representing a stability boundary that makes the system more robust as the kinetic constant increases.

The MATLAB–Simulink implementation of this model is very simple. The river system is represented by the Simulink linear model in Figure 6.45, where a single linear state-space block incorporates the four matrices which represent the model, including the initial conditions borrowed from the steady-state initial profile. The upstream loading is supplied as an input by the from Workspace block, while the output concentration matrix is routed to the workspace by the to Workspace data sink.

The MATLAB launch script is listed in Box 6.2 for the data input part, including the estimation of the diffusion with the Fisher formula Equation 6.76. The subsequent Box 6.3 shows how the system matrices are formed prior to being transferred to the Simulink state-space block. This script

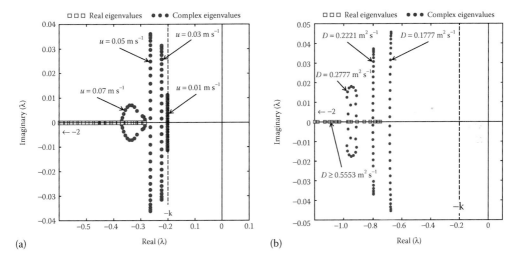

FIGURE 6.44 The eigenvalues of the system matrix Equation 6.109 are influenced by the velocity (a) and by the diffusion (b). In both cases, their negative real part is constrained by the kinetic constant (k), thus ensuring the system stability. These graphs were obtained with the MATLAB script Diffusive _ River _ Eigenvalues.m in the ESA _ MATLAB\Exercises\Chapter _ 6\Diffusive _ River folder.

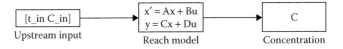

FIGURE 6.45 Simulink implementation of the river reach model Equation 6.109. The initial profile is passed as initial condition to the reach model state-space block.

BOX 6.2   PRELIMINARY OPERATIONS TO SIMULATE THE RIVER MODEL EQUATION 6.109: BOUNDARY CONDITIONS LOADING

```
clear
close all
% Load Initial Concentration profile
load Init_Conc
% Assign time and values of experimental observations
t_exp=Mis_exp(:,1);
C_exp=Mis_exp(:,2);
% Compute maximum reach length
L=ceil(t_exp(end));
% Set the number of cells
N_cell=30;
% Compute length of each cell
dz=L/(N_cell-1);
z=0:dz:L;
% Interpolate the initial profile to fit the cells
C_iniz=csaps(t_exp,C_exp,0.9,z);
% Load velocity data
load U_data
% Smooth the velocity profile to fit the cells
u=csaps(t_data,Y_data,0.9,z);
% Estimate the Diffusion with Fisher's formula
uo=mean(u);    % Average velocity (m/s)
w=15;          % Reach width (m)
s=0.0001;      % slope
h=1.5;         % average depth (m)
Dif=0.011*uo^2*w^2/(h*sqrt(9.81*s*h))
% Chemical kinetics
k=0.03; % (1/s)
% Load input upstream data
load Cin_data
t_in=Mis_exp(:,1);
% set the simulation time
tfin=t_in(end);
% Input upstream concentration
C_in=Mis_exp(:,2);
```

simply implements the system constants defined in Equation 6.107, launches the simulation, and plots the results.

A sample simulation is shown in Figure 6.46. The simulation starts with the specified initial profile, and since the upstream loading is zero for the entire simulation horizon, the concentration tends to flatten out and would eventually tend to zero for a sufficiently long simulation time. In this case, the velocity profile and the diffusivity are assumed to be the same for all the cells $\left(u = 0.1 \, \text{m s}^{-1}; D = 0.01 \, \text{m}^2\text{s}^{-1}\right)$.

Figure 6.47 compares the same simulation with the PF case, in which the diffusivity is set to zero. In both cases, the peak concentration travels downstream with the stream velocity, but while in the diffusive case (a) the pollutant spreads to adjacent cells as a result of diffusion, in the PF case (b) it remains confined to the water volume where it started, which travels downstream without mixing, in agreement with the PF hypothesis. In both cases, the slope of the peak in the time–space domain is given by the velocity $u = x/t$.

BOX 6.3 SET-UP OF MODEL EQUATION 6.109 AND SIMULATION LAUNCH

```
% Set-up the system matrix
A=zeros(N_cell,N_cell);
B_mat=zeros(N_cell,1);
C_mat=ones(1,N_cell);

for i=1:N_cell
  a(i)=u(i)/(2*dz)+Dif/(dz*dz);
  b(i)=-2*Dif/(dz*dz);
  c(i)=Dif/(dz*dz)-u(i)/(2*dz);

end
% First row
A(1,1)=b(1)-k;A(1,2)=c(1);
B_mat(1)=a(1);

% Middle rows
for i=2:N_cell-1
A(i,i-1)=a(i-1);
A(i,i)=b(i)-k;
A(i,i+1)=c(i-1);
end

% Last row
A(N_cell,N_cell-1)=a(N_cell-1);
A(N_cell,N_cell)=b(N_cell)-k+c(N_cell);

% Simulate the model
[t, x,y]=sim('Reach_model',tfin);
% Plot the results
figure(2)
surf(t,z,x')
axis tight
xlabel('time (h)')
ylabel('river length (km)')
shading interp
view([140 58])
grid off
zlabel('Conc. (mg/L)')
rotate3d on
```

6.3.6.2 Comparison with the CSTR Chain

As a concluding remark, we are now going to compare the space-discretized diffusive model Equation 6.109 to the comparatively simpler model in which each cell of Figure 6.42 is replaced by a CSTR. The two structures are shown in Figure 6.48 with the pertinent dynamical equation for the generic cell.

Recall that the CSTR chain model of Equation 6.24 is

$$
\begin{bmatrix} \dot{C}_1 \\ \dot{C}_2 \\ \vdots \\ \dot{C}_n \end{bmatrix} = \begin{bmatrix} -q_1-k & 0 & \cdots & 0 \\ q_2 & -q_2-k & \cdots & 0 \\ \vdots & \vdots & \vdots & 0 \\ 0 & 0 & q_n & -q_n-k \end{bmatrix} \times \begin{bmatrix} C_1 \\ C_2 \\ \vdots \\ C_n \end{bmatrix} + \begin{bmatrix} q_1 \\ 0 \\ \vdots \\ 0 \end{bmatrix} C_i \quad q_i = \frac{F}{V_i} \qquad (6.110)
$$

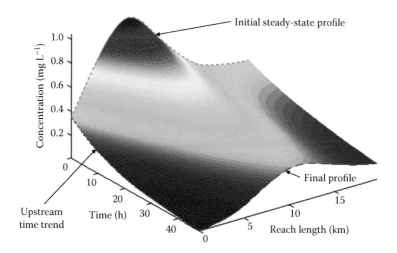

FIGURE 6.46 Simulation of the space–time evolution of a river reach with the model Equation 6.109, starting with a given steady-state profile, with given constant velocity ($u = 0.1$ m s^{-1}) and diffusivity ($D = 0.01$ m^2 s^{-1}). Since the input concentration is zero for the whole simulation period, the final profile flattens out and would eventually tend to zero, given enough simulation time. This graph was obtained with the MATLAB script `Go_Diffusive_River.m` in the `ESA_MATLAB\Exercises\Chapter_6\Diffusive_River` folder.

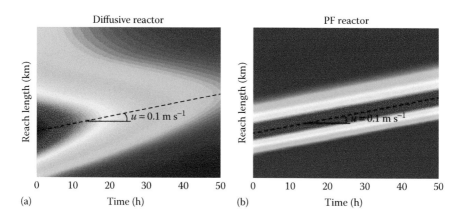

FIGURE 6.47 (a) Same simulation of Figure 6.46, presented as a contour graph, showing that the initial peak travels downstream with the stream velocity ($u = 0.1$ m s^{-1}). The concentration spread is due to diffusion. By comparison, in (b) the same reach is simulated with PF conditions ($D = 0$) and no pulse spreading occurs. These graphs were obtained with the MATLAB script `Go_Diffusive_River.m` in the `ESA_MATLAB\Exercises\Chapter_6\Diffusive_River` folder.

Here, the system matrix is sub-diagonal and its eigenvalues are given by the diagonal elements, as shown by Equation 6.25. In the CSTR chain, the diffusive term is clearly missing, as the only coupling between adjacent cells is given by advection through the dilution rate q. Clearly, the latter model Equation 6.110 is more robust, lacking the destabilizing effect of diffusion, but is also less accurate, omitting the diffusive term that plays an important role in the dynamics of river quality. To make the two models equivalent, the dilution rate q of the CSTR chain, used in Equation 6.110, was computed from the cell partitioning as follows:

$$q = \frac{F}{V} = \frac{u \cdot S}{\Delta x \cdot S} = \frac{u}{\Delta x} \tag{6.111}$$

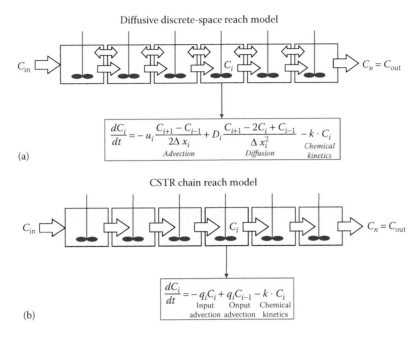

Diffusive discrete-space reach model

(a)

$$\frac{dC_i}{dt} = \underbrace{-u_i \frac{C_{i+1} - C_{i-1}}{2\Delta x_i}}_{Advection} + \underbrace{D_i \frac{C_{i+1} - 2C_i + C_{i-1}}{\Delta x_i^2}}_{Diffusion} \underbrace{- k \cdot C_i}_{\substack{Chemical \\ kinetics}}$$

CSTR chain reach model

(b)

$$\frac{dC_i}{dt} = \underbrace{-q_i C_i}_{\substack{Input \\ advection}} + \underbrace{q_i C_{i-1}}_{\substack{Onput \\ advection}} \underbrace{- k \cdot C_i}_{\substack{Chemical \\ kinetics}}$$

FIGURE 6.48 Comparison between the space-discretized diffusive model Equation 6.109 (a) and the CSTR chain Equation 6.110 (b). In the latter, the feedback diffusive term is missing and the cells communicate only by advection.

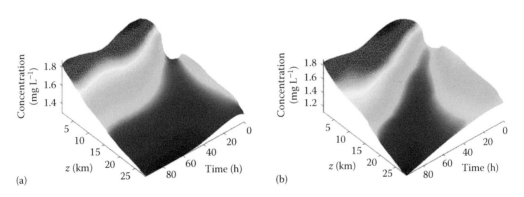

(a)

(b)

FIGURE 6.49 Comparison of diffusive river simulation using the two model schemes of Figure 6.48: (a) diffusive discrete-space model response, (b) CSTR chain model response. This comparison was produced by the MATLAB script Go_Diffusive_River.m in the ESA_MATLAB\Exercises\Chapter_6\Diffusive_River folder.

All the other parameters and boundary conditions, such as the initial profile $C_{in}(x)$ and the time-varying input $C_{in}(t)$, are the same. Figure 6.49 compares the two model responses, showing the better control that the diffusive model (a) has over the variable river conditions. Instead, the CSTR chain model (b) has less control over the domain, especially in the downstream reaches, where the input propagates with more delay along the CSTR chain, given the lack of diffusion.

7 Microbial Kinetics Modelling

We talk about life as being dull as ditchwater, but is ditchwater dull? Naturalists with microscopes have told me that it teems with quiet fun.

G.K. Chesterton

The focus of the preceding chapter was on the hydraulics of the system, and we assumed for simplicity that in the previous examples, the chemical transformations were governed by a first-order kinetics. This assumption made the model equations linear, facilitating its development and analysis. In this chapter, we are going to acknowledge that, in contrast, a wide variety of chemical transformations occurring in nature follow differing kinetics, and, in particular, we are going to focus on the role of microorganisms, which are involved in almost all of the natural chemical transformations. In Section 1.10.4.1, we introduced the Monod kinetics to have a benchmark model for our early exercises. This kinetics is the cornerstone of microbial transformation, and it is now time to put it into a broader context, explore its biochemical background and to take a wider look at the biochemical component of environmental models.

7.1 ENZYMATIC REACTION KINETICS

A brief review of chemical kinetics is now presented to introduce the biochemical reactions that will be encountered in environmental systems.

7.1.1 A SUMMARY OF ELEMENTARY CHEMICAL KINETICS

Let us begin by recalling the basic notions of chemical equilibrium and reaction kinetics (Alberty, 1979; Houston, 2006; House, 2007; Marin and Yablonsky, 2011). Consider the reversible equilibrium between reagents and products,

$$\underbrace{a[A]+b[B]}_{\text{reagents}} \underset{k_{\text{inv}}}{\overset{k_{\text{dir}}}{\rightleftharpoons}} \underbrace{c[C]+d[D]}_{\text{products}} \tag{7.1}$$

where:

$[A],[B]$ represent the molar concentrations of the reagents
$[C],[D]$ represent those of the products
(a,b,c,d) are stoichiometric constants that balance the equilibrium

By the law of mass–action, the forward and reverse reaction rates can be defined as

$$\text{Forward reaction rate} \quad r_{\text{dir}} = -\frac{1}{a}\frac{d[A]}{dt} = -\frac{1}{b}\frac{d[B]}{dt}$$

$$= k_{\text{dir}} \cdot [A]^a \cdot [B]^b$$

$$\text{Backward reaction rate} \quad r_{\text{inv}} = -\frac{1}{c}\frac{d[C]}{dt} = -\frac{1}{d}\frac{d[D]}{dt} \tag{7.2}$$

$$= k_{\text{inv}} \cdot [C]^c \cdot [D]^d$$

and thus the overall reaction rate R is given by

$$R = r_{\text{dir}} - r_{\text{inv}} = k_{\text{dir}} \cdot \left[A\right]^a \cdot \left[B\right]^b - k_{\text{inv}} \cdot \left[C\right]^c \cdot \left[D\right]^d$$

$$= k_{\text{dir}} \cdot \left(\left[A\right]^a \cdot \left[B\right]^b - \frac{\left[C\right]^c \cdot \left[D\right]^d}{k_{\text{eq}}} \right) \qquad (7.3)$$

where the equilibrium constant k_{eq} is defined as

$$k_{\text{eq}} = \frac{k_{\text{dir}}}{k_{\text{inv}}} = \frac{\left[C\right]^c \cdot \left[D\right]^d}{\left[A\right]^a \cdot \left[B\right]^b} \qquad (7.4)$$

From the previous definitions, it might appear that the reaction rate could be written simply by inspecting the chemical equilibrium (7.1) and using the stoichiometric constants as the rate exponents in Equation 7.3, but this is not so, unless the chemical equilibrium (7.1) is *elementary*, that is, it is not the sum of several reaction steps. In the latter case, only the dynamically relevant steps, that is, the rate-limiting steps, should be considered for the exponents. As an example of this lack of correspondence when the equilibrium is the sum of several reaction steps, consider the following irreversible chemical equilibrium, where the brackets have been omitted for clarity

$$A + 2B \xrightarrow{k} C \qquad (7.5)$$

If the law of mass–action is directly applied, the reaction rate should be

$$r = -k \cdot A \cdot B^2 \qquad (7.6)$$

In contrast, a chemist who knows about this reaction would tell us that the *experimentally observed* reaction rate is in fact

$$r_{\text{obs}} = -k \cdot A \cdot B \qquad (7.7)$$

How can we resolve this conflict? Obviously, it is the theory that should conform to the data and not the other way around, so we ask our chemist again, who will tell us that the stoichiometric equilibrium (7.5) is actually the sum of two steps: a slow one in which one mole of A reacts with one mole of B to form an intermediate AB^*, followed by a fast step in which the intermediate reacts with the second mole of B to yield the product C. From a kinetic viewpoint, the second step is unimportant, being much faster than the first, which in turn limits the reaction rate. Thus, before actually writing the kinetic rate equation by using the stoichiometric weights, the equilibrium (7.5) should be decomposed into its elementary steps, if we suspect there is any. In the present case,

$$
\begin{aligned}
\text{First step (slow)} \quad & A + B \xrightarrow{k} AB* \\
\text{Second step (fast)} \quad & AB* + B \rightarrow C
\end{aligned}
\qquad (7.8)
$$

thus, considering only the slow step yields the correct kinetics of Equation 7.7. This simple example illustrates that care must be taken in using the stoichiometric constants as kinetic exponents, as this can be done only if the chemical equilibrium is elementary and not the sum of several steps, in which case only the slow steps should be considered as kinetically relevant.

7.1.2 ENZYME-MEDIATED REACTION KINETICS

Now, we move on to consider an enzyme-mediated chemical reaction, which transforms a substrate (S) into a product (P) due to the catalytic action of an enzyme (E). The enzyme is a complex molecule that acts as a catalyst by making the reaction possible by lowering the activation energy barrier, but it does not take part in the reaction itself (Roels, 1983; Bailey and Ollis, 1986). The enzymatic reaction occurs when the substrate binds to a specific spot on the enzyme to form an activated complex, which then reacts to yield the product and eventually releases the enzyme. Because most of the enzymes have only one binding site, it makes sense to assume that one mole of enzyme combines with one mole of substrate. At the end of the reaction, the product will be released, and the enzyme will still be available to catalyse more substrate molecules. The enzyme-mediated reaction can then be described by the following chemical equilibrium:

$$S + E \underset{k_2}{\overset{k_1}{\rightleftarrows}} ES * \xrightarrow{k_3} P + E \tag{7.9}$$

supposing that the unstable intermediate (**ES***) can either react back to the reagents or proceed forward to yield the product (*P*), the latter path being considered here as irreversible, eventually releasing the enzyme (*E*). Knowing that the equilibrium (7.9) is elementary, we can write the reaction rates by inspection:

$$\begin{cases} \text{Substrate} & \dfrac{dS}{dt} = -k_1 \cdot E \cdot S + k_2 \cdot ES * \\[2mm] \text{Activated complex} & \dfrac{dES *}{dt} = k_1 \cdot E \cdot S - \left(k_2 + k_3\right) \cdot ES * \\[2mm] \text{Product} & \dfrac{dP}{dt} = k_3 \cdot ES * \end{cases} \tag{7.10}$$

The system of equations (7.10) describes the enzymatic reaction in any setting, but to derive the well-known Michaelis–Menten model for an enzymatic reaction, and then the subsequent Monod microbial kinetics, some simplifying assumptions must be made:

- The enzyme is not consumed during the reaction and is released on its completion. Therefore, during the reaction, the sum of the free (*E*) and complexed (*ES**) enzyme is constant and equal to the initial enzyme quantity E_o. Thus, $E_o = E + ES *$.
- The activated complex *ES** is at steady state, because its formation and dissociation balance at any one time; thus $dES */dt \cong 0$.

Applying the steady-state assumption to the second equation yields

$$\frac{dES *}{dt} \cong 0 \Rightarrow k_1 E \cdot S = \left(k_2 + k_3\right) ES* \Rightarrow E \cdot S = \frac{\left(k_2 + k_3\right)}{k_1} ES* = K_s ES* \tag{7.11}$$

Now, from the first assumption, we can express the free enzyme as $E = E_o - ES *$ and substitute it in Equation 7.11 to yield

$$\left(E_o - ES *\right)S = K_s ES * \quad \Rightarrow \quad E_o \cdot S = \left(K_s + S\right)ES * \quad \Rightarrow \quad ES* = \frac{E_o \cdot S}{\left(K_s + S\right)} \tag{7.12}$$

This allows the elimination of the second equation, so that $-dS/dt = dP/dt$. Thus, the system of equations (7.10) reduces to a single equation

$$-\frac{dS}{dt} = \frac{dP}{dt} = \mu_{min} = \mu_{max} \frac{S}{K_s + S} \tag{7.13}$$

where

$$\mu_{max} = k_3 \cdot E_o \quad \text{and} \quad K_s = \frac{k_2 + k_3}{k_1} \tag{7.14}$$

Several important conclusions can be drawn from Equation 7.13:

- The maximum reaction rate μ_{max} is linearly related to the total enzyme in the system, as per Equation 7.14.
- The nonlinear reaction rate (Equation 7.13) is modulated by the constant K_s, which depends on the ratio between the forward and backward equilibrium constants. K_s is also called *half-velocity* constant because when $K_s = S$, the reaction rate is $\mu_{max}/2$.

Figure 7.1 shows the Michaelis–Menten kinetics as a function of the available substrate. The maximum reaction rate μ_{max} is reached for very large values of the substrate, while for $S = K_s$ the reaction rate is half of that value. The initial slope of the kinetics is μ_{max}/K_s, and therefore for small values of the substrate, the Michaelis–Menten kinetics can be approximated by a first-order reaction:

$$\frac{dS}{dt} = -\frac{\mu_{max}}{K_s} \cdot S \tag{7.15}$$

while for large values of S, it tends to the constant value μ_{max}, thus becoming a zero-order kinetics.

We have seen that the formation of the activated complex (ES^*) and its reversibility are the key elements of the enzymatic reactions. The onset of the steady state depends on the relative abundance of the enzyme with respect to the substrate. If relatively little enzyme is available, it will immediately bind to the substrate, so that the steady-state hypothesis is satisfied, and the reaction rate follows Equation 7.13, but if the enzyme concentration is of the same order as the substrate, then the binding will involve only a fraction of the enzyme and the reaction dynamics will be different. This latter case is illustrated in Figure 7.2a, in which the initial substrate/initial enzyme ratio is one, resulting in wide variations of ES^*. Conversely, Figure 7.2b shows that if the enzyme is scarce, the onset of the steady state is almost immediate (inset), and the reaction proceeds as predicted by the Michaelis–Menten theory. Figure 7.3 shows the Simulink® model, launched by the MATLAB® script in Box 7.1, used to produce these simulations.

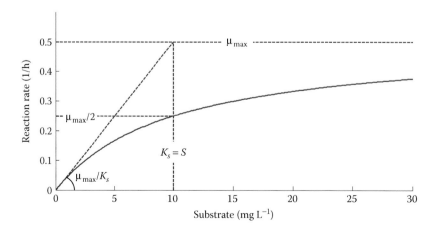

FIGURE 7.1 The Michaelis–Menten kinetics as a function of the available substrate (S). The initial slope is equal to μ_{max}/K_s, and when $K_s = S$, the reaction rate is $\mu_{max}/2$. The reaction rate tends to μ_{max} for large values of the substrate.

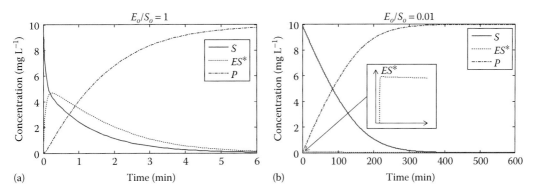

(a) Time (min)

(b) Time (min)

FIGURE 7.2 Time evolution of the Michaelis–Menten enzymatic reaction with differing initial enzyme concentrations. In (a), the enzyme is so abundant that the ES^* is never at steady state, thus violating the second hypothesis, while in (b), the shortage of enzyme almost immediately establishes the steady-state regime for the activated complex (see the inset). The graphs were produced with the MATLAB script listed in Box 7.1 and the Simulink model of Figure 7.3.

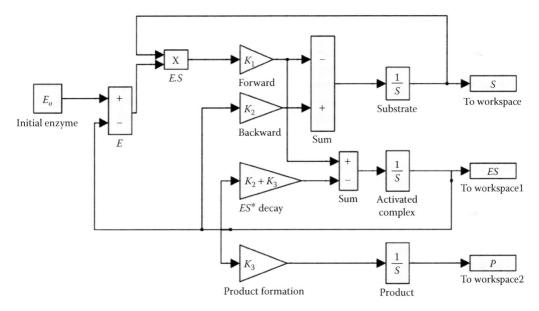

FIGURE 7.3 Simulink diagram to study the effect of relative enzyme abundance on the Michaelis–Menten reaction.

Differing hypotheses can be formulated regarding the reversibility of the reaction paths in the equilibrium (7.9). If, for example, the second transformation is assumed to be reversible

$$E + S \underset{k_2}{\overset{k_1}{\rightleftharpoons}} ES * \underset{k_4}{\overset{k_3}{\rightleftharpoons}} P \tag{7.16}$$

then by similar reasoning that led to Equation 7.13, the following kinetics can be obtained:

$$\mu_{rr} = \frac{\dfrac{\mu_{max}}{K_s} S - \dfrac{\mu_p}{K_p} P}{1 + \dfrac{S}{K_s} + \dfrac{P}{K_p}} \quad \text{with} \quad \begin{aligned} \mu_p &= \frac{k_2}{k_3} \mu_{max} \\[2mm] K_p &= \frac{k_1}{k_4} K_s \end{aligned} \tag{7.17}$$

BOX 7.1   LAUNCH SCRIPT TO SIMULATE THE MICHAELIS-MENTEN ENZYMATIC REACTION WITH THE POSSIBLE ONSET OF THE STEADY STATE OF THE ACTIVATED ENZYME. THE SCRIPT REFERS TO THE MM_BASIC SIMULINK MODEL OF FIGURE 7.3

```
% Select the enzyme initial concentration Eo to avoid (1) or
% reproduce (2) the steady-state condition
ESS=1;
switch ESS
    case 1
% Initial conditions to simulate the lack of steady-state for ES*
Eo=10;  % Initial enzyme
So=10;  % Initial substrate
tfin=6;
    case 2
% Initial conditions to simulate the steady-state for ES*
Eo=0.1; % Initial enzyme
So=10;  % Initial substrate
tfin=600;
end
% Other initial conditions
ESo=0;Po=0;
% Kinetic constants
K1=1;
K2=4;
K3=1;
% Note that the half velocity constant Ks=(K2+K2)/K1=5
% Launch the simulation
[t,x,y]=sim('MM_basic',tfin);
figure(1)
set(1,'Position',[279 244 557 323])
plot(t,S,t,ES,':b',t,P,'-.b')
set(gca,'FontName','Arial','FontSize',14)
title(['Eo/So = ',num2str(Eo/So)])
ylabel('Conc. (mg/l)')
xlabel('time (min)')
legend('S','ES*','P')
```

7.1.3   SELF-INHIBITING KINETICS

An important case of enzymatic reaction is the one that involves two active sites and the generation of two activated complexes, only one of which generates the final product, while the other is stable. The second active site is often referred to as the *inhibition* site, because it does not release the enzyme. This reaction scheme can be described by the following equilibria:

$$E + S \underset{k_2}{\overset{k_1}{\rightleftharpoons}} ES * \begin{array}{l} \rightarrow ES * + S \underset{k_5}{\overset{k_4}{\rightleftharpoons}} ES_2 \\ \xrightarrow{k_3} P + E \end{array} \tag{7.18}$$

While the lower branch is the same as in Equation 7.9, the upper one leads to a stable compound (ES_2) that sequesters the enzyme, which is no longer able to perform its catalytic function. As a result, the reaction rate slows down due to lack of enzyme, and because of the two-sided role of the substrate, which contributes to the product formation when it is scarce and binds to the active site, but slows the reaction when it is abundant and binds to the enzyme in the secondary site as well. For this reason, this mechanism is referred to as *self-inhibitory*. By applying the same basic hypotheses of enzyme conservation (including the sequestered fraction) and steady state of the activated complex, the Haldane kinetics (Bailey and Ollis, 1986) is obtained:

$$\mu_{hald} = \mu_{max} \frac{1}{1 + \dfrac{K_s}{S} + \dfrac{S}{K_i}} \quad \text{with} \quad \begin{aligned} K_s &= \frac{k_2 + k_3}{k_1} \\ K_i &= \frac{k_5}{k_4} \\ \mu_{max} &= k_3 \cdot E_o \end{aligned} \tag{7.19}$$

Figure 7.4 shows the reaction rate of the Haldane kinetics, which unlike Michaelis–Menten is not monotonic. In fact, the reaction rate increases with the substrate up to S_{max}, after which it decreases because of self-inhibition. The peak concentration S_{max} can be found by setting the derivative of the kinetics with respect to S to zero, to yield

$$\frac{d}{dt}\left(\frac{S}{K_s + S + \dfrac{S^2}{K_i}}\right) = \frac{K - \dfrac{S^2}{K_i}}{\left(K_s + S + \dfrac{S^2}{K_i}\right)^2} = 0 \quad \Rightarrow \quad S_{max} = \sqrt{K_s \cdot K_i} \tag{7.20}$$

Thus, the substrate concentration producing the maximum reaction rate S_{max} is the geometric mean of the two kinetic constants. The reader is referred to Bailey and Ollis (1986) for a thorough survey of many other inhibitory mechanisms, which are beyond the scope of this book.

7.1.4 Other Kinetics of Interest

So far, we have considered a limiting substrate, meaning that its concentration controls (limits) the reaction kinetics. An alternate, and very common, case is when, in addition to the one just examined, the reaction is based on another substrate. The second substrate in this case is not inhibiting, in

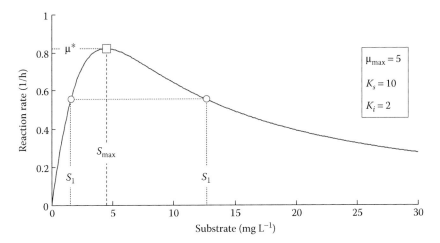

FIGURE 7.4 Haldane kinetic reaction rate as a function of the substrate. Unlike the Michaelis–Menten kinetics, the reaction rate increases with the substrate up to S_{max}, after which it decreases because of self-inhibition. Thus, there are two substrate values (S_1 and S_2) straddling S_{max} and producing the same reaction rates. This graph was obtained with the MATLAB script Haldane.m in the ESA _ Matlab\Exercises\ Chapter _ 7\Enzyme _ Kinetics folder.

the Haldane sense, because it does not bind with the substrate, but its presence decreases the overall reaction rate. This mechanism can be modelled as

$$\mu\left(S_1, S_2\right) = \mu_{max} \underbrace{\frac{S_1}{K_{s1} + S_1}}_{\substack{\text{Internally}\\\text{limiting}\\\text{substrate}}} \cdot \underbrace{\frac{K_{s2}}{K_{s2} + S_2}}_{\substack{\text{Externally}\\\text{limiting}\\\text{substrate}}} \tag{7.21}$$

In Equation 7.21, the first fraction is the Michaelis–Menten term, with S_1 acting as the internally limiting substrate, while in the second term S_2 is the externally limiting factor. In fact, the reaction rate is inversely proportional to the concentration of S_2, meaning that its presence acts as a restraining term on the reaction. A typical example of this mechanism will be observed with denitrifying microorganisms that are 'distracted' from their role (reducing nitrogen oxides into molecular nitrogen) by the oxygen availability. These facultative microorganisms prefer oxygen as their electron acceptor, as it allows them to activate a more robust adenosine triphosphate (ATP)-producing metabolism. If oxygen is absent, however, they settle for NO_3^- or NO_2^- as electron acceptors, hence the term 'facultative', resulting in a less energetic metabolism. In fact, in most activated sludge models (Henze et al., 2000), the μ_{max} for the denitrification reaction is about 80% of the same rate in aerobic conditions. Returning to Equation 7.21, in that case oxygen is the 'distracting' substrate S_2, because its concentration would make the second term smaller and slow down the entire kinetics.

Another very common occurrence is when the microorganisms need more than one substrate for their growth. In 1840, J. Liebig proposed his Law of the Minimum, suggesting that the growth of organisms that require multiple substrates is limited by the single substrate that is in shortest supply relative to demand (Palmeri et al., 2014). This can be expressed as the product of several Michaelis–Menten factors:

$$\mu\left(S_1, ..., S_n\right) = \mu_{max} \frac{S_1}{K_{s1} + S_1} \cdot \frac{S_2}{K_{sL2} + S_2} \cdots \frac{S_n}{K_{Sn} + S_n} \tag{7.22}$$

In this way, the single smallest substrate in the product affects the overall growth rate.

7.2 MICROBIAL METABOLISM OF CARBONACEOUS SUBSTRATES

The natural cycle of carbon is well described by Cloete and Muyima (1997), and it is sketched in Figure 7.5, which can be viewed as four intertwined cycles. From top to bottom, the upper part describes how organic carbon, briefly indicated as $\left(CH_2O\right)_n$, is synthesized from inorganic carbon $\left(CO_2\right)$ through either an anaerobic (left) or aerobic (right) path. The lower part of the diagram shows the two main paths through which the organic carbon thus generated is again oxidized to inorganic carbon, via an anaerobic (left) or an aerobic (right) path.

Figure 7.6 shows the main difference in carbon transformations through the aerobic (a) or the anaerobic (b) metabolism. In the aerobic pathway, nearly half of the organic carbon entering the system is assimilated into the biomass, whereas in the anaerobic pathway, most of it is converted into a gas mix of carbon dioxide (CO_2) and methane (CH_4). This is due to the very different microorganisms involved, with their differing metabolic mechanisms and yield factors, as will later be explained.

7.2.1 BASIC MOLECULE/MICROORGANISM INTERACTIONS OF ENVIRONMENTAL INTEREST

There exists a vast microbiological literature describing the biology of the microorganisms, both in natural (Campbell, 1983; Battley, 1987; Madigan et al., 2009) and man-made environments (Orhon and Artan, 1994; Cloete and Muyima, 1997; Henze et al., 2000; Gijzen, 2001) so only a brief

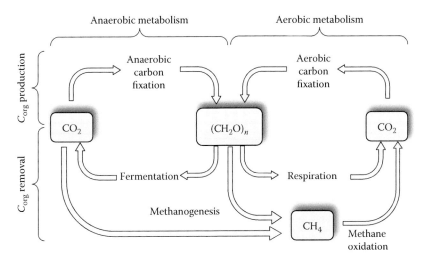

FIGURE 7.5 Natural carbon cycle can be viewed, from top to bottom, as the production/removal of organic carbon, or, from left to right, as carbon transformations that depend on the electron acceptor. Either way, the diagram hinges on the organic carbon (centre) and its inorganic counterpart (CO_2). The left branch is based on the anaerobic metabolism, and the result is methane production, whereas the right branch is based on aerobic metabolism.

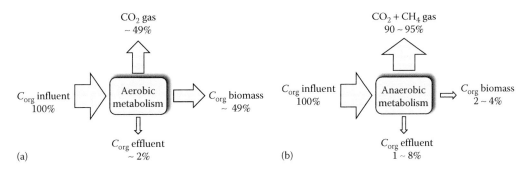

FIGURE 7.6 Differing partitioning of carbon in the aerobic (a) and in the anaerobic (b) metabolism. In (a), nearly half the available organic carbon is metabolized into cell biomass, whereas in (b), most of the organic carbon is transformed into gas, with very little biomass growth.

review of the biochemical foundations of cell metabolism will be included here, as many excellent textbooks, first and foremost the Lehninger textbook (Nelson and Cox, 2008), are available. The energy and material flows in an ecosystem are illustrated in Figure 7.7, with the solar radiation being the prime energy source, which sustains the entire food chain through the anabolic and catabolic pathways.

There are two complementary processes, through which the organisms transform the biodegradable materials:

- *Assimilative (or anabolic) processes* that transform the biodegradable substances (the substrate) into cell components, thus contributing to the biomass growth, and releasing metabolites such as H_2O, CO_2, NH_4, etc. in the process.
- *Dissimilative (or catabolic) processes*, multi-stage reduction–oxidation (redox) reactions that release energy that the microorganisms use for their assimilative processes.

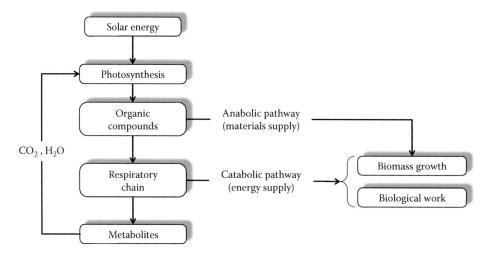

FIGURE 7.7 Energy and material flows in an ecosystem are initiated by the solar energy and they sustain the entire food chain through the anabolic and catabolic pathways.

The metabolism of microorganisms is based on four basic ingredients:

- An energy source, providing energy for the assimilative (anabolic) processes.
- A carbon source, the prime material for the build-up of biomass.
- An electron donor, to sustain the oxidation half reaction in the redox process.
- An electron acceptor, to sustain the reduction half reaction in the redox process.

Depending on the nature of these four factors, the metabolic pathway can have differing aspects, as summarized in Table 7.1, with the main distinction being between autotrophic and heterotrophic microorganisms. The former use inorganic carbon (CO_2) as their carbon source, while the latter utilize a wide variety of organic compounds for their growth.

From the modelling viewpoint, it is important to remember that *both* metabolic pathways (anabolic and catabolic) must be considered in a proper mass and energy balance.

TABLE 7.1
Summary of Microbial Metabolic Factors

Energy Source	Metabolism	Carbon Source	Electron Donor	Electron Acceptor	Process
Solar energy (reduction half-reaction missing)	Photo-autotrophic	CO_2	H_2O	CO_2	Photosynthesis
Chemical energy	Chemo-autotrophic	CO_2	Inorganic molecules H_2, Fe^{2+}, NH_4^+, NO_2^-	O_2 NO_3^-	Oxidation + Assimilation (e.g. Nitrification)
	Chemo-heterotrophic	Organic carbon	Organic carbon	O_2 NO_3^-	Aerobic respiration Anoxic respiration
		Organic carbon	Organic carbon	Organic carbon	Anaerobic fermentation

7.2.2 THE ENERGY OF MICROBIAL GROWTH

Depending on the kind of metabolism, Figure 7.8 illustrates the interactions between the microorganisms and the substrate, showing that for photo-autotrophic organisms, the energy is directly available as solar radiation (a). Conversely, for the chemo-trophic organism in (b) and (c), catabolic reactions provide the energy necessary for biosynthesis, the main difference being in the nature of the carbon source.

The basic energy storage/release mechanism in chemo-trophic microorganisms is based on the nucleoside *adenosine* and its derivatives obtained by adding two or three phosphoric acid groups to form adenosine diphosphate (ADP) or adenosine triphosphate (ATP). Figure 7.9 shows the closed cycle through which ATP is formed by phosphorylation at the expense of the energy released from the substrate metabolic reactions. Conversely, the enzymatic hydrolysis releases energy inside the cell to sustain its anabolic activities. The two complementary reactions illustrated in Figure 7.9 are (Bailey and Ollis, 1986; Orhon and Artan, 1994)

$$
\begin{aligned}
ADP + P_i \xrightarrow{\text{Phosphorilation}} ATP \qquad & \Delta G = -7.3 \text{ kcal mol}^{-1} \\
ATP + H_2O \xrightarrow{\text{Hydrolysis}} ADP + P_i \qquad & \Delta G = 7.3 \text{ kcal mol}^{-1}
\end{aligned}
\tag{7.23}
$$

While phosphorylation requires energy to incorporate one P_i, its mirror reaction (hydrolysis) releases the same amount of energy. For this reason, ATP/ADP can be regarded as the 'energy shuttle in the cell' (Bailey and Ollis, 1986).

The respiratory chain is sketched in Figure 7.10 to visualize the electron 'waterfall' from the energy level of the donor to the comparatively lower energy of the acceptor. In the process, some ΔG may be large enough to generate one or more ATP through oxidative phosphorylation.

Figure 7.11 shows the respiratory chains of environmental interest. Depending on the kind of electron acceptor, the magnitude of the energy gap ΔG may produce a varying number of ATP molecules, thus making the metabolism more or less lively. In fact, anaerobes are rather sluggish bacteria with very slow dynamics, whereas aerobic metabolism is comparatively faster. But then, it

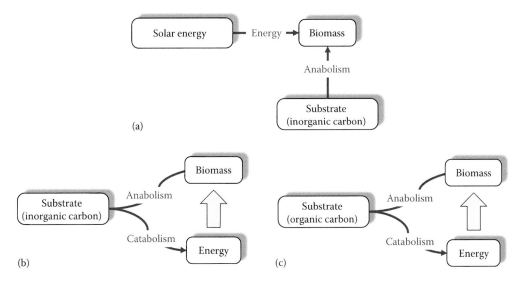

FIGURE 7.8 The three types of microbial metabolism: (a) photo-autotrophic metabolism; (b) chemo-autotrophic metabolism; and (c) chemo-heterotrophic metabolism. Photo-autotrophic organisms use the solar energy to metabolize inorganic carbon (a), whereas lithotrophs use chemical energy drawn from inorganic carbon for biosynthesis (b). Chemo-heterotrophs follow a similar pathway, but use organic carbon both as an energy and carbon source.

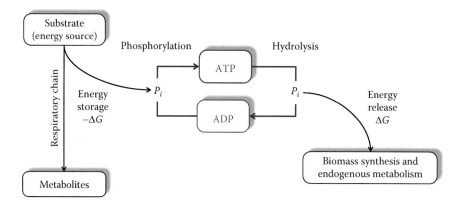

FIGURE 7.9   The cell wheel of energy revolves around the ATP/ADP couple: substrate degradation reactions provide the energy to attach one inorganic phosphorus (P_i) to ADP (phosphorylation), thus forming the energy-rich ATP. The reverse process of hydrolysis occurs in the respiratory chain and releases the energy required for the anabolic processes.

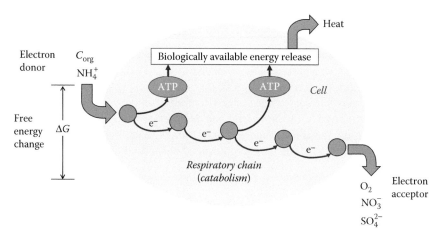

FIGURE 7.10   In the respiratory chain, electrons 'fall down' from the higher energy level of the donor to the relatively lower level of the acceptor, and in the process release the energy required to form one or more ATP molecules, depending of the magnitude of the free energy change ΔG.

should be reminded that anaerobes were the earliest forms of like on Earth, well before an atmosphere was established, so they made do with what they had, which was not much. Anyone should be sympathetic to these tiny forerunners of life on our planet, which are still with us, doing an obscure but vital job.

There are many alternate electron acceptors in nature (Hemond and Fechner, 1999), and their sequence of redox potentials is illustrated in Figure 7.12, which shows an increasing trend from the most sluggish reaction (methanogenesis) to the brisk aerobic respiration.

Of course, each respiration process is typical of one or more types of microorganisms, most of which are obligate users of certain electron acceptors, with some exceptions, like the already mentioned *facultative* heterotrophs, which can perform aerobic respiration if oxygen, their preferred electron acceptor, is available, but in its absence can settle for nitrate-N or nitrite-N as alternates. In this case, however, the comparatively smaller ΔG, reflected in the smaller redox potential, yields a lesser ATP production in the respiratory chain.

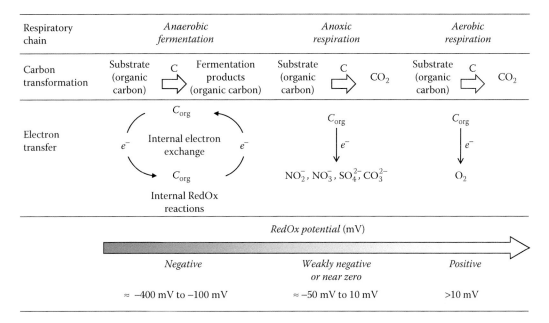

FIGURE 7.11 Microbial heterotrophic metabolism is determined by the extent of the respiratory chain and its ability to produce more or less energy. From left to right, the energy change increases, as does the ATP-producing potential. From the macroscopic viewpoint, the differing metabolisms can be related to the redox potential.

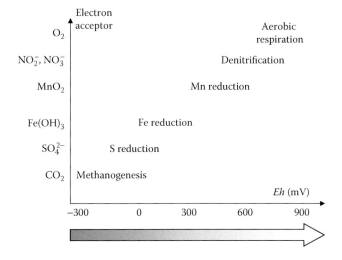

FIGURE 7.12 The RedOx sequence in natural processes. (Data from Hemond, H. F. and Fechner, E. J., *Chemical Fate and Transport in the Environment*, Academic Press, San Diego, CA, 433pp, 1999.)

7.2.3 MODELLING THE MICROORGANISM/SUBSTRATE INTERACTIONS: THE MONOD MODEL

In Section 7.1, we have introduced the enzymatic reactions assuming that the enzyme was an external agent artificially supplied to start the reaction. This is what happens in many industrial processes producing drugs or food. In nature, however, the availability of the enzymes is directly related to the organisms that produce these substances for their metabolism. Consequently, in transferring the enzymatic reaction from the lab to the environment, the dynamics of the living organism producing the enzyme must be accounted for, side by side with the enzymatic reaction. There are several hypotheses on how the microorganisms produce the enzyme they need. The first fundamental findings are due to Jacques Monod (1949), and it is widely acknowledged that the enzyme production is directly proportional to the biomass (Campbell, 1983). Let us recall that the maximum reaction rate $\mu_{max} = k_3 \cdot E_o$ is directly proportional to the quantity of available enzyme so, to transfer the enzyme kinetics into the natural environment, we must include the dynamics of the enzyme-producing biomass. This implies that we now have two coupled equations: one for the substrate utilization, and another for the biomass growth. To some extent, this situation is reminiscent of the prey–predator model encountered in Section 5.5, but with some important differences. The substrate can be considered as the prey, although it does not have a growth dynamics of its own. Likewise, the biomass, viewed as the predator, cannot grow without the nourishment provided by the substrate, though its decay (endogenous metabolism) is far more complex than the simple linear extinction usually assumed for the predator. Going back to the microbial version of the enzymatic reaction, the maximum growth rate μ_{max} now becomes biomass-specific. The following model uses the activated sludge models (ASM) notation (Henze et al., 2000), where the dissolved components are expressed as S and the particulate components as X. Here, we consider a dissolved substrate S that can be directly utilized by the biomass, while the particulate substrate should be previously hydrolysed. This distinction is somewhat different from the purely physical classification that considers soluble any component that can filter through a membrane with 0.45 μm porosity. In our case, the distinction is based on the biological availability of the substances, rather than on their physical dimensions, as sketched in Figure 7.13.

Several new concepts and parameters must be introduced before the model is fully specified:

- We must specify the fraction of new biomass (ΔX) synthesized through the utilization of a fraction of soluble substrate (ΔS). The *yield factor* term is defined as $Y = \Delta X/\Delta S$. Going back to Figure 7.6, the aerobic and anaerobic metabolisms greatly differ in the values of Y, which determines which fraction of the available organic carbon is assimilated as new biomass.
- As mentioned earlier, the biomass is subject to decay through *endogenous metabolism*, a comprehensive term including all of the cell processes that do not result in growth. Decay is generally assumed to affect a proportion of the biomass, and is thus expressed as $b \cdot X$.

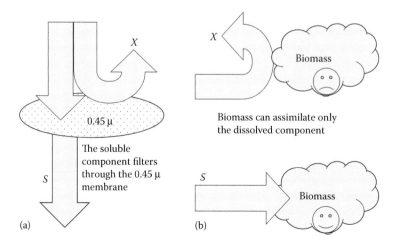

FIGURE 7.13 Physical (a) and biological (b) definitions of particulate matter.

Thus, the two equations that define the Monod model for the substrate/biomass interaction can be written as

$$\begin{cases} \text{Substrate utilization} & \dfrac{dS}{dt} = -\dfrac{1}{Y}\mu_{max}\dfrac{S}{K_s+S}X \\[2em] \text{Biomass growth} & \dfrac{dX}{dt} = \mu_{max}\dfrac{S}{K_s+S}X - b\cdot X \end{cases} \qquad (7.24)$$

which are exactly the pair of equations that we have already encountered in Chapters 1 and 2, if only as a mathematical object. Now, we must regard them as much more than just that, considering the legacy brought to us by several decades of dedicated microbiological research.

7.2.3.1 Which Units for the Carbon-Based Quantities?

Equations 7.24 relate two very different carbonaceous quantities: substrate, including a wide variety of biodegradable organic carbon molecules, and biomass, encompassing an equally vast inventory of microorganisms. These two heterogeneous variables can coexist in Equations 7.24 and be related to one another *because both are expressed in the same units*, precisely mg O_2/L. In fact, both are *measured by their oxygen equivalent*, that is, the amount of oxygen required to oxidize those substances into inert products. The chemical oxygen demand (COD) is in fact a global parameter expressing the amount of oxygen consumed in the chemical oxidation of all the carbonaceous substances, regardless of their precise nature and composition (Orhon and Artan, 1994; Metcalf & Eddy Inc., 2003; Sawyer et al., 2003; Brezonik and Arnold, 2011; Leslie Grady et al., 2011). Thus, COD is a unifying parameter, allowing very different substances to be compared and related, as in Equations 7.24. The oxygen-equivalent assumption, or the COD common denominator, will be used throughout and is comparable to the energy equivalent that was used in Chapters 1 and 5 to relate differing species in the same food chain on the basis of energy transfer, rather than on specific biomass weight. COD is a very precise and reproducible measure of the carbonaceous substances in the system, and accounts for the electron exchange in redox reactions (Orhon and Artan, 1994), but unfortunately, it does not reflect the biological mechanisms through which the bacteria metabolize those substrates. To reproduce these reactions, the biochemical oxygen demand (BOD) must be used, consisting of measuring the oxygen uptake of a batch in which substrate and bacteria interact like in their natural environment (Sawyer et al., 2003). The simple BOD-measuring equipment setup is sketched in Figure 7.14, while Figure 7.15 shows a typical time plot obtained from a BOD test modelled by a substrate/biomass interaction model (Marsili-Libelli, 1986). In the beginning, there is an incubation phase, during which the biomass grows enough to produce a measurable oxygen consumption. Then, the rapidly biodegradable substances are consumed and, when they are fully metabolized, the biomass begins to hydrolyse the slowly degradable fraction (plateau) before being able to assimilate it.

The BOD bottle method reproduces the natural conditions and the metabolic capabilities of bacteria, and hence it yields exactly the amount of oxygen consumed by the bacteria in metabolizing the biodegradable fraction of the carbonaceous substrates contained in the bottle. It has, however, the drawbacks common to all biological methods: poor reproducibility and the length of the test, normalized to five days, because the rate at which the bacteria degrade the substrates may depend on many uncontrollable factors, and in any case it cannot be accelerated at will. On the other hand, many substances that the COD test can oxidize are not biodegradable. Hence, for a given sample of water or wastewater, the BOD will be always less, often much less, than COD. Figure 7.16 shows a typical partitioning of a municipal wastewater used in the ASM models (Henze et al., 2000). The entire lot can be measured as COD, but only the dissolved components and part of the particulate (X_s) can be measured by the BOD biological test. Several components can be isolated by specific analytical methods, shown in the rightmost column.

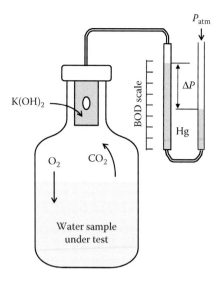

FIGURE 7.14   Setup scheme for the BOD test. The water sample contains both the substrate and the micro-organisms, whose metabolism consumes oxygen and releases carbon dioxide. This is captured by the reagents, usually $K(OH)_2$, so that the net gas pressure gap with respect to the atmospheric pressure (P_{atm}) can be measured by the mercury-filled U-tube. The scale is marked directly in BOD units (mg O_2/L).

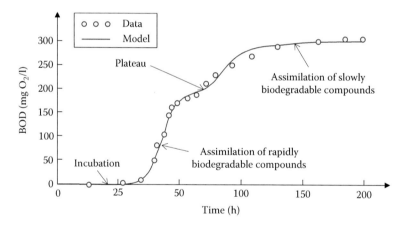

FIGURE 7.15   A sample BOD test. At the beginning, the biomass is too small to produce any appreciable oxygen consumption, but as it grows (incubation), it consumes first the rapidly biodegradable fraction of the waste. The plateau is likely due to the hydrolysis of slowly biodegradable compounds that are consumed last. The figure shows that a kinetic model can explain this observed behaviour. (Reproduced with permission from Marsili-Libelli, S., *Environ. Technol. Lett.*, 7, 341–350, 1986.)

7.2.4   THE CATABOLIC PATHWAY IN THE HETEROTROPHIC AEROBIC METABOLISM

Only half of the story, however, is told by the now familiar equations (7.24). In fact, they describe the *anabolic* part of the microbial metabolism, though we know from Section 7.2.2 that no cell anabolism is possible without an energy source, exploited by the catabolic metabolism. From Equations 7.24, we also notice that only the Y fraction of the substrate is utilized for growth, so if the substrate is totally biodegradable, the remaining $1-Y$ must follow the *catabolic* pathway. We will now consider this other pathway to set up the remaining part of the microbial metabolism considering, with reference to Figure 7.11, *aerobic respiration,* thus assuming that oxygen is the available electron acceptor.

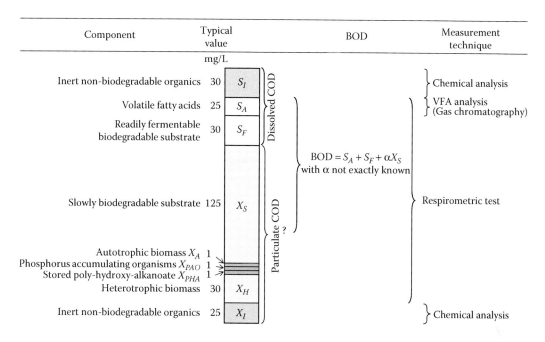

FIGURE 7.16 Components of a municipal wastewater adopted as a reference in the ASM models. The blue boxes indicate the particulate components. The whole lot can be expressed as COD, while the extent of the BOD equivalence includes the dissolved component and a variable part of the particulate components. The analytical method to measure each component is indicated on the right. (After Henze, M. et al., *Activated Sludge Models: ASM1, ASM2, ASM2d and ASM3*, IWA Scientific and Technical Report no. 9, London, 121pp, 2000.)

Now, we can specify Figure 7.8c better as Figure 7.17, in which the rate of substrate utilization (r_s) must balance the rate of biomass synthesis (r_X) and the rate of oxygen utilization (r_o)

$$r_s = r_X + r_o \quad \Rightarrow \quad \begin{cases} r_x = Y\dfrac{dS}{dt} = \dfrac{dX}{dt}\bigg|_s \\[3mm] r_o = (1-Y)\dfrac{dS}{dt} = -\dfrac{1-Y}{Y}\dfrac{dX}{dt} \end{cases} \tag{7.25}$$

In Equations 7.25, the subscript 's' refers to the synthetic (growth) part of the biomass dynamics in Equations 7.24, that is, neglecting the $b \cdot X$ term.

Now, regarding the endogenous respiration term, there are two hypotheses, shown in the two schemes of Figure 7.18, which imply differing oxygen utilizations. In the growth–decay model (a), the endogenous metabolism is considered as an oxygen-consuming process, while in the death–regeneration case (b), it does not consume oxygen, representing just the undoing of the cell, though part of the recycled material will be again metabolized, and hence will consume oxygen, as already explained. Therefore, the catabolic part of the biomass dynamics, with S_o indicating the dissolved oxygen (DO) concentration, can be written in the two cases as

$$\begin{cases} \text{Growth}-\text{decay} & \dfrac{dS_o}{dt} = -\dfrac{1-Y}{Y}\mu_{max}\dfrac{S}{K_s+S}X - b\cdot X \\[4mm] \text{Death}-\text{regeneration} & \dfrac{dS_o}{dt} = -\dfrac{1-Y}{Y}\mu_{max}\dfrac{S}{K_s+S}X \end{cases} \tag{7.26}$$

In the death–regeneration model, the oxygen consumption is related only to synthesis.

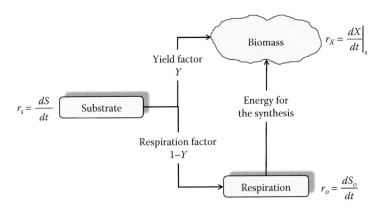

FIGURE 7.17 The substrate utilization rate must balance the synthetic rate r_X in the anabolic pathway and the oxygen utilization rate r_o in the catabolic pathway. S_o represents the oxygen concentration.

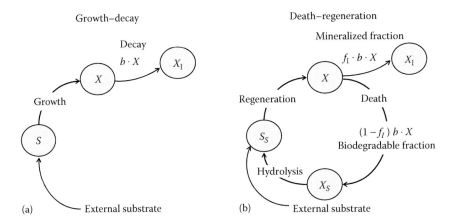

FIGURE 7.18 Two possible approaches to endogenous respiration modelling. The classical growth–decay approach (a) has given way to the death–regeneration mechanisms (b), adopted in the ASM models. X_S indicates particulate biodegradable material that can be used by the biomass (X) only after conversion into soluble substrate (S_S) through hydrolysis.

7.2.5 ENDOGENOUS METABOLISM

A further note on endogenous metabolism is necessary. As noted, it includes all the metabolic processes that do not result in biomass growth, but nevertheless require energy that the cell produces through the dissimilative pathway, which we have already examined. These processes may include the utilization of previously sequestered substrate, the use of the cell own protoplasm and eventually the release of cell material when the cell walls break down (lysis). There are alternative schools of thought regarding endogenous metabolism, whose multi-faceted nature is well described in van Loosdrecht and Henze (1999), while an interesting assessment of the evolution of activated sludge (the 'wastewater' version of microbial dynamics) modelling is provided by Gujer (2006). The evolution of endogenous metabolism modelling is illustrated in Figure 7.18, recalling the very wise suggestion (van Loosdrecht and Henze, 1999) that *a good-fitting model is not necessarily correct from a microbiological viewpoint*, and that not all the processes mentioned in the ASM models are microbiologically correct. Initially, the growth–decay model, Figure 7.18a, was used, in which a fixed portion of the biomass $\left(b \cdot X\right)$ decayed, and this process was considered to be oxygen consuming. Later, the death–regeneration process has been replaced by the 'death–regeneration' hypothesis, Figure 7.18b, in which a fraction f_I of the material released during the cell decay becomes inert, thus no longer utilizable, $f_I \cdot b \cdot X$, while the remaining part $(1 - f_I) \cdot b \cdot X$ is recycled as particulate biodegradable substrate (X_s), which can still be used by the active cell, provided it is previously hydrolysed into soluble substrate (S_S). While in the old scheme, the decay was considered as an oxygen-consuming process, in the death–regeneration model no oxygen consumption is involved, or rather it is shifted to the reuse of the recycled material in the regeneration phase. Incorporating the death–regeneration model into the basic substrate–biomass basic model (7.24) yields the new model

$$
\begin{cases}
\text{Substrate utilization} & \dfrac{dS}{dt} = -\dfrac{1}{Y} \mu_{\max} \dfrac{S}{K_s + S} X + \left(1 - f_I\right) \cdot b_{\mathrm{dr}} \cdot X \\[4mm]
\text{Biomass growth} & \dfrac{dX}{dt} = \mu_{\max} \dfrac{S}{K_s + S} X - b_{\mathrm{dr}} \cdot X
\end{cases}
\tag{7.27}
$$

where the second term in the substrate equation represents the regeneration term provided by the recycle of still biodegradable cell material. For simplicity, the intermediate hydrolysis stage $(X_S \rightarrow S_S)$ has been omitted, so it is assumed that the released material is already in soluble form. Jeppsson (1996) pointed out that, as a consequence of the death–regeneration hypothesis, the decay constant b_{dr} in Equations 7.27 is larger than the corresponding parameter in the growth–decay model (7.24), their relationship being

$$
b_{\mathrm{dr}} = \frac{b_{\mathrm{gd}}}{1 - Y\left(1 - f_I\right)}
\tag{7.28}
$$

where:

b_{dr} indicates the endogenous constant in the death–regeneration hypothesis
b_{gd} is the corresponding parameter in the traditional growth–decay model

The mineralized part of the substrate cannot be utilized by the biomass. Its asymptotic value can be found by setting the substrate derivative in Equations 7.27 to zero to yield

$$
\frac{dS}{dt} = -\frac{1}{Y} \mu_{\max} \frac{S}{K_s + S} X + \left(1 - f_I\right) b_{\mathrm{dr}} \cdot X = 0 \quad \Rightarrow \quad \bar{S} = \frac{K_s \left(1 - f_I\right) b_{\mathrm{dr}}}{\dfrac{\mu_{\max}}{Y} - \left(1 - f_I\right) b_{\mathrm{dr}}}
\tag{7.29}
$$

7.2.6 OTHER KINETICS OF CARBONACEOUS SUBSTANCES

The Monod kinetics not only represents the fundamental interaction model between carbonaceous substances and the heterotrophic biomass, but has also been extended to other contexts, such as the nitrogenous species and the autotrophic biomass, as described in Section 7.3. In special cases, this dynamics can be approximated with a first-order kinetics, as in Equation 7.15. This simplification is at the basis of the Streeter–Phelps model, already introduced in Section 1.10.5, describing the fate of BOD in water-quality models (Rinaldi et al., 1979; Chapra, 1997). In that model, the coupling between the two variables, BOD and DO, was made possible by the use of common units (mgO_2/L) for the carbonaceous substance and for the DO, resulting in a very simple model:

$$
\begin{aligned}
\text{BOD} \quad & \frac{dB}{dt} = -k_b \cdot B \\[3mm]
\text{DO} \quad & \frac{dS_o}{dt} = k_r \left(S_o^{\mathrm{sat}} - S_o\right) - k_b \cdot B
\end{aligned}
\tag{7.30}
$$

The consumption term $-k_b \cdot B$ is the same in both equations as a result of the common units of mgO_2/L. The first term in the DO equation is the re-aeration rate and will be considered next in Section 7.2.9.1.

7.2.6.1 Unbalanced Growth

Having considered the simplest form of carbonaceous substance dynamics, let us now turn to the other end of the modelling spectrum and explain the observed behaviour of delayed growth. In fact, in practice, it may happen that the disappearance of substrate from the medium is not immediately matched by a corresponding microbial growth. This is often referred to as *unbalanced growth*. The reason for this behaviour is in the storage of carbonaceous material inside the cell, to be later metabolized. The ASM3 model (Henze et al., 2000) introduces a new variable with respect to ASM1 to take into account this storing property of microorganisms. This new variable (X_{STO}) is an intermediate between the external soluble substrate and the cell material. It represents an internal reserve of organic storage polymers, mostly polyhydroxyalkonates (PHA), glycogen and other polysaccharides and lipids. This additional variable does not directly correspond to the chemical analysis of those molecules, but it was introduced as a component of X_H to describe the delayed growth (Gujer et al., 1999). Figure 7.19 shows the main differences between the ASM1 and ASM3 models regarding the heterotrophic biomass. The ASM1 is based on the death-regeneration assumption of Figure 7.18b, so that decay does not consume oxygen, and part of the decayed cell material is recycled as particulate substrate. Conversely, in the ASM3, endogenous respiration produces only mineralized particulate, but is oxygen consuming. In general, the ASM3 tends to decouple the autotrophic and heterotrophic metabolisms and reintroduces the concept of oxygen-consuming endogenous respiration. A full assessment of the two models and their differences can be found in Gujer et al. (1999) and Gernaey et al. (2004).

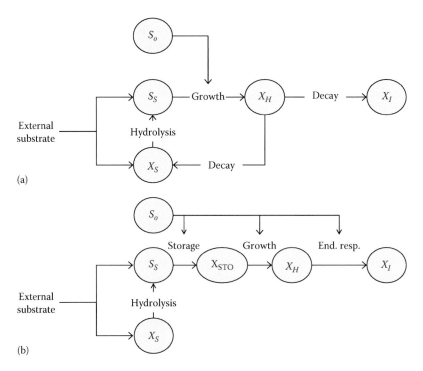

FIGURE 7.19 Main differences between (a) ASM1 and (b) ASM3 models regarding the heterotrophic biomass. The ASM1 adopts the death-regeneration assumption of Figure 7.18b, whereas in ASM3 endogenous respiration does not produce any recycled particulate matter, but is oxygen consuming.

7.2.7 THE MATRIX FORMAT OF MICROBIAL MODELS

Microbial models describing complex substrate/biomass interactions involve both kinetic processes and stoichiometry. A compact notation to describe both of these aspects is the matrix format, used in the ASM models (Henze et al., 2000). The matrix consists of as many rows as rate processes (ρ_j) and as many columns as components. The empty elements of the matrix should be interpreted as zeros, while the non-zero entries are the stoichiometric coefficients (v_{ji}) relating the ith component to the jth process. The total reaction rate (r_i) for the ith component is obtained by the weighted sum of the rate processes for the related coefficients, that is,

$$r_i = \sum_j v_{ji} \cdot \rho_j \qquad (7.31)$$

Apart from the ease of assembling the kinetics for each variable, the matrix format allows a quick continuity check, provided, as discussed in 7.2.3.1, that consistent units are used for all the variables. The sign convention is that the oxygen equivalent of COD subtracted from the variable is negative, while that added to the variable is positive. Immediate verification of the continuity of COD, electric charge, nitrogen and phosphorus is an additional benefit of the matrix format. The generic continuity equation encompassing all the j processes and the c materials is

$$0 = \sum_j v_{ji} \cdot i_{ci} \quad \forall i \qquad (7.32)$$

where the coefficient i_{ci} represents the fraction of material c in component i. As an example, $i_{N,SS}$ indicates that every gram of COD in the S_s component contains $i_{N,SS}$ grams of nitrogen. Figure 7.20 illustrates the correspondence between the matrix format and the dynamical model equations. For each variable (column), the stoichiometric coefficient corresponding to every process in which the variable is involved multiplies the pertinent rate (row). The weighted sum of the rates defines the dynamical equation for that variable. The solid and dashed paths in Figure 7.20 refer to the S_1 variable only. It is left to the reader to verify the second equation.

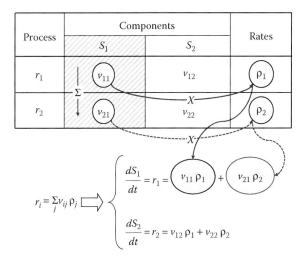

FIGURE 7.20 From the matrix format to the dynamical equations, illustrated with a hypothetical example. For each variable (column), each stoichiometric coefficient multiplies the pertinent rate (row). The weighted sum of the rates defines the dynamical equation for that variable. The solid and dashed paths refer to the S_1 variable only (hatched grey area).

TABLE 7.2

Matrix Format of a Simple Hydrolysis–Growth–Endogenous Metabolism Model

	Components				
Process	X_s	S_s	S_o	X_H	**Rates**
Hydrolysis	-1	1			$\rho_1 = \mu_h \dfrac{X_s/X_H}{K_X + X_s/X_H} X_H$
Growth		$-\dfrac{1}{Y_H}$	$-\dfrac{1-Y_H}{Y_H}$	1	$\rho_2 = \mu_{max} \dfrac{S_s}{K_s + S_s} \dfrac{S_o}{K_o + S_o} X_H$
Endogenous metabolism	$1-f_I$			-1	$\rho_3 = b_H X_H$

Moving a step further, the matrix format of a simple hydrolysis–growth–decay model is shown in Table 7.2, while the flow diagram is sketched in Figure 7.21. In this model, there are four variables (X_s, S_s, S_o, X_H) and three processes (ρ_1, ρ_2, ρ_3) representing the reaction rates of hydrolysis, growth and decay. The dynamical equation for each variable can be obtained by selecting the pertinent column, and multiplying the non-zero stoichiometric coefficients for the corresponding process, namely

$$\frac{dX_s}{dt} = -\mu_h \frac{X_s/X_H}{K_X + X_s/X_H} X_H + (1-f_I) b_H X_H$$

$$\frac{dS_s}{dt} = \mu_h \frac{X_s/X_H}{K_X + X_s/X_H} X_H - \frac{1}{Y_H} \mu_{max} \frac{S_s}{K_s + S_s} \frac{S_o}{K_o + S_o} X_H$$

$$\frac{dX_H}{dt} = \mu_{max} \frac{S_s}{K_s + S_s} \frac{S_o}{K_o + S_o} X_H - b_H X_H \qquad (7.33)$$

$$\frac{dS_o}{dt} = -\frac{1-Y_H}{Y_H} \mu_{max} \frac{S_s}{K_s + S_s} \frac{S_o}{K_o + S_o} X_H$$

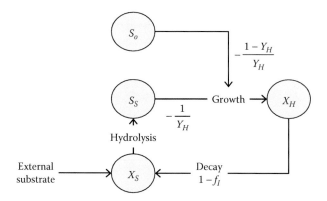

FIGURE 7.21   The flow diagram of the simple model expressed in matrix format in Table 7.2, producing the set of Equations 7.33. The stoichiometric coefficients are shown next to the pertinent link.

7.2.8 MATLAB Implementation of the Aerobic Heterotrophic Model

The full microbial metabolic model is obtained by combining the substrate utilization with the two microbial pathways, anabolic (growth) and catabolic (oxygen utilization). The complete batch model in the growth–decay hypothesis then becomes

$$
\begin{cases}
\text{Substrate utilization} & \dfrac{dS}{dt} = -\dfrac{1}{Y}\mu_{\max}\dfrac{S}{K_s+S}X \\[2mm]
\text{Biomass growth} & \dfrac{dX}{dt} = \mu_{\max}\dfrac{S}{K_s+S}X - b_{\mathrm{gd}}\cdot X \\
\text{(anabolism)} & \\[2mm]
\text{Oxygen utilization} & \dfrac{dS_o}{dt} = -\dfrac{1-Y}{Y}\mu_{\max}\dfrac{S}{K_s+S}X - b_{\mathrm{gd}}\cdot X \\
\text{(catabolism)} &
\end{cases}
\tag{7.34}
$$

while in the death–regeneration case, we have the additional regeneration term in the substrate equation and no endogenous respiration term in the oxygen equation

$$
\begin{cases}
\text{Substrate utilization} & \dfrac{dS}{dt} = -\dfrac{1}{Y}\mu_{\max}\dfrac{S}{K_s+S}X + \left(1-f_I\right)\cdot b_{\mathrm{dr}}\cdot X \\[2mm]
\text{Biomass growth} & \dfrac{dX}{dt} = \mu_{\max}\dfrac{S}{K_s+S}X - b_{\mathrm{dr}}\cdot X \\
\text{(anabolism)} & \\[2mm]
\text{Oxygen utilization} & \dfrac{dS_o}{dt} = -\dfrac{1-Y}{Y}\mu_{\max}\dfrac{S}{K_s+S}X \\
\text{(catabolism)} &
\end{cases}
\tag{7.35}
$$

The two models were implemented in the Simulink diagram of Figure 7.22, while Figure 7.23 shows the simulation of a batch in which the same amount of initial biomass and substrate interact until all the available substrate is consumed. While in the growth–decay assumption (a), all the substrate is assimilated, in the death–regeneration case (b) a residual inert fraction of the substrate remains, according to Equation 7.29. The substrate/biomass evolution is very similar in the two cases, with the only exception that in the death–regeneration model the inert fraction of the substrate is not metabolized. The oxygen utilization plots are the absolute value of the third equation of each model and represent the cumulative consumed oxygen throughout the batch. They correspond to the evolution of the BOD measurement obtained with the BOD bottle test of Figure 7.14, conventionally lasting five days, and resulting in the cumulative oxygen utilization similar to Figure 7.15, though the plateau related to slowly degradable material is now absent. The equivalence between the two models requires that the decay coefficients b_{gd} and b_{dr} be related by Equation 7.28.

7.2.9 The Oxygen Transfer Rate

Figure 7.24 shows the two basic mechanisms through which oxygen dissolves into the water, either through the free liquid surface or from gas bubbles. In both cases, the driving force of diffusion, according to Fick's law, is the concentration gradient, which in this case is represented by the oxygen deficit $\left(S_o^{\mathrm{sat}}-S_o\right)$. Thus, oxygen tends to penetrate into the liquid phase where it has a concentration $\left(S_o\right)$, which is lower than the saturated concentration $\left(S_i^{\mathrm{sat}}\right)$ it has in the gaseous phase. The efficiency of the oxygen transfer is determined by the oxygen transfer rate coefficient k_r, whose dimension is the reciprocal of time $\left[T^{-1}\right]$. The oxygen transfer rate (r_{rear}), which is controlled by several factors, is given by the product between the deficit and the transfer rate coefficient (Tchobanoglous

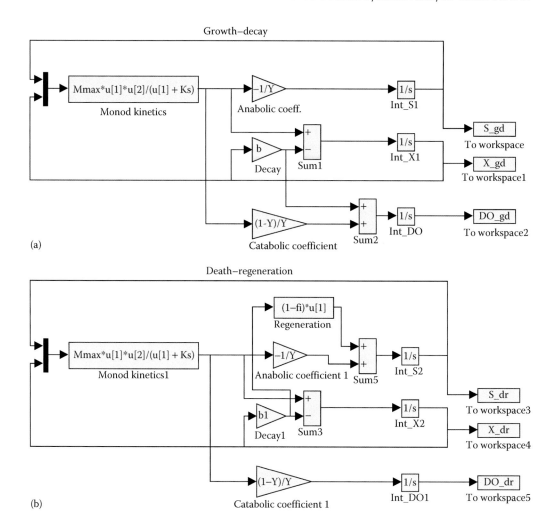

FIGURE 7.22 Simulink model to simulate the batch Monod kinetics with the two differing endogenous metabolism hypotheses: (a) growth–decay; (b) death–regeneration. This Simulink model can be retrieved as `Monod _ Batch _ GD _ DR.mdl` in the `ESA _ Matlab\Exercises\Chapter _ 7\Enzyme _ Kinetics` folder.

and Schroeder, 1985; Orhon and Artan, 1994; Chapra, 1997; Metcalf & Eddy Inc., 2003; Leslie Grady et al., 2011).

$$r_{rear} = k_r \left(S_o^{sat} - S_o \right) \qquad \left(mgO_2/L\ h \right) \tag{7.36}$$

In Figure 7.24, two differing notations are used for this coefficient, which is usually referred to as k_r in the case of natural surface re-aeration, while K_La is generally used in the wastewater literature to indicate forced aeration. The oxygen saturation concentration S_o^{sat} is proportional to the gas partial pressure P_{O_2} according to Henry's law of gas solubility,

$$S_o^{sat} = K_h^{O_2} \cdot P_{O_2} \tag{7.37}$$

where Henry's constant for oxygen at 20°C is $42.88 \left(mg\ L^{-1}\ atm^{-1} \right)$. By comparison, the solubility constant for nitrogen is only $19.012 \left(mg\ L^{-1}\ atm^{-1} \right)$, while it is much larger for CO_2 $1667.6 \left(mg\ L^{-1}\ atm^{-1} \right)$. Because the oxygen percentage in the air is 20.9%, the maximum concentration of oxygen that can be dissolved in the water by air blowing is therefore

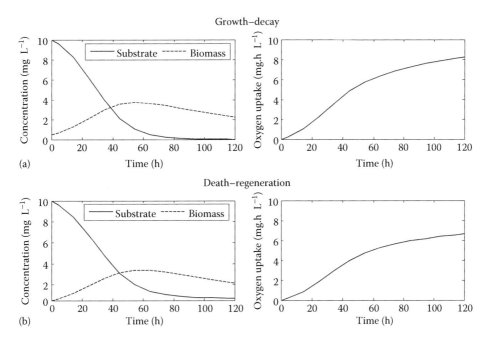

(a)

(b)

FIGURE 7.23 Simulation of the batch Monod model obtained with the Simulink diagram of Figure 7.22. In (a), the classic 'growth–decay' model implies that all the substrate is metabolized, while in (b), the 'death–regeneration' model results in a residual inert fraction of the substrate. These graphs were obtained with the MATLAB script `Go _ Monod _ Batch _ GD _ DR.m` in the `ESA _ Matlab\Exercises\Chapter _ 7\ Enzyme _ Kinetics` folder.

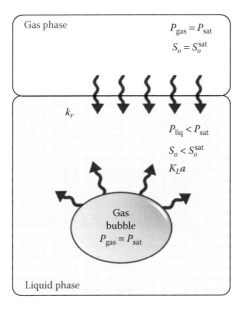

FIGURE 7.24 The two main oxygen diffusion mechanisms in the liquid phase. In natural systems, surface diffusion is the prevailing mechanism, while in wastewater plants, the air is injected through bottom diffusers. The wavy arrows indicate the direction of diffusion.

$$S_o^{\text{sat}}\Big|_{\text{air}} = 42.88 \times 0.209 = 8.9619 \text{ mg L}^{-1} \tag{7.38}$$

whereas if pure oxygen at the atmospheric pressure $\left(P_{O_2} = 1\right)$ is used, the saturation concentration increases almost fivefold to

$$S_o^{\text{sat}}\Big|_{O_2} = 42.88 \times 1.000 = 42.88 \text{ mg L}^{-1} \tag{7.39}$$

7.2.9.1   Oxygen Transfer in Natural Waters

In natural waters, diffusion occurs through the water surface, and several authors (Tchobanoglous and Schroeder, 1985; Chapra, 1997) have proposed to structure the k_r re-aeration coefficient in the second equation of the Streeter–Phelps model (7.30) as a function of the stream velocity, river depth and the state of the surface, that is,

$$k_r = \alpha \cdot u^\beta \cdot h^{-\gamma} \tag{7.40}$$

with the coefficients (α, β, γ) being positive quantities. Equation 7.40 indicates that the re-aeration increases with the water velocity and decreases with the river depth. In addition, k_r depends on the temperature through the Arrhenius law

$$k_r(T) = k_r(20) \cdot \theta^{(T-20)} \tag{7.41}$$

where

    T is the temperature (°C),
    $\theta = 1.0243$ is the Arrhenius constant.

A thorough survey of available re-aeration models and parameters can be found in Chapra (1997), who collected the findings of many researchers obtained in differing river morphologies. Regarding the influence of temperature and salinity, the following empirical relation holds for natural waters:

$$S_o^{\text{sat}}(T, Sal) = 14.6244 - 0.367134 \cdot T + 0.0044972 \cdot T^2$$
$$- 0.0966 \cdot Sal + 0.00005 \cdot T \cdot Sal + 0.0002739 \cdot Sal^2 \tag{7.42}$$

where

    T is the temperature (°C),
    Sal (‰) is the salinity.

7.2.9.2   Oxygen Transfer in Wastewater Treatment

In the wastewater literature, the oxygen transfer coefficient is usually indicated as $K_L a$. It expresses the *forced* oxygen transfer rate obtained by blowing air into the oxidation tank containing a mix of water and activated sludge (Metcalf & Eddy Inc., 2003; Rieger et al., 2006), while natural re-aeration is considered negligible compared to the aeration needs of the sludge, because the surface is often covered with bubbles or foam that hinders the natural atmospheric diffusion. Injecting gas from the bottom of the oxidation tank, which is the normal practice in wastewater engineering, requires that the pressure generated by the blower at the diffuser depth (P_{blw}) is greater than the hydraulic head, that is

$$P_{\text{blw}} = P_{\text{atm}}\left(1 + \frac{h}{10.44}\right) \tag{7.43}$$

where h (m) is the depth of the diffusers, as shown in Figure 7.25. This pressure increase is beneficial for oxygen solubility, which is proportionally increased according to Equation 7.37. As an example, if the diffusers are submersed at 4 m below the surface, the oxygen saturation becomes

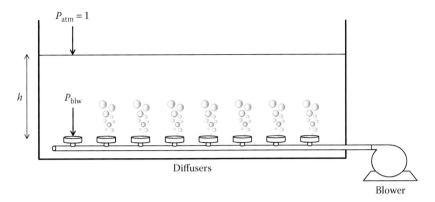

FIGURE 7.25 A submerged diffuser system powered by a blower is normally used to dissolve air into the water–sludge mixed liquor. To produce bubbles, the blower must provide a pressure (P_{blw}) greater than the one created by the hydraulic head (h).

$$S_o^{sat} = K_h^{O_2} \cdot P_{atm} \left(1 + \frac{h}{10.44}\right) = 42.88 \left(1 + \frac{4}{10.44}\right) = 20.36 \left(mg\ L^{-1}\right) \tag{7.44}$$

Conventionally, the oxygen saturation level of the entire tank is computed at half the depth of submergence; thus

$$\bar{S}_o^{sat} = K_h^{O_2} \cdot P_{atm} \left(1 + \frac{\dfrac{h}{2}}{10.44}\right) = 42.88 \times 0.209 \left(1 + \frac{2}{10.44}\right) = 10.68 \left(mg\ L^{-1}\right) \tag{7.45}$$

Thick sludge suspensions, like those encountered in wastewater oxidation tanks, hinder the diffusion of oxygen and the saturation value obtained by Equation 7.44 should be decreased by as much as 20%. Other factors determine the actual oxygen transfer efficiency, first of all the bubble diameter. Fine bubbles, with diameter less than 3 mm, have a high efficient surface/volume ratio and are generally preferred. The oxygen transfer coefficient $K_L a$ is a *global* parameter representing the rate of oxygen transfer from the gas phase to the liquid phase, and as such, it is often left unstructured. On the other hand, air flow rate is the primary manipulated variable in wastewater process control, and it would therefore be desirable to relate $K_L a$ to the air flow (U_a). Experimental data confirm that a linear relationship can be established between air flow and $K_L a$ by separating the natural re-aeration (K_o) from the forced oxygen transfer caused by gas flow,

$$K_L a = K_o + K_a \cdot U_a \tag{7.46}$$

Figure 7.26 shows that air flow data correlate well with the observed $K_L a$ values.

7.2.9.3 Complete DO Dynamics in Microbial Systems

Going back to the microbial metabolism of carbonaceous substrates, we can now complement the oxygen utilization equations (7.26) with an oxygen supply term to obtain the full DO dynamics. In fact, Equations 7.26 describe only the oxygen depletion due to the uptake by the biomass, while the complete DO dynamics is obtained by adding an oxygen supply term to balance the consumption. This term was already defined in Equation 7.36, and in the case of wastewater treatment models, the transfer rate coefficient can be further structured as in Equation 7.46. Thus, the oxygen supply term takes the form $K_L a \left(S_o^{sat} - S_o\right)$ and the full DO dynamics becomes

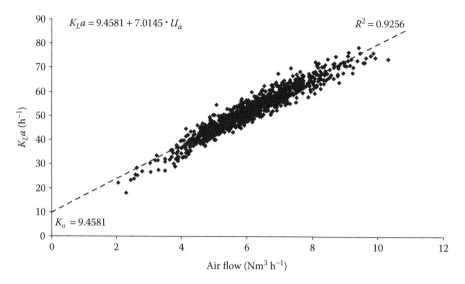

FIGURE 7.26 Correlation between air flow (U_a) and K_La obtained from a set of experimental data. (Courtesy of Nokia OY.)

$$
\begin{cases}
\text{Growth–decay} & \dfrac{dS_o}{dt} = K_La\left(S_o^{\text{sat}} - S_o\right) - \dfrac{1-Y}{Y}\mu_{\max}\dfrac{S}{K_s+S}X - b_{\text{gd}}\cdot X \\[3mm]
\text{Death–regeneration} & \dfrac{dS_o}{dt} = K_La\left(S_o^{\text{sat}} - S_o\right) - \dfrac{1-Y}{Y}\mu_{\max}\dfrac{S}{K_s+S}X
\end{cases}
\tag{7.47}
$$

depending on which metabolic model is adopted. In the death–regeneration case, it is assumed that decay is not an oxygen-consuming process.

7.2.10 A Naïve Wastewater Treatment Plant Model

A very simple, almost naïve, model can now be set up to illustrate the previous interactions between the carbonaceous substrate and the heterotrophic biomass. Because a continuous-flow system is considered, the batch equations must be complemented with the flow terms to obtain the process scheme of Figure 7.27.

The process model can be obtained by complementing the previous batch kinetics, based on the death–regeneration case of Equations 7.35, with the flow terms to yield

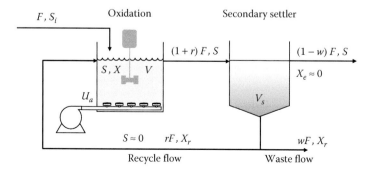

FIGURE 7.27 An extremely simplified wastewater treatment plant to demonstrate the previous substrate–biomass interactions.

$$\begin{cases} \dfrac{dS}{dt} = -\dfrac{1}{Y}\mu_{max}\dfrac{S}{K_s+S}X - q(1+r)S + (1-f_I)b_{dr}X + q \cdot S_i \\[3mm] \dfrac{dX}{dt} = \mu_{max}\dfrac{S}{K_s+S}X - b_{dr}X - q(1+r)X + qrX_r \qquad\qquad q = \dfrac{F}{V} \\[3mm] \dfrac{dS_o}{dt} = K_a \cdot U_a\left(S_o^{sat}-S_o\right) - \dfrac{1-Y}{Y}\mu_{max}\dfrac{S}{K_s+S}X - q(1+r)S_o \end{cases} \tag{7.48}$$

where the following assumptions have been made:

- The influent flow F carries an organic load with concentration S_i, but no biomass $\left(X_i \approx 0\right)$ nor oxygen $\left(S_{o_i} \approx 0\right)$.
- The recycle biomass (sludge) concentration from the settler underflow is X_r, to be defined later.
- The substrate concentration in the recycle flow is negligible $\left(S_r \approx 0\right)$.
- No solids (sludge) are discharged in the effluent $\left(X_e \approx 0\right)$.
- The recycle flow is expressed as a fraction r of the process flow $\left(F_r = r \cdot F\right)$.

The secondary settler is a very complex plant component, and a widely used model based on the hindered flux theory (Takács et al., 1991) was used in the standard benchmark model (Copp, 2002; Jeppsson et al., 2007). Simpler models based on the same theory can also be used in reduced-order models (Marsili-Libelli, 1993). In the present, extremely simplified, model, the settler can be assumed to be at steady-state, and with reference to Figure 7.28, the underflow concentration $\left(X_r\right)$ can be obtained from the mass balance

$$F(1+r)X = F(1-w)X_e + F(w+r)X_r \tag{7.49}$$

which, assuming that $X_e \approx 0$, yields

$$X_r = \frac{1+r}{w+r}X \tag{7.50}$$

Thus, the recycle concentration X_r is a function of the recycle and waste fraction of the process flow, and because $r < 1$ and w is usually very small $\left(w < 0.1\right)$, the recycle concentration X_r is usually greater than X. In fact, the primary function of the secondary settler is to separate the treated water from the sludge and to concentrate it in the underflow. Substituting Equation 7.50 into the biomass equation in Equations 7.48 yields

$$\begin{aligned} \frac{dX}{dt} &= \mu_{max}\frac{S}{K_s+S}X - b_{dr}\cdot X - q(1+r)X + rq\frac{1+r}{r+w}X \\[2mm] &= \mu_{max}\frac{S}{K_s+S}X - b_{dr}\cdot X - qw\frac{1+r}{r+w}X \end{aligned} \tag{7.51}$$

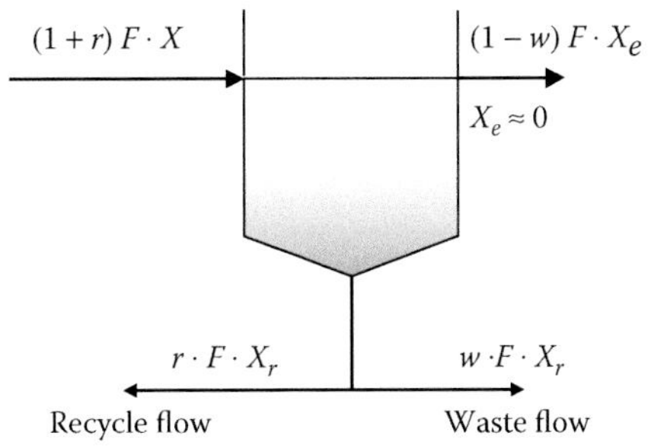

FIGURE 7.28 A steady-state mass balance around the secondary settler.

In this way, the internal variable X_r is eliminated, and the model becomes

$$
\begin{cases}
\dfrac{dS}{dt} = -\dfrac{1}{Y}\mu_{\max}\dfrac{S}{K_s+S}X - q(1+r)S + (1-f_I)b_{dr}X + qS_i \\[2ex]
\dfrac{dX}{dt} = \mu_{\max}\dfrac{S}{K_s+S}X - b_{dr}\cdot X - qw\dfrac{1+r}{r+w}X \\[2ex]
\dfrac{dS_o}{dt} = K_a\cdot U_a\left(S_o^{\text{sat}}-S_o\right) - \dfrac{1-Y}{Y}\mu_{\max}\dfrac{S}{K_s+S}X - q(1+r)S_o
\end{cases}
\tag{7.52}
$$

As an exercise, let us now compute the steady state of the model (7.52) by setting the derivatives of the first two equations to zero. The steady-state oxygen balance will be computed later, as this does not affect the steady state of the first two variables. Solving the steady-state equations

$$
0 = -\frac{1}{Y}\mu_{\max}\frac{S}{K_s+S}X - q(1+r)S + (1-f_I)b_{dr}X + qS_i
$$

$$
0 = \mu_{\max}\frac{S}{K_s+S}X - b_{dr}\cdot X - qw\frac{1+r}{r+w}X
\tag{7.53}
$$

for S and X yields

$$
\bar{S} = \frac{K_s(b_{dr}+a)}{\mu_{\max}-(b_{dr}+a)} \quad \text{with} \quad a = qw\frac{1+r}{r+w}
$$

$$
\bar{X} = q\frac{S_i-(1+r)\bar{S}}{\dfrac{1}{Y}(b_{dr}+a)-(1-f_I)b_{dr}}
\tag{7.54}
$$

Figure 7.29 shows the steady-state profiles of substrate and biomass as a function of the two control variables: the dilution rate q, depending on the process flow, and the waste ratio w, controlling the sludge age, or mean cell residence time (MCRT), defined as the total amount of sludge in the system $\left(X_{ox}\cdot V_{ox} + X_s\cdot V_s\right)$ divided by the amount of sludge wasted per day, that is,

$$
\text{MCRT} = \frac{X\cdot V + X_s\cdot V_s}{w\cdot X_r\cdot F}
\tag{7.55}
$$

where X_s is the sludge concentration in the settler, occupying a V_s volume.

Substituting these values into the DO equation and assuming a desired oxygen level $\bar{S}_o$, the steady-state aeration requirement is obtained as

$$
\bar{U}_a = \frac{\dfrac{1-Y}{Y}\mu_{\max}\dfrac{\bar{S}}{K_s+\bar{S}}\bar{X} + b_{gd}\cdot\bar{X} + q(1+r)\bar{S}_o}{K_a\left(S_o^{\text{sat}}-\bar{S}_o\right)}
\tag{7.56}
$$

Before we actually set up the Simulink model for Equations 7.52, a fourth equation could be added to introduce a very simple dynamics of the secondary settler, in addition to the steady-state balance of Equation 7.50. Suppose that the settler acts like a buffer with dilution rate $q_s = F/V_s$, where V_s is

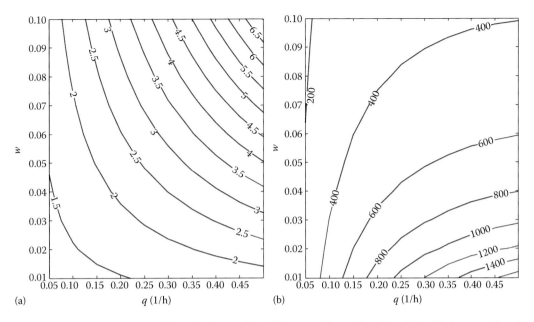

FIGURE 7.29 Steady-state profiles of substrate (a) and biomass (b) as a function of the dilution rate (q) and the waste fraction (w), computed with Equations 7.54.

the equivalent settler volume, already defined in Equation 7.55. A simple mass balance around this component yields

$$\frac{dX_r}{dt} = q_s(1+r)X - q_s(r+w)X_r \tag{7.57}$$

which is consistent with the steady-state mass balance equation (7.50). Adding this fourth equation to the previous model equations (7.52) yields the complete simple wastewater treatment plant (WWTP) model, whose Simulink implementation is shown in Figure 7.30.

This model produced the results of Figure 7.31. The simulation horizon is 720 h long, corresponding to 30 days of operation. During this period, the upper figure shows that the input substrate fluctuates with a fairly regular daily cycle, but at hours 221 and 558, a larger input peak occurs, producing a brief overload. The biomass increases almost immediately and returns to the normal level after a few days, being followed by the underflow biomass with a similar but slower trend $\left(X_r\right)$. The increased biological activity induces a larger oxygen utilization, and because the air flow is constant, the DO decreases to almost zero. In practice, a low DO level should be avoided because it adversely affects the biomass dynamics. The DO dependence of microbial dynamics is accounted for in the complex ASM models (Henze et al., 2000), to which the reader is referred for more details on wastewater treatment models.

7.2.11 ANAEROBIC HETEROTROPHIC METABOLISM

Let us now look up again at Figure 7.11 and concentrate on the leftmost instance, the case in which no external electron acceptor is available. Definitely, this is the most disadvantaged situation for the microbial metabolism, as the respiratory chain cannot use any external electron acceptor. This results in slow growth and limited assimilation, as shown in Figure 7.6b. The anaerobic

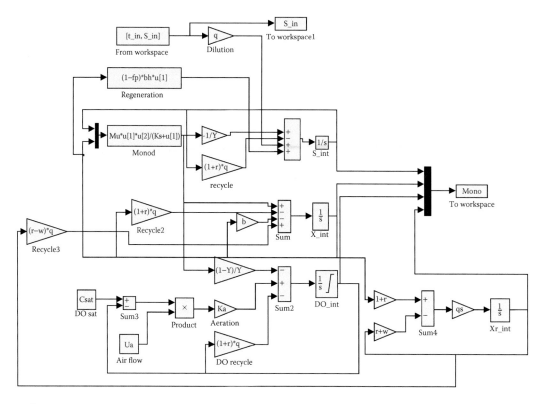

FIGURE 7.30 Simulink model implementing the WWTP equations (7.52). The Simulink model `ASTP_simple_DO_dr.mdl` can be retrieved from the `ESA_Matlab\Exercises\Chapter_7\WWTP` folder.

transformation of carbonaceous substrates is a very complex system of reactions and involves several coexisting microbial communities (Rozzi et al., 1985) with specific kinetics (Pavlostatis and Giraldo-Gomez, 1991). Figure 7.32 shows a very simplified scheme of reactions globally referred to as anaerobic digestion. There are three main groups of microorganisms: the acetoclastic bacteria, which transform the various organic molecules into short-chain volatile fatty acids (VFA), mostly acetic acid (CH_3COOH) but also propionic (CH_3CH_2COOH), butyric ($CH_3(CH_2)_2COOH$), valeric ($CH_3(CH_2)_3COOH$) in minor proportions. As indicated in the scheme of Figure 7.32, the volatile fatty acids follow differing paths: acetic acid on one side and the heavier VFAs (propionic, butyric, etc.) on the other. While acetic acid is directly metabolized by acetoclastic methanogens producing a gas mixture of methane (CH_4) and carbon dioxide (CO_2), the heavier VFAs are mediated by obligate hydrogen producers, which are inhibited by the hydrogen they produce and are dependent on its removal by methanogens, which are in turn inhibited by an excess of VFAs.

Several models have been developed to describe the anaerobic digestion process (Rozzi, 1984; Marsili-Libelli and Beni, 1996) before the anaerobic digestion model n. 1 (ADM1) was proposed (Batstone et al., 2002a, 2002b) and became the definitive reference model for this type of process, to be widely used in the applications (Shang et al., 2005; Batstone et al., 2006).

A simplified model (Marsili-Libelli and Beni, 1996) is briefly summarized here. It was developed to study the digester behaviour in organic shock load conditions, already analysed in Section 4.5.5 from the pattern recognition viewpoint. In this context, the goal of the model was the prediction of the bicarbonate alkalinity required to counterbalance the pH decrease caused by an increased level of VFAs. The model assumes that the substrate is available in dissolved form and is partly

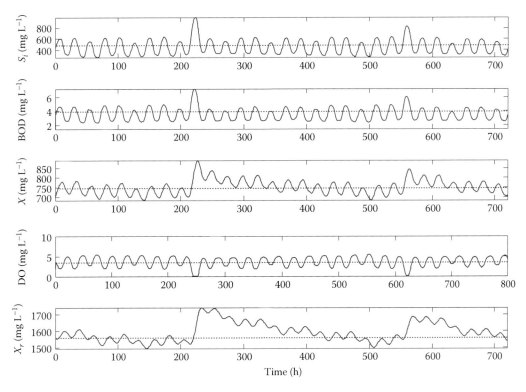

FIGURE 7.31 Simulation result of the simple WWTP model equations (7.52). The dotted lines represent the average values. This simulation was obtained with the MATLAB script Go_WWTP_DO_dr.m in the `ESA _ Matlab\Exercises\Chapter _ 7\WWTP` folder.

assimilated by the acetoclastic biomass with $y_{s,a}$ yield factor, and is partly converted into VFAs through Y_{vfa}, whereas the remaining fraction yields CO_2. Likewise, acetic acid yields both carbon dioxide and methane through $y_{co,m}$ and y_{ch}, respectively, and only part of it is taken up by methanogens through $y_{s,m}$. The interactions at the basis of the model, together with the related yield factors, are sketched in Figure 7.33.

The entire model cannot be described in detail here and the reader is referred to the original publication (Marsili-Libelli and Beni, 1996) for a full description. However, two interesting aspects of the model are worth considering in the context of microbial kinetics:

- A Monod kinetics was used for the substrate conversion into acetic acid, with an additional multiplying term depending on the input substrate concentration (S_{in}), as suggested by Pavlostatis and Giraldo-Gomez (1991).
- The methanogenic bacteria are very pH-sensitive, being inhibited by an excessive amount of the un-ionized fraction of the acetic acid, which plays the double role of substrate/inhibitor. Thus, they are modelled with a Haldane kinetics.

The difficulty in controlling an anaerobic digestion process rests primarily in the pH sensitivity of methanogenic bacteria. If the digester is overloaded with the organic substrate, the acetogenic bacteria produce a large quantity of VFA, which inhibits the methanogens and creates a VFA build-up. This leads to the acidification of the digester, and eventually to the destruction of the biomass. Therefore, pre-emptive remedial strategies have been proposed to correct the pH before an irreversible damage is created (Rozzi, 1984; Hawkes et al., 1992; Müller et al., 1997). The batch model equations of the digester are then

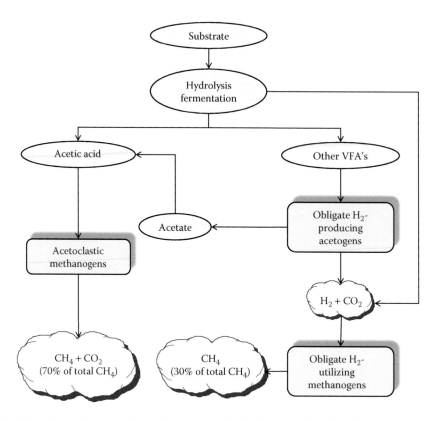

FIGURE 7.32 Main pathways and microbial communities in the anaerobic digestion process. (Reproduced with permission from Marsili-Libelli, S. and Beni, S., *Ecol. Model.*, 84, 215–232, 1996.)

Organic substrate

$$\frac{dS}{dt} = -\left(\frac{1}{y_{s,a}} + y_{vfa} + y_{co,a}\right)\frac{\mu_{a,\max}}{K_{s,a} + S} \cdot \frac{S_{in} + K_{s,a}}{S_{in}} \cdot X_a$$

Acetogenic bacteria

$$\frac{dX_a}{dt} = \frac{\mu_{a,\max}}{K_{s,a} + S} \cdot \frac{S_{in} + K_{s,a}}{S_{in}} \cdot X_a - k_{d,a} \cdot X_a$$

Acetic acid

$$\frac{dV_a}{dt} = \frac{\mu_{a,\max}}{K_{s,a} + S} \cdot \frac{S_{in} + K_{s,a}}{S_{in}} \cdot X_a - k_{d,a} \cdot X_a$$

(7.58)

$$-\left(\frac{1}{y_{s,m}} + y_{co,m} + y_{ch}\right)\frac{\mu_{m,\max}}{1 + \dfrac{K_{s,m}}{HAc} + \dfrac{HAc}{K_{i,m}}}$$

Methanogenic bacteria

$$\frac{dX_m}{dt} = \frac{\mu_{m,\max}}{1 + \dfrac{K_{s,m}}{HAc} + \dfrac{HAc}{K_{i,m}}} - k_{d,m} \cdot X_m$$

where $\left(\mu_{a,\max}, K_{s,a}, k_{d,a}\right)$ are the Monod parameters of the acetogenic bacteria, $\left(\mu_{m,\max}, K_{s,m}, k_{d,m}\right)$ those of the methanogenic bacteria, and the various yield factors are illustrated in Figure 7.33. *HAc* is the un-ionized fraction of acetic acid, which can be computed from the chemical equilibrium of acetic acid $\left(S_a\right)$,

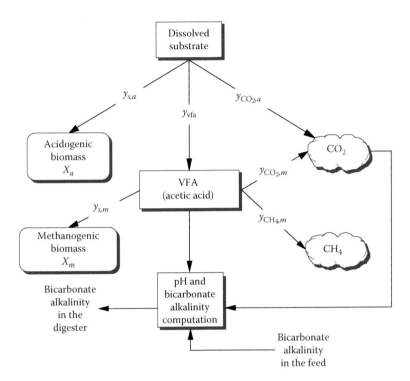

FIGURE 7.33 Interactions that form the basis of the simplified model, showing the yield factors controlling the conversion from one stage to another. (Reproduced with permission from Marsili-Libelli, S. and Beni, S., *Ecol. Model.*, 84, 215–232, 1996.)

$$HAc = \frac{S_a}{1 + \dfrac{K_a}{H^+}} \tag{7.59}$$

with dissociation constant K_a. Equations 7.58 must then be complemented with the dynamics of carbon dioxide (CO_2) and methane (CH_4) gas production and the electrochemical equilibrium required to determine the H^+ concentration used in Equation 7.59. The full description of the pH model can be found in Marsili-Libelli and Beni (1996).

7.3 MICROBIAL METABOLISM OF NITROGENOUS SUBSTRATES

Nitrogen is present in each and every compartment of the environment, given its property to cycle globally in various forms. Figure 7.34 provides a sketch of the nitrogen cycle in the environment, where it cycles between its organic form (N_{org}) in living organisms and their litter, and its inorganic form, which is either utilized as NH_4^+ and NO_3^- by the primary producers, or eventually returned to the atmosphere as dinitrogen (N_2).

The anthropogenic nitrogen cycle (Gijzen and Mulder, 2001) illustrated in Figure 7.35 shows a worrying imbalance between the large use of nitrogen-based fertilizer and inefficient food production methods.

The consequence of this imbalance is a vast release of unused nitrogen, which contaminates both groundwater and surface waterbodies. A further difference between the natural and anthropogenic nitrogen cycles is shown in Figure 7.36, where the role of the decomposers has been largely replaced by the artificial—and expensive—wastewater treatment processes.

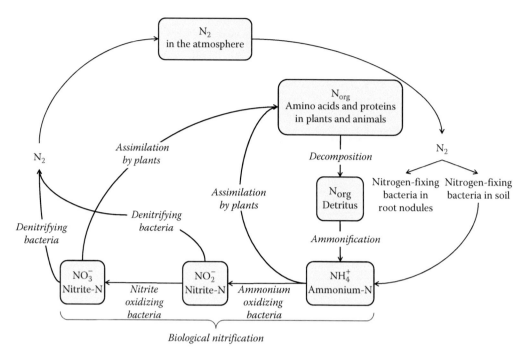

FIGURE 7.34 Nitrogen is present in all the compartments of the environment, given its ability to cycle globally. Nitrogen in the atmosphere cannot be directly utilized, but specific nitrogen-fixing bacteria in the soil convert it to ammonium-N, which is also produced by the decomposition of waste organic-N. The biological nitrification chain provides inorganic nitrogen in the forms that the plants can utilize (NH_4^+ and NO_3^-), while the denitrifying bacteria produce dinitrogen (N_2) to be released into the atmosphere, thus closing the cycle.

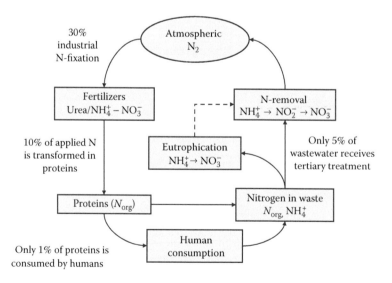

FIGURE 7.35 The anthropogenic nitrogen cycle (Gijzen, 2001; Gijzen and Mulder, 2001) shows worrying imbalances between the high amount of fertilized being applied to agricultural soil, and the inefficiency of food production.

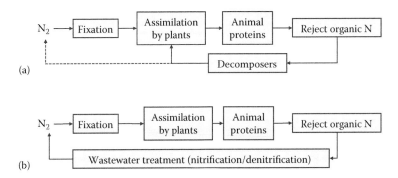

FIGURE 7.36 Differing nitrogen cycles: (a) in the natural N cycle the reduction to dinitrogen is minimized by the decomposers shortcut. The anthropogenic N cycle (b), instead, goes to great length (and costs) to close the loop to dinitrogen.

The forms of nitrogen present in the various environmental compartments are

- Dinitrogen (N_2) in the atmosphere.
- Reduced inorganic nitrogen, such as ammonia and ammonium-N $\left(NH_3, NH_4^+\right)$, in the pedosphere and hydrosphere.
- Oxidized inorganic nitrogen, as nitrite and nitrate $\left(NO_2^-, NO_3^-\right)$, in the pedosphere and hydrosphere.
- Organic nitrogen (N_{org}), in the biosphere (food web), pedosphere and hydrosphere.

The main transformations in the nitrogen cycle are

- Biological nitrogen fixation by bacteria in the pedosphere and hydrosphere.
- Assimilation and biosynthesis in the biosphere.
- Decomposition by bacteria and fungi in the pedosphere and hydrosphere.
- Hydrolysis of particulate biodegradable organic nitrogen.
- Ammonification, converting organic nitrogen into ammonia.
- Biological ammonium-N oxidation by autotrophic bacteria in the pedosphere and hydrosphere.
- Biological denitrification by heterotrophic bacteria in the pedosphere and hydrosphere.

The relations among these processes were already outlined in Figure 7.34 and are now be briefly reviewed.

Nitrogen fixation is a process in which dinitrogen is converted into ammonium-N according to the enzymatic reaction

$$N_2 + 8H^+ + 6e^- \xrightarrow{\text{nitrogenase}} 2NH_4^+ \tag{7.60}$$

This process is performed in both the soil and the water by nitrogen-fixing bacteria. The common factor in both cases is a specific enzyme (*nitrogenase*). In the soil, the genus *Rhizobium* performs fixation by forming root nodules in symbiosis with legumes. This can fix from 50 to 200 kg N ha^{-1} yr^{-1}. Non-symbiotic bacteria, like *Azotobacter* and *Clostridium*, can fix between 5 and 20 kg N ha^{-1} yr^{-1}. In the water, blue-green algae (*Oscillatoris princeps* and *Gloeocapsa sp.*) can fix between 10 and 50 kg N ha^{-1} yr^{-1}.

Ammonification is part of the biological degradation of organic nitrogen compounds, like proteins, that act as carbon source for heterotrophic bacteria, which uptake some of the nitrogen for their growth and release the excess as ammonium-N. The two most important ammonification processes are *deamination*, an enzymatic reaction consisting of the removal of an amine group, which is then converted to ammonium-N, and the *hydrolysis of urea*, catalysed by a specific enzyme (urease),

$$\left(NH_2\right)_2 CO + H_2O \xrightarrow{\text{urease}} CO_2 + 2NH_3 + \text{energy} \tag{7.61}$$

Both processes are energy producing, and the bacteria use this energy for their growth.

Nitrification is the biological process in which specialized autotrophic bacteria oxidize the ammonium-N, first into nitrite by ammonium oxidizers, and then into nitrate, by nitrite oxidizers, in the presence of oxygen. The biological oxidation of ammonium-N is also an energy-producing process, which provides the autotrophs with the energy required to assimilate CO_2, their carbon source. The nitrification consists of the following two steps:

$$NH_4^+ + \frac{3}{2}O_2 \xrightarrow[\text{oxidizers}]{\text{ammonium-N}} NO_2^- + 2H^+ + H_2O + \text{energy}$$

$$\tag{7.62}$$

$$NO_2^- + \frac{1}{2}O_2 \xrightarrow[\text{oxidizers}]{\text{nitrite}} NO_3^- + \text{energy}$$

The first nitrification step lowers the pH, because for each NH_4^+ oxidized two H^+ are generated. Hence, to maintain a constant pH, two moles of bicarbonate should be added for each mole of oxidized ammonium-N. However, because nitrification requires oxygen, and an air flow is usually present, at least in artificial nitrification, the CO_2 stripping due to the air flow may partly offset the acidifying effect of nitrification. The amount of oxygen required for the complete nitrification of one mole of ammonium-N can be computed by inspection of the stoichiometry of Equations 7.62. Because 4 g of oxygen is required to oxidize 1 g of ammonium-N, the amount of oxygen required in the two nitrification steps is

$$\text{First step} \qquad \frac{1.5 O_2}{1 N} = \frac{1.5 \times 32}{14} = 3.43 \quad \frac{g O_2}{g N}$$

$$\tag{7.63}$$

$$\text{Second step} \qquad \frac{0.5 O_2}{1 N} = \frac{0.5 \times 32}{14} = 1.14 \quad \frac{g O_2}{g N}$$

The combined oxygen requirement for the oxidation of 1 g of ammonium-N is thus $3.43 + 1.14 = 4.57$ g of oxygen. We shall see later that the actual amount of consumed oxygen is less than the stoichiometric equivalent of Equations 7.63, because part of the nitrogen is assimilated by the autotrophs, rather than oxidized.

Denitrification is the reduction of the oxidized forms of nitrogen $\left(NO_2^- \text{ and } NO_3^-\right)$ to dinitrogen (N_2) by means of heterotrophic facultative bacteria, *facultative* meaning that they can use nitrite/nitrate as alternate electron acceptors when oxygen is not available. Being heterotrophs, they require organic carbon both as a carbon source and an electron donor. The reduction chain is

$$NO_3^- \rightarrow NO_2^- \rightarrow NO \rightarrow N_2O \rightarrow N_2 \tag{7.64}$$

and the stoichiometry of denitrification from nitrate or nitrite is

$$\frac{1}{5}NO_3^- + \frac{6}{5}H^+ + \underset{\text{from } C_{\text{org}}}{e^-} \xrightarrow{X_H} \frac{1}{10}N_2 + \frac{3}{5}H_2O + \text{energy} \tag{7.65}$$

$$\frac{1}{3}NO_2^- + \frac{4}{3}H^+ + \underset{\text{from } C_{\text{org}}}{e^-} \xrightarrow{X_H} \frac{1}{6}N_2 + \frac{2}{3}H_2O + \text{energy} \tag{7.66}$$

Recall the definition of the yield coefficient, where Y_D is the substrate fraction utilized in the denitrification anabolic pathway and $(1 - Y_D)$ is the remaining part utilized in the catabolic pathway. From Equation 7.65, for one-fifth of nitrogen reduced one mole of electron is required, whereas in the case of nitrite, from Equation 7.66, for one-third of nitrogen reduced one mole of electron is required. Because one mole of electron is equivalent to 8 g of oxygen (Orhon and Artan, 1994), the removal of 1 g of COD will provide the reduction of

$$(1-Y_D)\frac{14}{5}\frac{1}{8} = \frac{1-Y_D}{2.857} \frac{g\left(NO_3^- - N\right)_{reduced}}{g\left(COD\right)_{removed}} \tag{7.67}$$

grams of nitrate-N. Likewise, in the reduction of nitrite, one-third of nitrogen is required, so that the corresponding grams of nitrite reduced for each gram of COD is

$$(1-Y_D)\frac{14}{3}\frac{1}{8} = \frac{1-Y_D}{1.714} \frac{g\left(NO_2^- - N\right)_{reduced}}{g\left(COD\right)_{removed}} \tag{7.68}$$

It should be recalled that even though the heterotrophic bacteria performing denitrification are the same that degrade COD by oxidation, if oxygen is available, the yield coefficients may be different in the two cases, thus $Y \neq Y_D$. As a consequence of denitrification, the total amount of carbonaceous substrate (S_S) removed by anoxic respiration on either nitrite or nitrate is, including also the oxygen uptake for aerobic respiration,

$$S_s = \frac{2.857}{1-Y_D}S_{NO_3} + \frac{1.714}{1-Y_D}S_{NO_2} + \frac{1}{1-Y}S_o \tag{7.69}$$

where the last term represents the amount of substrate removed by aerobic respiration if DO is available, for which the yield coefficient is Y. When oxygen is not available, the facultative micro-organisms switch to anoxic respiration and metabolize the organic substrate in the proportions indicated by the first two terms.

Though the natural nitrification/denitrification chain normally includes all of the steps just examined, in man-made nitrogen removal processes, it is sometime profitable to limit the extent of the reaction to nitrite, avoiding the extension to nitrate (Giusti et al., 2011c). The so-called nitrite shortcut reaction scheme is illustrated in Figure 7.37, which shows the possible savings in oxygen and organic carbon:

NH_4^+ oxidation to NO_3^- $NH_4^+ + 2O_2 \rightarrow NO_3^- + H_2O + 2H^+$
NH_4^+ oxidation to NO_2^- $NH_4^+ + 1.5O_2 \rightarrow NO_2^- + H_2O + 2H^+$

$$\Rightarrow 25\% \ O_2 \ saving \tag{7.70}$$

NO_3^- denitrification $6NO_3^- + 5CH_2OH + CO_2 \rightarrow 3N_2 + 6HCO_3^- + 7H_2O$
NO_2^- denitrification $6NO_2^- + 3CH_2OH + 3CO_2 \rightarrow 3N_2 + 6HCO_3^- + 3H_2O$

$$\Rightarrow 40\% \ C_{org} \ saving$$

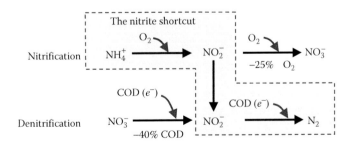

FIGURE 7.37 The nitrite shortcut (blue area) results in a considerable saving in oxygen and organic carbon. (Reproduced with permission from Giusti, E. et al., *Environ. Modell. Softw.*, 26, 938–949, 2011c.)

7.3.1 Nitrification Modelling

As we have seen, the biological nitrification process includes two successive oxidations performed by specialized autotrophic bacteria. A simplified one-step kinetics is often sufficient to describe the nitrification process in natural environments, whereas more complex two-step models are required when a detailed description of the nitrification process is needed (Orhon and Artan, 1994; Leslie Grady et al., 2011). Even in the context of complex models, the ASM family (Henze et al., 2000) considers nitrification as a single-step reaction, thus neglecting the intermediate nitrite-N. On the other hand, in many environmental studies and in the wastewater engineering practice, monitoring this intermediate component is mandatory, because of its adverse effects on many living organisms, and for human health in particular. Further, as pointed out in Equations 7.70, nitrite can become the pivotal component in specialized nitritation processes, based on the nitrite shortcut of Figure 7.37, with obvious savings in oxygen and organic carbon. Specialized nitrogen removal processes, such as SHARON, which is an acronym for Single reactor High activity Ammonia Removal Over Nitrite (Hellinga et al., 1998, 1999; Mulder et al., 2001), or the sequencing batch reactors (SBRs) treating special waste (Spagni et al., 2008; Spagni and Marsili-Libelli, 2009) are technological implementations of the nitrite shortcut, to name but a few. Several models of increasing complexity will now be presented that can be used in differing contexts, from simplified river water-quality models, to detailed wastewater process models.

7.3.1.1 Simplified Nitrification Modelling in Natural Waters

In the QUAL2E (Brown and Barnwell, 1987) and in QUAL2K (Chapra and Pelletier, 2003; Pelletier et al., 2006) water-quality models, nitrification is described by the single-step reaction $NH_4^+ + 2O_2 \rightarrow NO_3^- + H_2O + 2H^+$ with an oxygen-limited first-order kinetics. A simple ammonium-N dynamics thus takes the form

$$\frac{dS_{NH4}}{dt} = k_{hn} \cdot S_{N_{org}}^{in} - \frac{S_o}{K_o + S_o} \cdot k_n(T) \cdot S_{NH4} - \delta \frac{S_{NH4}}{K_v + S_{NH4}} X_{veg} \tag{7.71}$$

where the first term represents the organic nitrogen ammonification, the second describes the oxidation into nitrate and the last term accounts for the ammonium-N uptake by the submerged vegetation (X_{veg}) with a preference factor (δ) of NH_4 over NO_3 (Di Toro et al., 1975; Brown and Barnwell, 1987; Asaeda and Van Bon, 1997; Asaeda et al., 2000, 2001). This term can be modelled as

$$\delta = \frac{S_{NH4}}{S_{NO3} + S_{NH4}} \tag{7.72}$$

According to Equation 7.72, the preferred form of nitrogen utilized by the submerged vegetation is ammonium-N, because its assimilation requires less energy than that of nitrate-N (Britto and Kronzucker, 2001), but when its concentration falls below that of the nitrate-N (S_{NO3}), the primary producers switch to this alternate form of inorganic nitrogen, as plotted in Figure 7.38 for a nitrate concentration of 5 mg N L^{-1}.

The nitrate-N produced by Equation 7.71 is then consumed by denitrification and plant uptake:

$$\frac{dS_{NO3}}{dx} = \frac{S_o}{K_o + S_o} \cdot k_n(T) \cdot S_{NH4} - (1-\delta) \cdot \frac{S_{NO3}}{K_v + S_{NO3}} X_{veg} - k_{dn} \cdot S_{NO3} \tag{7.73}$$

In Equation 7.73, the first term is the counterpart of the second term of Equation 7.71 describing the oxidation of ammonium-N to nitrate. The second term accounts for the utilization of nitrate by the vegetation, with a preferential factor $(1-\delta)$, and the third term is the denitrification rate.

An example of natural nitrification in rivers can be observed in Figure 7.39. In October 1989, there was a considerable ammonium-N spillage in the Arno River, caused by dredging of the dams upstream of the city of Florence. This massive upheaval of the sediments brought a considerable

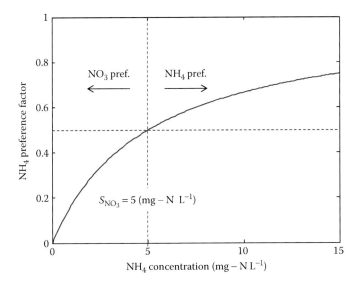

FIGURE 7.38 Ammonium-N preference factor described by Equation 7.72. When the concentration of ammonium-N falls below the nitrate half-saturation, in this case 5 mg N L^{-1}, the uptake is shifted to this form of inorganic N. This graph was obtained with the MATLAB script `Ammonia_Pref.m` in the `ESA_Matlab\ Exercises\Chapter _ 7` folder.

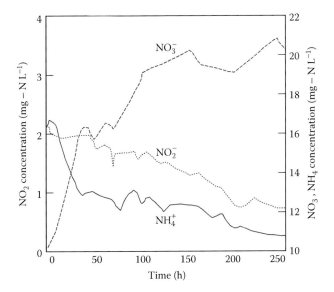

FIGURE 7.39 An ammonia spillage event in the River Arno. The recording started at 8:00 am on 7 October 1989. The plots show the natural ammonium oxidation to nitrite and nitrate.

amount of detritus to the surface, with subsequent resuspension of reduced material containing vast quantities of ammonium-N, which flowed downstream. This caused an alarm at the water potabilization plant of the city of Florence, located just 30 km downstream from the dredged area, and the nitrogen in the river was closely monitored for the next few days after the beginning of the event. Figure 7.39 shows the nitrification process observed just upstream of the abstraction point of the potabilization plant, with the ammonium-N being gradually oxidized to nitrite, and then to nitrate-N. It is left as an exercise for the reader to digitize the data and approximate them with a first-order nitrification model.

7.3.1.2 Detailed Two-Step Nitrification Modelling

The single-step and first-order kinetic are simplifying assumptions in modelling the nitrogen cycle in natural waters, given an inability to obtain detailed information about the autotrophic populations. However, in wastewater processes, a two-step description of the nitrification process is possible, because the autotrophs are abundant enough to be reliably measured, and advisable, because nitrites are a source of concern and there are specialized processes that are focussed on this nitrogen form. As already mentioned SHARON is one of them, being based on the advantage of ammonium oxidizers over nitrite oxidizers depending on the differing temperature response of the two genera (Hellinga et al., 1998). SHARON is often coupled with the ANAMMOX process (Van Dongen et al., 2001), whose name is an acronym for anoxic ammonium oxidation. Other examples of two-step nitrification models have been proposed (Nowak et al., 1995; Ossenbruggen et al., 1996; Hellinga et al., 1999).

A basic two-step nitrification model includes two autotrophic microbial populations: the ammonium oxidizers (X_{aob}) and the nitrite oxidizers (X_{nob}), whose growth rates can be structured as Monod kinetics with two limiting substrates: DO and the pertinent nitrogen species, NH_4^+ in the first step, and NO_2^- in the second step. In this model, the nitrogen assimilation by the autotrophic biomass is represented by the yield factors Y_{aob} and Y_{nob}.

First step $NH_4^+ \rightarrow NO_2^-$

$$\frac{dS_{NH_4}}{dt} = -\frac{1}{Y_{aob}} \mu_{max}^{aob} \cdot \alpha(pH) \cdot \frac{S_o}{K_o + S_o} \cdot \frac{S_{NH_4}}{K_{NH_4} + S_{NH_4}} X_{aob} \tag{7.74}$$

$$\frac{dX_{A1}}{dt} = \mu_{max}^{aob} \cdot \alpha(pH) \cdot \frac{S_o}{K_o + S_o} \cdot \frac{S_{NH_4}}{K_{NH_4} + S_{NH_4}} X_{aob} - b_{aob} \cdot X_{aob}$$

Second step $NO_2^- \rightarrow NO_3^-$

$$\frac{dS_{NO_2}}{dt} = -\frac{1}{Y_{nob}} \mu_{max}^{nob} \cdot \alpha(pH) \cdot \frac{S_o}{K_o + S_o} \cdot \frac{S_{NO_2}}{K_{NO_2} + S_{NO_2}} \cdot \frac{K_{NH_i}}{K_{NH_i} + S_{NH}} X_{nob} \tag{7.75}$$

$$\frac{dX_{A2}}{dt} = \mu_{max}^{nob} \cdot \alpha(pH) \cdot \frac{S_o}{K_o + S_o} \cdot \frac{S_{NO_2}}{K_{NO_2} + S_{NO_2}} \cdot \frac{K_{NH_i}}{K_{NH_i} + S_{NH}} X_{nob} - b_{nob} \cdot X_{nob}$$

In the first step, described by Equation 7.74, the kinetics is controlled by two limiting substrates: DO (S_o) and ammonium-N (S_{NH_4}), with half-velocity constants K_o and K_{NH_4}, while the second step described by Equation 7.75 is more complex because, in addition to the two limiting substrates, ammonium-N acts as an inhibitor with constant K_{NH_i}. The nitrate-N production rate is the opposite of the nitrite-N decay; thus $dS_{NO_3}/dt = -dS_{NO_2}/dt$.

The oxygen requirements for nitrification follow the stoichiometry already computed in Equations 7.63, and thus the rate of oxygen consumption for nitrification is

$$\frac{dS_o}{dt} = -\frac{3.43 - Y_{aob}}{Y_{aob}} \cdot \frac{dS_{NH_4}}{dt} - \frac{1.14 - Y_{nob}}{Y_{nob}} \cdot \frac{dS_{NO_2}}{dt} \tag{7.76}$$

assuming that the oxygen consumption due to autotrophic decay is negligible. The nitrification kinetics is very sensitive to both pH and temperature and can be inhibited by an excess of nitrite, which lowers the pH, or by an excess of ammonium-N, which increases the pH, in addition to the

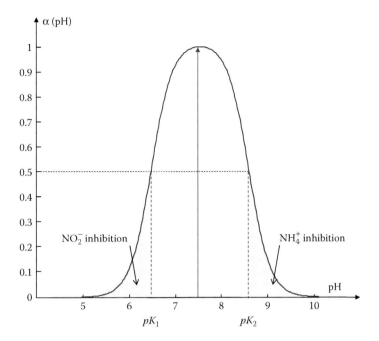

FIGURE 7.40 Sensitivity of nitrification to pH, whose optimal value is around 7.5. The factor α indicates the maximum growth rate reduction of both autotrophic groups at low pH, the inhibiting agent is the nitrite, while at high pH the ammonium-N acts as an inhibitor. Ammonium-N inhibition is also explicitly accounted for in the second nitrification step of Equations 7.74.

inhibiting effect already accounted for in Equations 7.75. The rate reduction factor $\alpha(pH)$ multiplying the maximum growth rates in Equations 7.74 and 7.75 can be expressed as a function of the two limits (pK_1, pK_2) shown in Figure 7.40,

$$\alpha(pH) = \frac{1}{1 + 10^{pK_1 - pH} + 10^{pH - pK_2}} \tag{7.77}$$

The temperature influences both maximum growth rates as follows:

$$\mu(T) = \mu(15) \cdot e^{0.059(T-15)} \tag{7.78}$$

Table 7.3 shows the broad range of values for the parameters in the two-step nitrification model (7.74)–(7.75).

It is interesting at this point to establish a link between heterotrophic and autotrophic metabolisms, as shown in Figure 7.41.

Here, the autotrophic metabolism (blue box) and the heterotrophic metabolism (grey boxes) are linked together by the nitrogen transformations. Both nitrification and C_{org} assimilation require oxygen and release CO_2, which is metabolized by the autotrophs. Instead, the facultative heterotrophs performing denitrification require C_{org} for the reduction of oxidized nitrogen.

7.3.1.3 More Advanced Nitrification Models

A more specialized two-step nitrification and denitrification model, derived from the ASM3 kinetics, is the ASM3_2N model summarized here and fully described elsewhere (Iacopozzi et al., 2007). Here, the ammonium oxidizers and the nitrite oxidizers have separate growth kinetics:

TABLE 7.3

Possible Range of Values of the Parameters in the Two-Step Nitrification Model Equations (7.74 and 7.75)

Parameter	Range of Values	Units
μ_{max}^{aob}	0.46–2.2	d^{-1}
μ_{max}^{nob}	0.28–1.45	d^{-1}
Y_{aob}	0.03–0.16	–
Y_{nob}	0.02–0.08	–
K_o	0.5–1.3	$mg\,O_2\,L^{-1}$
K_{NH_4}	0.06–5.6	$mg\,N\,L^{-1}$
K_{NO_2}	0.06–8.4	$mg\,N\,L^{-1}$
K_{NH_i}	1.00–2.1	$mg\,N\,L^{-1}$
b_{aob}	0.05–0.15	d^{-1}
b_{nob}	0.01–0.02	d^{-1}

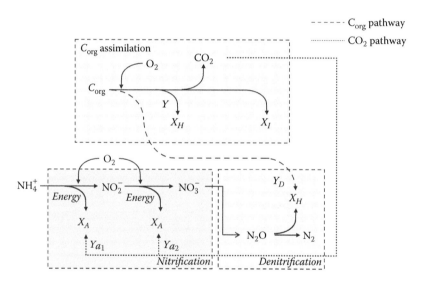

FIGURE 7.41 The autotrophic metabolism (blue box) and the heterotrophic metabolism (grey boxes) are linked together by the nitrogen transformations. Both nitrification and C_{org} assimilation require oxygen and release CO_2, which is metabolized by the autotrophs. The facultative heterotrophs performing denitrification require C_{org} for the reduction of oxidized nitrogen. For each anabolic transformation, the yield factor is indicated.

$$r_{aob} = \mu_{max}^{aob} \cdot \frac{S_o}{K_o + S_o} \cdot \frac{S_{NH_4}}{K_{A,NH_4} + S_{NH_4}} \cdot \frac{S_{ALK}}{K_{A,ALK} + S_{ALK}} \cdot X_{aob}$$

$$r_{nob} = \mu_{max}^{nob} \cdot \frac{S_o}{K_o + S_o} \cdot \frac{K_{I,NH_4}}{K_{I,NH_4} + S_{NH_4}} \cdot \frac{S_{ALK}}{K_{A,ALK} + S_{ALK}} \cdot \frac{S_{NO_2}}{K_{NO_2} + S_{NO_2}} \cdot X_{nob}$$

(7.79)

which are similar to the previous model (7.74)–(7.75), with the introduction of alkalinity as a further limiting factor. As in the previous model, the second step is inhibited by the presence of ammonium-N. In addition to these growth rates, the model also considers the endogenous respiration, both aerobic and anoxic:

$$\text{Aerobic end. resp. of } X_{\text{aob}} \qquad r_{\text{end,aob}}^{\text{Ox}} = b_{\text{aob,O}_2} \cdot \frac{K_o}{K_o + S_o} \cdot X_{\text{aob}}$$

$$\text{Aerobic end. resp. of } X_{\text{nob}} \qquad r_{\text{end,nob}}^{\text{Ox}} = b_{\text{nob,O}_2} \cdot \frac{K_o}{K_o + S_o} \cdot X_{\text{nob}}$$

$$\text{Anoxic end. resp. of } X_{\text{aob}} \qquad r_{\text{end,aob}}^{\text{Anox}} = b_{\text{aob,NO}_2} \cdot \frac{K_o}{K_o + S_o} \cdot \frac{S_{\text{NO}_2}}{K_{\text{NO}_2} + S_{\text{NO}_2}} \cdot X_{\text{aob}}$$

$$\text{Anoxic end. resp. of } X_{\text{nob}} \qquad r_{\text{end,nob}}^{\text{Anox}} = b_{\text{nob,NO}_3} \cdot \frac{K_o}{K_o + S_o} \cdot \frac{S_{\text{NO}_3}}{K_{\text{NO}_3} + S_{\text{NO}_3}} \cdot X_{\text{aob}}$$

(7.80)

The stoichiometry of these kinetics have been recomputed with respect to the ASM3 model, as fully described in Iacopozzi et al. (2007) and further extended to accommodate the nitritation process of Figure 7.37 (Giusti et al., 2011c).

7.3.2 Denitrification Modelling

Denitrification is the inverse process, whereby oxidized forms of nitrogen $\left(\text{NO}_2^- \text{ and NO}_3^-\right)$ are enzymatically reduced to dinitrogen (N_2) through a complex chain of biochemical reactions performed by heterotrophic facultative bacteria. In the simplest form, which is frequently used in water-quality models, denitrification can be described by a first-order kinetics, as in Equation 7.73. More detailed descriptions, used in wastewater treatment, consider that the facultative heterotrophic biomass $\left(X_H\right)$ can reduce either nitrite or nitrate by anoxic respiration with double limitation Monod kinetics involving organic carbon $\left(S_s\right)$ and each oxidized nitrogen species, with DO acting as an external inhibitor, because the presence of oxygen adversely affects this process:

$$r_{dn,\text{NO}_2} = \mu_H \cdot \eta_{dn} \frac{K_o}{K_o + S_o} \cdot \frac{S_{\text{NO}_2}}{K_{\text{NO}_2} + S_{\text{NO}_2}} \cdot \frac{S_s}{K_s + S_s}$$

$$r_{dn,\text{NO}_3} = \mu_H \cdot \eta_{dn} \frac{K_o}{K_o + S_o} \cdot \frac{S_{\text{NO}_3}}{K_{\text{NO}_3} + S_{\text{NO}_3}} \cdot \frac{S_s}{K_s + S_s}$$

(7.81)

The factor $\eta \approx 0.8$ reduces the maximum growth rate μ_H with respect to aerobic growth as a consequence of the smaller biochemical energy (ATP) produced by anoxic respiration. The organic substrate consumption is given by the first two terms of Equation 7.69, so that the nitrate and nitrite consumption rates and the heterotrophic growth due to denitrification are

$$\frac{dS_{\text{NO}_2}}{dt} = -\frac{1 - Y_D}{Y_D \times 1.713} r_{dn,\text{NO}_2} X_H$$

$$\frac{dS_{\text{NO}_3}}{dt} = -\frac{1 - Y_D}{Y_D \times 2.857} r_{dn,\text{NO}_3} X_H$$

(7.82)

$$\left. \frac{dX_H}{dt} \right|_{dn} = \left(r_{dn,\text{NO}_2} + r_{dn,\text{NO}_3} \right) X_H - b_H \cdot X_H$$

where the yield factors for denitrification are given by Equations 7.67 and 7.68. In the already mentioned ASM3_2N model, the anoxic growth on nitrite and nitrate (denitrification) is based on the stored particulate material (X_{STO}), while additional limiting factors are ammonium-N and alkalinity:

$$r_{dn,\text{NO}_2} = \mu_H \cdot \eta_{dn} \frac{K_o}{K_o + S_o} \cdot \frac{S_{\text{NO}_2}}{K_{\text{NO}_2} + S_{\text{NO}_2}} \cdot \frac{S_{\text{NH}_4}}{K_{\text{NH}_4} + S_{\text{NH}_4}} \cdot \frac{X_{\text{STO}}/X_H}{K_{\text{STO}} + \left(X_{\text{STO}}/X_H \right)} \cdot X_H$$

$$r_{dn,\text{NO}_3} = \mu_H \cdot \eta_{dn} \frac{K_o}{K_o + S_o} \cdot \frac{S_{\text{NO}_3}}{K_{\text{NO}_3} + S_{\text{NO}_3}} \cdot \frac{S_{\text{NH}_4}}{K_{\text{NH}_4} + S_{\text{NH}_4}} \cdot \frac{X_{\text{STO}}/X_H}{K_{\text{STO}} + \left(X_{\text{STO}}/X_H \right)} \cdot X_H$$

(7.83)

These kinetics were further modified to consider the nitrogen removal through the nitrite shortcut illustrated in Equations 7.70 and in Figure 7.37, as described in Giusti et al. (2011c).

A last remark on denitrification regards the importance of hydrolysis. It can be seen from Equations 7.69 that denitrification requires considerable quantities of soluble organic substrate C_{org} due to an inability to use particulate substrate directly. If soluble substrate is depleted, then hydrolysis may become the bottleneck of the entire process, limiting the availability of C_{org}. The kinetics involved in anoxic hydrolysis is the same as in the aerobic counterpart, save for a correction of $\eta_{anox} \approx 0.4$ for the maximum rate k_h, which accounts for the lower reaction speed in anoxic conditions. Thus, the anoxic hydrolysis rate producing soluble organic substrate can be written as

$$\frac{dX_s}{dt} = -k_h \cdot \eta_{anox} \frac{X_s/X_H}{K_X + X_s/X_H} X_H \tag{7.84}$$

7.3.3   MATLAB/Simulink Implementation of ASM3_2N

The already cited ASM3_2N model (Iacopozzi et al., 2007) was implemented in the MATLAB/Simulink platform following the Benchmark guidelines (Copp, 2002). This initiative, supported by the European Union under (Cooperation in Science and Technology) COST 682 and 624 actions, defined a simulation protocol for wastewater treatment processes and control strategies (Gernaey et al., 2004; Vrecko et al., 2006; Jeppsson et al., 2007; Nopens et al., 2009). The full Benchmark protocol is available at the web URL http://apps.ensic.inpl-nancy.fr/benchmarkWWTP/, while the MATLAB script implementing the ASM3_2N model is available in the companion software, in the ESA _ Matlab\Exercises\Chapter _ 7\ASM3 _ 2N folder.

7.3.3.1   Creating and Using MEX Files

The software organization of ASM3_2N is shown in Figure 7.42. Given the complexity of the model structure, including structured kinetics and stoichiometric matrix, it was decided to implement the most computationally intensive parts as MEX files, where the acronym stands for MATLAB Executable. MEX modules are dynamically linked subroutines, which may be written in C, C++ or Fortran, that can be invoked from within the MATLAB environment and executed as if they were built-in functions. They use both MATLAB runtime libraries, language-specific libraries or specialized runtime libraries, hence the dynamic linking. To integrate non-MATLAB code in the MATLAB or Simulink framework, the external language source code, for example, a C code, is required, and a compatible compiler must be installed; then it is possible to build the MEX file simply by typing mex filename.c in the command window, and the corresponding filename.mexw64 will be created. In this case, the extension means that the file will be compatible with a Windows 64 bit operating system. This arrangement preserved the advantages of both platforms (MATLAB and C), making the superior execution speed of an executable code available in the Simulink block diagram.

The implementation of the full ASM3_2N model is based on four custom S-function MEX blocks shown in Figure 7.43, which are now briefly described. The input/output dimensions of the blocks are consistent with the six soluble and eight particulate wastewater components in the ASM3 model, as depicted in Figure 7.16.

7.3.3.1.1   CSTR

This block implements the set of reactions in either the anoxic or aerobic tank, depending on the presence/absence of the air flow. Its inputs are the soluble and particulate components, together with water and air flows, while its outputs are the same quantities at the next timestep.

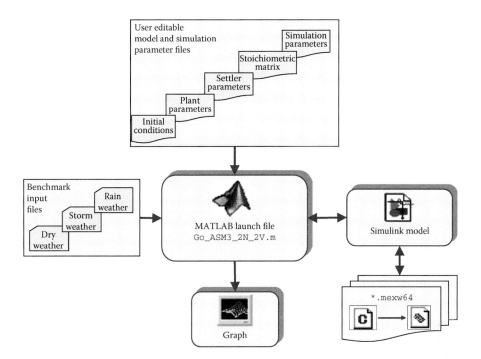

FIGURE 7.42 Software organization of ASM3_2N operating in the Benchmark protocol. The MATLAB launch file `Go _ ASM3 _ 2N _ 2V.m` supervises the entire simulation: after loading the pertinent input file and the specified model parameters, it invokes the Simulink module, which performs the simulation using the precompiled `*.mexw64` routines as model blocks. The Graph module performs the graphical post-processing. The relevant launch script, Simulink model and other routines can be retrieved from the `ESA _ Matlab\Exercises\Chapter _ 7\ASM3 _ 2N` folder. (Reproduced with permission from Iacopozzi, I. et al., *Environ. Modell. Softw.*, 22, 847–861, 2007.)

7.3.3.1.2   Settler

This block implements the Takásc settler model (Takács et al., 1991) using 10 horizontal layers, with the feed placed at the height of the sixth layer. This model is the one normally adopted in all of the benchmark studies (Copp, 2002).

7.3.3.1.3   Mixer

This utility block combines two mixing flows, diluting matching input components of each input to generate the output, consistent with the ASM3 wastewater composition of Figure 7.16.

7.3.3.1.4   Splitter

This utility block generates the COD partition required by the ASM3 model from unstructured COD data, according to Figure 7.16, whereas the other inputs (ammonia, alkalinity and flow) are simply fed through.

The ASM3_2N Simulink model is shown in Figure 7.44. It is a simplified version of the Benchmark configuration, with only one anoxic and one aerated biological reactors, followed by the secondary Takásc settler model. The input is processed by the splitter to produce the required COD partition, then two mixers merge the internal and external flows with the input flow, and then the following blocks implement the anoxic and aerated reaction tanks. The two blocks in the return paths introduce a transport delay required to avoid algebraic loops and to add realism by taking into account the actual flow delays introduced by the recycle flow routing. The simulation is launched by the MATLAB script `Go _ Plant3 _ 2step _ 2V.m`, which loads the initial conditions, plant

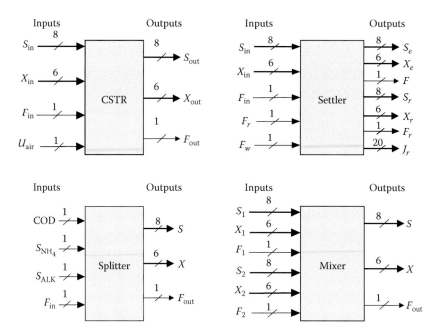

FIGURE 7.43 The four main model blocks implementing the ASM3_2N model in the Simulink platform. The connection stubs indicate which variables are supposed to serve as an input or output for that particular connector. Thick arrows denote a data vector, whose dimension is indicated above the line. Each block is implemented as a precompiled S-function. (Reproduced with permission from Iacopozzi, I. et al., *Environ. Modell. Softw.*, 22, 847–861, 2007.)

configuration, stoichiometric matrix and then launches the Simulink model `plant3 _ 2step _ base _ 2v.mdl` shown in Figure 7.44 with the Benchmark input data for the dry weather condition, shown in Figure 7.45.

Some of the corresponding outputs are shown in Figure 7.46, while the cumulative output statistics for the nitrogen species in the output flow are shown in Figure 7.47.

They were computed with the methods used in Chapter 3 (see e.g. Figures 3.7, 3.8 and 3.15) to obtain a cumulative distribution of an experimental time series.

7.4 MICROBIAL METABOLISM OF PHOSPHORUS

The natural phosphorus cycle is characterized by a vast storage compartment, releasing phosphorus on a geological time scale through atmospheric erosion and slow dissolution of mineral phosphorus (e.g. apatite). In the water cycle, the phosphorus cycle of interest is much faster. It is also simpler than the nitrogen cycle because there is no gaseous form of phosphorus, which can basically exist in two forms: organic P and inorganic P, as orthophosphate $\left(PO_4^{3-}\right)$. The other difference with nitrogen is the lack of a phosphorus equivalent of nitrification, because P can only be absorbed and released by the vegetation and stored in the sediment.

A simplified P cycle in the aquatic environment is shown in Figure 7.48, where the particulate organic phosphorus (POP) pool is fed by several inputs (P_{input}) from various kinds of waste or detritus. It is then hydrolysed into dissolved organic phosphorus (DOP) and converted back into inorganic phosphorus $\left(PO_4^{3-}\right)$ through regeneration. In this form, it is stored in the aquatic vegetation (P_V) and partly utilized for growth. Part of POP settles as detritus in the upper sediment layers, and its subsequent resuspension and dissolution depend on the oxido-reduction state of the sediment, while the part that buries deep into the sediment is transformed back into inorganic by diagenesis.

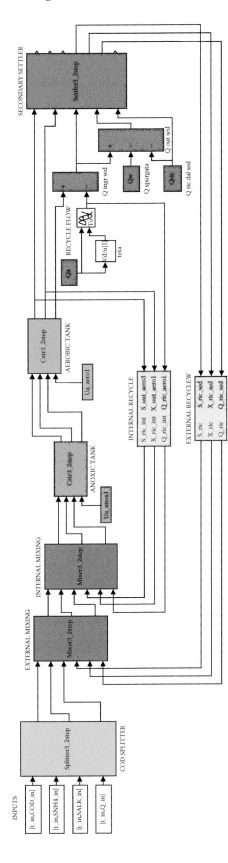

FIGURE 7.44 Complete Simulink diagram implementing the ASM3_2N model, connecting the main S-function blocks. The two blocks in the return paths introduce a transport delay required to avoid algebraic loops and to add realism by taking into account the actual flow delays introduced by the internal and external recycles. This Simulink model plant3 _ 2step _ base _ 2v.mdl can be retrieved from the ESA _ Matlab\Exercises\Chapter _ 7\ASM3 _ 2N folder. (Reproduced with permission from Iacopozzi, I. et al., *Environ. Modell. Softw.*, 22, 847–861, 2007.)

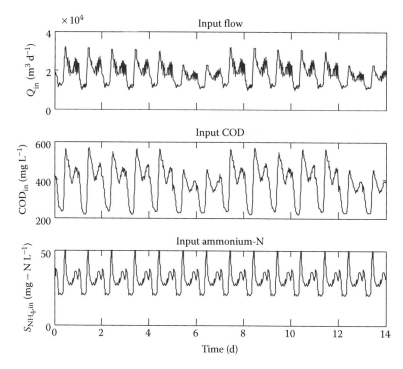

FIGURE 7.45   The Benchmark input files for the dry weather conditions.

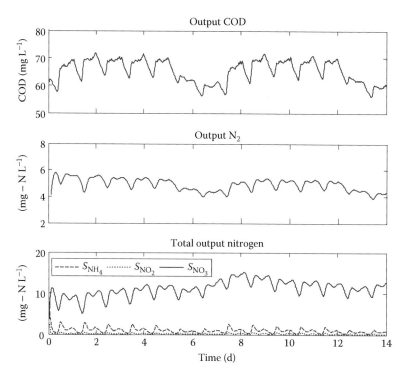

FIGURE 7.46   A sample of ASM3_2N model output regarding total output COD, the amount of output dinitrogen produced by the denitrification process and the various nitrogen forms in the liquid output. These graphs were obtained with the MATLAB launch script Go _ ASM3 _ 2N _ 2V.m in the ESA _ Matlab\ Exercises\Chapter _ 7\ASM3 _ 2N folder.

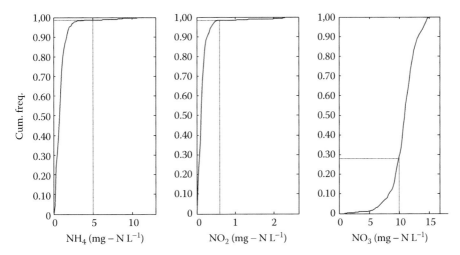

FIGURE 7.47 Cumulative distributions of the three output nitrogen species shown in Figure 7.46 during the 2-week simulation period. The dashed lines indicate some relevant discharge limit, and the corresponding exceedance statistics. In the plots above, it can be seen that the ammonium-N and nitrite are well below their thresholds, while only 30% of the samples are below the nitrate limit. These graphs were obtained with the MATLAB launch script `percentiles.m` in the `ESA _ Matlab\Exercises\Chapter _ 7\ ASM3 _ 2N` folder on completion of the previous simulation of Figure 7.46.

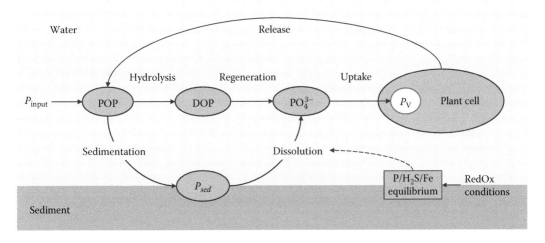

FIGURE 7.48 Basic phosphorus cycle in the aquatic environment. The POP is hydrolysed into dissolved organic phosphorus (DOP), and then regenerated (mineralized) into orthophosphate $\left(PO_4^{3-}\right)$. In this form (P_V), it is utilized by the plant cell. Subsequent release as decaying material is recycled into POP and partly settled in the upper sediment layer. Its eventual dissolution depends on the oxidation/reduction condition of the sediment.

If the detritus is sufficiently oxidized, phosphorus is normally sequestered as $FePO_4$, but if reducing conditions prevail, then phosphorus can be released according to the reaction

$$FePO_4 + HS^- + e^- \rightarrow FeS + HPO_4^{2-} \tag{7.85}$$

releasing phosphorus in biologically available form. The change from an oxidized to a reduced sediment condition may be caused by a low-DO condition (hypoxia), which can trigger eutrophication

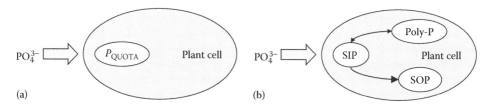

(a)   (b)

FIGURE 7.49 Two models of phosphorus utilization in the plant cell. (a) The cell-quota model (Droop, M. R., *Bot. Mar.*, 26, 99–112, 1983) considers only one growth-controlling storage compartment inside the cell, while in (b), the phosphate interaction model (PIM) has three separate internal compartments: soluble inorganic P (SIP), polyphosphate (Poly-P) and structural organic P (SOP) (John, E. H. and Flynn, K. J., *Ecol. Model.*, 125, 145–157, 2000.)

by the release of inorganic P according to the reaction (7.85). The iron, sulphur and phosphorus cycling, and their relation to possible eutrophication, is thoroughly described in the literature (Azzoni et al., 2001; Melia et al., 2003; Nizzoli et al., 2006; Giordani et al., 2009). As to the role of phosphorus in the vegetation, Droop (1983) introduced the notion of cell quota as the growth-controlling factor and the concept of a nutrient pool inside the cell was further developed by John and Flynn (2000). The common feature of both phosphorus models is that the uptake is not immediately related to growth, which may start well after the nutrient is absorbed from the water. Growth is sustained by the nutrient reserves in the cell until a lower limit is reached, below which this pool cannot be used anymore. The two approaches are sketched in Figure 7.49, where the cell quota (a) receives and stores the external nutrient, while in the PIM hypothesis (b), a more complex storage mechanisms is envisaged.

7.4.1 DROOP CELL-QUOTA MODEL

Droop (1983) formulated the cell-quota hypothesis as 'the quantity of substrate to produce a given biomass'. In this sense, the cell quota can be regarded as a yield factor, but it also represents an internal nutrient pool on which growth depends, but which cannot be completely depleted. On the basis of these considerations, the following three-variable cell-quota model can be set up:

P uptake
$$\frac{dP}{dt} = -\frac{u_m P}{k_p + P} + P_{\text{in}}$$

Cell quota
$$\frac{dQ}{dt} = \frac{u_m P}{k_P + P} - \mu_Q \left(1 - \frac{k_q}{Q}\right) Q \qquad (7.86)$$

Biomass growth
$$\frac{dX}{dt} = \frac{1}{Y} \mu_Q \left(1 - \frac{k_q}{Q}\right) Q - k_d X$$

Comparing model (7.86) with the previous Monod-based substrate/biomass interactions, it can be seen that growth is mediated by the cell quota, which acts as an internal reservoir. The Simulink model of Figure 7.50 implements the cell-quota model (7.86), where the uptake and the quota consumption are implemented with two `fcn` block. The three integrators produce the required state variables of the model (P, Q, X).

The launch MATLAB script is listed in Box 7.2, and it is very straightforward. After defining the model parameters, it invokes the Simulink model named `Cell _ quota.mdl` and plots the results. Given the large amount of simulation data, a decimation factor of 10 was introduced in the `To Workspace` blocks to limit the number of plotted data, hence the need to store a decimated simulation time `ts`. The results plotted in Figure 7.51 clearly show the delayed growth with respect to the phosphorus uptake and the cell quota settling at its lower limit equal to k_q. Because no further

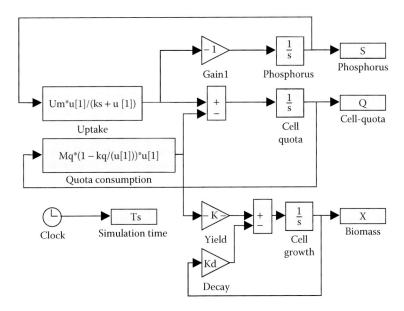

FIGURE 7.50 Simulink diagram implementing the cell-quota phosphorus uptake and cell growth model. This Simulink model is available as `Cell _ Quota.mdl` in the `ESA _ Matlab\Exercises\Chapter _ 7\ Phosphorus` folder.

BOX 7.2   LAUNCH SCRIPT FOR THE CELL-QUOTA MODEL (DROOP, 1983) INVOKING THE SIMULINK MODEL OF FIGURE 7.50

```
clear
close all
Po=15; % Initial nutrient concentration
% Parameter vector bundle
Um=0.042;    % Uptake maximum rate
Mq=0.001;    % Cell-quota maximum rate
kq=1.5;      % Cell-quota constant
ks=15;       % Uptake half-velocity
kd=0.00013;  % Biomass decay rate
Xo=0.3;      % Initial biomass
Qo=0.0015;   % Initial cell-quota
Y=0.75;      % Yield factor
% Simulation
tfin=10000; % Final simulation time
Dec=10; % Simulation data decimation factor
[t,x,y]=sim('Cell _ Quota',tfin);
% Plot simulation results
figure(1)
plot(ts,S,'-.b',ts,Q,'b',ts,X,':b','Linewidth',2)
set(gca,'FontName','Arial','FontSize',14)
legend('Ext. Substrate','Cell-Quota','Biomass','location','northeast')
xlabel('time (h)')
ylabel('Concentration (mg/L)')
```

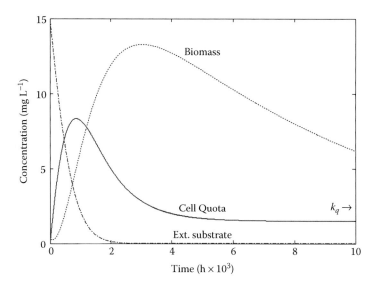

FIGURE 7.51 Simulation of the cell-quota model of Figure 7.50 in the case of no continuous external phosphorus input. The cell quota settles to the k_q constant value, while growth is delayed with respect to phosphorus uptake, and then decays because of endogenous metabolism with rate k_d. This simulation was produce by the MATLAB script `Go_CQ.m` in the `ESA_Matlab\Exercises\Chapter_7\Phosphorus` folder.

nutrient was supplied after the initial condition $P_o = 15 \, (\text{mg P} / \text{L})$, the biomass decays with rate k_d after having reached its peak development by exploiting the reserves provided by the cell quota.

7.4.2 PIM MODEL

An improvement to the cell-quota concept was later provided by John and Flynn (2000), who proposed a PIM corresponding to the scheme of Figure 7.49b. In this model, phosphorus is shared by three internal storage pools: Poly-P, SIP and SOP. The authors claim that PIM can simulate the decoupling of phosphate transport and assimilation, especially in transient regimes and P-limited conditions. Another important aspect of introducing Poly-P as a phosphorus pool is that in this way inorganic polyphosphate can be accumulated in very large quantities to support prolonged growth, even in the absence of external P. For a complete description of the PIM, the reader is referred to the original paper (John and Flynn, 2000), where the model is fully described in Table 1 (model equations) and Table 2 (model parameters). Here, we concentrate on its Simulink implementation, which differs from the previous Simulink models examined so far, with the exception of the ASM3_2N model. In fact, in the latter case, the complex dynamics of the microbial metabolism were implemented with the S-function block hosting a pre-compiled C-code. In the implementation of PIM, an intermediate approach was used: the S-function block was still used, but instead of embedding an external code, a MATLAB script was inserted into the S-function template according to the scheme presented in Figure 7.52.

The detailed Simulink model is shown in Figure 7.53 and the internal structure of the S-function is listed in Box 7.3, using the template available in the Simulink library.

The advantage of PIM over the cell-quota model is the flexibility in simulating differing levels of poly-P accumulation, and to decouple P transport from assimilation. In fact, comparing the external P exhaustion in Figure 7.55 with the poly-P accumulation in Figure 7.54, it can be seen that the latter resource is used after the external P has been completely exhausted.

Differing levels of Poly-P accumulation can be simulated through the PIM, by acting on the `PolyPmax` constant. It can be shown that the decline in carbon growth is slower in species with a

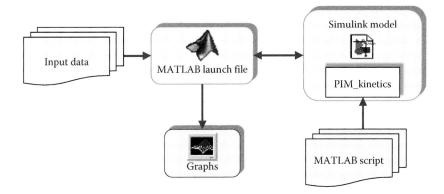

FIGURE 7.52 Software organization to incorporate an S-function MATLAB script. The MATLAB code is written in the S-function template and then incorporated into the Simulink model.

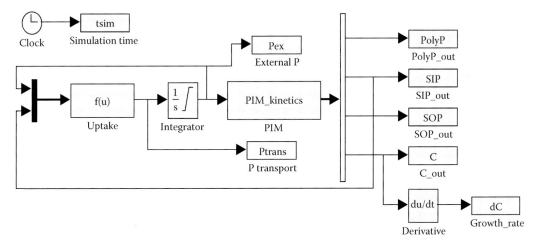

FIGURE 7.53 Simulink diagram implementing the PIM model (John and Flynn, 2000). The PIM kinetics is implemented with an S-function incorporating MATLAB code, as described in Figure 7.52, while the uptake function is a function block requiring external integration. The thick arrows indicate transmission of a data vector. This Simulink model is available as PIM.mdl in the ESA _ Matlab\Exercises\Chapter _ 7\ Phosphorus folder. The PIM_Kinetics block is an S-function incorporating the PIM _ Kinetics.m in the same folder.

BOX 7.3   MATLAB SCRIPT INCORPORATED IN THE
S-FUNCTION TO IMPLEMENT THE PIM KINETICS

```
function  [sys, x0] = PIM _ Kinetics(t,x,u,flag)
%S-FUNCTION implementation of the PIM
% Phosphorus kinetics according to John and Flynn,(2000).
%
global P IC
% Parameter values and initial conditions
% are passed as external objects via a global statement
if abs(flag) == 1,
% Parameter vector unbundle                          Box 7.3 continued
```

Box 7.3 continued

```
KqP=P(1);
PolyPmax=P(2);
Ksip =P(3);
Kps =P(4);
Kpt1=P(5);
Kpt2=P(6);
PTkSIP=P(7);
PTKs=P(8);
QoP=P(9);
SIPmax=P(10);
Ksop1=P(11);
Ksop2=P(12);
SOPmax=P(13);
Umax=P(14);
% Intermediate variables
PAmax=SOPmax*Umax;
PTmax=4*PAmax;
PolyP=x(1);SIP=x(2);SOP=x(3);C=x(4);
PT=PTmax*(u/(u+PTKs))*(1-SIP/SIPmax)/((1-SIP/SIPmax)+PTkSIP);
PolyPs=2*PAmax*(SIP/(SIP+Ksip))*((1-PolyP/PolyPmax)^4)/(((1-PolyP/
PolyPmax)^4)+Kps);
PolyPt=PAmax*(PolyP/(PolyP+Kpt1))*((1+SIP/SIPmax)^4)/(((1+SIP/
SIPmax)^4)+Kpt2);
SOPs=PAmax*(SIP/(SIP+Ksop2))*((1-SOP/SOPmax)^4)/(((1-SOP/
SOPmax)^4)+Ksop1);
Cmu=Umax*(SOP-QoP)/(SOP-QoP+KqP);
dC=C*Cmu;
dPolyP=PolyP*Cmu;
dSIP=SIP*Cmu;
dSOP=SOP*Cmu;
% Model equations
dx(1)=PolyPs-PolyPt-dPolyP;
dx(2)=PolyPt+PT-SOPs-PolyPs-dSIP;
dx(3)=SOPs-dSOP;
dx(4)=dC;
  sys = dx;
end
if flag == 3,
  % If FLAG==3, then SIMULINK wants to know what the next output is.
  % **** In this template system, y gets the current state x ****
  % (SYS =) Y = X
  for i=1:4
      if x(i)<0.001, x(i)=0.001;
      end
  end
  sys = x;
end
```

Box 7.3 continued

Box 7.3 continued

```
if flag == 0,
  % This part takes care of all initialization; it is used only once.
  % The sizes vector is six elements long, and it looks like this:
  sizes(1) = 4; %   number of continuous states
  sizes(2) = 0; %   number of discrete states
  sizes(3) = 4; %   number of system outputs
  sizes(4) = 1; %   number of system inputs
  sizes(5) = 0; %   number of   discontinuous   roots;   unused
                    feature, set to zero
  sizes(6) = 0;
  % Set the initial conditions on the states
  x0 = IC;
  sys = sizes';
end
if flag==2 |flag==4
  % Flags not considered here are treated as unimportant.
  % Notice that since there are no discrete states in this system,
  % there is no need to deal with FLAG==2 or FLAG==4.
  % Output is set to [].
  sys = [];
end      % if abs(flag) == ...
```

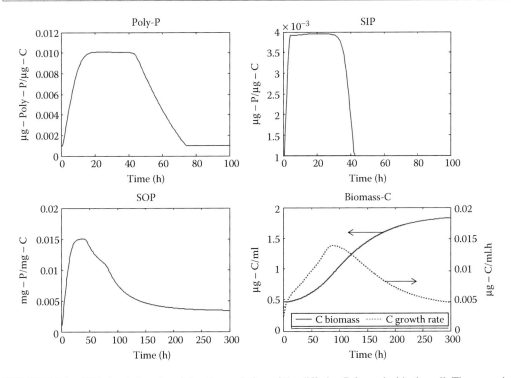

FIGURE 7.54 PIM simulation describing the evolution of the differing P forms inside the cell. These graphs were obtained with the MATLAB script Go_PIM.m in the ESA_Matlab\Exercises\Chapter_7\ Phosphorus folder.

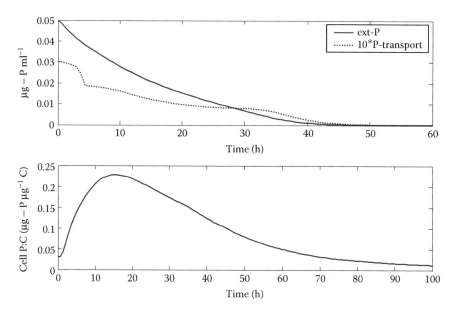

FIGURE 7.55 PIM simulation describing the external P transport and P:C ratio inside the cell. These graphs were obtained with the MATLAB script `Go _ PIM.m` in the `ESA _ Matlab\Exercises\Chapter _ 7\ Phosphorus` folder.

higher Poly-P accumulation, representing a competitive advantage of this species over others with a more limited Poly-P storage capacity.

7.4.3 Phosphorus Removal in Wastewater Treatment Processes

Phosphorus has no gaseous form in the natural environment, apart from the phosphine poisonous gas (PH_3) (Han et al., 2010), and thus it cannot be disposed of through a chain of biochemical reactions, such as the nitrification–denitrification nitrogen counterpart. The only possibility for removing phosphorus from the environment is to accumulate it in some living organism, and then remove that organism with its P pool. In fact, we have just seen that many plant cells, especially microalgae and some specialized bacteria, have the ability to build internal P reserves. In wastewater technology, polyphosphate accumulating organisms (PAOs) have been isolated and can remove phosphorus from the wastewater if kept in adequate conditions, such as the Dephanox process (Bortone et al., 1999) and in SBRs (Marsili-Libelli et al., 2001; Wilderer et al., 2001; Artan and Orhon, 2005).

PAOs are bacteria that, under certain environmental conditions (*feast/famine*), can accumulate phosphorus inside the cell, removing it from wastewater (Mino et al., 1998; Hesselmann et al., 1999; Seviour et al., 2003; Oehmen et al., 2007). As we have seen, the capability of accumulating intracellular phosphorus is common to many plant cells and bacteria. What makes PAOs different from other P-accumulating organisms is the ability to store phosphorus as poly-β-hydroxybutyrate (PHB), and to consume simple carbon compounds by generating energy from internally stored polyphosphate and glycogen (Beun et al., 2000, 2002; Carta et al., 2001) without the availability of an external electron acceptor. The glycogen metabolism is described in Dircks et al. (2001), who showed that formation and consumption of glycogen appears to be much faster than for PHB. Competition for carbonaceous substrate between PAOs and denitrifiers during simultaneous nitrogen and phosphorus removal depends on the nature of the carbon source (Lopez-Vazquez et al., 2009; Guerrero et al., 2011) and may lead to competition between the two groups, PAOs and the glycogen-accumulating organisms (GAOs), as investigated by Lanham et al. (2014). PAOs can operate in three differing conditions with strikingly differing behaviours.

1. In anaerobic conditions and with external substrate available, they metabolize acetate (HAc) using the stored Poly-P as energy source. In the process, phosphorus is released and internal PHB reserves are built up. In these conditions, they have a competitive advantage over other heterotrophs, being able to metabolize the substrate without external electron acceptors (Figure 7.56a).
2. In anoxic $\left(NO_3^-\right)$ or aerobic $\left(O_2\right)$ conditions, and *without* external substrate, they utilize the previously produced phosphorus and produce new cells equipped with their own PHB reserves. The amount of phosphorus accumulated in the process is greater than that previously released, resulting in a net P removal from the wastewater (Figure 7.56b).
3. In anoxic $\left(NO_3^-\right)$ or aerobic $\left(O_2\right)$ conditions, and *with* external substrate, this is converted into PHB, and at the same time phosphorus is released, thus achieving the opposite result (Figure 7.56c).

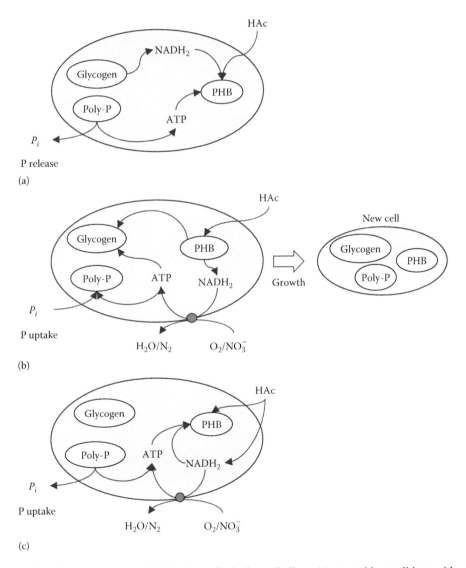

FIGURE 7.56 The three states of PAOs 'feast–famine' metabolism: (a) anaerobic conditions with external substrate; (b) anoxic/aerobic conditions without external substrate; and (c) anoxic/aerobic conditions with external substrate.

Thus, PAOs are characterized by an unbalanced growth (feast–famine regime), during which they can assimilate the external substrate in anaerobic conditions to build internal reserves, while they grow in anoxic or aerobic conditions using their internal reserves, if external substrate is not available. Given their peculiar metabolism, PAOs are disadvantaged in normal activated sludge processes with aerobic conditions and external carbon availability, while they have a competitive advantage in processes alternating anaerobic/anoxic to aerobic phases, where they accumulate more phosphorus in the anaerobic phase than they release during the aerobic phase. In addition, they can also perform denitrification during the anoxic phase.

Figure 7.56 illustrates the three metabolic states of PAO metabolism. In (a), the external substrate (HAc) is metabolized into PHB under anaerobic conditions. The required energy is provided by ATP, which is phosphorylated by the Poly-P reserves, while the reducing power ($NADH_2$) is provided by the glycogen and phosphorus is released in the process. In (b), no external substrate is available. Therefore, PHB is used as a substrate for the oxidative phosphorylation of ATP and the glycogen is replenished, thanks to the external electron acceptors (O_2 or NO_3^-). External phosphorus is taken up to resupply the Poly-P chains and new cells are generated, with their own supplies of Poly-P. In this way, the P-uptake is greater than the previous P release and a net P removal is obtained. In (c), both external substrate and electron acceptors are available. The PHB reserves are replenished using this carbon source, and phosphorus is released in the process. From the wastewater engineering viewpoint, this is an undesirable situation, because phosphorus is released instead of being removed. The P-removal processes are conceived to alternate between the (a) and (b) situations, avoiding the occurrence of (c). In these conditions, the phosphorus cycling between the two phases is illustrated in Figure 7.57. During the anaerobic phase, PAOs build up PHB reserves using the external carbon source (HAc) and release P according to the mechanism of Figure 7.56a. After all the substrate has been consumed, aeration is switched on. The PAO metabolism shifts to the growth condition of Figure 7.56b, uptaking P from the water. The combined cycle results in a net P removal.

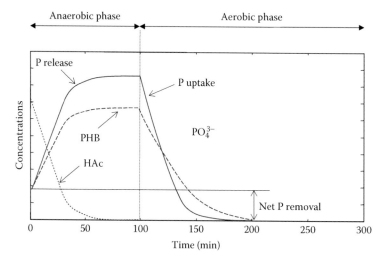

FIGURE 7.57 Phosphorus exchange during the anaerobic/aerobic cycling. During the anaerobic phase, PAOs build up PHB reserves using the external carbon source (HAc) and release P according to the mechanism of Figure 7.56a. When aeration is switched on, their metabolism shifts to the growth condition of Figure 7.56b and uptake P from the water. The combined cycle results in a net P removal. This graph was obtained with the MATLAB script Go _ SBR.m in the ESA _ Matlab\Exercises\Chapter _ 7\SBR folder.

7.4.4 THE SEQUENCING BATCH REACTOR PROCESS FOR NUTRIENT REMOVAL

Given the unbalanced growth regime of PAOs, alternating periods of phosphorus uptake and cell growth in differing periods, a specific process configuration has been conceived to provide these bacteria with the appropriate feast/famine conditions in which they can perform P removal. The sequencing batch reactor (SBR), first introduced by Irvine and Ketchum (1989), is a simple and effective engineering solution to provide the required alternating conditions for P-removal by PAOs, and also the anoxic/oxic conditions for the denitrification by facultative heterotrophs and the same PAOs. SBR is now a well-established engineering solution for nutrient removal from wastewater (Wilderer et al., 2001; Artan and Orhon, 2005). The engineering aspects of SBRs are thoroughly described in specialized textbooks (Metcalf & Eddy Inc., 2003; Leslie Grady et al., 2011).

The phases of the SBR are sequentially organized as in Figure 7.58. In phase 1, the reactor is filled with the raw effluent. In phase 2, the reactor is kept in anoxic/anaerobic conditions by stirring without aeration. In phase 3, the air is switched on so that the conditions become aerobic. After the aeration is switched off, the mixed liquor is allowed to settle, and then the treated water is extracted from the clarified zone (5) and eventually the excess sludge is extracted (6), after which the SBR is ready to start a new cycle. Traditionally, the SBR timing is fixed, with the length of each phase shown in Figure 7.59 so that a full cycle is completed in 6 h, but a considerable improvement can be obtained by adapting the switching sequence to the actual state of the reactions, as will be illustrated in the next section.

7.4.4.1 MATLAB Implementation of the SBR Model

The established model describing the microbial reactions related to nutrient (N and P) removal is the ASM2d model (Henze et al., 2000) that, unlike ASM1 and ASM3, contains the P-removal reactions described in the previous section. The model involves twenty state variables and a large

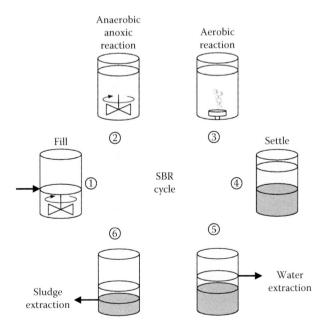

FIGURE 7.58 The six phases of the SBR cycle. In 1, the reactor is filled with the effluent; in 2, the reactor is stirred and kept in anaerobic/anoxic conditions; in 3, the air is switched on; in 4, the mixed liquor is allowed to settle; in 5, the treated water is extracted; and in 6, the excess sludge is extracted. Then, the reactor is ready for another cycle.

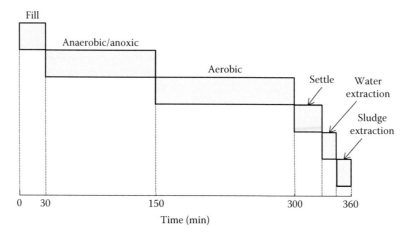

FIGURE 7.59 Typical length of each SBR phase, with a full cycle duration of 6 h.

number of reaction processes and numerical parameters. For this reason, a block Simulink implementation of the SBR model would be too complex and inefficient, and again the MEX S-function approach has been adopted (Marsili-Libelli et al., 2001). In fact, calling a compiled C-code from an S-function block improves the computation speed by at least a factor of 100 and makes the source code more compact. The same software engineering approach, already used to code the ASM3_2N model of Section 7.3.3, was followed here. A single S-function suffices to implement all of the SBR reactions, while the control inputs (fill, aeration, extraction, and sludge waste) are implemented as pulse generators. The simulation is launched by the MATLAB script `Go_SBR.m`, which first passes the parameters and initial conditions to the Simulink model and then plots the simulation results. It was decided to pass the parameters to the S-function through a `global` statement, so that they could be manipulated outside the SBR S-function. This is necessary during the model calibration (see Chapter 2), where the parameters are changed by the optimization algorithm at each simulation run. The model was calibrated, as shown in Figure 7.60, with data from an experimental SBR treating an effluent containing both nitrogen and phosphorus.

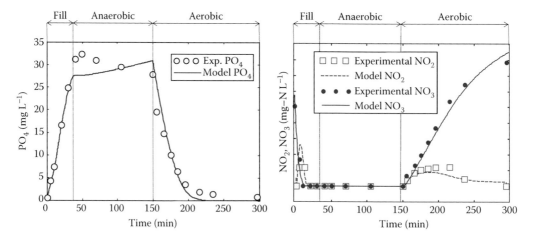

FIGURE 7.60 Comparison between the experimental data and the calibrated SBR model output, considering an effluent with both phosphorus and nitrogen. These graphs were obtained with the MATLAB script `Go_SBR.m` in the `ESA_Matlab\Exercises\Chapter_7\SBR` folder. (Redrawn from Marsili-Libelli, S. et al., *Water Sci. Technol.*, 43, 69–76, 2001.)

The experimental setup was located in the ENEA Laboratories in Bologna, Italy (www.enea.it) and is described elsewhere (Marsili-Libelli et al., 2001; Spagni et al., 2001; Luccarini et al., 2002). A closer look at Figure 7.60 shows that the during the fill phase, and the subsequent anaerobic period, there is a progressive phosphorus release, which is later re-absorbed during the aerobic phase, after the carbonaceous substrate has been eliminated. The nitrogen transformations, instead, are limited to denitrification during the fill/anaerobic phases, while during the aerated phase, the intermediate nitrite is oxidized to nitrate, to be denitrified in the anoxic phase of the next cycle.

7.4.5　Controlling the SBR Switching with Fuzzy Logic

As previously mentioned, normally the SBR cycle has a fixed duration, with each phase being timed regardless of the process requirements. This may result in an inefficient operation and a waste of energy. Because the SBR process is often used in low-cost applications and/or with aggressive effluents, the key control factor is the use of simple and cheap online process measurements to infer the concentration of the chemical variables $\left(NH_4^+, NO_2^-, NO_3^+, PO_4^{2-}\right)$, while it would be difficult and expensive to measure them directly. There has long been a general consensus on two main issues: that SBR control aims to adapt the switching sequence to the actual load and that physico-chemical parameters such as pH, oxido-reduction potential (ORP) and DO may be used as indirect process indicators.

This section summarizes the results obtained by using fuzzy logic to control the SBR switching sequence. First, the relevant process transitions are defined; then a switching fuzzy logic is defined to terminate each reactive phase as soon as the relevant chemical transformations are completed.

7.4.6　Indirect Detection of Process Transitions

If all the relevant process quantities $\left(NH_4^+, NO_2^-, NO_3^+, PO_4^{2-}\right)$ could be easily measured in real time, switching could be performed according to the following simple rules:

1. The anaerobic/anoxic phase ends when both the nitrate-N has been denitrified and all the available organic carbon is used in the process, as in Figure 7.61. At this time the aeration could be switched on.
2. The aerobic phase ends when all the phosphate-P is taken up by the PAO or all the ammonium-N is converted into nitrate-N, whichever takes longer, as illustrated in Figure 7.62. At this time the aeration could be switched off.

Though ion-specific probes are becoming cheaper and more reliable, still their use in the SBR technology is still limited for economical and maintenance considerations. In their place, indirect quantities such as pH, ORP and DO, which can be measured cheaply and reliably, are used as indirect pattern indicators to signal the end of each phase and switch the process accordingly. The correspondence between such indicators and the process state can be established by inspecting Figure 7.61 for the anaerobic/anoxic phase, and Figure 7.62 for the aerobic phase. In both cases, the slope of these physico-chemical measurements can be related to significant events in the SBR reaction.

7.4.6.1　Anaerobic/Anoxic Patterns

During this phase, the pH-influencing processes are denitrification and P-release, generally occurring in this order. Figure 7.61 shows that the end of denitrification produces a brief pH increase, after which it decreases as a consequence of phosphorus release. The ORP further decreases as the process gets deeper into anaerobic conditions. Eventually, when all the available nitrate is reduced and all the phosphorus is released, both pH and ORP level off, with the remaining fraction of the phase becomes redundant. In summary, the relevant transitions indicating the end of this phase are the changes in ORP slope and the 'kick' and levelling of pH.

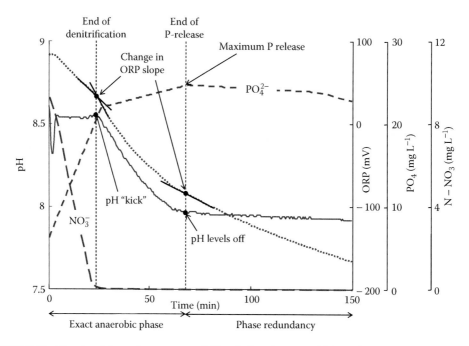

FIGURE 7.61   Trends of the main process variables during the anaerobic/anoxic phase. When all the phosphorus is released the pH levels off and the ORP changes slope, while the end of denitrification is signalled by a pH kick. After these transitions, whichever comes last, no other transformation occurs and the rest of the phase is redundant. (Reproduced with permission from Marsili-Libelli, S., *Water Res.*, 40, 1095–1107, 2006.)

7.4.6.2   Aerobic Patterns

During this phase ammonia is oxidized into nitrate, whereas phosphorus is utilized by the PAO provided that no organic carbon is available. Figure 7.62 shows that the relevant process variables are pH and DO, while ORP carries little information when positive. The pH exhibits an inflexion at the end of phosphorus uptake, while DO levels off when all the ammonium-N has been oxidized. However, because during the aerobic phase nitrification and P-uptake are concurrent processes, pH may exhibit differing patterns depending on which ends first. This complex behaviour has been thoroughly examined by Spagni et al. (2001), who observed differing pH patterns as a consequence of the relative duration of ammonia oxidation and P-uptake, in addition to CO_2 stripping, responsible for a rapid increase of pH at the beginning of the aerobic phase. According to this analysis two differing behaviours may occur, depending on whichever process ends first. If nitrification ends before P-uptake, the end of ammonia oxidation is marked by a change in the slope of pH, but it is difficult to detect the so-called ammonia valley (Kim et al., 2004; Akin and Ugurlu, 2005; Hu et al., 2005). Conversely, if P-uptake ends before nitrification, the 'ammonia valley' is clearly identifiable, confirming the effect of P-uptake on pH. In the first part of the aeration phase, the pH increase is due to CO_2 stripping and P-uptake, when the latter ends pH starts decreasing due to ammonia oxidation. The 'aerobic pH apex' can be related to the end of P-uptake.

7.4.7   Design of the Fuzzy Switching Logic

From the behaviours shown in Figures 7.61 and 7.62 the relevant patterns in Table 7.4 were defined as phase-end indicators.

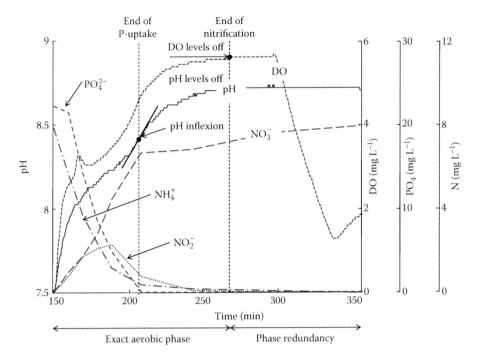

FIGURE 7.62 Trends of the main process variables during the aerobic phase. The previously released phosphorus is absorbed and the pH inflexion indicates its elimination. The levelling off of the DO signals indicates that all the ammonium-N is oxidized to nitrate-N. After these transitions, whichever comes last, no other transformation occurs and the rest of the phase is redundant. (Reproduced with permission from Marsili-Libelli, S., *Water Res.*, 40, 1095–1107, 2006.)

Because the information is contained in the signal variations, these signals must be carefully denoised before numerical differentiation is applied. For this, the dyadic wavelet denoising of Sections 3.4.2 and 3.4.3 is used in the scheme of Figure 3.33, and the denoised pH and ORP signal are shown in Figure 7.63.

Once sufficiently smooth numerical derivatives have been obtained, they were classified with respect to the patterns of Table 7.4 by clustering a set of training data as in Section 4.5.4 using the Gustafson–Kessel algorithm of Section 4.5.8.

The use of the clusters is the same as that already demonstrated in the case of the diagnosis of an anaerobic digester in Section 4.5.5.1 and further described elsewhere (Marsili-Libelli and Müller, 1996; Mueller et al., 1997). The partition entropy criterion of Equation 4.52 showed that two clusters $\left(C_{AA}^{1}, C_{AA}^{2}\right)$ represent the optimal partition for the anaerobic/anoxic phase, as illustrated in Figure 7.64, while three clusters $\left(C_{OX}^{1}, C_{OX}^{2}, C_{OX}^{3}\right)$ best describe the data during the aerobic phase, as shown in Figure 7.65. These figures also show the trajectory of a training and a validation experiment, and the consequent centroid displacement induced by the adaptation mechanisms described in Section 4.5.6.

7.4.8 DESIGN OF THE FUZZY INFERENCE SYSTEM FOR THE PHASE SWITCHING

The fuzzy inference system capable of recognizing these indicators requires the steps illustrated in the overall diagram of Figure 7.66, which includes preliminary filtering, numerical differentiation, pattern recognition and feature extraction.

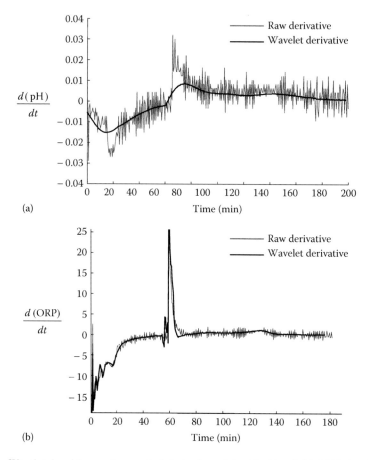

FIGURE 7.63 Wavelet denoising and numerical derivative of the (a) pH and (b) ORP signals. (Reproduced with permission from Marsili-Libelli, S., *Water Res.*, 40, 1095–1107, 2006.)

TABLE 7.4

Process Indicators for SBR Switching

Phase	Terminated Process	Indicator
Anaerobic anoxic	Denitrification	pH kicks up and then falls
		ORP negative slope increases
	P-release	pH slope levels off
Aerobic	Nitrification	DO slope levels off
	P-uptake	pH inflexion point

Source: Marsili-Libelli, S., *Water Res.*, 40, 1095–1107, 2006. Reproduced with permission.

Using the cluster structure just determined and using the approach of Section 4.5.4, two FIS are now designed, one for the anaerobic/anoxic phase, and one for the aerobic phase. Another external variable, time, was been added for safety, to prevent both premature termination and excessive duration of the phase. The lower time limit was specified through a set of fuzzy memberships assuming that the aerobic phase should last long enough to ensure consistent ammonia oxidation, whereas the minimum anoxic/anaerobic duration is related to loading and denitrification.

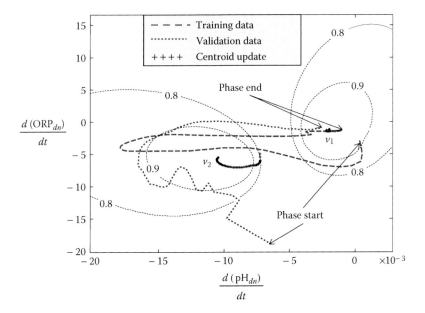

FIGURE 7.64 Two clusters, with centroids v_1 and v_2, describe the transitions during the anaerobic/anoxic phase in the space of the denoised derivatives of pH and ORP. The trajectories relative to two experiments are shown, together with the centroid adaptation during training. (Reproduced with permission from Marsili-Libelli, S., *Water Res.*, 40, 1095–1107, 2006.)

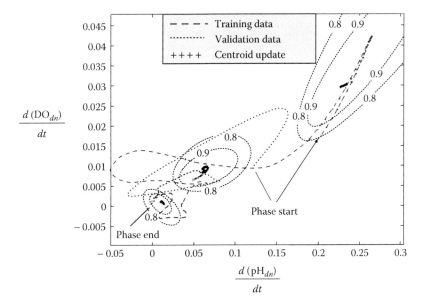

FIGURE 7.65 Three clusters, with centroids v_1, v_2 and v_3, describe the critical transitions during the aerobic phase in the space of the denoised derivatives of pH and DO. The trajectories relative to two experiments are shown, together with the centroid adaptation during training. (Reproduced with permission from Marsili-Libelli, S., *Water Res.*, 40, 1095–1107, 2006.)

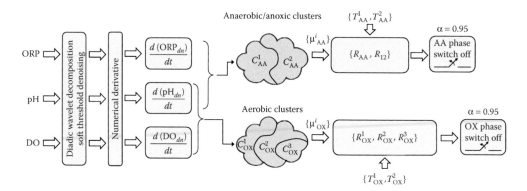

FIGURE 7.66 Structure of the fuzzy inference system to switch the reactive SBR phases. The derivative of the process signals are denoised in real-time, then their values are classified according to the predefined clusters to form the antecedents to the fuzzy rules of Equations 7.87 and 7.89. The defuzzified outputs are compared with a threshold $(\alpha = 0.95)$ and when exceeded, switching is activated.

7.4.8.1 Anoxic/Anaerobic Phase

The following two Sugeno rules are defined:

$$R_{AA}^1 : \text{IF} \left[\frac{d\left(pH_{dn}\right)}{dt}, \frac{d\left(ORP_{dn}\right)}{dt} \right] \subset C_{AA}^1 \text{ AND } t \subset T_{AA}^1 \text{ THEN } y_{AA}^1 = 0$$

$$R_{AA}^2 : \text{IF} \left[\frac{d\left(pH_{dn}\right)}{dt}, \frac{d\left(ORP_{dn}\right)}{dt} \right] \subset C_{AA}^2 \text{ AND } t \subset T_{AA}^2 \text{ THEN } y_{AA}^2 = 1$$

(7.87)

where the time thresholds $\left(T_{AA}^1, T_{AA}^2\right)$ are the safeguards against the abnormal phase duration. The indicator of the anaerobic phase-end is obtained by defuzzification:

$$y_{AA} = \frac{\displaystyle\sum_{i=1}^{2} y_{AA}^i \cdot \mu_{AA}^i}{\displaystyle\sum_{i=1}^{2} \mu_{AA}^i}$$

(7.88)

with y_{AA} representing the likelihood that anaerobic/anoxic phase is approaching its end.

7.4.8.2 Aerobic Phase

The rules for the end of the aerobic phase are similar to the previous equations (7.87), but because three clusters represented the optimal partition, this inference engine is composed of three Sugeno rules:

$$R_{OX}^1 : \text{IF} \left[\frac{d\left(pH_{dn}\right)}{dt}, \frac{d\left(DO_{dn}\right)}{dt} \right] \subset C_{OX}^1 \text{ AND } t \subset T_{OX}^1 \text{ THEN } y_{OX}^1 = 0$$

$$R_{OX}^2 : \text{IF} \left[\frac{d\left(pH_{dn}\right)}{dt}, \frac{d\left(DO_{dn}\right)}{dt} \right] \subset C_{OX}^2 \text{ AND } t \subset T_{OX}^2 \text{ THEN } y_{OX}^2 = 0.7$$

(7.89)

$$R_{OX}^3 : \text{IF} \left[\frac{d\left(pH_{dn}\right)}{dt}, \frac{d\left(DO_{dn}\right)}{dt} \right] \subset C_{OX}^3 \text{ AND } t \subset T_{OX}^2 \text{ THEN } y_{OX}^3 = 1$$

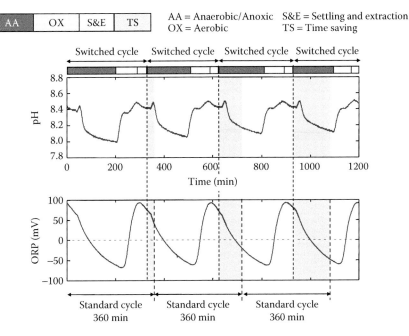

FIGURE 7.67 Experimental record of the fuzzy switching logic of Equations 7.87 through 7.89, applied to a bench scale SBR in the ENEA Laboratories, Bologna, Italy. This logic produces an early switching in both phases and the time saved for each cycle, compared to the standard duration, is indicated by the grey boxes.

from which the defuzzified aerobic output is obtained as

$$y_{OX} = \frac{\sum_{i=1}^{3} y_{OX}^{i} \cdot \mu_{OX}^{i}}{\sum_{i=1}^{2} \mu_{AOX}^{i}} \tag{7.90}$$

which, like the previous y_{AA} of Equation 7.88, represents the likelihood that anaerobic/anoxic phase is about to end. So, the output of the combined FIS is represented by the two mutually exclusive defuzzified values $y = \begin{bmatrix} y_{Aa} & y_{OX} \end{bmatrix}$. To be operational, this output must be converted into a crisp On/Off switching command, which is obtained from either y_{AA} or y_{OX}, whichever is active, via thresholding with an α-cut = 0.95, that is, whichever variable is greater than 0.95 is assumed to be 'true' and drives the controller. Figure 7.67 shows the testing of the fuzzy inference algorithm just described with a pilot SBR operated in the ENEA Bologna Laboratories, confirming that the FIS-driven switching can save a considerable time over the conventional cycle, thus allowing more wastewater to be treated each day. Further applications of this method and more experimental results can be found in the relevant literature (Marsili-Libelli et al., 2008; Spagni et al., 2008; Spagni and Marsili-Libelli, 2009, 2010).

8 Analysis of Aquatic Ecosystems

I hear babies cryin', I watch them grow
They'll learn much more than I'll ever know
And I think to myself, what a wonderful world

Louis Armstrong

In this final chapter, we will review all of the techniques presented in the previous chapters and put them to work with real environmental analysis problems. The emphasis here will be on the importance of combining two or more approaches, and it will be shown that such joint usage can achieve results that could not be obtained by employing a single technique alone. The range of applications presented here covers the various aspects of the aquatic environment. First, an application of an extended Streeter & Phelps (S&P) model will be presented and applied to some small Italian rivers. Then, the analysis will move on to the estimation of bioavailable nutrients in natural waters by controlled microalgae growth. Next, we shall consider the modelling of constructed wetlands, involving both structural modelling and parameter estimation, and, finally, a lagoon ecosystem will be considered from several viewpoints, including modelling and management.

8.1 THE OXYGEN CYCLE IN THE AQUATIC ENVIRONMENT

In this section, a simple dynamic balance for dissolved oxygen (DO) will be introduced, utilizing the concepts of general modelling (Chapter 1), model identification (Chapter 2), and microbial metabolism and oxygen transfer (Chapter 7). As Figure 8.1 shows, this model consists of two consumption terms (sediment and aquatic oxygen demand to oxidize the carbonaceous and nitrogenous biodegradable compounds) and two re-supply terms (diffusion from the atmosphere and photosynthesis). While the former has already been considered in Section 7.2.9 in regard to the oxygen transfer and solubility notions, a model of photosynthesis will now be introduced to describe the other re-supply terms in the oxygen balance. However, first let us consider a more detailed spatial DO description along the water column.

8.1.1 DO Distribution along the Water Column

So far, we have considered the average DO concentration in the water body as it if were evenly distributed from the surface to the bottom. A closer look, though, reveals that for several reasons, its distribution may change considerably with depth, as shown in Figure 8.2. At the surface, the DO concentration equals the saturation value because of atmospheric diffusion, while at shallow depths it may become supersaturated as a consequence of photosynthesis, as this process produces pure oxygen, while the saturation is referred to the air gas mixture, according to Equation 7.37. As light becomes scarcer with depth, photosynthesis decreases, and respiration and sediment oxygen demand (SOD) consume oxygen, lowering its concentration. In the upper layer of the sediment, decomposition processes consume the remaining oxygen until, after a short distance, the DO is fully depleted.

Of course, it makes sense to consider the vertical DO profile only when the depth of the water body is considerable and a euphotic (well-lit) upper zone can be distinguished from a low-light deep zone. As an example, Figure 8.3 shows the seasonal variations of the vertical DO profile in the Bilancino Lake, whose location will be given later in Figure 8.31, where during the summer

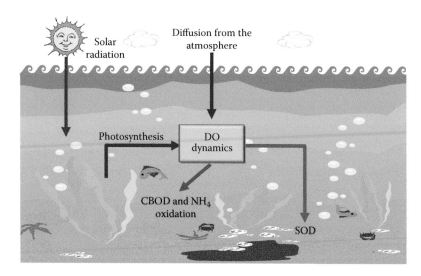

FIGURE 8.1 Pictorial view of the oxygen cycle in natural waters. There are two consumption terms: oxygen demand from the aquatic environment and from the sediment, and two supply factors: photosynthesis and diffusion from the atmosphere.

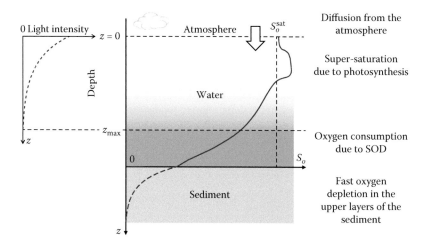

FIGURE 8.2 DO profile along the water column. At the surface, the DO is at the saturation value because of atmospheric diffusion, while at shallow depths it may become supersaturated as a consequence of photosynthesis. As light becomes scarcer with depth, photosynthesis decreases, while the SOD depletes the DO concentration even more. In the upper layer of the sediment, the decomposition processes consume the remaining oxygen.

stratification, a stable gradient is created, which is disrupted by mixing during the cold season. The winter DO profiles (14 February and 30 December) show a well-mixed lake with an almost constant DO profile. The high DO value can be explained by the reduced metabolic activities, given the low temperature. The two summer profiles (23 June and 3 October) denote a strongly stratified lake, with supersaturation in the Epilimnion (upper layer), and a severe oxygen depletion in the Hypolimnion (deep layer) due to the fact that the oxygen consumed by the oxidation processes is no longer resupplied by mixing from the upper layer.

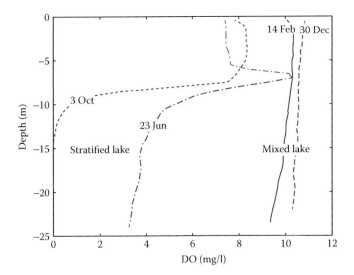

FIGURE 8.3   DO profiles along the water column in the Bilancino Lake, as a consequence of thermal stratification. The winter DO profiles on the right (14 February and 30 December) show that during the winter the lake is well mixed, with no DO gradient. The DO profile is almost constant at a high level due to reduced metabolic activities caused by the low temperature. Instead, the two profiles on the left (23 June and 3 October) describe a strongly stratified lake, with supersaturation in the Epilimnion (upper layer), and a severe oxygen depletion in the Hypolimnion (deep layer) due to consumption processes no longer balanced by mixing (Publiacqua SpA is gratefully acknowledged for the data availability).

8.1.2   The Role of the Aquatic Vegetation

Of the two oxygen supply terms, diffusion from the atmosphere into a natural stream has been treated in Section 7.2.9.1. The other fundamental process of *photosynthesis* is briefly considered here, while more details can be found in specialized textbooks (Jorgensen and Bendoricchio, 2001; Falkowski and Raven, 2007; Kirk, 2011). Photosynthesis is the single most important process in the biosphere. Using the energy supplied by the sun, it converts carbon dioxide into glucose and its polymers, thus providing the primary food source for the consumers. It also closes the carbon and oxygen cycles by reducing the carbon oxidized by respiration (CO_2) and producing oxygen, as shown in the cyclic path of Figure 8.4. The basic photosynthetic reaction can be expressed as

$$6\,CO_2 + 6\,H_2O + h\nu \xrightarrow{\text{Chl-a}} C_6H_{12}O_6 + 6\,O_2 \tag{8.1}$$

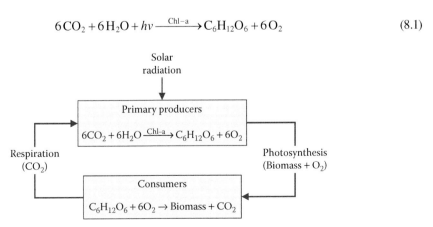

FIGURE 8.4   Energy and matter cycling between primary producers and consumers through the complementary processes of respiration and photosynthesis.

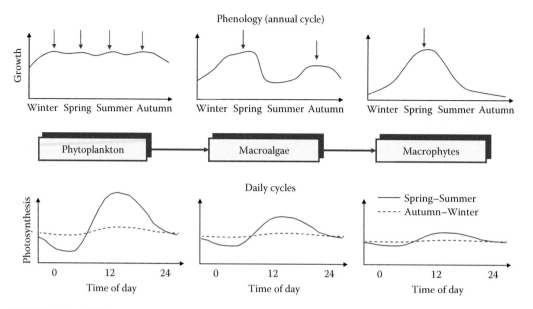

FIGURE 8.5 Differing annual and daily cycles of the submerged vegetation. As a general rule, the smaller the organism, the faster is its response to the environmental conditions. Phytoplankton grows almost all year round and has a marked diel oxygen production, while macroalgae have two growth peaks in spring and autumn, and the macrophytes have only one peak, with rather limited diel variations.

where hv represents the photon energy provided by the solar radiation, which is a function of its wavelength (v).

Special organelles, called *chloroplasts* and containing chlorophyll (Chl-a), are able to absorb the photosynthetically active radiation (PAR) in the waveband between the 690 and 430 nm wavelengths, thus providing the energy required for the reaction (8.1). The huge variety of photosynthetic organisms in the aquatic environment have differing growth rates, depending on several factors that will be examined shortly, but as a general rule it is observed that the smaller the organism, the quicker its response to environmental conditions. As an example, Figure 8.5 shows the differing response of the three major groups of submerged vegetation to the annual and daily cycles.

The growth of the aquatic vegetation depends on three factors: temperature, nutrients, and solar radiation (Chapra, 1997; Chapra and Pelletier, 2003; Cole and Wells, 2015). Each species has a specific optimal range or a limitation zone for these factors, as shown in Figure 8.6, whose effects will now be examined in detail.

8.1.3 SOLAR RADIATION MODELLING

Not all of the incoming solar radiation stimulates photosynthesis, but only the radiation in the 700–400 nm wavelength range. This fraction of the broadband radiation is referred to as PAR and represents the main input to the photosynthesis model that we are going to develop. Measurements and models to calculate PAR have been proposed (Alados et al., 1996; Knyazikhin et al., 1998; Rosati et al., 2004; Sudhakar et al., 2013), and a rough estimate considers that PAR represents about 56% of the incoming radiation. If radiation data are available from a meteorological station near the site of interest, they can be used, after suitable smoothing and interpolation. However, in many cases, such data are difficult to find, so they can be replaced by synthetic data computed on the basis of the following simple astronomical formulas, more details of which can be found

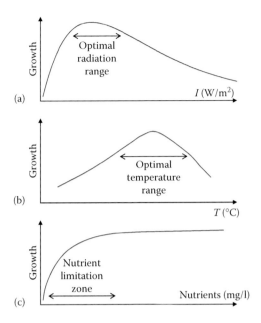

FIGURE 8.6 The three factors controlling the growth of aquatic vegetation. In (a) and (b), there is an optimal range for temperature and solar radiation, while the nutrients availability (c) has a limitation zone.

in France and Thornley (1984). To compute the total daily radiation for a given location, the *year angle* y_n must be defined as

$$y_n = 360 \times \left(\frac{n-80}{365} \right) \tag{8.2}$$

where n is the calendar day beginning on January 1st. The year angle y_n is zero at the vernal equinox, while $n = 80$ corresponds to March 21st.

The other preliminary quantity that must be defined is the *solar declination* δ, defined as the angle (in degrees) between the equatorial plane and the line joining the sun to the earth. This quantity can be computed as a function of the year angle y_n with the following empirical expression adapted from France and Thornley (1984), where $\delta = 0°$ at the equinoxes, $\delta = +23.45°$ on 21 June and $\delta = -23.45°$ on 23 December

$$\delta = 0.38092 - 0.76996\cos(y_n) + 32.265\sin(y_n) + 0.36958\cos(2y_n) + 0.10868\sin(2y_n) \tag{8.3}$$

From Equations 8.2 and 8.3, the *photoperiod f*, defined as the fraction of light hours in the 24-h day length, given the day number n and the latitude φ, can be computed as

$$f = \frac{2\cos^{-1}(-\tan\varphi \cdot \tan\delta)}{360} \tag{8.4}$$

where the constraint $|\tan\varphi \cdot \tan\delta| < 1$ must hold for any n. Because the maximum values of δ is 23.45° and $\tan(23.45) = 0.4348124$, it follows that $|\tan\varphi| < 2.29984$, which yields a maximum φ value of 66.5°, corresponding to the latitude of the polar circles, beyond which a 24-h daylight spell is possible.

The total daily radiation throughout the year can be estimated by fitting a sine function

$$J_n = J_1 + J_2 \sin(J_3 \cdot n + J_4) \tag{8.5}$$

where the four parameters (J_1, J_2, J_3, J_4) are location dependent. On a daily basis the instantaneous radiation $I(t)$ is related to the total radiation J_n according to

$$\begin{cases} I(t) = \dfrac{J_n}{f}\left[1 + \cos\left((t-0.5)\dfrac{360°}{f}\right)\right] & \text{for } 0.5 - \dfrac{f}{2} \le t \le 0.5 + \dfrac{f}{2} \\ I(t) = 0 & \text{otherwise} \end{cases} \tag{8.6}$$

where the 24-h daytime has been normalized between 0 and 1. Obviously, the time outside the interval $0.5 \pm f/2$ represents the dark portion of the day. Figure 8.7 shows the results of the previous astronomical computations for a given latitude (43.7° N). The average daily radiation is computed in (b), recalling that the *solar constant*, that is, the incoming radiation measured outside the atmosphere, is about 1361 W/m², so the most commonly used unit for the solar radiation is watt per square metre (W/m²). Of course, Equations 8.5 and 8.6 do not consider the cloud cover. Figure 8.7a shows the photoperiod computed with Equation 8.4, while (c) displays the trend of the instantaneous radiation at the same latitude and on the summer solstice. The shaded area in (c) represents the total daily radiation and coincides with the dot in (b). Equations 8.5 and 8.6 give the required radiation data, provided

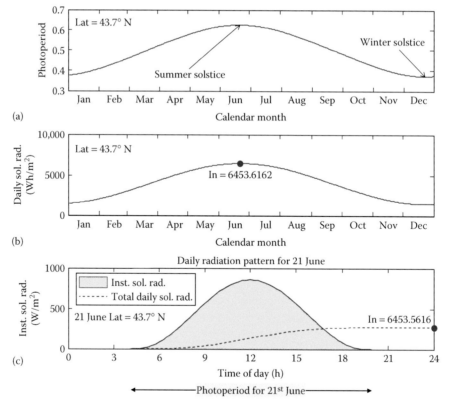

FIGURE 8.7 Astronomical computation of photoperiod (a) through Equation 8.4 and solar radiation (b) for a given latitude (43.7° N) by fitting Equation 8.5 to the data provided by the ENEA Solar Energy computation web site (http://www.solaritaly.enea.it/CalcRggmmOrizz/Calcola1.php). The yellow-filled area shown in (c) represents the daily radiation pattern on the summer solstice, whereas the dotted line shows that its integral eventually equals the total daily radiation for that day (dot in b), within the tolerance of numerical integration. These graphs were obtained with the MATLAB script Go _ Sol _ Rad.m in the A _ ESA _ Matlab\ Exercises\Chapter _ 8\Sol _ Rad folder.

that Equation 8.5 is calibrated for the location of interest. As noted, these equations do not consider light attenuation by the cloud cover, or by other factors that may attenuate the solar radiation actually reaching the surface. These correction factors may be introduced as a multiplicative factor. Their variability can be modelled as random series on the basis of statistical observations, following the method used in Section 3.3.4.1. for synthetizing the wind series in the Orbetello lagoon.

8.1.4 PHOTOSYNTHESIS MODELLING

On a daily basis, the photosynthetic rate is a function of the incoming solar radiation provided by Equation 8.6. Terrestrial and aquatic plants have very different responses, with the former exhibiting a monotonic response

$$\varphi_T(t) = \frac{I(t)/I_k}{\sqrt{1+\left(I(t)/I_k\right)^2}} \tag{8.7}$$

while aquatic plants have their maximum photosynthetic rate confined to a range of radiation intensities

$$\varphi_A(t) = \frac{I(t)}{I_s} \cdot e^{1-\frac{I(t)}{I_s}} \tag{8.8}$$

implying that photosynthesis is inhibited at high light intensities. The two shape parameters, I_k in Equation 8.7 and I_s in Equation 8.8, have differing values and meanings and are species specific. These light response functions are shown in Figure 8.8. Equation 8.8 is, in principle, applicable to all kinds of aquatic plants, from phytoplankton (Steele, 1962; Bannister, 1974; Jassby and Platt, 1976) to macrophytes (Lapointe et al., 1984).

The photosynthetic oxygen production can be integrated over the optical depth of the water body. In this operation, we should take into account the light attenuation down the water column, until a depth is reached (optical depth) defined as the depth z_{max} where its intensity is too weak to sustain photosynthesis. The Beer–Lambert law is used to describe the light attenuation with the depth z from the water surface

$$I(z) = I_o \cdot e^{-k_e \cdot z} \tag{8.9}$$

where k_e is the attenuation factor, which may depend on both inert suspended material and vegetation density (self-shading). In the case of free floating vegetation, such as algae and phytoplankton, the total photosynthesis can be computed over the entire water column, from the surface to the

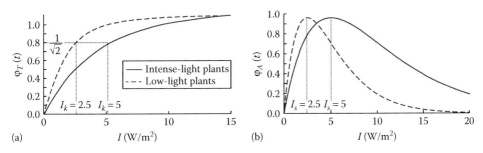

FIGURE 8.8 Photosynthetic response to the solar radiation of terrestrial plants (a), described by Equation 8.7, and of aquatic plants (b), described by Equation 8.8.

optical depth z_{max}. Integrating the photosynthetic rate (8.8) over the optical depth and taking into account the light attenuation (8.9) yields

$$\Phi(t) = \int_0^{z_{max}} \left(\frac{I_o(t)}{I_s} \cdot e^{-k_e \cdot z} \times e^{1 - \frac{I_o(t)}{I_s} \cdot e^{-k_e \cdot z}} \right) dz \tag{8.10}$$

thus providing the total instantaneous photosynthesis

$$\Phi(t) = \frac{1}{k_e} \left(e^{1 - \frac{I_o(t)}{I_s} \cdot e^{-k_e \cdot z_{max}}} - e^{1 - \frac{I_o(t)}{I_s}} \right) \tag{8.11}$$

If we limit ourselves to considering the average daily radiation I_m over the photoperiod f, Equation 8.11 simplifies into

$$\Phi_m = \frac{f}{k_e} \left(e^{1 - \frac{I_m}{I_s} \cdot e^{-k_e \cdot z_{max}}} - e^{1 - \frac{I_m}{I_s}} \right) \tag{8.12}$$

Numerical methods for the integration of photosynthesis can be found in Walsby (1997).

8.1.4.1 Temperature Limitation

Each aquatic species has an optimal range of temperatures, which can be approximated with one of the following expressions:

$$f_T = \theta^{(T - T_o)} \tag{8.13}$$

$$f_T = \frac{1}{1 + \left[(T - T_o)/a \right]^b} \tag{8.14}$$

Equation 8.13 is the typical Arrhenius temperature dependence $(\theta \simeq 1.066)$, while Equation 8.14 defines a bell-shaped range of temperatures around the optimal value T_o, with a and b as shape parameters. At temperate latitudes the solar radiation is saturating, so the main limiting factor becomes the photoperiod f, and many species have a 'window' of optimal combinations of temperature and photoperiod, whose influence on growth can be described by the following function

$$f_f = 1 - \frac{1}{1 + c \cdot e^{g(f - f_o)}} \tag{8.15}$$

Combining Equation 8.14 with Equation 8.15, a joint temperature–photoperiod limiting function is obtained

$$f_{T,p} = \frac{1}{1 + \left[(T - T_o)/a \right]^b} \times \left[1 - \frac{1}{1 + c \cdot e^{g(p - p_o)}} \right] \tag{8.16}$$

Considering the daily temperature and the photoperiod recorded at the Orbetello lagoon (see Figure 8.31) in 2002, Figure 8.9 is obtained, showing that the May–July period is the most favourable for the development of the algae and the macrophytes in the lagoon.

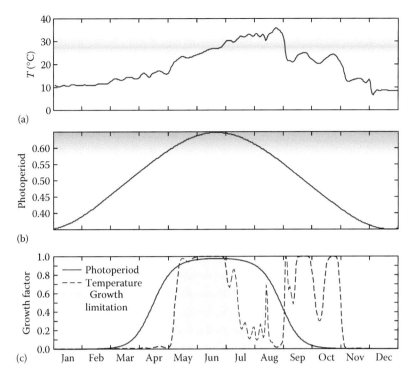

FIGURE 8.9 Combined effect of temperature and photoperiod, according to Equation 8.16, in determining the joint growth factor. The coloured bands indicate the optimal ranges for temperature (a) and photoperiod (b). Their combined effect is shown in (c), where the shaded area indicates the most favourable combination. The data refer to the year 2002 in the Orbetello lagoon.

8.1.4.2 Nutrients Limitation

Inorganic nitrogen and phosphorus limit the development of the aquatic vegetation, and various expressions of the Michaelis–Menten kind have been proposed to model this limitation, that is, with a saturation for large concentrations. The most commonly used expressions are

Minimum law:

$$\Phi_N^{\min} = \min\left(\frac{N_1}{K_1 + N_1}, \frac{N_2}{K_2 + N_2}, \cdots, \frac{N_m}{K_m + N_m}\right) \tag{8.17}$$

Product law:

$$\Phi_N^{\text{prod}} = \frac{N_1}{K_1 + N_1} \times \frac{N_2}{K_2 + N_2} \times \cdots \times \frac{N_m}{K_m + N_m} \tag{8.18}$$

Harmonic mean law:

$$\Phi_N^{\text{harm}} = \frac{1}{\dfrac{1}{\dfrac{N_1}{K_1 + N_1}} + \dfrac{1}{\dfrac{N_2}{K_2 + N_2}} + \ldots + \dfrac{1}{\dfrac{N_m}{K_m + N_m}}} \tag{8.19}$$

Combining the previous limiting terms yields the dynamics of photosynthesis, which can be expressed as the increase in chlorophyll-a (Chl-a) inside the primary biomass, from which the carbon growth rate and the oxygen production rate can be deduced. Approximately, the average ratio (α_o) of organic matter (OM) to Chl-a is of the order of 145 mg OM dry weight per μg Chl-a (Cole and Wells, 2015), and the oxygen production (α_1) and uptake for respiration (α_2) are of the order of 1.4–1.8 and 1.6–2.3, respectively (Brown and Barnwell, 1987). Consequently, the basic Chl-a dynamics can be written as

$$\frac{d\text{Chl}}{dt} = \underset{\text{growth}}{k_g \cdot \varphi_A(t) \cdot f_T \cdot f_p \cdot \Phi_N} - \underset{\text{decay}}{k_a \cdot \text{Chl}} \tag{8.20}$$

which translates into the OM dynamics

$$\frac{dA}{dt} = \underset{\text{growth}}{\alpha_o \cdot k_g \cdot \varphi_A(t) \cdot f_T \cdot f_p \cdot \Phi_N} - \underset{\text{decay}}{\alpha_o \cdot k_a \cdot \text{Chl}} \tag{8.21}$$

and into the DO dynamics

$$\frac{dS_o}{dt} = \underset{\text{re-aeration}}{K_r \left(S_o^{\text{sat}} - S_o \right)} + \underset{\text{photosynthesis}}{\alpha_1 . \alpha_o . k_g \cdot \varphi_A(t) \cdot f_T \cdot f_p \cdot \Phi_N} - \underset{\substack{\text{CBOD and SOD} \\ \text{oxidation}}}{R(t)} - \underset{\text{algal respiration}}{\alpha_2 . \alpha_o \cdot k_a \cdot \text{Chl}} \tag{8.22}$$

8.2 A SHORT-TERM OXYGEN DYNAMICAL BALANCE

The three factors controlling photosynthesis in Equation 8.20 operate on widely differing time-scales, and their influence will be examined in later sections of this chapter. While the temperature influence is predominantly seasonal and the nutrients operate on a monthly to weekly time-scale, the solar radiation determines the daily fluctuations in DO, which is the most important supply term, at least during the spring and summer seasons. In this section, we are going to develop and test a short-term (daily) oxygen balance that can be set up by considering only the influence of the solar radiation and assuming that the other factors (nutrients and temperature) remain constant over a daily time horizon.

Figure 8.10 illustrates this point by showing the large day DO swings due to daytime photosynthesis that are frequently observed in rivers during summer, as analysed in Marsili-Libelli (1991). The fact that the DO exceeds the saturation value during the time of maximal solar radiation is due to photosynthesis, which produced pure oxygen, while the saturation is referred to the air gas mixture, hence DO concentrations above the saturation value are frequently observed during the light hours. During these periods, oxygen production is greater than its consumption $(P > R)$, while the reverse is true during the night and low-light hours, when photosynthesis is weak or absent, but respiration continues all the same $(P < R)$. A simple model can now be proposed to describe the daily DO fluctuations, considering the river reach as a CSTR whose oxygen balance can be written from Equation 8.22 in a simpler form

$$\frac{dS_o}{dt} = K_r \left(S_o^{\text{sat}} - C \right) + a \frac{I(t)}{I_s} e^{1 - \frac{I(t)}{I_s}} - R_t \tag{8.23}$$

where the first term represents the oxygen transfer rate from the atmosphere, and the second term is the photosynthetic oxygen production rate. The last term (R_t) is representative of all the respiration processes, including algae respiration, carbonaceous biochemical oxygen demand (CBOD) oxidation and SOD. Although the equivalence coefficient $(a = \alpha_o \cdot \alpha_1)$ between photosynthesis and oxygen production rate is, in principle, known, its variability with the prevailing algal species and the environmental conditions suggests estimating it from the data, rather than using literature values.

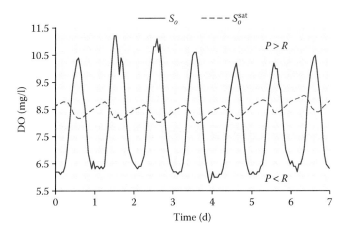

FIGURE 8.10 A week-long record of diurnal DO fluctuations in the Arno River upstream of the city of Florence during the last week of August 2006. During the daytime, the photosynthetic oxygen production drives the DO concentration well above the saturation level, whose value varies as a consequence of the temperature variations. In these conditions, production (P) prevails over respiration (R), while during the night hours, the reverse is true ($P < R$).

The same considerations apply to the other constants in Equation 8.23, such as the re-aeration coefficient (K_r) and the 'knee radiation' (I_s). The global respiration rate (R_t) incorporates a large number of differing time-varying processes, but on the short time-scale it can be considered as a constant and estimated along with the other parameters. Therefore, Equation 8.23 is a typical case of a simple model whose validity heavily depends on a robust identification. This will be demonstrated along the guidelines outlined in Chapter 2. First, the error functional is defined as follows:

$$E = \sum_{k=1}^{24} \left(S_o^{mod}(k) - S_o^{obs}(k) \right)^2 \tag{8.24}$$

where:
S_o^{mod} are the hourly DO values obtained from the model
S_o^{obs} are the corresponding observations

Then, the estimation problem consists of minimizing this error functional with respect to the four parameters, that is,

$$\min_{(K_r, R_t, a, I_s)} E \tag{8.25}$$

Before undertaking the actual minimization of the functional (8.24), the parameter sensitivity, both static and dynamical, will be considered, and after the optimal parameters have been obtained, a validation with the regression line F-test will follow. The data from two consecutive days (3 and 4 August 2006) were considered, collected by the regional environmental authority (ARPAT) in the Arno River upstream of the city of Florence. The first day was used for the calibration, and the following day for the validation. First, the static sensitivities, shown in Figure 8.11, were computed and did not reveal any critical situation, because all the parameters appear to have regular sensitivity plots. Then, the trajectory sensitivities were computed (Figure 8.12), selecting an incremental perturbation $\delta = 0.001$. The maximal sensitivity intervals, highlighted by the shaded areas, are almost all located in the central daytime hours, save for the re-aeration coefficient K_r, which has a

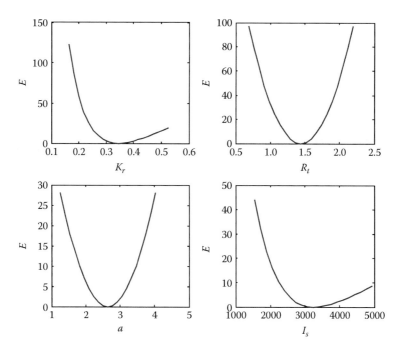

FIGURE 8.11 Static sensitivity of the model parameters in Equation 8.23. These graphs were obtained with the MATLAB script Go _ River _ Resp.m in the ESA _ Matlab\Matlab _ Examples _ Home\ Chapter _ 8\River _ Respiration folder.

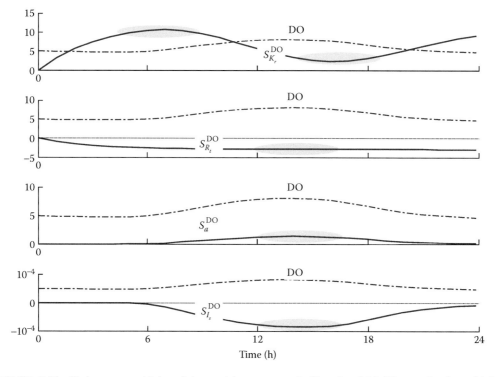

FIGURE 8.12 Trajectory sensitivity of the model parameters in Equation 8.23. The maximal sensitivity intervals are shaded. These graphs were obtained with the MATLAB script Go _ River _ Resp.m in the ESA _ Matlab\Matlab _ Examples _ Home\Chapter _ 8\River _ Respiration folder.

TABLE 8.1
Estimated Parameters of the DO Model (Equation 8.23)

Parameter	Units	Sensitivity Ranking	Estimated Value	Confidence Interval
K_r	1/h	34.6373	0.3446	±0.1397
R_t	mgO$_2$/l h	13.0315	1.4402	±0.098782
a	mgO$_2$/l h	3.8282	2.6408	±1.222
I_s	W/m^2	0.0024	3244	±2.2538

maximum around dawn and another in the afternoon, because in these intervals it is comparatively the most important oxygen supply term. However, having hourly DO data covering the entire period, it was decided to use them all, though their efficiency in the estimation depends on the time of day.

The parameters were then ranked according to their sensitivities by computing the following quantity:

$$S_p = \sqrt{\sum_{k=1}^{24} \left(S_p^{DO}(k)\right)^2} \qquad (8.26)$$

The sensitivity ranking is shown in Table 8.1, together with the estimated parameter values and their confidence intervals. It appears that the sensitivities are distributed along a wide range of values, with the re-aeration coefficient being the most sensitive and the 'knee radiation' the least sensitive. However, given the lack of critical aspects in the estimation, it was decided to estimate the whole lot, even if the number of data is only just sufficient. The Fisher Information Matrix (FIM) and the covariance matrix were computed to determine the parameters confidence intervals of Table 8.1, while Figure 8.13 shows the contours of the error functional (8.24) in the subspace of the two most sensitive parameters (K_r, R_t), whose 95% confidence region is indicated by the dotted contour. Finally, Figure 8.14 shows the calibration and validation results, together with the corresponding solar radiation data, representing the model input.

To validate the estimated parameters, a regression line F-test is performed, for both the calibration and the validation runs. The results, shown in Figure 8.15, confirm that in both cases there is no reason to reject the hypothesis that the model response is in agreement with the observed data.

8.3 ANALYSIS OF THE CIRCADIAN DO CYCLES BY WAVELET FILTERING AND FUZZY CLUSTERING

The validity of the previous DO model is limited to a few consecutive days, beyond which its simplifications are no longer valid. Any attempt to extend its time horizon is doomed to failure because it would conflict with the assumption of considering respiration terms as constant and neglecting the external inputs, while that model considered the river reach as a virtual batch. Furthermore, with that model, little progress is made, both into identifying the causes that produce the fluctuations and in interpreting their ecological significance. In this section, we take a step further in the diagnostic aspect, considering the causes responsible for the DO daily swings, which represent an environmental indicator. However, rather than attempting to build a dynamical model, this section will describe a method for diagnosing the state of the aquatic ecosystem from the observed DO variations, using a method based on wavelet denoising, described in Section 3.4, and fuzzy clustering, illustrated in Section 4.5. This study was carried out in the Orbetello lagoon, which is located on Italy's west coast, along the Tyrrhenian sea, and consists of two shallow coastal ponds with a combined surface of approximately 27 km^2 and an average depth of 1 m. Two water quality monitoring

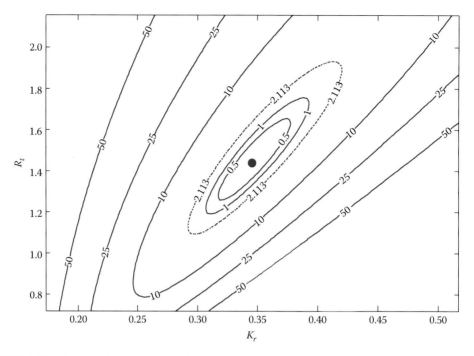

FIGURE 8.13 Contour of the error functional (8.24) for the two most sensitive parameters in Equation 8.23. Its regular shape, together with the absence of local minima or of narrow valleys, rules out numerical problems for the Simplex search. The dotted contour line indicates the 95% confidence region of the estimated parameters. These graphs were obtained with the MATLAB script Go _ River _ Resp.m in the ESA _ Matlab\Matlab _ Examples _ Home\Chapter _ 8\River _ Respiration folder.

stations, indicated by the two circles in Figure 8.16, communicate hourly physico-chemical data to the Orbetello Lagoon Managerial Office headquarters.

Around 2003, a scientific committee was set up to study the lagoon ecosystem and take remedial actions to improve the deteriorating environmental quality caused by the accumulating nutrients in the sediment and ensuing eutrophication. The main problem in the Orbetello Lagoon is the control of the submersed vegetation, given the critical coexistence between macroalgae (*Chaetomorpha linum, Cladophora vagabunda, Gracilaria verrucosa, Ulva rigida*) and rooted macrophytes, mostly *Ruppia maritima* (de Biasi et al., 2003; Lenzi et al., 2003). Macroalgae are of epiphytic origin, but, after reaching their maturity, they float in dense mats and absorb a large quantity of nutrients, eventually producing sudden blooms that cause dystrophic crises (Park and Jaffé, 1996; Christian et al., 1998; Azzoni et al., 2001; Zeng et al., 2006). In fact, when the macroalgae decompose after their bloom, the oxygen consumption due to their decomposition exceeds the photosynthetic production, and the imbalance may cause hypoxia or anoxia. The causes of the dystrophic crises are well understood and documented (Christian et al., 1996; Azzoni et al., 2001; Melia et al., 2003; Nizzoli et al., 2006; Giordani et al., 2009), and, by analysing the DO fluctuations, we are able to assess the quality of the ecosystem and predict possible dystrophic crises determined by an abnormally low DO level. Thus, the shape of the DO daily cycle contains important season-dependent ecological information, which this method attempts to extract from noisy data. As we have seen, daytime DO has a well-defined afternoon peak due to photosynthesis, often well above the saturation level, because as explained in Sections 7.2.9 and 8.2, the DO saturation is referred to the oxygen percentage in the air (20.9%, see Equation 7.38), while photosynthesis produces *pure* oxygen. Thus, the saturation concentration may increase, as predicted by Equation 7.39, from about 9 to over 42 mg/l. When the growth phase ends, the fast anoxic decomposition enriches the sediment with reduced organic nitrogen (Christian et al., 1998). These reducing conditions can be detected by low, almost constant DO

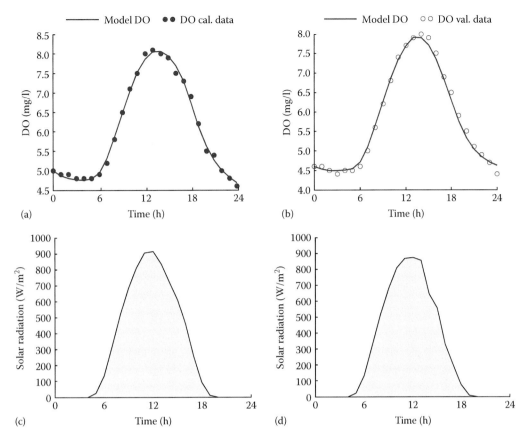

FIGURE 8.14 Calibration and validation of the parameters in Equation 8.23. The calibration data and the calibrated model response are shown in (a), while the input solar radiation for the same day is shown in (c). The same quantities for the validation data are shown in (b) and (d). These graphs were obtained with the MATLAB script `Go _ River _ Resp.m` in the `ESA _ Matlab\Matlab _ Examples _ Home\ Chapter _ 8\River _ Respiration` folder.

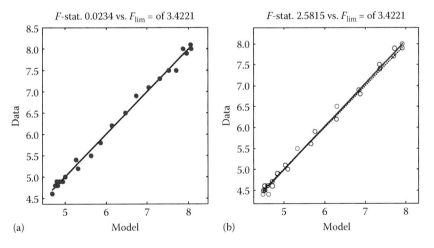

FIGURE 8.15 F-test for the calibration (a) and validation (b) data for the model (Equation 8.23). In both cases, the regression line F-test is passed. These graphs were obtained with the MATLAB script `Go _ River _ Resp.m` in the `ESA _ Matlab\Matlab _ Examples _ Home\Chapter _ 8\River _ Respiration` folder.

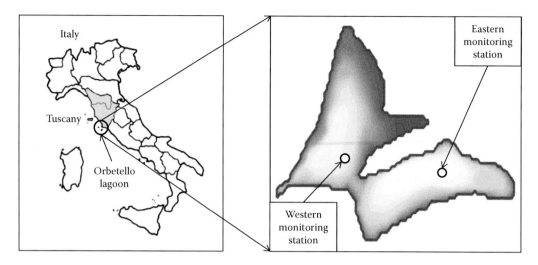

FIGURE 8.16   Location of the Orbetello lagoon, at the southern border of Tuscany (left). It is composed (right) of two communicating shallow coastal ponds.

and low daytime oxidation–reduction potential (ORP) in the water, which corresponds to near-zero or negative ORP at the sediment surface, as reported by Azzoni et al. (2001).

The analysis of the diel DO cycles is based on wavelet filtering and fuzzy clustering, as described in Marsili-Libelli and Arrigucci (2004). The reasons for applying this two-step procedure are the following:

1. DO data are normally affected by noise in many ways, and a dyadic wavelet decomposition (Section 3.4.2), with its multi-scale property, can reveal data trends that other signal analyses can miss. It can also denoise the data without appreciable signal degradation by retaining low-frequency components, which are usually ecologically meaningful, and rejecting high-frequency disturbances.
2. Once the basic patterns are isolated, the filtered data are sorted into a number of meaningful behaviours through fuzzy clustering. In particular, the Fuzzy Maximum Likelihood Estimator (FMLE) algorithm (Section 4.5.9) is used for its variable metric and cluster volume adaptability.

Figure 8.17 shows the algorithm organization used to extract the relevant information from the DO daily cycles: first, the daily data are denoised by a dyadic wavelet decomposition, and then the relevant features are extracted and clustered to determine the relevant prototypes. The correspondence between such prototypes and specific ecological conditions is then assessed.

8.3.1   DYADIC DECOMPOSITION DENOISING

Recalling the notions explained in Section 3.4.4, the dyadic wavelet decomposition of a signal at the k level is fully represented by the coefficients of the *Approximation* cA_k and of the *Detail* cD_k. The latter contains most of the noise component, which *adapts* to the signal behaviour. Denoising can be obtained by limiting the details by thresholding, that is, compressing the coefficients with magnitude below the threshold towards zero. Then, the denoised signal is obtained by combining the original *Approximation* coefficients cA_k and the *modified Details* $(cD_1', cD_2', \ldots, cD_k')$, as shown in Figure 3.32. The resulting signal is smooth enough to provide a reliable extraction of the three features defined in Figure 8.18. For this application, the Meyer wavelet was found to be the best

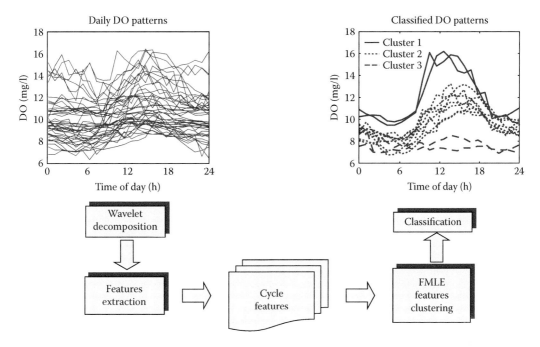

FIGURE 8.17 Algorithm organization to extract circadian patterns from the DO data in the Orbetello lagoon. The various daily DO data (top-left graph) are filtered by a dyadic wavelet decomposition, then the features describing each cycle are extracted and clustered by similarity, isolating significant patterns. The upper-right corner graph shows the various patterns, identified by differing line styles.

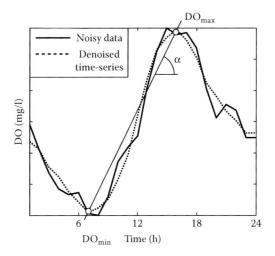

FIGURE 8.18 Features identifying the denoised DO daily cycle: the minimum and maximum DO value and the angle of the line connecting these extremes. The denoised signal is compared to the original noisy data.

performer, and a level 2 decomposition proved sufficient to remove the unwanted noise without excessively decreasing the number of available data per cycle.

8.3.2 DIURNAL DO CYCLE CHARACTERIZATION

Some distinctive features must now be defined to characterize each daily DO cycle. The features' selection must be parsimonious yet exhaustive at the same time, and after some trial and error the triplets of features illustrated in Figure 8.18, minimum DO (DO_{min}), maximum DO (DO_{max}), and the angle of the line connecting these two extremes (α), provided an efficient description of the daily DO cycle. After clustering the DO data, the corresponding temperature, pH and ORP patterns were associated to them for further processing.

8.3.2.1 Fuzzy Clustering

The triplets of data corresponding to the DO cycles were clustered using the FMLE algorithm, given its cluster volume adaptability. The optimal number of clusters $(c = 3)$ was determined with the partition entropy criterion (Section 4.5.3.1), and the patterns that received a low membership, near the theoretical minimum of 1/3, to all the three clusters, were labelled as unclassified. Several years between 2000 and 2005 were analysed. As an example, Figure 8.19 shows the results for 2001, where three clusters could classify the majority of the observed cycles, with the following characteristics:

- *Cluster 1* groups typical late summer behaviours, where the DO takes almost any value with little diurnal variation. This behaviour is typical of a mature algal population, which is less sensitive to the solar radiation.

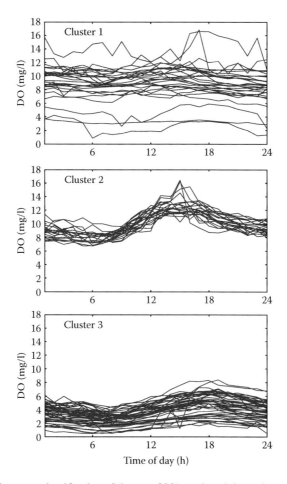

FIGURE 8.19 The DO pattern classification of the year 2001 produced three clusters, with definite seasonal characters.

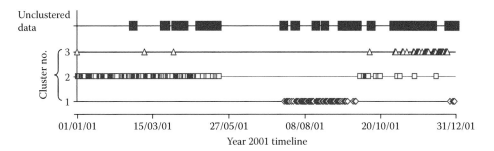

FIGURE 8.20 Seasonal distribution of the classified and unclassified patterns of Figure 8.19. The data that receive a membership to all clusters near to 1/3 are classified as 'unclustered'.

- *Cluster 2* represents typical end-of-winter and spring situations, with very close cycles characterized by strong afternoon peak due to intense photosynthesis.
- *Cluster 3* is typical of the winter situation, with low oxygen due to a declining and decaying algal population.
- *Unclustered cycles* present none of the previous characteristics. They may occur almost randomly throughout the year, and may be caused by adverse weather conditions, which prevent the development of the full diel cycle.

Figure 8.20 shows the distribution of the three clusters, plus the unclustered data, over the year 2001. It can be seen that there is a considerable match between the seasonality and each classified cluster, with the unclustered cycle distributed evenly throughout the year.

Repeating this procedure for the data collected during 2003, the entropy criterion indicates that four clusters are the best choice for these data, and the timeline of these patterns is shown in Figure 8.21. However, a closer look into cluster 2 reveals that a further partition is possible into clusters 2a and 2b.

As shown in Figure 8.22, this subpartition is obtained by clustering the ORP data that exhibit two differing patterns. This discrimination is particularly relevant because it isolates the pattern associated with a dystrophic crisis. This is characterized by very low daytime DO and large negative ORP, both indicative of strongly reducing chemical conditions at the sediment surface, which is conducive to nutrient release and subsequent eutrophication. The dystrophic patterns of DO and ORP during the crisis are shown in Figure 8.23, where the DO is very close to zero, and with almost no daytime rise, while the ORP has deep night time negative peaks.

8.4 EXTENDING THE SPATIAL–TEMPORAL VALIDITY OF DO MODELS

The two previous sections have considered the analysis of the diel DO variations from fundamentally differing viewpoints. In Section 8.2, we have developed a simple dynamical model with a very limited time horizon, while in Section 8.3 the time horizon of the fluctuations was extended to the full yearly cycle and analysed from the pattern recognition viewpoint, but the dynamical nature of the fluctuation was disregarded. In this section, we shall attempt a more comprehensive description of the DO dynamics, reintroducing its dynamical nature in the short time-scale, and yet retaining the seasonal characterization of a long time horizon. The motivation for this study was the already-mentioned environmental remediation project for the Orbetello lagoon, for which a complex ecological model was developed (Giusti and Marsili-Libelli, 2005, 2006). This section summarizes the results presented in Giusti and Marsili-Libelli (2009) and explains how the short and long time-scales can be combined in a bank of models with a fuzzy activation depending on the circumstances.

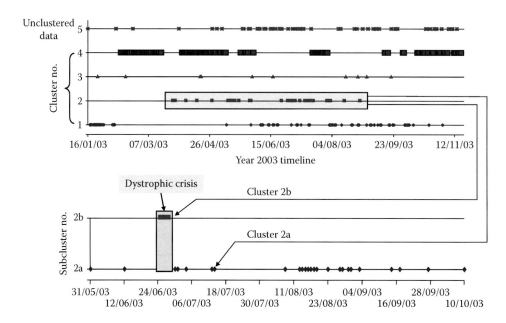

FIGURE 8.21 The timeline of the daily patterns during 2003. The inner structure of cluster 2 isolates the dystrophic crisis. The data which have a membership to all clusters near to 1/4 are classified as 'unclustered'.

To some extent, this study complements and extends the previous case study of Section 8.3, in which the circadian DO fluctuations were just regarded as behavioural classes. Now the classification is made on models, rather than on behaviours, and we shall see that this has important implications for the model identification. The structure of the dynamical DO model describing the diel fluctuations as a function of the supply and consumption terms is shown in Figure 8.24, while its details can be found in the relevant publication (Giusti and Marsili-Libelli, 2009; Giusti et al., 2010). The DO module integrates into a comprehensive lagoon model, initially termed 'LaguSoft 1.0' (Giusti and Marsili-Libelli, 2005), and later enhanced into 'LaguSoft 2.0' (Giusti and Marsili-Libelli, 2006). Here, we use the integrated model shown in Figure 8.25, obtained by the integration of LaguSoft 2.0, and the DO model of Figure 8.24, to explain the daily DO cycles on a yearly basis.

Using the hourly DO data collected in the years 2001–2004, the most representative daily patterns were extracted with the same techniques described in the previous section, obtaining the four clusters of ephemeral parameters $(DO_{min}, DO_{max}, \alpha)$ shown in Figure 8.26, together with their centroids whose numerical values are listed in Table 8.2.

Using the DO patterns corresponding to the prototypical diel cycles, four models were identified, as shown in Figure 8.27.

In these models, all of the parameter values were drawn from the literature or from previously calibrated models, save for four parameters that were specifically calibrated for this application, as described in Giusti and Marsili-Libelli (2009). Now, the problem is how to blend these four models to find the best cycle approximation for any calendar day. The most obvious solution could be the fuzzy combination of all the model responses, according to the membership of the day with respect to the four clusters, but this approach produced disappointing results. A much better result was obtained, as described in Giusti and Marsili-Libelli (2009), by using a single model whose parameters are a fuzzy combination of the four prototypical parameter sets, weighted by the degree of activation of the corresponding cluster, as illustrated in Figure 8.28. A validation run of the DO model with fuzzy patched parameters is shown in Figure 8.29, where the model was tested against the observations in the most critical spring period. Between 1 and 3 June the data acquisition was disrupted by a DO probe failure, but the model still provided a consistent output.

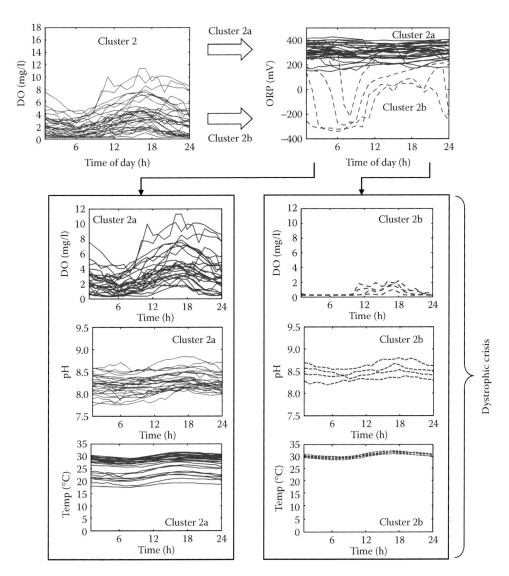

FIGURE 8.22 The finer cluster structure of the 2003 data, revealing a subcluster (2b) containing the typical features of the dystrophic crisis, with strongly negative ORP and high pH.

The integrated DO model, thus calibrated and tested, was used to predict the DO concentrations over the lagoon, given the calendar day and the previously modelled synthetic solar radiation data. As an example, Figure 8.30 shows the DO spatial distribution at four differing days, considered sufficiently representative of the four seasonal behaviours. Spring is confirmed as a highly transitional period, with clusters 3 and 4 being activated almost equally, whereas summer, autumn and winter are well represented by clusters 2, 4, and 1, respectively. In agreement with observations, summer has the lowest DO distribution, whereas winter has the highest, thanks to a diminished vegetation density and lower temperatures. In the eastern lagoon, the DO concentrations are usually lower, given the higher macroalgae concentration, but the widgeon grass prairie in the southern part produces a considerable oxygenation. In general, the DO-depleted areas as indicated by the model were confirmed by direct observations.

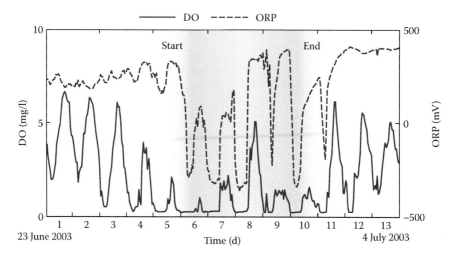

FIGURE 8.23 Time portrait of the dystrophic crisis in the Orbetello lagoon at the end of June 2003, indicated by the shaded area, whose density is proportional to the 2b cluster membership. The onset of the crisis is foretold by the lowering DO values, followed by the ORP plummeting to large negative values. The end of the episode is first indicated by a sharp ORP rising, followed by the sluggish DO increasing fluctuations. (Courtesy of the Orbetello Lagoon Scientific Committee, Grosseto, Italy.)

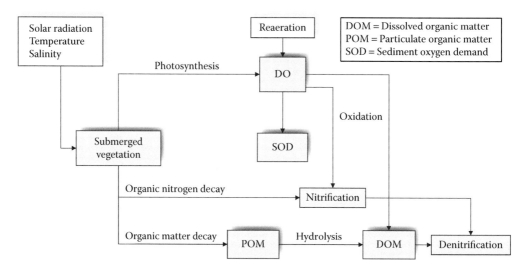

FIGURE 8.24 Structure of the dynamical DO module to reproduce the diel DO fluctuations in the Orbetello Lagoon: The DO dynamics has two supply terms (re-aeration and photosynthesis) and three consumption terms (nitrification, oxygen demand from DOM and SOD). (Reproduced with permission from Giusti, E. and Marsili-Libelli, S., *Ecol. Model.*, 220, 2415–2426, 2009.)

8.5 BEYOND STREETER & PHELPS: A WATER QUALITY MODEL CASE STUDY

The case study proposed in this section draws heavily from the notions explained in Chapter 2 (parameters estimation, FIM), Chapter 6 (Flow dynamics, follow-the-plug), and Chapter 7 (reaction kinetics, nitrogen and phosphorus dynamics in rivers). The discussion that follows is a résumé of the material presented in Marsili-Libelli and Giusti (2008). The pioneering S&P model considered in the previous chapters is a very effective pedagogical tool to introduce the concepts of river

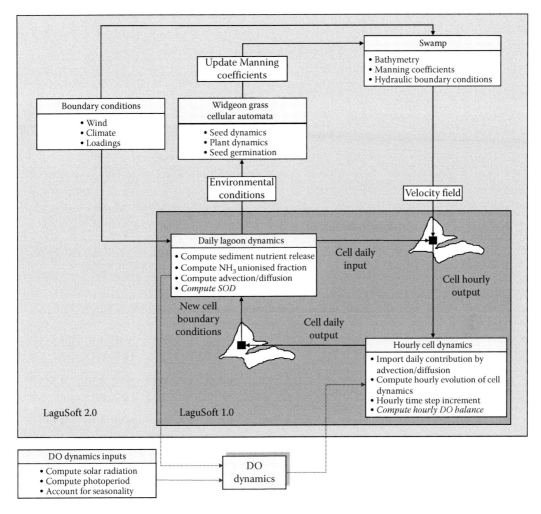

FIGURE 8.25 Integration between LaguSoft 2.0 (Giusti and Marsili-Libelli, 2006) and the DO dynamics of Figure 8.24 to develop the present DO cycle modelling. (Reproduced with permission from Giusti, E. and Marsili-Libelli, S., *Ecol. Model.*, 220, 2415–2426, 2009.)

quality illustrated in the previous chapters (Rinaldi et al., 1979). Furthermore, its structural simplicity allows a quick and reliable identification of its parameters. On the other hand, it oversimplifies the reactions that determine the water quality in a river, and for this reason it has been superseded by more detailed models, which include not only the fate of carbonaceous substances, but also the kinetics of nutrients, the contribution of algae to DO balance through photosynthesis, and the fate of toxicants. The most comprehensive model of all these processes is the IWA River Model N. 1, RWQM1 (Reichert et al., 2001), which was produced by a task group set up by the International Water Association (IWA) along the same successful guidelines that resulted in the already cited activated sludge models (ASM) (Henze et al., 2000), described in Chapter 7. Subsequent studies were aimed at assessing the structural properties of RWQM1 and the identifiability of its parameters (Reichert and Vanrolleghem, 2001). Before RWQM1 was published, other public domain water quality models were made available by the United States Environmental Protection Agency (US EPA), such as QUAL2E (Brown and Barnwell, 1987), later improved to become QUAL2K (Chapra and Pelletier, 2003), and eventually QUAL2Kw (Pelletier et al., 2006), equipped with a parameter calibration functionality based on genetic algorithms, while a more general two-dimensional model

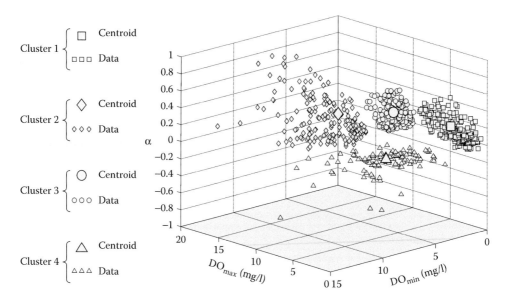

FIGURE 8.26 Four clusters of daily DO patterns were defined on the basis of the three DO features of Figure 8.18. The larger grey shapes indicate the cluster prototypes, whose numerical values are listed in Table 8.2. (Reproduced with permission from Giusti, E. and Marsili-Libelli, S., *Ecol. Model.*, 220, 2415–2426, 2009.)

TABLE 8.2

Prototypical Values of Daily DO Patterns in the Orbetello Lagoon DO Model

Cluster	DO_{min}	DO_{max}	α
1	7.86	10.71	0.31345
2	2.88	6.61	0.45162
3	3.67	6.62	0.40291
4	1.16	2.65	0.08602

Source: Giusti, E. and Marsili-Libelli, S., *Ecol. Model.*, 220, 2415–2426, 2009.

---

with enhanced hydrodynamic and water quality capabilities has been released from the Portland State University and the US Army Corps of Engineers (Cole and Wells, 2015). AQUATOX (Clough, 2014; Park and Clough, 2014) is a more specialized EPA software package dealing with toxic materials. It is a general ecological risk model that gives the combined fate and effects of toxic chemicals, with the aid of an extensive internal library of chemical molecules, plants and animals. In this book, however, we strive to use MATLAB®/Simulink® as a tool for translating ideas into working computer codes, so we shall confine our development to this domain, while it is left to the reader to explore the vast and varied world of public domain water quality models just described.

While the QUAL2 model family is appropriate for rivers of considerable length and detailed ecological structure, an ad hoc approach may be preferable for smaller rivers, which pose specific problems due to data scarcity, lack of major investments as a consequence of their minor importance and the large number of diverse inputs, especially if they flow through densely populated areas.

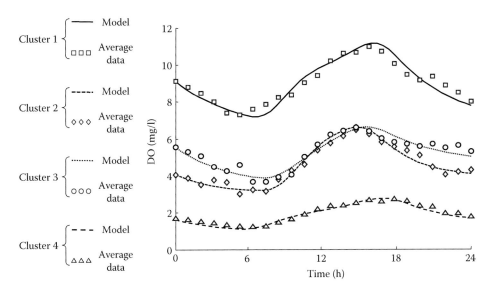

FIGURE 8.27 Identification of the four prototypical DO models corresponding to the four prototypes of Figure 8.26. (Reproduced with permission from Giusti, E. and Marsili-Libelli, S., *Ecol. Model.*, 220, 2415–2426, 2009.)

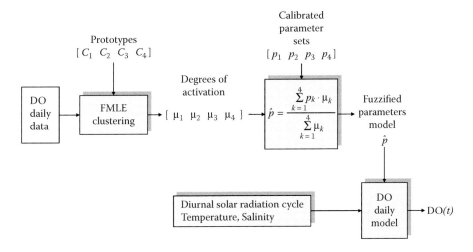

FIGURE 8.28 The DO model is obtained by a fuzzy combination of the prototypical parameters, weighted by the degree of activation of the corresponding cluster in Table 8.2. (Reproduced with permission from Giusti, E. and Marsili-Libelli, S., *Ecol. Model.*, 220, 2415–2426, 2009.)

Furthermore, it is questionable whether model complexity always pays off in terms of prediction accuracy vis-à-vis the modelling effort, evaluated in terms of money and manpower to gather the data and develop the model (Pearl, 1978; Halfon, 1983; Lindenschmidt, 2006). For these reasons, it is difficult to adapt major water quality models, such as QUAL2 or RWQM1, which would require a considerable budget and data-gathering effort to minor river systems. This paragraph describes the development of a model to implement a water quality study in two small, but ecologically important, rivers in Tuscany (Italy), whose location is illustrated in Figure 8.31.

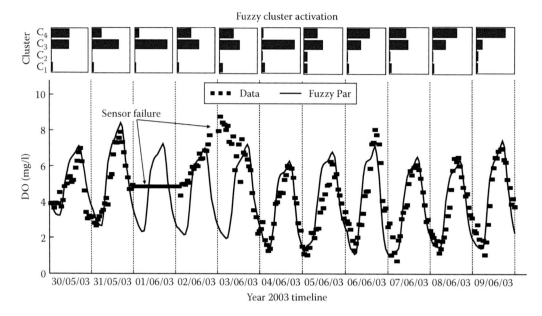

FIGURE 8.29 DO model validation, including an episode of DO sensor failure. The upper boxes show the daily activation of the four prototypes of Table 8.2. (Reproduced with permission from Giusti, E. and Marsili-Libelli, S., *Ecol. Model.*, 220, 2415–2426, 2009.)

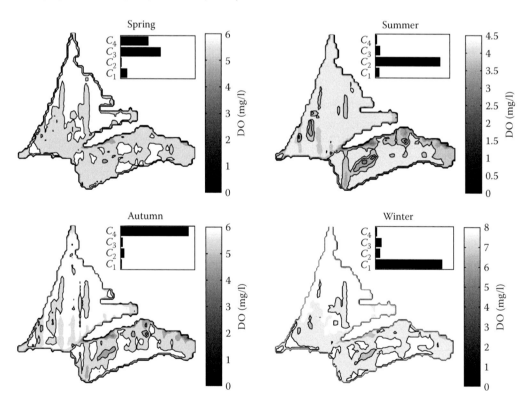

FIGURE 8.30 DO distribution predicted by the integrated model shown in Figure 8.25. For each day in the example, the upper horizontal bars indicate the degree of activation of the prototypical behaviours, showing the good correspondence between clusters and seasons. The vertical bars show the scale of the DO concentration. (Reproduced with permission from Giusti, E. and Marsili-Libelli, S., *Ecol. Model.*, 220, 2415–2426, 2009.)

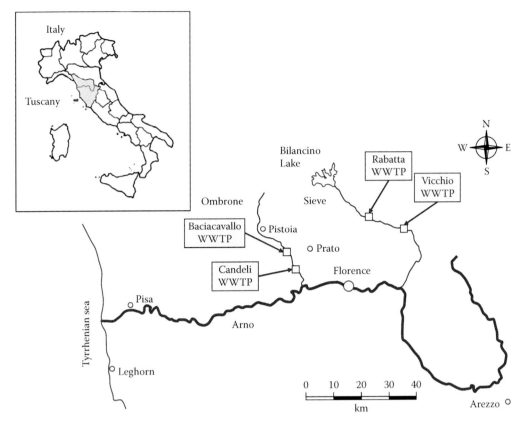

FIGURE 8.31 Location of the Ombrone and Sieve rivers, representing two of the main right bank tributaries to the Arno River, in Tuscany, outlined as a shaded area in the upper-left inset. (Reproduced with permission from Marsili-Libelli, S. and Giusti, E., *Environ. Model. Softw.*, 23, 451–463, 2008.)

This study was partly supported by the regional environmental protection agency (ARPAT) in the context of a region-wide project of river water quality awareness, and the developed model is fully described elsewhere (Marsili-Libelli and Giusti, 2008). The study was based on an ad hoc data-collection campaign, for which the most effective sampling spots were selected by a preliminary sensitivity study. With the collected data, the model was then calibrated and further structural changes were made on the basis of the information acquired on the field, specifically regarding the parameter and distributed-inputs structuring. Thus, in developing this study, several notions treated in the previous chapters were used: the Streeter & Phelps model (Rinaldi et al., 1979) was the basis on which the actual model was built, sensitivity analysis was used to find the most efficient sampling locations in terms estimation accuracy, the parameter estimation and validation techniques of Chapter 2 were used to determine the best model parameters, and finally a set of Monte Carlo simulations were run to test the model robustness, using the PEAS package (Checchi et al., 2007).

8.5.1 A Tale of Two Rivers

The two rivers considered in this study, Sieve and Ombrone, are both part of the Arno River system and have widely differing characteristics. Hydraulically, both rivers are subject to massive winter and autumn floods, while they reduce to almost a trickle during the summer months. In this period the river quality becomes critical and a water quality model is needed. The models were developed for typical low-flow conditions, which were determined by averaging the daily flows during the summer months to obtain a typical flow of 2.9 m^3/s for the Sieve, and 1.7 m^3/s for the Ombrone.

8.5.1.1   Sieve Basin

The quality of the Sieve River is definitely the best in the Arno basin, and a great deal of efforts is being made to conserve its pristine conditions. The reach considered here extends from the upstream Bilancino reservoir down to the confluence with the Arno, for a total length of 47 km. The water released from the Bilancino reservoir, normally 3 m³/s, is meant to provide the minimum sustainable flow during the summer months, and to feed adequate water to the Florence potabilization plant, placed at about 20 km downstream of the confluence into the Arno. The slope of the reach is less pronounced in the upstream part and increases in the final part. This, together with the major tributaries located in the downstream portion, makes the final part of the river much richer in flow than the upstream part. However, the increased depth in this portion counterbalances the higher flow, and no increase in re-aeration rate was observed. There are two wastewater treatment plants discharging into the Sieve River, located at Rabatta and downstream of Vicchio, as, shown in Figure 8.31, but many untreated discharges and nonpoint agricultural pollution sources, detected during this study, make a considerable contribution to the river pollution. In setting up the model, information about these sources was obtained through direct interviews with local municipal officers, or was estimated in the model calibration phase, as explained later. The river description in terms of characteristics and sources (point and nonpoint) is shown in Figure 8.32.

8.5.1.2   Ombrone Basin

Unlike the Sieve, the Ombrone River flows through a heavily industrialized area dominated by the textile centre of Prato and is subject to many discharges, some of which are untreated, in addition to the output of the major wastewater treatment works of Baciacavallo, with a capacity exceeding 750,000 PE. A smaller wastewater treatment plant is located at Candeli, 3 km downstream of Baciacavallo, with a capacity of 5000 PE. The reach under study is 15 km in length and can be divided into two parts with widely differing characteristics. The upstream part, 10 km long, flows through a densely populated area, with artificial riverbanks and many pollution sources, whereas

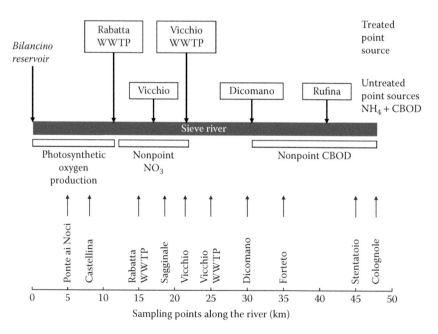

FIGURE 8.32   Distribution of measuring points (upward arrows), point and nonpoint sources along the Sieve River. (Reproduced with permission from Marsili-Libelli, S. and Giusti, E., *Environ. Model. Softw.*, 23, 451–463, 2008.)

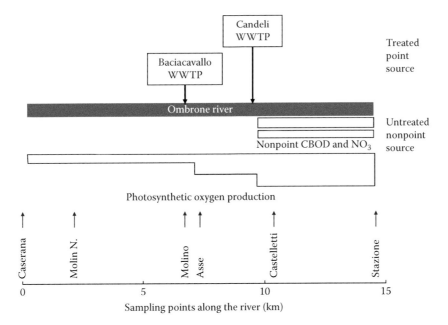

FIGURE 8.33 Distribution of measuring points (upward arrows), point and nonpoint sources along the Ombrone River, which presents a considerable photosynthetic oxygen production, shown as the lower box. (Reproduced with permission from Marsili-Libelli, S. and Giusti, E., *Environ. Model. Softw.*, 23, 451–463, 2008.)

the downstream reach, of about 5 km, is relatively unspoiled, with densely vegetated riparian zones and sparse human settlements. In this reach, the photosynthetic activity is considerable, given the high nutrient content of the incoming water, and self-purification is very active. The river description, in terms of loadings and sampling stations, is shown in Figure 8.33.

8.5.2 Model Structure

The model is intended to describe the water quality along each river in a hydraulic steady-state condition, and therefore the locations along the river length are identified through their flow-times relative to the stationary dominant flow regime. In this condition, the quality variables along the river are referred to the physical location (length from the upstream end) for an easy visualization, but are related by the underlying 'follow-the-plug' assumption. Carbonaceous and nitrogenous substances are the main pollutants in both rivers, so the model includes the degradation of the CBOD and the two main nitrogen species: ammonium-N and nitrate-N. An important source term is given by the photosynthetic oxygen production, which is particularly strong in the downstream part of the Ombrone. A single general structure for the model was set up, and then adapted to the specific characteristics of the two rivers. In this way, the generality of the model and its flexibility can be appreciated. The model equations and the parameters' structures are grouped in Table 8.3, while Table 8.4 indicates by a check mark ✓ which process in the block diagram of Figure 8.34 is included in each river model.

Furthermore, by comparing the model response to the data, it was concluded that sometimes a constant kinetics worked better than a structured one, and vice versa. To provide the best parametrization in each situation, a switching block was introduced, as in Figure 8.35, where the most suitable parametrization was activated, depending on the location along the river. This figure refers to the ammonium-N kinetics, for which a first-order reaction was found appropriate in the upstream part, while a Monod kinetics performed better in the downstream part. The value x_s represents the downstream distance controlling the switching.

TABLE 8.3

Processes and Structured Parameters Included in the Sieve–Ombrone River Water Quality Model

Dynamics

$$\frac{dB}{dx} = -\frac{1}{u(x)}K_b(x)B + B_d(x)$$

$$\frac{d\text{NH}_4}{dx} = -\frac{1}{u(x)}K_a(x)\cdot\text{NH}_4 - \frac{1}{u(x)}\delta\cdot K_{al}\frac{\text{NH}_4}{K_f+\text{NH}_4}$$

$$\frac{d\text{NO}_3}{dx} = \frac{1}{u(x)}K_a(x)\cdot\text{NH}_4 - \frac{1}{u(x)}(1-\delta)\cdot K_{al}\frac{\text{NO}_3}{K_f+\text{NO}_3} - \frac{1}{u(x)}K_o\cdot\text{NO}_3 + \text{NO}_{3d}(x)$$

$$\frac{d\text{DO}}{dx} = \frac{1}{u(x)}K_r(u)(\text{DO}_{sat}-\text{DO}) - \frac{1}{u(x)}K_b(x)B - \frac{1}{u(x)}\frac{(4.57-Y_a)}{Y_a}K_a(x)\cdot\text{NH}_4 + \text{DO}_{ph}(x)$$

<table>
<tr><td align="center">Sieve</td><td align="center">Ombrone</td></tr>
</table>

$$K_a(x) = \begin{cases} K_{a\_max}\dfrac{\text{NH}_4}{K_{sa}+\text{NH}_4} & x>21 \\ K_{aa} & x<21 \end{cases}$$

$$K_a(x) = \begin{cases} K_{a\_max1}\cdot f(T,\text{pH},\text{DO}) & 0<x<10.5 \\ K_{a\_max2}\cdot f(T,\text{pH},\text{DO}) & x>10.5 \end{cases}$$

$$f(T,\text{pH},\text{DO}) = e^{C_I(T(x)-15)}\cdot\frac{1}{1+10^{pk_1-pH(x)}+10^{pH(x)-pK_2}}\cdot\left(\frac{\text{DO}}{K_{os}+\text{DO}}\right)$$

$$K_b(x) = K_{bm}\cdot\begin{cases} 1.802 & 0<x\le12.492 \\ 2.479 & 12.492<x<16.859 \\ 0.451 & 16.859<x<48.454 \end{cases}$$

$$K_b(x) = \begin{cases} K_{b1} & 0<x<10.5 \\ K_{b2} & x>10.5 \end{cases}$$

$$K_r(u) = K_c\sqrt{\frac{u(x)}{3.6}}$$

Source: Marsili-Libelli, S. and Giusti, E., *Environ. Model. Softw.*, 23, 451–463, 2008.

TABLE 8.4

Processes Included in Each Model

Process	Sieve	Ombrone
BOD degradation	✓	✓
BOD nonpoint source	✓	
NO₃-N nonpoint source	✓	
Nitrification (first order)		✓
Nitrification (half-saturation)	✓	
NH₄-N assimilation by submerged vegetation		✓
NO₃-N assimilation by submerged vegetation		✓
Denitrification	✓	✓
Re-aeration	✓	✓
Photosynthetic oxygen production	✓	✓

Source: Marsili-Libelli, S. and Giusti, E., *Environ. Model. Softw.*, 23, 451–463, 2008.

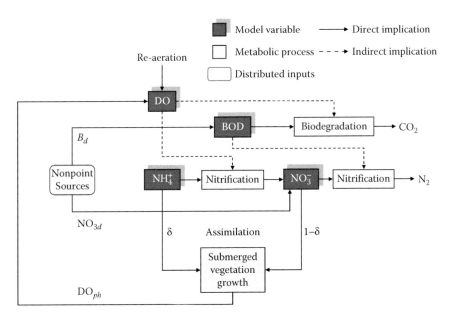

FIGURE 8.34 Water quality variables and transformation included in the river model structure. (Reproduced with permission from Marsili-Libelli, S. and Giusti, E., *Environ. Model. Softw.*, 23, 451–463, 2008.)

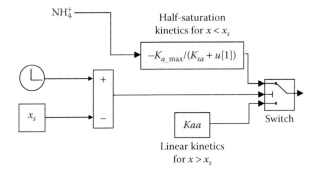

FIGURE 8.35 Mechanising a variable structure parameter in Simulink requires a switch block which routes either input to the output, depending on the sign of the central input. In this case, when this is negative, that is, $x < x_s$, the half-saturation kinetics is used, while for $x > x_s$ this is replaced by the linear kinetics. x_s represents the switching time, or rather the switching length, because the independent variable is downstream length.

8.5.2.1 Analysis of the Reactive Terms

With reference to the diagram of Figure 8.34 and to Table 8.3, the reactive processes included in the model are degradation of dissolved carbonaceous substances, ammonium oxidation, algal uptake and denitrification and DO balance, including depletion by degradation and supply by physical re-aeration and photosynthetic production. Because the model is space-referenced through the 'follow-the-plug' assumption, the variation of flow and velocity along the course was included in the model. Given the differing morphology of the two rivers, the general model of Table 8.3 had to be adapted to each river including the appropriate processes as shown in Table 8.4. To relate this model structure with the complexity grades introduced by Lindenschmidt (2006), the BOD was differentiated between carbonaceous (CBOD) and nitrogenous (NBOD) components, with a loose coupling between the DO-BOD cycle and the algae–nitrogen interaction. Phosphorus was not included in the model because, in both rivers, its concentration was too low to be observed and modelled reliably. Furthermore, almost the entire algae population is composed of N-limited species, and its interaction with dissolved inorganic nitrogen is described by the preferential absorption coefficient δ introduced by Brown and Barnwell (1987) in the QUAL2E model. The relatively simple model structure is partially offset by structuring some parameters as a function of the varying river morphology. In fact, it was observed that only by relating some parameters to the river characteristics, a good data fit could be obtained. The main examples of this approach are the CBOD biodegradation rate K_b, which depends on the river characteristics (water flow, river bed morphology, depth), as pointed out by several studies (Thomann and Mueller, 1987; Chapra, 1997; Chapra and Pelletier, 2003; Pelletier et al., 2006). In the model of Table 8.3, this parameter has a variable structure, depending on the location, and is implemented with a switching block, such as the one shown in Figure 8.35 for the kinetics of the ammonium-N oxidation, which varies between a first-order or a half-velocity kinetics depending on the river conditions. Furthermore, the photosynthetic oxygen production $DO_{ph}(x)$ varies widely on a daily and seasonal basis, as we have seen in Sections 8.1.4 and 8.2. Because this model has a spatial reference (x) originating from the flow-time steady-state assumption, Section 6.3.4, an average value will be identified for the photosynthetic term on the basis of in situ inspections to assess the level of vegetative development. All these structured parameters, in addition to requiring a careful study of the river morphology, introduce a large number of switching functions into the model, which not only increases its complexity but also makes its parameter estimation more difficult (Checchi and Marsili-Libelli, 2005; Checchi et al., 2007), as already pointed out in Chapter 2.

8.5.2.2 Introduction of the Nonpoint Pollution Sources

Detecting the effects of nonpoint sources can be a daunting task, involving indirect estimation of basin characteristics and land use, as shown by Azzellino et al. (2006). In this study, each point source is modelled by diluting the discharge with the upstream flow, and restarting the integration with the new initial conditions. Nonpoint sources of CBOD, NH_4-N and the average photosynthetic oxygen production rate $DO_{ph}(x)$, which are very difficult to detect, were introduced as unknown piecewise constant inputs and estimated together with the other model parameters. This increases the difficulty of the parameter estimation and adversely affects the global identifiability by introducing even more switching functions. The pros and cons of introducing the point and nonpoint pollution sources into the estimation are discussed in Marsili-Libelli and Giusti (2008), while Heathwaite (2003) analyses the processes whereby nutrients are exported to the rivers and how effective and economically viable mitigation decisions can be made.

8.5.3 Parameter Estimation

The relatively simple model structure of Table 8.3 produces good results only if its parameters are accurately estimated. The techniques illustrated in Chapter 2 are applied here, starting with a sensitivity analysis, which has the double purpose of ranking the parameters according to their sensitivity, and hence their expected estimation accuracy, and of selecting the most efficient sampling spots characterized by a high sensitivity. For the practical model identification, the PEAS software (Section 2.4.5) was used.

8.5.3.1 Sensitivity Analysis and Parameter Identifiability

The problem of selecting the best compromise in terms of data availability, computational complexity, and model accuracy limits the number of parameters that can be reliably estimated. In this case, the problem is made more difficult by the need to include the unknown nonpoint sources as additional parameters. The selection of the identifiable parameter subset is a particularly relevant subject in environmental system identification, particularly in wastewater treatment models (Weijers and Vanrolleghem, 1997; Petersen, 2000; Dochain and Vanrolleghem, 2001; Brun et al., 2002; Gernaey et al., 2004; De Pauw, 2005; Sin et al., 2005; Ruano et al., 2007). Almost all of these studies are based on trajectory sensitivity functions, and ultimately refer to the FIM, which can be used as a figure of merit to assess the identifiability of a parameter subset, and to design the best experiment, that is, sampling campaign planning, in terms of estimation accuracy (Fedorov, 1972; Atkinson and Donev, 1992; Versyck et al., 1998; Checchi and Marsili-Libelli, 2005). As already discussed in Chapter 2, maximizing the FIM implies minimizing the estimation error covariance matrix C, because $C = F^{-1}$ (Equation 2.81).

The identifiable parameter subset for this exercise was determined with a two-step procedure. First, the parameters were ranked according to their global sensitivity index, similar to other indexes already in use (Reichert and Vanrolleghem, 2001; De Pauw, 2005), and defined as

$$\xi_i = \sqrt{\frac{1}{N}\sum_{j=1}^{q}\sum_{k=1}^{N}\left(S_k^{ij}\right)^2} \quad i = 1,\ldots,n_p \tag{8.27}$$

where:

$S_k^{ij} = \dfrac{\partial y_j(k)}{\partial p_i}$ is the jth output sensitivity with respect to the ith parameter computed at the kth sampling instant

n_p is the number of model parameters

N is the number of experimental data for the q output variables

The parameters of the model in Table 8.3 are then ranked in Table 8.5 according to their sensitivity scores, computed according to Equation 8.27.

TABLE 8.5
Sensitivity Scores of the Model Parameters for the Two Rivers Computed with Equation 8.27

River	Parameter	ξ	
Sieve	K_{sa}	791.25	
	K_{a_max}	376.76	Core parameters
	K_b	181.65	
	K_o	69.79	
	B_d	49.49	Nonpoint sources
	NO_{3d}	25.29	
	DO_{ph}	10.77	
	K_c	5.28	Literature or heuristic values
	K_{ao}	0.43	
Ombrone	K_{a_max}	313.68	
	K_b	198.46	Core parameters
	K_{al}	150.98	
	K_o	44.61	
	DO_{ph}	42.38	Nonpoint source
	K_c	17.29	Literature or heuristic values
	K_f	13.68	
	K_{os}	11.12	
	K_{ao}	4.90	

Note: On the basis of the identifiability results shown in Figure 8.36 for the Sieve River and in Figure 8.37 for the Ombrone River, only the 'core' parameters and the nonpoint sources were actually calibrated. The other parameter values were either adapted from the literature, or obtained from direct sources.

The second step in the development is to build a sequence of sensitivity matrices $S_m \in \mathbb{R}^{q \times m}$ limited to the first m parameters in the sensitivity ranking

$$
S_m = \begin{bmatrix}
\dfrac{\partial y_1}{\partial p_1} & \dfrac{\partial y_1}{\partial p_2} & \cdots & \dfrac{\partial y_1}{\partial p_m} \\[2mm]
\dfrac{\partial y_2}{\partial p_1} & \dfrac{\partial y_2}{\partial p_2} & \cdots & \dfrac{\partial y_2}{\partial p_m} \\[2mm]
\vdots & \vdots & \vdots & \vdots \\[2mm]
\dfrac{\partial y_q}{\partial p_1} & \dfrac{\partial y_q}{\partial p_2} & \cdots & \dfrac{\partial y_q}{\partial p_m}
\end{bmatrix} \quad \text{with} \quad m = 2, \dots, n_p \tag{8.28}
$$

from which a sequence of FIMs $F_m \in \mathbb{R}^{m \times m}$ of increasing order (m) are formed

$$
F_m = \sum_{k=1}^{N} \left(S_m \right)^T Q_k \left(S_m \right) \quad \text{with} \quad m = 2, \dots, n_p \tag{8.29}
$$

In this way, F_2 considers the two most sensitive parameters, F_3 considers the previous couple plus the third most sensitive parameter, and so on, until all the parameters have been included. The Optimal Experiment Design criteria A, E and mod E, defined in Table 8.6, are then applied to the FIM sequence $\left(F_2, \dots, F_{n_p} \right)$ to detect any significant changes in the indicators while the FIM order is increased.

Figure 8.36 shows the identifiability test for the Sieve River. As expected, as far as the core parameters only are considered, all the criteria of Table 8.6 remain almost constant at a safe value, but as soon as the nonpoint sources are added, they exhibit a considerable variation, denoting a significant degradation in the identification performance. The same behaviour is obtained for the Ombrone River, as shown in Figure 8.37.

8.5.3.2 Calibration Results

The set of estimated parameters are listed in Table 8.7 for the Sieve River, and in Table 8.8 for the Ombrone River.

Each table reports the estimated value of the parameter in the form, $\left(\hat{p} \pm \sigma_p \right)$, with its confidence intervals computed with Equation 2.89, which takes into account the limited number of experimental data. It can be seen in both tables that while the parameters are estimated with a relatively small confidence interval, denoting a high estimation accuracy, the nonpoint source estimates come with a fairly large confidence interval, as predicted by the identifiability indicators of Figures 8.36 and 8.37. Figures 8.38 and 8.39 compare the model output with the experimental observations, represented by its average and standard error, for DO/CBOD and NH_4/NO_3, respectively.

TABLE 8.6

Optimal Experiment Design Criteria Based on FIM (F)

Criteria	Method	Effect
A	$\min(\text{tr } F^{-1})$	Minimization of the arithmetic mean of parameter errors
E	$\max(\lambda_{\min})$	Maximizes the minimum eigenvalues of F, which is proportional to the length of the largest axis of the confidence ellipsoid
mod E	$\min(\lambda_{\max}/\lambda_{\min})$	Minimizes the condition number of F

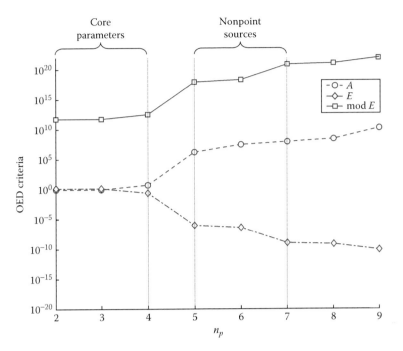

FIGURE 8.36 Identifiability test for the Sieve River as a function of the number of estimated parameters n_p. The core parameters (K_{sa}, K_{a_max}, K_{bm}, K_o) are indexed from 1 to 4, whereas indexes from 5 to 7 refer to the nonpoint sources B_d, NO_{3d}, DO_{ph}. (Reproduced with permission from Marsili-Libelli, S. and Giusti, E., *Environ. Model. Softw.*, 23, 451–463, 2008.)

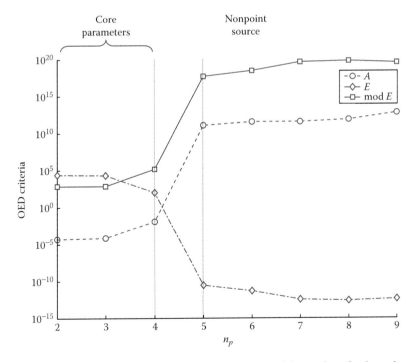

FIGURE 8.37 Identifiability test for the Ombrone River as a function of the number of estimated parameters n_p. The core parameters (K_{a_max}, K_b, K_{al}, K_o) are indexed from 1 to 4, whereas index 5 refers to the nonpoint source DO_{ph}. In this case also, the introduction of nonpoint source identification makes the entire estimation more difficult. (Reproduced with permission from Marsili-Libelli, S. and Giusti, E., *Environ. Model. Softw.*, 23, 451–463, 2008.)

TABLE 8.7
Calibrated Parameters of the Sieve River Model

Parameter	Estimated Value and Monte Carlo Confidence Limits $\hat{p} \pm \sigma_p$	Length (km)
K_{sa}	0.037 ± 0.032	$0 < x < 48.4$
K_{a_max}	0.048 ± 0.089	$0 < x < 48.4$
K_{bm}	0.029 ± 0.008	$0 < x < 48.4$
K_o	0.139 ± 0.073	$0 < x < 48.4$
B_d	0.321 ± 0.282	$35 < x < 48.4$
NO_{3d}	0.385 ± 0.158	$8.2 < x < 21.4$
DO_{ph}	2.11 ± 0.281	$0 < x < 10$
	Constant value	*Reference*
Y_a	0.24	Brown and Barnwell (1987); Chapra (1997)
K_c	1.5	Thomann and Mueller (1987); Chapra (1997)
K_{aa}	0.075	Chapra (1997)

TABLE 8.8
Calibrated Parameters of the Ombrone River Model

Parameter	Estimated Value and Monte Carlo Confidence Limits $\hat{p} \pm \sigma_p$	Length (km)
K_b	0.109 ± 0.043	$0 < x < 10.5$
	0.009 ± 0.001	$10.5 < x < 14.6$
K_{al}	0.512 ± 0.106	$0 < x < 10.5$
	0.093 ± 0.092	$10.5 < x < 14.6$
K_{a_max}	0.117 ± 0.031	$0 < x < 10.5$
	0.137 ± 0.064	$10.5 < x < 14.6$
K_o	0.117 ± 0.077	$0 < x < 10.5$
	0.137 ± 0.093	$10.5 < x < 14.6$
DO_{ph}	1.656 ± 0.279	$0 < x < 7.1$
	1.760 ± 0.684	$7.1 < x < 10.4$
	3.847 ± 0.632	$10.4 < x < 14.6$
	Constant value	*Reference*
Y_a	0.24	Brown and Barnwell (1987); Chapra (1997)
K_f	0.026	Lindenschmidt (2006)
K_c	$0.85 \; 0 < x < 10.5$	Thomann and Mueller (1987); Chapra (1997)
	$0.67 \; 10.5 < x < 14.6$	
δ	0.2	Brown and Barnwell (1987); Chapra (1997)
C_t	0.1	Jorgensen and Bendoricchio (2001)
pK_1	5	Jorgensen and Bendoricchio (2001)
pK_2	8.8	Jorgensen and Bendoricchio (2001)
K_{os}	1.34	Lindenschmidt (2006)

The thin grey lines are Monte Carlo model trajectories generated by perturbing the estimated parameter $\hat{p}$ with a value drawn from the Gaussian distribution with a standard deviation equal to the parameter confidence interval, that is, $N\left(\hat{p}, \sigma_p\right)$. The same comparison for the Ombrone River is presented in Figure 8.40 for the DO/CBOD couple, and in Figure 8.41 for the nitrogen species. It is interesting to notice that, in both rivers, the family of perturbed trajectories never exceeds the measurement error bands.

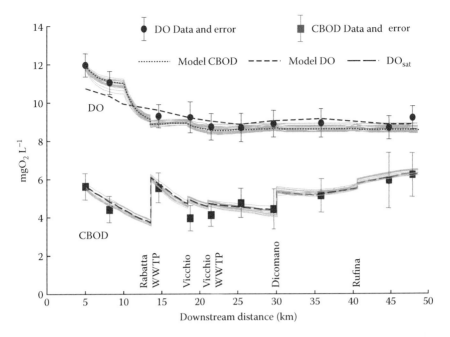

FIGURE 8.38 Performance of the calibrated model for the Sieve River for DO and CBOD. The experimental data are represented by their mean (dot or square) and error (bar), while the model response is represented by the dotted/dashed lines. The thin grey lines represent Monte Carlo simulations generated with parameter values drawn from a Gaussian distribution $N\left(\hat{p}, \sigma_p^2\right)$. (Reproduced with permission from Marsili-Libelli, S. and Giusti, E., *Environ. Model. Softw.*, 23, 451–463, 2008.)

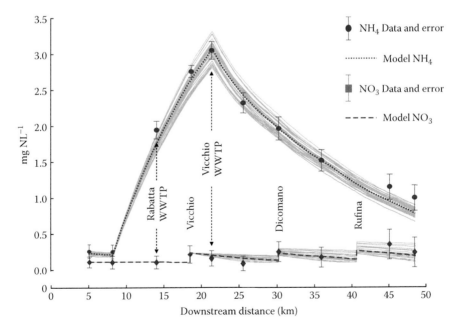

FIGURE 8.39 Performance of the calibrated model for the Sieve River for ammonium-N and nitrate-N. The experimental data are represented by their mean (dot or square) and error (bar), while the model response is represented by the dotted/dashed lines. The thin grey lines represent Monte Carlo simulations generated with parameter values drawn from a Gaussian distribution $N\left(\hat{p}, \sigma_p^2\right)$. (Reproduced with permission from Marsili-Libelli, S. and Giusti, E., *Environ. Model. Softw.*, 23, 451–463, 2008.)

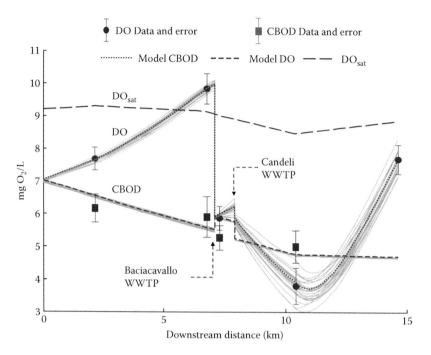

FIGURE 8.40 Performance of the calibrated model for the Ombrone River for DO and CBOD. The experimental data are represented by their mean (dot or square) and error (bar), while the model response is represented by the dotted/dashed lines. The thin grey lines represent Monte Carlo simulations generated with parameter values drawn from a Gaussian distribution $N\left(\hat{p}, \sigma_p^2\right)$. (Reproduced with permission from Marsili-Libelli, S. and Giusti, E., *Environ. Model. Softw.*, 23, 451–463, 2008.)

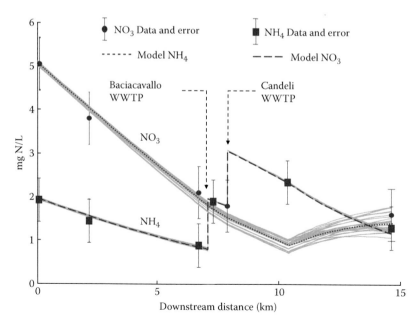

FIGURE 8.41 Performance of the calibrated model for the Ombrone River for ammonium-N and nitrate-N. The experimental data are represented by their mean (dot or square) and error (bar), while the model response is represented by the dotted/dashed lines. The thin grey lines represent Monte Carlo simulations generated with parameter values drawn from a Gaussian distribution $N\left(\hat{p}, \sigma_p^2\right)$. (Reproduced with permission from Marsili-Libelli, S. and Giusti, E., *Environ. Model. Softw.*, 23, 451–463, 2008.)

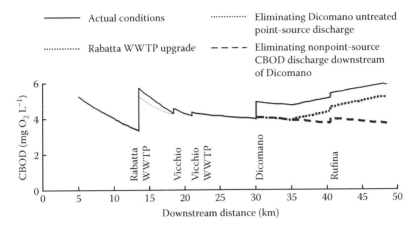

FIGURE 8.42 Scenarios generation on the basis of the calibrated model for the Sieve River. The effects of the various management decisions are clearly shown by the differing model behaviours. As expected, the largest improvement is obtained by eliminating the unchecked nonpoint sources in the downstream part of the river. (Reproduced with permission from Marsili-Libelli, S. and Giusti, E., *Environ. Model. Softw.*, 23, 451–463, 2008.)

8.5.3.3 Lessons to Be Drawn from the Exercise

Considering the Sieve River, while the DO/CBOD levels are satisfactory along the entire river length, the nitrogen behaviour requires some explanation. In the upstream part, the river flows through a cattle farming area, which explains the rising values of ammonium-N up to the Vicchio wastewater treatment plant (WWTP). Downstream of this point, the land use is predominantly agricultural, and nitrate from fertilizers is predominant, while nitrate discharged by the small WWTPs is negligible. The Sieve River study was required (and funded) by the Regional Environmental Agency (ARPAT) to acquire a tool for the assessment of various management options. The results of the possible environmental actions are shown in Figure 8.42, which compares the various scenarios generated by introducing one hypothesis at a time. It is found that upgrading the Rabatta WWTP brings only a small, local contribution, while eliminating the unchecked nonpoint sources downstream of Dicomano would result in the most important improvement.

From the environmental viewpoint, there are two major differences between the Sieve and Ombrone. While the Sieve River flows through a basically unspoilt territory and receives very limited treated discharges, the Ombrone River is more exposed to organic pollution and receives a considerable flow and load contribution from the major WWTP of Baciacavallo, which in the summer months makes up almost entirely for the river flowrate. The asset of this river, however, is the considerable photosynthetic oxygen production of the downstream reach, which results in a fair quality improvement. The deep oxygen sag downstream of Baciacavallo is indicative of the major load discharged by this plant, but also of a strong river recovery due to the self-purification mechanisms.

8.6 ESTIMATION OF NUTRIENT BIOAVAILABILITY IN RIVERS

The case examined in this section draws heavily from the notions explained in Chapter 2 regarding parameter estimation, while its aim is to determine the bioavailable fraction of nutrients in rivers. This is important for both water quality assessment and for understanding the cycling of nutrients in the aquatic ecosystem. In many ecological models of complex aquatic ecosystems, the concentration of nutrients has been based on chemical analyses (Crispi et al., 1998; Kiirikki and Haapama, 1998; Jamu and Piedrahita, 2002; Arhonditsis and Brett, 2005; Kuo et al., 2006; Zhang and Jørgensen, 2005), but the difference between bulk concentration and its bioavailable fraction is often overlooked, though only the latter is relevant for the characterization of water quality from a biological viewpoint. In fact, the biologically available fraction of nitrogen and phosphorus may be significantly lower than the gross chemically measured concentration of these elements, and it has a fundamental importance in terms of eutrophication and habitat assessment. This fraction cannot be determined by standard chemical analyses, but it requires a biological assay, which relates the nutrient abundance to the growth of a known microalgae species in a controlled environment. *Selenastrum capricornutum,* recently renamed *Raphidocellis subcapitata,* has been selected as the test organism for its widespread presence in natural streams and for its sensitivity to different nutrients. For this reason, this species has been extensively reported in the literature as a significant water quality indicator (Chiaudani and Vighi, 1976; Nyholm, 1978, 1985; Nyholm and Lingby, 1988). To relate the trophic potential of a river to the observed variations in the growth characteristics of *S. capricornutum,* the following method was developed:

1. A growth model for the *S. capricornutum* was set up as a function of nutrient concentrations. This model was then calibrated with laboratory data, growing several batches of previously starved algae with known initial nutrient concentrations. A sample calibration run of the Richards model, Section 5.1.4, against optical density data from a *S. capricornutum* in vitro batch is shown in Figure 8.43. The population carrying capacity was modelled by Equation 8.30 as a function of the nutrient initial concentration in the batch.
2. The unknown bioavailable nutrient concentrations was inferred by growing new starved batches in river water samples until the carrying capacity was reached.

8.6.1 A Steady-State Model Relating *S. capricornutum* Growth to Nutrients

The controlled growth of microalgae has been extensively researched. Multispecies assays with organic nitrogen sources were considered (Elgavish et al., 1980; Berman and Chava, 1999), and

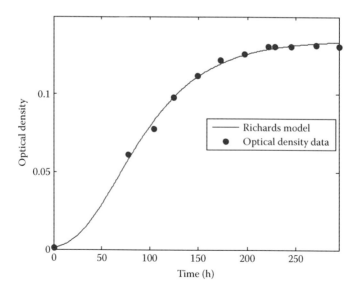

FIGURE 8.43 Calibration of the Richards model, Section 5.1.4, with the optical density data relative to an in vitro batch of *S. capricornutum.* This calibration was obtained with the MATLAB script `Selen _ Cal.m` in the `ESA _ Matlab\Exercises\Chapter _ 8\Selenastrum\Selen` folder.

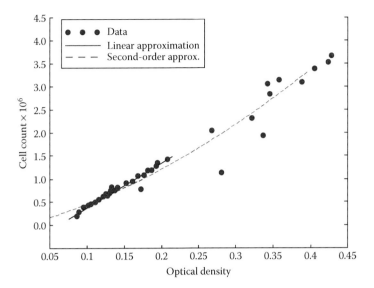

FIGURE 8.44 The correspondence between cell count and optical density can be approximated with a linear relationship in a limited range of cell density.

the experimental conditions[*] are detailed in Marsili-Libelli et al. (1997). It was experimentally observed that a batch of *S. capricornutum* grown in a controlled medium fully develops in about 14 days (300 h), following the Richards growth function as shown in Figure 8.43, and that the steady-state eventually reached depends on the initial nutrients concentration.

To calibrate the model, particularly its nutrient-limited carrying capacity, nine batches of previously starved *S. capricornutum* cells were grown in controlled conditions, and their number was related to the optical density of the medium as shown in Figure 8.44.

To model this double limitation mechanism a Liebig's law was proposed (Nyholm and Lingby, 1988; Sterner and Grover, 1998) consisting of a double Monod kinetics. This approach was initially tried, but resulted in poor fitting of the experimental data. Instead, the mathematical model which best described the relationship between steady-state algal density and initial nutrient concentration was the following

$$G = \frac{\alpha}{\dfrac{1}{\dfrac{N}{K_n + N}} + \dfrac{1}{\dfrac{P}{K_p + P}}} \tag{8.30}$$

where N and P represent the nutrient concentrations (mg/l), and the constants $\left(\alpha, K_n, K_p\right)$ are to be estimated by minimizing the following error functional

$$E\left(\alpha, K_n, K_p\right) = \sum_{i=1}^{3} \sum_{j=1}^{3} \left(\frac{G_{obs}\left(i, j\right) - G_{mod}\left(i, j\right)}{G_{obs}\left(i, j\right)}\right)^2 \tag{8.31}$$

where G_{obs} are the observed cell densities of Table 8.9, inferred from the optical density according to the linear approximate relationship shown in Figure 8.44, and G_{mod} are the optical densities generated by Equation 8.30.

---

[*] The cooperation of the ARPAT Prato Provincial Laboratory is gratefully acknowledged.

TABLE 8.9

Steady-State Cell Density (10^6 cell/ml) as a Function of Initial Nutrient Concentration

Nutrient Concentration	$N = 0.22$ (mg/l) $i = 1$	$N = 0.44$ (mg/l) $i = 2$	$N = 0.88$ (mg/l) $i = 3$
$P = 0.02$ (mg/l) $j = 1$	0.450	0.716	0.820
$P = 0.04$ (mg/l) $j = 2$	0.442	0.922	1.186
$P = 0.08$ (mg/l) $j = 3$	0.470	0.918	1.432

Note: The indexes i and j refer to the nutrient combinations in each batch.

TABLE 8.10

Estimated Values of the Model Parameters in Equation 8.30

Parameter	Units	Value and Confidence Interval
K_n	(mg/l)	5.62789 ± 0.107768
K_p	(mg/l)	0.168063 ± 0.0033893
α	–	1.50984 ± 0.0265657

The indexes (i, j) refer to the initial nutrient concentrations corresponding to the headings of Table 8.9. The estimated parameter values are listed in Table 8.10, and their cross-correlation matrix is

$$C = \begin{bmatrix} 1.00001 & -0.0276 & 0.2009 \\ -0.0276 & 1.0000 & 0.0768 \\ 0.2009 & 0.0768 & 1.0000 \end{bmatrix} \qquad (8.32)$$

showing that the cross-correlations are fairly small, thus confirming that the model (8.30) is not overparametrized. The surface of the observed cell density distribution as a function of the initial nutrients and its estimate according to Equation 8.31 are compared in Figure 8.45.

8.6.2 ESTIMATION OF THE BIOAVAILABLE NUTRIENT CONCENTRATIONS

The calibrated model (8.30) can now be used *in reverse* to determine the bioavailable fraction of nitrogen and phosphorus in a water sample by observing the ultimate algal growth in that medium, from which the unknown bioavailable nutrient quantities are inferred. However, there is a mathematical snag, because the nonlinear equation (8.30) cannot be inverted to compute two unknown quantities $(N_x$ and $P_x)$ from one observed quantity (algal final density). To make the problem mathematically determined, we adopted the following procedure: three more samples were generated from the initial water sample by adding to the raw water, containing unknown nutrient concentrations (N_x, P_x), either an increment of one nutrient (ΔP or ΔN) or the same increment of both (ΔP and ΔN) in separate batches. The steady-state cell densities $\left(G_i^{\text{obs}}, \quad i = 0, \ldots, 3\right)$ in these four conditions are related to the initial nutrient concentrations in the following way:

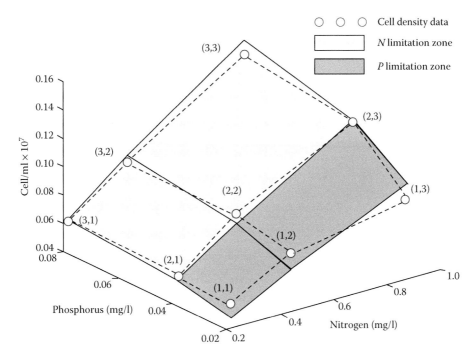

FIGURE 8.45 Calibration of the steady-state *S. capricornutum* development as a function of nutrient concentrations, as described by Equation 8.30. The numbers in parentheses refer to the entries of Table 8.9. The domain is clearly divided in two zones, where either *N* or *P* is the limiting nutrient.

$$
G_0^{obs} = \cfrac{1}{\cfrac{1}{\cfrac{N_x}{K_n + N_x}} + \cfrac{1}{\cfrac{P_x}{K_p + P_x}}} \qquad
G_1^{obs} = \cfrac{1}{\cfrac{1}{\cfrac{N_x + \Delta N}{K_n + N_x + \Delta N}} + \cfrac{1}{\cfrac{P_x}{K_p + P_x}}}
$$

$$
G_2^{obs} = \cfrac{1}{\cfrac{1}{\cfrac{N_x}{K_n + N_x}} + \cfrac{1}{\cfrac{P_x + \Delta P}{K_p + P_x + \Delta P}}} \qquad
G_3^{obs} = \cfrac{1}{\cfrac{1}{\cfrac{N_x + \Delta N}{K_n + N_x + \Delta N}} + \cfrac{1}{\cfrac{P_x + \Delta P}{K_p + P_x + \Delta P}}}
$$

(8.33)

These four values generate a unique four-edge polyhedron on the nonlinear growth surface of Equation 8.30, as shown in Figure 8.46.

The problem of finding the unknowns (N_x, P_x), now overdetermined by the four conditions of Equation 8.33, can be solved in the least-squares sense by minimizing the error functional

$$
(N_x, P_x) = \underset{(N,P)}{\arg\min} \left(\sum_{i=0}^{3} \frac{G_i^{obs} - G_i}{G_i^{obs}} \right)^2
$$

(8.34)

where the G_is are generated by the search algorithm initialized with a problem-compatible set of initial concentration (N_x, P_x). It is important to notice that the model parameters (K_n, K_p, α) are algal specific, so as far as *S. capricornutum* is used for the assay, there is no need to repeat step 1 and

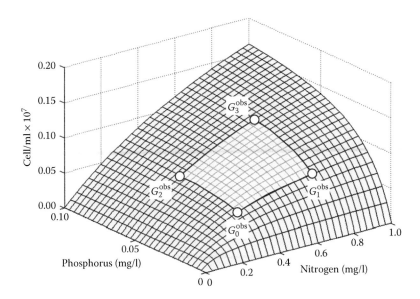

FIGURE 8.46   The observed cell densities form a four-edge polyhedron on the nonlinear growth surface of Equation 8.30. This graph was obtained with the MATLAB script Go _ Selen.m in the ESA _ Matlab\ Matlab _ Examples _ Home\Chapter _ 8\Selenastrum folder.

recalibrate the model, which can be directly applied to differing water samples, provided starved algal inoculum are used. A typical search pattern is shown in the contour map of Figure 8.47, where the minimum is correctly located by the simplex algorithm. The polyhedron obtained by differential addition of nutrients is shown as a shaded area.

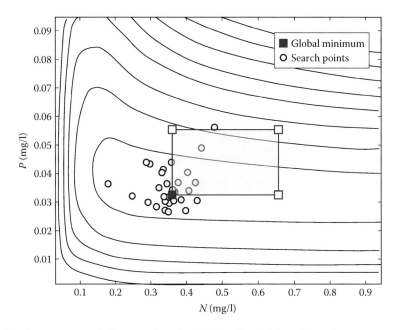

FIGURE 8.47   Contour map of the error functional (Equation 8.34) and search pattern to estimate the unknown bioavailable nutrient concentrations N_x and P_x. The shaded rectangle represents the polyhedron formed by the four samples defined by Equation 8.33. This graph was obtained with the MATLAB script Go_ Nutrients.m in the ESA _ Matlab\Matlab _ Examples _ Home\Chapter _ 8\Selenastrum folder.

8.6.3 FIELD APPLICATION OF THE METHOD

The method just described was applied to the assessment of nutrient bioavailability along the course of the Arno River, using samples routinely taken by ARPAT for the normal water quality monitoring. The samples were taken across the mouth of the main tributaries, as indicated by the circled numbers in Figure 8.48.

Points 1 and 2 refer to the most upstream tributary, the Chiana channel, an artificial water channel flowing through an area with intensive cattle farming. Points 3 and 4 are placed across the Bisenzio confluence and account for the loading generated by the textile district. Point 5 is located downstream of the Ombrone tributary, which is shortly downstream of the Bisenzio confluence, and carries the load from the Pistoia area, where a significant portion of pollution is generated by domestic and greenhouse farming. The last two points (6 and 7) straddle the Usciana confluence, the most downstream tributary, carrying pollution from the leather district. All the samples were taken on the same day during the summer low-flow conditions, and thus, given the high temperature and the low dilution, refer to a high-pollution situation.

From each sample, the associated triplet was obtained by adding 1 mg/l of nitrogen and 50 µg/l of phosphorus in bioavailable form. Algae were then grown in these media, and the final cell densities were recorded and used in the previous algorithm to determine the bioavailable nutrient concentrations, as shown in Table 8.11 and graphed in Figure 8.49, showing the major contributions of Bisenzio, Ombrone, and Usciana. The impact of the Chiana is rather limited, because it flows into the relatively unpolluted upstream part of the Arno, where self-purification is very active. The substantial nitrogen contribution from the Bisenzio tributary is later diluted with the flow coming from the Ombrone tributary, which, in turn, is rich in phosphorus. The cumulative

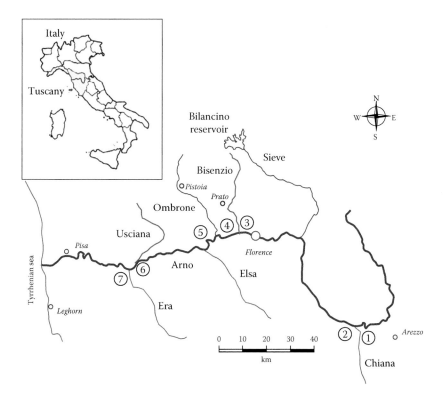

FIGURE 8.48 The Arno River and its tributaries. The sampling points for the bioavailable nutrients campaign were selected upstream and downstream of the major confluences. The numbers refer to the locations listed in Table 8.11.

TABLE 8.11

Estimation of Bioavailable Nutrients at the Sampling Points Shown in Figure 8.48

Sampling Point	Location	N (mg/l)	P (mg/l)
1	Upstream of Chiana	0.6267	1.6
2	Downstream of Chiana	0.6454	4.0
3	Upstream of Bisenzio	2.6263	23.1
4	Downstream of Bisenzio	7.7922	76.9
5	Downstream of Ombrone	2.9156	114.4
6	Upstream of Usciana	5.7252	59.2
7	Downstream of Usciana	8.4411	67.8

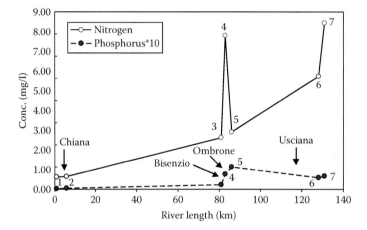

FIGURE 8.49 Bioavailable nutrient concentrations along the Arno River. The numbers refer to the sampling locations shown in Table 8.11 and in Figure 8.48. The contribution of some tributaries is considerable.

effect along the stream, due to additional nonpoint pollution sources, is represented by the relatively high value at point 6, further augmented by the nitrogen load delivered by the Usciana tributary.

8.7   MODELLING AN HORIZONTAL SUBSURFACE CONSTRUCTED WETLAND

The study illustrated in this section involves the basic notions of structural modelling treated in Chapter 1 and in Chapter 6 concerning the flow modelling. It also involves the model identification techniques explained in Chapter 2. The aim of this section is the development of a simple dynamical model for a horizontal subsurface constructed wetland (HSSCW). This type of wastewater treatment represents an important low-impact alternative to conventional wastewater treatment processes for low strength domestic sewage, or for tertiary treatment. The design of constructed wetlands has evolved from early empirical rules to advanced models, which try to explain the complexity of hydraulics in a porous medium combined with the many processes involved in pollution reduction (Kadlec and Wallace, 2009). Still, recent studies suggest that the interdependence between hydraulics and kinetics is so strong, and influenced by such a large number of factors, that even the early complex models (Wynn and Liehr, 2001; Mashauri and Kayombo, 2002; Langergraber, 2003; Kincanon and McAnally, 2004) proved inadequate to fully explain the observed behaviours and provide a reliable estimation of their parameters. Only recently the numerical aspects of this complex man-made ecosystem were thoroughly addressed (Langergraber et al., 2009a) and biokinetic models based on the ASM approach were proposed (Langergraber et al., 2009b; Pálfy and Langergraber, 2014). A comprehensive review of existing models was recently published (Meyer et al., 2015), while more details about the flow model described in this section can be found in Marsili-Libelli and Checchi (2005).

8.7.1 HYDRAULIC MODEL IDENTIFICATION

We are going to follow a different approach, and present a model with a very simple, but easily identifiable structure, to describe the hydraulics and carbon removal of HSSCW. The model structure proposed here is fully described in Marsili-Libelli and Checchi (2005) and extends the distributed parameters approach described in Chapter 6 as an extension of the tank-in-series model (Levenspiel, 1972; Marsili-Libelli, 1997; Rousseau et al., 2004). Figure 8.50 shows the flow routing scheme used to model the horizontal subsurface flow in the porous media, while Table 8.12 lists the model equations.

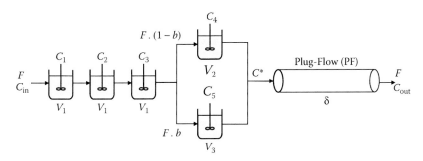

FIGURE 8.50 Basic structure of the subsurface constructed wetland hydraulic model, consisting of a front chain of three CSTRs, followed by a split into two parallel branches, and a plug-flow section accounting for the deadtime. (Reproduced with permission from Marsili-Libelli, S. and Checchi, N., *Ecol. Model.*, 187, 201–218, 2005.)

TABLE 8.12
Model Equations for the Composite Flow Structure of Figure 8.50

Component	Model Equations
Series branch, first CSTR, volume V_1	$\dfrac{dC_1}{dt} = \dfrac{F}{V_1}C_{in} - \dfrac{F}{V_1}C_1$
Series branch, second CSTR, volume V_1	$\dfrac{dC_2}{dt} = \dfrac{F}{V_1}C_1 - \dfrac{F}{V_1}C_2$
Series branch, third CSTR, volume V_1	$\dfrac{dC_3}{dt} = \dfrac{F}{V_1}C_2 - \dfrac{F}{V_1}C_3$
First parallel branch, flow partition b, volume V_2	$\dfrac{dC_4}{dt} = \dfrac{bF}{V_2}C_3 - \dfrac{bF}{V_2}C_4$
Second parallel branch, flow partition $(1-b)$, volume V_3	$\dfrac{dC_5}{dt} = \dfrac{(1-b)F}{V_3}C_3 - \dfrac{(1-b)F}{V_3}C_5$
Output dilution	$C* = bC_4 + (1-b)C_5$
Plug-flow dead-time	$C_{out}(t) = C*(t-\delta)$

Source: Marsili-Libelli, S. and Checchi, N., *Ecol. Model.*, 187, 201–218, 2005.

The flow model consists of a front chain of three CSTRs of equal volume V_1, followed by two parallel branches of differing volumes V_2 and V_3, and a trailing plug-flow section to account for the input–output dead-time δ. In diffusive reactors, the latter can still be defined as the time delay between a pulse injection and the rising edge of the output concentration (Marsili-Libelli, 1997), as discussed in Section 6.3.3.2. In this sense, dead-time should not be confused with hydraulic residence time as it is normally defined (Levenspiel, 1972; Chazarenc et al., 2003; Persson and Wittgren, 2003). To estimate the parameters $\boldsymbol{p} = (V_1, V_2, V_3, b, \delta)$ of the model of Figure 8.50, the following error functional was defined

$$E(\boldsymbol{p}) = \frac{1}{N - n_p} \sum_{i=1}^{N} \left(C_{\text{out}}^{(i)} - C_{\text{obs}}^{(i)} \right)^2 \tag{8.35}$$

where:
$C_{\text{obs}}^{(i)}$ is the set of observed output tracer concentrations
$C_{\text{out}}^{(i)}$ is the corresponding model output

The shape of this error functional poses several numerical problems, shown in Figure 8.51.

The identification appears difficult because the parameter couples (V_1, V_2) and (V_3, b) present local minima (Figure 8.51a and c), so the simplex should be carefully initialized within the domain of attraction of the global minimum. A further problem is represented in Figure 8.51b by

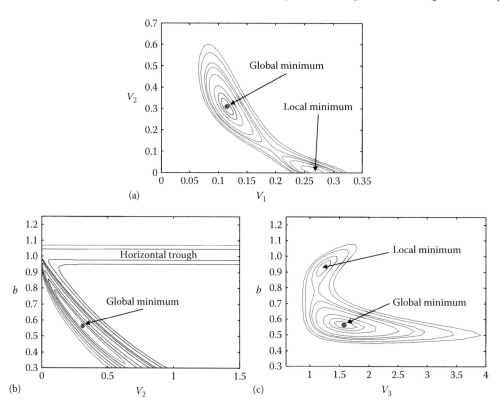

FIGURE 8.51 Contours of the error surface for estimating the parameters of the wetland flow structure of Figure 8.50. The estimation is made difficult by the presence of local minima (a and c) and a long horizontal trough (b) representing an indeterminate region of convergence. The location of the global minimum is indicated by the dot. (Reproduced with permission from Marsili-Libelli, S. and Checchi, N., *Ecol. Model.*, 187, 201–218, 2005.)

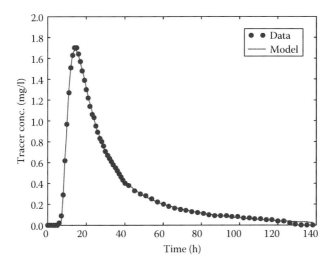

FIGURE 8.52   Fitting the model of Table 8.12 and Figure 8.50 with experimental data using lithium as a tracer. (Reproduced with permission from Marsili-Libelli, S. and Checchi, N., *Ecol. Model.*, 187, 201–218, 2005.)

the horizontal trough in the (V_2, b) subspace, where the search could be impaired by the lack of gradient. This is a consequence of the obvious fact that diverting the entire flow to V_3 $(b = 1)$ makes the model insensitive to V_2. This behaviour, however, is not symmetrical between V_2 and V_3, because V_3 is the dominant branch. This problem can be avoided by a proper simplex initialization. The agreement between the tracer data and the hydraulic model of Figure 8.50 and Table 8.12 is shown in Figure 8.52.

8.7.2   Pollution Abatement Model

In the subsurface constructed wetlands, a rich microbial colony soon becomes established as a biofilm around the porous media and the root zone, providing a self-purification action and reducing the incoming polluting load (Werner and Kadlec, 1996; Brix, 1997; Werner and Kadlec, 2000; Kadlec, 2003; Kadlec and Wallace, 2009). The best kinetics to describe these purification processes is still open to debate (Meyer et al., 2015) and should be decided on a case-by-case basis. In this example, two hypotheses have been tested, comparing a first-order and a Monod kinetics, as suggested by Mitchell and McNevin (2001), for the abatement of the carbonaceous substances, expressed as biodegradable COD. To augment the hydraulic model of Table 8.12 with the pollution abatement terms, a first-order or a Monod kinetics were added, assuming that no reaction occurred in the last stage plug-flow (PF). In this way, assuming that C_i is the COD concentration in the ith tank, the dynamics of each CSTR takes the form

$$\text{First-order kinetics} \quad \frac{dC_i}{dt} = \frac{F_i}{V_i} C_{i-1} - \frac{F_i}{V_i} C_i - k_1 C_i$$

$$\text{Monod kinetics} \quad \frac{dC_i}{dt} = \frac{F_i}{V_i} C_{i-1} - \frac{F_i}{V_i} C_i - \frac{R C_i}{K_s + C_i}$$

(8.36)

In the Monod kinetics case, R is the equivalent of the maximum growth rate μ_{max} divided by the yield coefficient Y, that is, $R = \mu_{max}/Y$. Being both unknown and difficult to estimate, they were lumped into a single parameter, as suggested by Dochain and Vanrolleghem (2001). A relationship may be established between the two models (Equation 8.36) by considering that for COD

TABLE 8.13

Estimated Parameters and Confidence Intervals for the Constructed Wetland Model of Figure 8.50 and Tracer Data of Figure 8.52

Parameter and Units	$\hat{p}$	δp_{FIM} (%)	δp_H (%)
Volume V_1 (m³)	0.1226	±2.246	±2.271
Volume V_2 (m³)	0.2263	±23.514	±25.049
Volume V_3 (m³)	1.5545	±4.313	±4.313
Flow split b	0.6268	±7.881	±8.891
Dead time δ (h)	5.98	–	–
Tracer mass M_o (g)	3.1011	–	–
Estimated tracer M_{est} (g)	3.1578	±1.789	±1.820
k_1 (1/d)	0.0265	±48.792	±51.094
R (1/d)	204.9	±4521.179	±1525.342
K_s (mg/l)	7737.7	±4605.091	±1536.262

concentrations well below the half-velocity constant $(C_i \ll K_s)$, the Monod kinetics may be approximated by a first-order kinetics with rate coefficient $k_1 \simeq R / K_s$. However, given the already large number of parameters vis-à-vis the limited number of available data, the first-order kinetics looks attractive, given the difficulties of estimating the Monod parameters. In fact, looking at the percentage confidence intervals for the kinetic coefficients in Table 8.13, it can be concluded that a much more robust model is obtained using a first-order kinetic approximation.

8.7.3 MODEL IDENTIFICATION

Direct samples were collected at a constructed wetland site, with a surface of 12 m² that was planted with *Phragmites australis* two years before the experiment. The filling medium had an average diameter of 8 mm and a porosity of 35%. More details can be found in Marsili-Libelli and Checchi (2005). The parameters of the kinetic models (Equation 8.36) were estimated by minimizing the error functional of Equation 8.35, and their confidence intervals were computed as percentages of the estimated values as

$$\delta p_{FIM}(i) = \pm \frac{t_{N_{exp}-n_p}^{1-\frac{\alpha}{2}} \sqrt{C_{FIM}(i,i)}}{\hat{p}_i} \times 100$$

(8.37)

$$\delta p_H(i) = \pm \frac{t_{N_{exp}-n_p}^{1-\frac{\alpha}{2}} \sqrt{C_H(i,i)}}{\hat{p}_i} \times 100$$

where

$t_{N_{exp}-n_p}^{1-\frac{\alpha}{2}}$ is the *t*-Student distribution for $(N_{exp} - n_p)$ degrees of freedom,
α is the confidence level,
C is the covariance matrix, computed by either approximation (FIM or Hessian).

As explained in Section 2.4.3.3, comparing these two approximations, FIM and Hessian, provides an accuracy check, depending on their agreement or disagreement. It must be recalled that the FIM approximation C_{FIM} is based on the sensitivity trajectories, whereas the Hessian approximation C_H

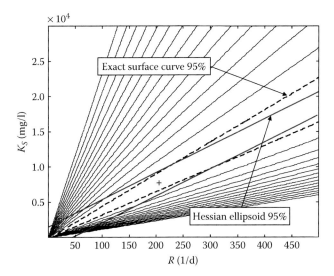

FIGURE 8.53 Comparison of approximate ellipses in the (R, K_s) space for the identification of the Monod term of Equation 8.36. The cross indicates the estimated point, and the shape of the contours reveals an extremely high parameter correlation, indicating a poor identifiability of the Monod kinetics. For this reason, the first-order approximation was preferred.

depends on the shape of the error surface. For nonlinear systems, this may be heavily influenced by the *curvature*, reflecting the degree of nonlinearity in the parametrization (Bates and Watts, 1988; Seber and Wild, 1989), so neglecting this term may have a significant effect. Comparing the confidence intervals obtained with C_{FIM} or C_H represents a way to assess the curvature effect, and hence the extremal nature of the estimated parameters $\hat{p}$. It has been shown with practical examples (Marsili-Libelli et al., 2003; Checchi and Marsili-Libelli, 2005) that when the two intervals are very near, the curvature effect is negligible and the identification can be relied upon. Conversely, if the search terminates at a point differing from the minimum, where the curvature may cause the two approximations to disagree, this should be considered as a warning that the estimation is not reliable. In this example, the first-order kinetic estimation proved more reliable than the Monod kinetics, as the confidence intervals in Table 8.13 show. It is interesting to notice that the ratio of the Monod parameters $R / K_s = 204.9 / 7737.7 = 0.02648$ is in excellent agreement with the calibrated value of $k_1 = 0.0265$. This confirms the validity of the assumption that when the substrate is considerably lower than the half-saturation constant, the first-order kinetics is a good approximation of the Monod kinetics. Furthermore, the inadequacy of the Monod kinetics is confirmed by the very high correlation between R and K_s, as shown by the very narrow and elongated shape of the confidence ellipses in Figure 8.53.

This confirms that the only quantity that can be reliably estimated is their ratio, equal to k_1, rather than their separate values. The low value of the kinetic constant k_1 is also in agreement with the conclusion of Wynn and Liehr (2001), who reported heterotrophic maximum growth rates as three orders of magnitude smaller than the values used for conventional wastewater treatments. The model with the calibrated first-order kinetic is compared to the data in Figure 8.54.

8.8 ENVIRONMENTAL ASSESSMENT OF A COASTAL LAKE AND SURROUNDING WETLANDS

This section describes the environmental assessment of a particularly sensitive ecosystem in Tuscany: the Massaciuccoli Lake on the Tyrrhenian coast, central Italy and the surrounding areas, shown in Figure 8.55.

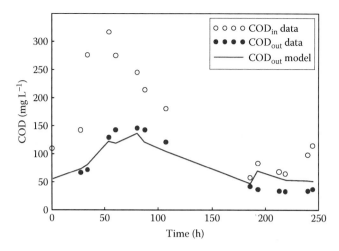

FIGURE 8.54 Performance of the fitted model obtained by augmenting the fitted hydraulic model of Table 8.12 with the first-order kinetics of Equation 8.36. (Reproduced with permission from Marsili-Libelli, S. and Checchi, N., *Ecol. Model.*, 187, 201–218, 2005.)

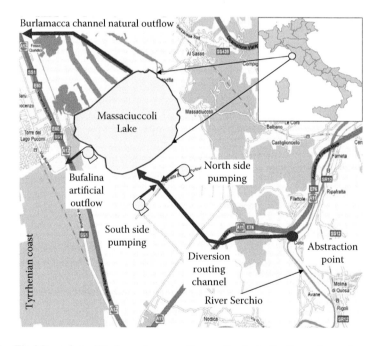

FIGURE 8.55 The Massaciuccoli Lake and surrounding wetlands. In the figure the main components of the water augmentation project are outlined: the diversion route from the Serchio River is outlined, together with the existing Burlamacca and Bufalina outgoing channels.

The Massaciuccoli Lake is a coastal lake receiving water from a web of draining channels from the surrounding cultivated plots, whose soil is mainly composed of peat. The main threats to this fragile ecosystem come from saltwater intrusion, subsidence and nonpoint agricultural pollution, resulting from a difficult coexistence between agricultural land use and environmental protection. The land subsidence around the lake and the considerable water usage for irrigation produce a severe water shortage and saltwater intrusion, as shown in the scheme of Figure 8.56.

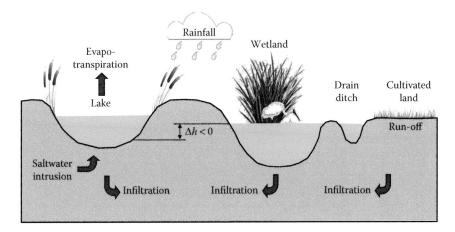

FIGURE 8.56 The water imbalance between the lake and the surrounding areas. Because of the subsidence, the lake is at a higher level than the adjacent areas and a considerable amount of water is lost to infiltration. Because the lake level is often lower than the mean sea level, saltwater intrusion is also possible.

To alleviate this water imbalance, a diversion from the nearby Serchio River is envisaged, as part of the environmental improvement scheme of the Massaciuccoli Lake area. The water flow around the Massaciuccoli Lake is shown in Figure 8.55, where the route of the possible diversion from the Serchio River is shown, together with the existing input and output channels and the pumping station drawing water from the wetlands and delivering it back into the lake.

8.8.1 MODELLING THE MASSACIUCCOLI LAKE WATER BALANCE

The main lake problem is the water conservation during the summer months, when the water level falls below the sea level. This, in addition to representing a water shortage for irrigation, favours the intrusion of saltwater along the Burlamacca channel bottom. The Serchio River abstraction is primarily conceived to mitigate this level drop, but this model demonstrates that not all the water diverted from the river contributes to the volume enhancement because of infiltration, an effect that should be accounted for in defining the abstraction policy. Before defining the lake model, its inputs were synthesized by using the time-series techniques of Chapter 3, as shown in Figure 8.57, combining the water inflow from the ditches and pumps with the rainfall. More details and the rainfall/runoff model and the input flow synthesis can be found in the specific publications (Giusti et al., 2011a, b). The lake

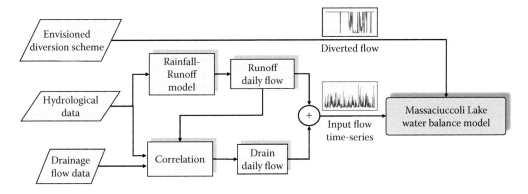

FIGURE 8.57 The input flow to the lake is modelled as a combination of rainfall and surface drain flow. The envisioned diversion is also accounted for. (Adapted with permission from Giusti et al., *Water Sci. Technol.*, 63, 2061–2070, 2011b.)

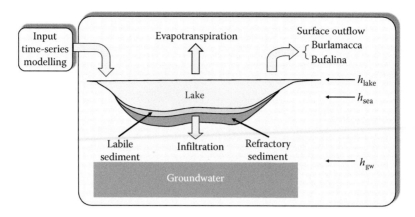

FIGURE 8.58 Massaciuccoli Lake water balance modelling, using the input model of Figure 8.57 and considering the groundwater infiltration.

volume model is described by the schematic diagram of Figure 8.58, where the evapotranspiration rate and the outflow rating curves (Burlamacca and Bufalina) were experimentally determined from local data provided by the Serchio River Water Authority.

The following flow balance can be written

$$\frac{dV}{dt} = Q_{\text{lake}}^{\text{in}} + Q_{\text{Serchio}}^{\text{in}} - Q_{\text{wdr}}^{\text{out}} - Q_{\text{et}}^{\text{out}} - Q_{\text{Bur}}^{\text{out}} - Q_{\text{Buf}}^{\text{out}} - Q_{\text{gw}}^{\text{out}}$$

$$h = \frac{V}{S(V)}$$

(8.38)

where the right-hand terms have the following meanings:

V	Lake volume (m³)
h	Lake mean depth (m)
$S(V)$	Lake variable surface (m²)
$Q_{\text{Buf}}^{\text{out}}$	Bufalina outflow (m³/d)
$Q_{\text{lake}}^{\text{in}}$	Lake surface inflow (m³/d)
$Q_{\text{et}}^{\text{out}}$	Gross evapotranspiration (m³/d)
$Q_{\text{Serchio}}^{\text{in}}$	Serchio diversion input (m³/d)
$Q_{\text{wdr}}^{\text{out}}$	Lake water withdrawal (m³/d)
$Q_{\text{Bur}}^{\text{out}}$	Burlamacca outflow (m³/d)
$Q_{\text{gw}}^{\text{out}}$	Groundwater infiltration (m³/d)

The groundwater seepage was modelled with the following equations:

$$Q_{\text{gw}}^{\text{out}} = \begin{cases} -\alpha \cdot \sqrt{|h - h_{\text{gw}}|} & \text{for } h > h_{\text{gw}} \\ 0 & \text{otherwise} \end{cases}$$

(8.39)

$$h_{\text{gw}} = -\beta - \gamma \cdot f(d)$$

where

$Q_{\text{gw}}^{\text{out}}$ is the estimated flow of lake water lost to groundwater seepage,

h_{gw} is the depth of the underground water table,

$f(d)$ is the function describing variable depth of the water table (Di Toro, 2001).

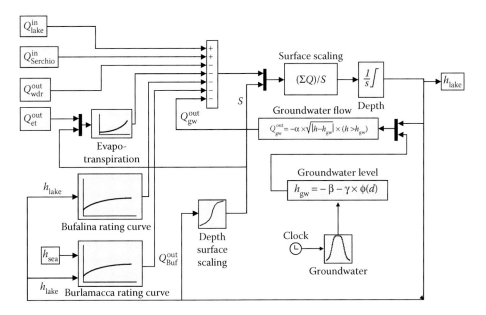

FIGURE 8.59 Simulink model of the Massaciuccoli Lake water balance.

The parameter α is the gauge factor for the groundwater head, β represents the offset value of the water table depth and γ is a scaling factor of the function $f(d)$. These two dimensionless parameters modulate the seasonal variations of the h_{gw} term. The Simulink model describing the water budget model of Equations 8.38 and 8.39 is shown in Figure 8.59. It was then calibrated by minimizing the sum of squared errors between the observed $\left(h^{obs}\right)$ and simulated $\left(h^{sim}\right)$ lake levels observed daily in the years 2000–2006 with each year representing a separate calibration horizon and thus yielding a separate parameter set.

$$\min_{(\alpha,\beta,\gamma,Q_{in},Q_{out})} \sum_{i=1}^{365} \left(h_i^{obs} - h_i^{sim}\right)^2 \qquad (8.40)$$

The calibrated model is compared to the observed lake levels throughout the year 2006 in Figure 8.60.

The parameters of the infiltration submodel of Equation 8.39 were assessed with the sensitivity analysis described in Marsili-Libelli and Giusti (2008), and they were then estimated with reasonable accuracy from the level data (h_i^{obs}) using the PEAS toolbox (Checchi et al., 2007). Table 8.14 shows the estimation results with their 95% confidence intervals. These parameters showed considerable variations over the years, reflecting the influence of climate changes on the groundwater seepage mechanism. As an example, 2003 was an exceptionally warm and dry year, and the corresponding α value is the highest of the set. The relation with the cumulative hydrograph is equally interesting, showing the saturation of β.

8.8.2 Using the Calibrated Model to Assess the Efficiency of the Diversion

The calibrated model was used to compare the present situation with the envisioned policy of routing part of the water from the Serchio River into the lake whenever the lake level falls below a prescribed threshold, now set at 10 cm below sea level. Instream flow considerations required that the river flow downstream of the abstraction should not be less than 6 m³/s; therefore, the maximum diverted flow was set at 3 m³/s. The simulation of Figure 8.61 shows that with this routing scheme the water imbalance would be greatly reduced and that the lake level would be maintained at the

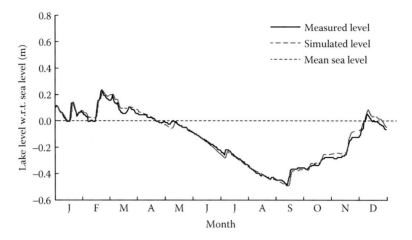

FIGURE 8.60 Calibration of the Massaciuccoli Lake water balance model with the level data of 2006. The water depletion during the summer month is quite evident. The level data were kindly provided by the Serchio River Water Authority.

TABLE 8.14

Calibrated Parameters of the Groundwater Seepage (Equation 8.39) in the Massaciuccoli Lake Water Balance Model

	Estimated Values and 95% Confidence Interval		
Year	$\alpha \times 10^{-4}$	β	γ
2000	1.9945 ± 0.0135	3.8182 ± 0.0182	0.3321 ± 0.0534
2001	4.5665 ± 0.0079	1.2401 ± 0.0173	0.0195 ± 0.0404
2002	4.8954 ± 0.0181	0.6719 ± 0.0163	0.4029 ± 0.0728
2003	5.5628 ± 0.0066	0.6811 ± 0.0163	0.2639 ± 0.0271
2004	4.8667 ± 0.0118	0.7177 ± 0.0146	0.2128 ± 0.0472
2005	3.9185 ± 0.0102	0.8953 ± 0.0173	0.2682 ± 0.0361
2006	3.8167 ± 0.0054	0.9037 ± 0.0192	0.3826 ± 0.0214

prescribed level, except for a short period in September when drought is most acute. However, as Figure 8.62 shows, much of the diverted water would be lost to infiltration in the summer months when the water table is lowest and the soil porosity highest, as a consequence of the increased head. So the routing policy should be carefully reviewed to limit the additional water loss.

8.9   A FISH HABITAT SUITABILITY ASSESSMENT BASED ON FUZZY LOGIC

In this section, we describe a method for assessing the environmental quality of the aquatic habitat based on synthetic hydraulic and water quality parameters. The method is meant to overcome the limitations of IFIM (Instream Flow Incremental Methodology, Bovee et al., 1998) based on the suitability response of target fish species to river parameters through a fuzzy inference system (FIS), as described in Chapter 4. Through the use of simulated river conditions, a wide range of scenarios can be generated to detect potentially critical situations. The FIS was then adapted to produce a fuzzy index for the suitability of the habitat (FISH) by determining the range of admissible flows for

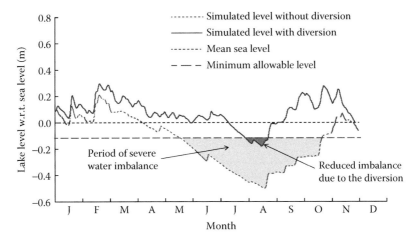

FIGURE 8.61 Comparison of yearly lake level with (solid line) and without (dashed line) the Serchio River diversion, showing that this can greatly reduce the summer imbalance.

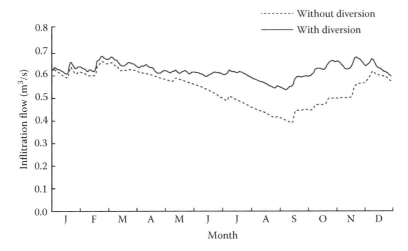

FIGURE 8.62 Comparison of seepage flow in the present situation (dotted line) and with the diversion. It can be seen that as a consequence of the increased head, the quantity of water lost by infiltration is increased.

which the habitat suitability is preserved to a prescribed degree. The approach and the main results will be described here, while the full description of the FISH method can be found in the specific publications (Marsili-Libelli et al., 2008a, 2013) whose results are described here.

The fuzzy approach to the definition of habitat suitability had been previously used (Young et al., 2000) for its ability to incorporate nonnumerical information into its rule base, often complementing the scarcity of field data. Furthermore, the inherent uncertainty of ecological variables can be accounted for in the definition of rules and membership functions (Prato, 2007), and results can be easily grasped by nontechnical parties, such as decision makers and stakeholders (Van Broekhoven et al., 2006). A fuzzy approach had already been used to develop the MesoCASiMiR method (Schneider et al., 2001), which was later applied at differing levels in the trophic chain, from macro-invertebrates to fish (Mouton et al., 2006, 2007, 2008; Fukuda and Hiramatsu, 2008; Fukuda, 2009; Fukuda et al., 2011). Habitat indicators based on fuzzy logic were also developed and compared with expert knowledge based systems (Marchini et al., 2009; Mouton et al., 2009), indicating that the combination of both approaches can improve model reliability, especially in site extrapolation of local results. However, the importance of expert judgment in deciding the final fuzzy model acceptability has recently been reaffirmed (Fukuda et al., 2011; Mouton et al., 2011; Wieland et al., 2011).

8.9.1 Development of the FISH Algorithm

The goal of this paper was to extend the habitat assessment from the analysis of an existing situation to a variety of hypothetical scenarios involving differing hydraulic and water quality data. This study cannot be compared to much more complex habitat suitability methods (Mouton et al., 2009) or species distribution models (Fukuda et al., 2011), because it only aims to provide a simple instream flow analysis tool for river management. Given the scarce data availability often encountered in instream flow analysis, this method makes minimum use of field data and emphasizes the importance of the expert in supplementing the missing data, in specifying the habitat characteristics, and in the interpretation of results.

The FISH algorithm is based on synthetic hydraulic and ecological parameters. Its aim is the generation of a whole paradigm of possible scenarios based on combinations of flow and quality variations obtained by simulation. Its aims are threefold:

- Incorporating expert-knowledge into the FIS as a complement to the available field data
- Extending the analysis to a combination of synthetic flow/quality scenarios generated by HEC-RAS (Brunner, 2010) and QUAL2K (Chapra and Pelletier, 2003; Pelletier et al., 2006), thus enabling the generation of a grid of scenarios in the flow/water quality 2D domain
- Providing integration into a Decision Support System for the management of regulated rivers by extending the old notion of hydraulic instream flow into a broader concept of ecological instream flow

The preliminary knowledge on which the FISH method is based is now considered. First, a suitable model must be available for the hydraulics and the quality of the river under study. Then, the target fish species must be selected. In the following, the algorithm will be applied to the Arno River, whose catchment is shown in Figure 8.63, and for which a HEC-RAS/QUAL2K hydraulic/quality model was developed and calibrated, as shown in Figure 8.64.

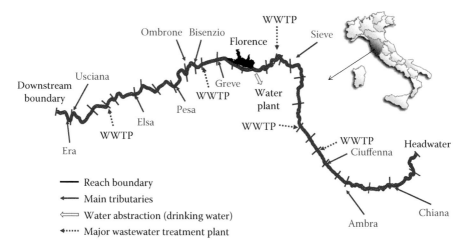

FIGURE 8.63 Catchment of the Arno River, for which the FISH algorithm was applied. The portion of the river course considered in this study (thick line) was divided into 30 reaches. There are 11 main tributaries and 5 major wastewater treatment plants, together with 1 large water treatment plant providing drinking water to the city of Florence. (Reproduced with permission from Marsili-Libelli, S. et al., *Environ. Model. Softw.*, 41, 27–38, 2013.)

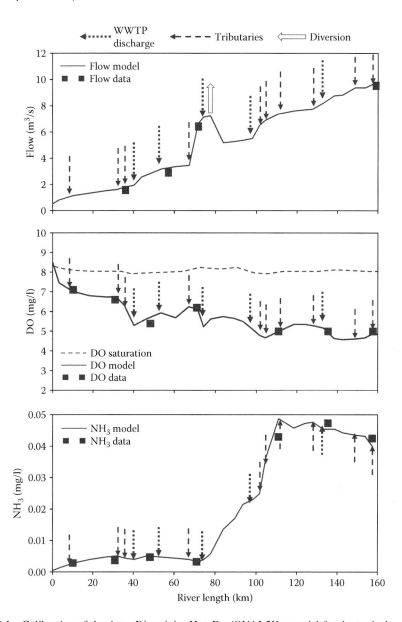

FIGURE 8.64 Calibration of the Arno River joint Hec-Ras/QUAL2Kw model for the typical summer low-flow regime, used to generate the FISH scenarios. The variables used in this study (Flow, DO, and NH₃) are shown with the arrows indicating the position of the main tributaries of the wastewater treatment plants, and of the diversion of the Florence potabilization plant. The calibration data were kindly provided by the regional environmental protection authority (ARPAT). (Reproduced with permission from Marsili-Libelli, S. et al., *Environ. Model. Softw.*, 41, 27–38, 2013.)

Regarding the target species, Barbel (*Barbus tyberinus*) and European chub (E. chub) (*Leuciscus cephalus*) were selected, these being the most widespread autochthonous and reophylic species in the region. Their average response to the environmental parameters is described by the suitability curves shown in Figure 8.65. These were determined by combining literature studies (McMahon, 1982; McMahon et al., 1984; Raleigh et al., 1986; Svobodova et al., 1993; Lamouroux and Capra, 2002; Lamouroux and Souchon, 2002; Moir et al., 2005; Ayllón et al., 2012; Macura et al., 2012) with

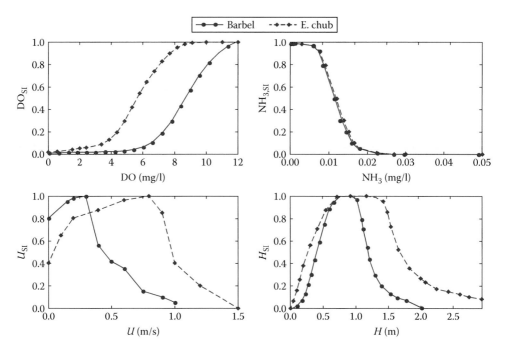

FIGURE 8.65 Suitability curves for the target species Barbel (*Barbus tyberinus*) and E. chub (*Leuciscus cephalus*). The suitability index for each parameter varies between 0 (totally unsuitable) and 1 (perfectly suitable). (Reproduced with permission from Marsili-Libelli, S. et al., *Environ. Model. Softw.*, 41, 27–38, 2013.)

direct observations obtained with the IFIM sampling protocol (Nocita et al., 2009). Furthermore, the suitability curves of Figure 8.65 are fuzzified into three classes of merit (low, medium, and high) using the membership functions shown in Figure 8.66.

8.9.2 THE FISH ALGORITHM

With this preliminary information available (flow and quality river model and fuzzy suitability curves), the FISH algorithm can now be described. Its general structure is outlined in Figure 8.67. The fuzzy inference algorithm uses the knowledge described above to determine the composite FISH index through fuzzy inference (Algorithm 1). Critical reaches can then be detected via a low-flow sensitivity analysis, for which Algorithm 2 performs an instream flow analysis. Algorithm 1 consists of the four steps listed in Box 8.1.

The organization of Algorithm 1 is illustrated in Figure 8.68: the scenario information (river flow and quality) generates the antecedents of the FIS, while the consequent singletons are defined for both species as high (H = 1.0), middle high (MH = 0.77), middle (M = 0.5), middle low (ML = 0.33), and low (L = 0.1).

The inference rules were obtained by examining some direct fish abundance data in the summer low flow (SLF) and assuming that they represented a rough approximation of the habitat suitability. This comparison, shown in Figure 8.69, should by no means be considered a calibration in the strict sense, but was used nonetheless as a guideline to adjust the inference rules for the two species, together with expert judgement, eliminating the rules that were never activated by more than 2.5%

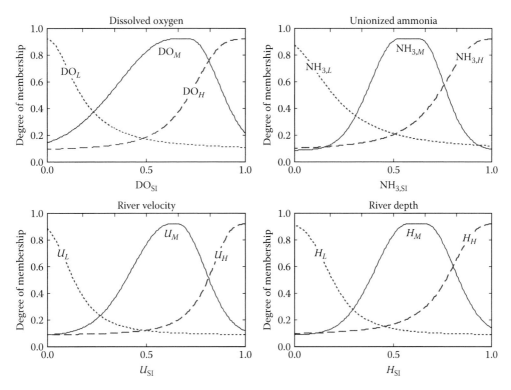

FIGURE 8.66 Fuzzification of the suitability curves of Figure 8.65 into three membership functions (L, Low; M, Medium; H, High). (Reproduced with permission from Marsili-Libelli, S. et al., *Environ. Model. Softw.*, 41, 27–38, 2013.)

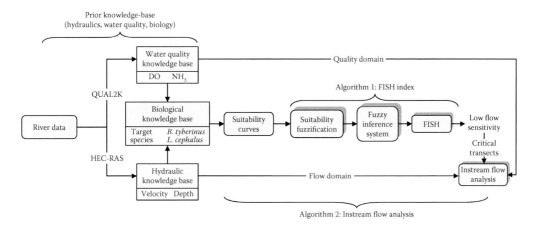

FIGURE 8.67 General outline of the FISH algorithm. The prior knowledge about the river hydraulics and ecology are used as input data to the suitability curves. Algorithm 1 computes the FISH index via the FIS, while Algorithm 2 performs an instream flow analysis based on the critical areas detected by Algorithm 1. (Reproduced with permission from Marsili-Libelli, S. et al., *Environ. Model. Softw.*, 41, 27–38, 2013.)

BOX 8.1   STEPS OF ALGORITHM 1 IN THE FISH METHOD

- *Step 1*: Generate the synthetic hydraulic/quality data for the prescribed flows and reaches
- *Step 2*: Define a set of membership functions for the fuzzification of the suitability curves
- *Step 3*: Compute the individual fuzzy suitability responses
- *Step 4*: Process the suitability indexes through the fuzzy inferential system to obtain the FISH along all the reaches for each computed flow.

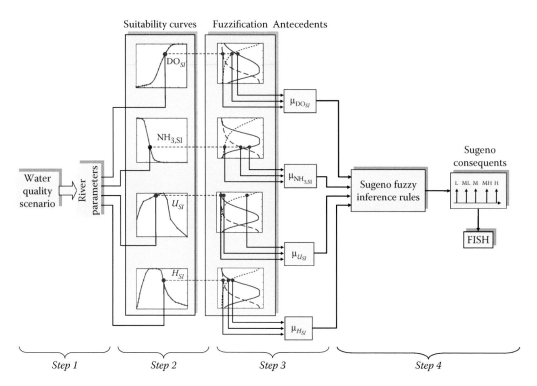

FIGURE 8.68   Detailed organization of Algorithm 1 for computing the FISH index. Step 1 generates the river parameters; Step 2 produces the relative suitability indexes. Step 3 fuzzifies this information producing the antecedents of the Sugeno inferential system; Step 4 computes the FISH index on the basis of the available knowledge base. (Reproduced with permission from Marsili-Libelli, S. et al., *Environ. Model. Softw.*, 41, 27–38, 2013.)

of the total, and to adjust the membership functions. After this pruning, the 14 rules were retained for the Barbel (Table 8.15), and 20 rules were defined for E. chub (Table 8.16).

8.9.3   INSTREAM FLOW ANALYSIS WITH FISH

Instream flow analysis is a part of the IFIM (Bovee et al., 1998) that assesses the impact of flow alterations on stream habitat. We are now going to use FISH to detect reaches where the habitat suitability is critically affected by flow. We define the sustainable instream flow $\{Q_{\text{instream}}\}$ as the

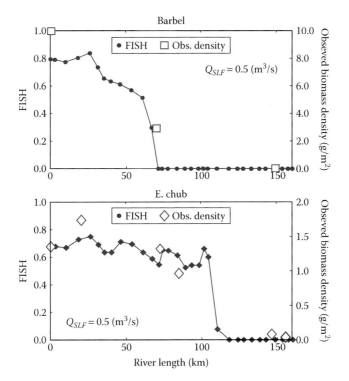

FIGURE 8.69 FISH computation (left scale) for the Arno River in the summer low-flow condition $Q_{SLF} = 0.5 \left(m^3/s \right)$. Some observed fish densities for the same conditions are shown for comparison (right scale). (Reproduced with permission from Marsili-Libelli, S. et al., *Environ. Model. Softw.*, 41, 27–38, 2013.)

TABLE 8.15

Inferential Fuzzy Rules for the Barbel

1.	If (DO is M)	and (NH_3 is H)	and (U is M)	and (H is L)	then (S is ML)
2.	If (DO is L)	and (NH_3 is M)	and (U is M)	and (H is L)	then (S is L)
3.	If (DO is H)	and (NH_3 is H)	and (U is M)	and (H is M)	then (S is MH)
4.	If (DO is M)	and (NH_3 is M)	and (U is H)	and (H is H)	then (S is H)
5.	If (DO is L)	and (NH_3 is L)	and (U is M)	and (H is M)	then (S is ML)
6.	If (DO is M)	and (NH_3 is M)	and (U is M)	and (H is H)	then (S is MH)
7.	If (DO is M)	and (NH_3 is M)	and (U is M)	and (H is L)	then (S is MH)
8.	If (DO is M)	and (NH_3 is H)	and (U is M)	and (H is M)	then (S is MH)
9.	If (DO is L)	and (NH_3 is H)	and (U is L)	and (H is L)	then (S is L)
10.	If (DO is M)	and (NH_3 is M)	and (U is H)	and (H is M)	then (S is H)
11.	If (DO is H)	and (NH_3 is H)	and (U is H)	and (H is H)	then (S is H)
12.	If (DO is H)	and (NH_3 is H)	and (U is M)	and (H is H)	then (S is MH)
13.	If (DO is L)	and (NH_3 is M)	and (U is L)	and (H is H)	then (S is L)
14.	If (DO is M)	and (NH_3 is M)	and (U is M)	and (H is M)	then (S is MH)

Source: Marsili-Libelli, S. et al., *Environ. Model. Softw.*, 41, 27–38, 2013.

TABLE 8.16

Inferential Fuzzy Rules for the E. Chub

1.	If (DO is H)	and (NH₃ is M)	and (U is M)	and (H is H)	then (S is H)
2.	If (DO is H)	and (NH₃ is M)	and (U is M)	and (H is L)	then (S is M)
3.	If (DO is L)	and (NH₃ is H)	and (U is M)	and (H is L)	then (S is ML)
4.	If (DO is H)	and (NH₃ is L)	and (U is H)	and (H is L)	then (S is MH)
5.	If (DO is H)	and (NH₃ is L)	and (U is M)	and (H is L)	then (S is H)
6.	If (DO is M)	and (NH₃ is L)	and (U is M)	and (H is L)	then (S is ML)
7.	If (DO is L)	and (NH₃ is M)	and (U is M)	and (H is L)	then (S is L)
8.	If (DO is H)	and (NH₃ is H)	and (U is M)	and (H is M)	then (S is MH)
9.	If (DO is M)	and (NH₃ is M)	and (U is H)	and (H is H)	then (S is H)
10.	If (DO is H)	and (NH₃ is M)	and (U is L)	and (H is L)	then (S is ML)
11.	If (DO is H)	and (NH₃ is H)	and (U is H)	and (H is L)	then (S is ML)
12.	If (DO is M)	and (NH₃ is M)	and (U is M)	and (H is H)	then (S is MH)
13.	If (DO is M)	and (NH₃ is M)	and (U is L)	and (H is M)	then (S is MH)
14.	If (DO is H)	and (NH₃ is M)	and (U is M)	and (H is M)	then (S is ML)
15.	If (DO is M)	and (NH₃ is H)	and (U is M)	and (H is M)	then (S is MH)
16.	If (DO is M)	and (NH₃ is M)	and (U is H)	and (H is M)	then (S is H)
17.	If (DO is H)	and (NH₃ is H)	and (U is H)	and (H is H)	then (S is H)
18.	If (DO is H)	and (NH₃ is H)	and (U is M)	and (H is H)	then (S is MH)
19.	If (DO is M)	and (NH₃ is M)	and (U is M)	and (H is M)	then (S is MH)
20.	If (DO is H)	and (NH₃ is L)	and (U is M)	and (H is H)	then (S is MH)

Source: Marsili-Libelli, S. et al., *Environ. Model. Softw.*, 41, 27–38, 2013.

BOX 8.2   STEPS OF ALGORITHM 2 IN THE FISH METHOD

- *Step 1*: Select a critical reach based on FISH
- *Step 2*: Run the hydraulic water quality model for the entire flow set at the selected reach
- *Step 3*: Evaluate FISH, according to Algorithm 1, as a function of flow
- *Step 4*: Compute the FISH cumulative distribution as a function of flow and determine the lower FISH limits
- *Step 5*: Determine the instream inflow range $\{Q_{instream}\}$, according to Equation 8.41.

---

flow range for which FISH remains above the fourth quartile ($FISH_{75}$) of its statistical distribution, or above a minimum threshold ($FISH_{min}$), whichever is greater

$$\{Q_{instream}\} = \underset{Q}{\arg}\left[\left(FISH > FISH_{75}\right) \wedge \left(FISH > FISH_{min}\right)\right]$$ (8.41)

Algorithm 2 develops along the five steps listed in Box 8.2. This second algorithm extends the capabilities of Algorithm 1 by determining the flow values between which the suitability never deteriorates below a set threshold. Repeated application of Algorithm 1 for a suitable set of flows,

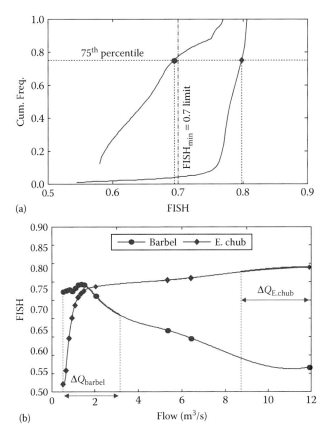

FIGURE 8.70 (a) Cumulative FISH distribution and selection of the suitability limits for the two target species evaluated for the Arno River at reach #5 around Q_{SLF}. (b) The admissible flow range is computed by considering the last quartile $FISH_{75}$ and an absolute $FISH_{min}$ of 0.7 (thick lines), whichever is greater (in this example, the latter holds for the Barbel). The dots and diamonds refer to the actual flow values used for the interpolation (thin solid lines). (Reproduced with permission from Marsili-Libelli, S. et al., *Environ. Model. Softw.*, 41, 27–38, 2013.)

and the subsequent statistical distribution of FISH, yields the flow intervals (ΔQ) in which the suitability is above a prescribed minimum level. This analysis can be useful in gaining further insight into critical reaches, or for assessing diversions in regulated rivers. Figure 8.70 shows the application of Algorithm 2 to the critical fifth reach in the Arno River, in which the habitats of the two species are widely separated. It also shows that the E. chub would be greatly disadvantaged by a flow decrease.

References

Abdi, H. and Williams, L. J., 2010. Principal component analysis. *Wiley Interdisciplinary Reviews: Computational Statistics*, **2**, 433–459.

Abonyi, J., 2003. *Fuzzy Model Identification for Control*, Birkhauser, Boston, MA, 273pp.

Abramowitz, M. and Stegun, I. A., 1970. *Handbook of Mathematical Functions with Formulas, Graphs, and Mathematical Tables*, Dover, New York, 1022pp.

Agren, G. and Bosatta, E., 1996. *Theoretical Ecosystem Ecology*, Cambridge University Press, Cambridge, 234pp.

Akçakaya, H. S., Burgman, M. A., and Ginzburg, L. R., 1999. *Applied Population Ecology: Principles and Computer Exercises Using RAMAS EcoLab 2.0*, Sinauer Associates, Sunderland, MA.

Akin, B. S. and Ugurlu, A., 2005. Monitoring and control of biological nutrient removal in a Sequencing Batch Reactor. *Process Biochemistry*, **40**, 2873–2878.

Alados, I., Foyo-Moreno, I., and Alados-Arboledas, L., 1996. Photosynthetically active radiation: Measurements and modelling. *Agricultural and Forest Meteorology*, **78**, 121–131.

Alberty, R. A., 1979. *Physical Chemistry*, John Wiley & Sons, New York, 682pp.

Alferes, J., Tik, S., Copp, J., and Vanrolleghem, P. A., 2013. Advanced monitoring of water systems using in situ measurement stations: Data validation and fault detection. *Water Science & Technology*, **68**, 1022–1030.

Arhonditsis, G. B. and Brett, M. T., 2005. Eutrophication model for Lake Washington (USA): Part I. Model description and sensitivity analysis. *Ecological Modelling*, **187**, 140–178.

Artan, N. and Orhon, D., 2005. *Mechanism and Design of Sequencing Batch Reactors for Nutrient Removal*, IWA Scientific and Technical Report no. 19, London, 100pp.

Asaeda, T., Trung, V. K., and Manatunge, J., 2000. Modeling the effects of macrophyte growth and decomposition on the nutrient budget in shallow lakes. *Aquatic Botany*, **68**, 217–237.

Asaeda, T., Trung, V. K., Manatunge, J., and Van Bon, T., 2001. Modelling macrophyte-nutrient-phytoplankton interactions in shallow eutrophic lakes and the evaluation of environmental impacts. *Ecological Engineering*, **16**, 341–357.

Asaeda, T. and Van Bon, T., 1997. Modelling the effects of macrophytes on algal blooming in eutrophic shallow lakes. *Ecological Modelling*, **104**, 261–287.

Astrom, K. J., 1970. *Introduction to Stochastic Control Theory*, Academic Press, New York, 299pp.

Athans, M., Dertouzos, M. L., Spann, R. D., and Mason, S. J., 1974. *Systems, Networks, and Computation: Multivariable Methods*, McGraw-Hill, New York, 552pp.

Atkinson, A. C. and Donev, A. N., 1992. *Optimum Experimental Designs*, Clarendon Press, Oxford, 352pp.

Ayllón, D., Almodóvar, A., Nicola, G. G., and Elvira, B., 2012. The influence of variable habitat suitability criteria on PHABSIM habitat index results. *River Research and Applications*, **28**, 1179–1188.

Azzellino, A., Salvetti, R., Vismara, R., and Bonomo, L., 2006. Combined use of the EPA-QUAL2E simulation model and factor analysis to assess the source apportionment of point and non point loads of nutrients to surface waters. *Science of the Total Environment*, **371**, 214–222.

Azzoni, R., Giordani, G. M., Bartoli, M., Welsh, D. T., and Viaroli, P. L., 2001. Iron, sulphur and phosphorus cycling in the rhizosphere sediments of a eutrophic Ruppia cirrhosa meadow (Valle Smarlacca, Italy). *Journal of Sea Research*, **45**, 15–26.

Babuska, R., 1998. *Fuzzy Modeling for Control*, Kluwer Academic Publishers, Boston, MA, 260pp.

Babuska, R., van der Veen, P. J., and Kaymak, U., 2002. Improved covariance estimation for Gustafson-Kessel clustering. In: *Proceedings of the IEEE International Conference on Fuzzy Systems*, Honolulu, HI, pp. 1081–1085.

Baggiani, F. and Marsili-Libelli, S., 2009. Real-time fault detection and isolation in biological wastewater treatment plants. *Water Science & Technology*, **60**, 2949–2961.

Bailey, J. E. and Ollis, D. F., 1986. *Biochemical Engineering Fundamentals*, McGraw-Hill, New York, 984pp.

Banks, S. P., 1986. *Control Systems Engineering*, Prentice Hall, New York, 614pp.

Bannister, T. T., 1974. Production equations in terms of chlorophyll concentration, quantum yield, and upper limit to production. *Limnology and Oceanography*, **19**, 1–12.

Bàrdossy, A. and Duckstein, L., 1995. *Fuzzy Rule-Based Modeling with Applications to Geophysical, Biological and Engineering Systems*, CRC Press, Boca Raton, FL, 232pp.

Basille, M., Calenge, C., Marboutin, É., Andersen, R., and Gaillard, J.-M., 2008. Assessing habitat selection using multivariate statistics: Some refinements of the ecological-niche factor analysis. *Ecological Modelling*, **211**, 233–240.

Bastianoni, S., 2008. Eco-exergy to emergy flow ratio. In: S. E. Jorgensen (ed.), *Encyclopedia of Ecology*. Elsevier B.V., Amsterdam, the Netherlands, pp. 979–983.

Bastin, G. and Dochain, D., 1990. *On-Line Estimation and Adaptive Control of Bioreactors*, Elsevier B.V., Amsterdam, the Netherlands, 394pp.

Batchelor, G. K., 2000. *An Introduction to Fluid Dynamics*, Cambridge University Press, Cambridge, 615pp.

Bates, D. M. and Watts, D., 1988. *Nonlinear Regression Analysis and its Applications*, John Wiley & Sons, New York, 392pp.

Batstone, D. J., Keller, J., Angelidaki, I., Kalyuzhnyi, S. V., Pavlostathis, S. G., Rozzi, A., Sanders, W. T., Siegrist, H., and Vavilin, V. A., 2002a. *Anaerobic Digestion Model No. 1*, IWA Scientific and Technical Report no. 13, London, 77pp.

Batstone, D. J., Keller, J., Angelidaki, I., Kalyuzhnyi, S. V., Pavlostathis, S. G., Rozzi, A., Sanders, W. T., Siegrist, H., and Vavilin, V. A., 2002b. The IWA anaerobic digestion model no 1 (ADM1). *Water Science & Technology*, **45**, 65–73.

Batstone, D. J., Keller, J., and Steyer, J. P., 2006. A review of ADM1 extensions, applications, and analysis: 2002–2005. *Water Science & Technology*, **54**, 1–10.

Battley, E. H., 1987. *Energetics of Microbial Growth*, John Wiley & Sons, New York, 450pp.

Begon, M. and Mortimer, M., 1986. *Population Ecology*, Blackwell, Oxford, 220pp.

Bellman, R. and Åström, K. J., 1970. On structural identifiability. *Mathematical Biosciences*, **7**, 329–339.

Benedini, M. and Tsakiris, G., 2013. Dispersion in rivers and streams. In: *Water Quality Modelling for Rivers and Streams, Water Science and Technology Library 70*. Springer-Verlag, Berlin, Germany, 288pp.

Bennett, N. D., Croke, B. F. W., Guariso, G., Guillaume, J. H. A., Hamilton, S. H., Jakeman, A. J., Marsili-Libelli, S. et al., 2013. Characterising performance of environmental models. *Environmental Modelling & Software*, **40**, 1–20.

Berger, R. D., 1981. Comparison of the Gompertez and logistic equations to describe plant disease progress. *Phytopathology*, **71**, 716–719.

Berman, T. and Chava, S., 1999. Algal growth on organic compounds as nitrogen sources. *Journal of Plankton Research*, **21**, 1423–1437.

Beun, J. J., Dircks, K., Van Loosdrecht, M. C. M., and Heijnen, J. J., 2002. Poly-beta-hydroxybutyrate metabolism in dynamically fed mixed microbial cultures. *Water Research*, **36**, 1167–1180.

Beun, J. J., Verhoef, E. V., Van Loosdrecht, M. C. M., and Heijnen, J. J., 2000. Stoichiometry and kinetics of poly-beta-hydroxybutyrate metabolism under denitrifying conditions in activated sludge cultures. *Biotechnology and Bioengineering*, **68**, 496–507.

Bezdek, J. and Pal, N. R., 1995. On cluster validity for the fuzzy c-means model. *IEEE Transactions on Fuzzy Systems*, **3**, 370–379.

Bezdek, J. C., 1981. *Pattern Recognition with Fuzzy Objective Function Algorithms*, Plenum Press, New York, 256pp.

Bezdek, J. C., 1993. Fuzzy models—What are they, and why? *IEEE Transactions on Fuzzy Systems*, **1**, 1–5.

Bortone, G., Marsili-Libelli, S., Tilche, A., and Wanner, J., 1999. Anoxic phosphate uptake in the DEPHANOX process. *Water Science & Technology*, **40**, 177–185.

Bovee, K. D., Lamb, B. L., Bartholow, J. M., Stalnaker, C. B., Taylor, J., and Henriksen, J., 1998. *Stream Habitat Analysis Using the Instream Flow Incremental Methodology*, U.S. Geological Survey, Biological Resources Division Information and Technology Report, Fort Collins, CO, 131pp.

Box, G. E. P., Jenkins, G. M., and Reinsel, G. C., 2008. *Time Series Analysis, Forecasting and Control*, John Wiley & Sons, New York, 784pp.

Brandt-Williams, S. L., 2002. *Handbook of Emergy Evaluation: A Compendium of Data for Emergy Computation*, Center for Environmental Policy, University of Florida, Gainesville, FL, 40pp.

Brauer, F. and Castillo-Chaves, C., 2014. *Mathematical Models in Population Biology and Epidemiology*, Springer Science & Business Media, New York, 508pp.

Brezonik, P. and Arnold, W., 2011. *Water Chemistry: An Introduction to the Chemistry of Natural and Engineered Aquatic Systems*, Oxford University Press, Oxford, 808pp.

Brillinger, D. R., 2013. Aligning some Nicholson sheep-blowfly data sets with system input. *Stat*, **2**, 9–21.

Britto, D. T. and Kronzucker, H. J., 2001. Constancy of nitrogen turn-over kinetics in the plant cell: Insights into the integration of subcellular N fluxes. *Planta*, **213**, 175–181.

Brix, H., 1997. Do macrophytes play a role in constructed treatment wetlands? *Water Resources Research*, **35**, 11–17.

Brown, L. C. and Barnwell, T. O., 1987. *The Enhanced Stream Water Quality Models QUAL2E and QUAL2E-UNCAS: Documentation and User Manual.* Environmental Research Laboratory, US Environmental Protection Agency, Athens, GA, 189pp.

Brown, M. and Harris, C., 1994. *Neurofuzzy Adaptive Modelling and Control*, Prentice Hall, London, 508pp.

Brown, M. T. and Bardi, E., 2001. *Handbook of Emergy Evaluation—A Compendium of Data for Emergy Computation*, Center for Environmental Policy, University of Florida, Gainesville, FL, 94pp.

Brun, R., Kühni, M., Siegrist, H., Gujer, W., and Reichert, P., 2002. Practical identifiability of ASM2d parameters—Systematic selection and tuning of parameter subsets. *Water Research*, **36**, 4113–4127.

Brunner, G., 2010. *HEC-RAS River Analysis System. Hydraulic Reference Manual*, Version 4.1, US Army Corps of Engineers Hydrologic Engineering Center, Davis, CA, 790pp.

Butcher, J. C., 2003. *Numerical Methods for Ordinary Differential Equations*, John Wiley & Sons, New York, 440pp.

Campbell, R., 1983. *Microbial Ecology*, Blackwell, Oxford, 191pp.

Carta, F., Beun, J. J., Van Loosdrecht, M. C. M., and Heijnen, J. J., 2001. Simultaneous storage and degradation of PHB and glycogen in activated sludge cultures. *Water Research*, **35**, 2693–2701.

Casti, J. L., 1979. *Connectivity, Complexity, and Catastrophe in Large-Scale Systems*, John Wiley & Sons, London, 203pp.

Causton, D. R. and Venus, J. C., 1982. *The Biometry of Plant Growth*, Edward Arnold, New York, 320pp.

Chapra, S. C., 1979. Applying phosphorus loading models to embayments. *Limnology and Oceanography*, **24**, 168–171.

Chapra, S. C., 1997. *Surface Water-Quality Modeling*, McGraw-Hill, New York, 844pp.

Chapra, S. C. and Canale, R., 1991. *Numerical Methods for Engineers with Software and Programming Applications*, McGraw-Hill, New York, 944pp.

Chapra, S. C. and Pelletier, G. J., 2003. *QUAL2K: A Modeling Framework for Simulating River and Stream Water Quality. Documentation and Users Manual*, Civil and Environmental Engineering Department, Tufts University, Medford, MA, 121pp.

Chazarenc, F., Merlin, G., and Gonthier, Y., 2003. Hydrodynamics of horizontal subsurface flow constructed wetlands. *Ecological Engineering*, **21**, 165–173.

Checchi, N., Giusti, E., and Marsili-Libelli, S., 2007. PEAS: A toolbox to assess the accuracy of estimated parameters in environmental models. *Environmental Modelling and Software*, **22**, 899–913.

Checchi, N. and Marsili-Libelli, S., 2005. Reliability of parameter estimation in respirometric models. *Water Research*, **39**, 3686–3696.

Chiaudani, G. and Vighi, M., 1976. Comparison of different techniques for detecting limiting or surplus nitrogen in batch cultures of *Selenastrum capricornutum*. *Water Research*, **10**, 725–729.

Christensen, V. and Pauly, D., 1992. Ecopath II—A software for balancing steady-state ecosystem models and calculating network characteristics. *Ecological Modelling*, **61**, 169–185.

Christensen, V. and Walters, C. J., 2004. Ecopath with ecosim: Methods, capabilities and limitations. *Ecological Modelling*, **172**, 109–139.

Christensen, V., Walters, C. J., Pauly, D., and Forrest, R., 2008. *Ecopath with Ecosim Version 6: User Guide*, Fisheries Centre, University of British Columbia, Vancouver, Canada. Available on the web at http://www.ecopath.org/.

Christian, R. R., Forés, E., Comin, F., Viaroli, P. L., Naldi, M. C., and Ferrari, I., 1996. Nitrogen cycling networks of coastal ecosystems: Influence of trophic status and primary producer form. *Ecological Modelling*, **87**, 111–129.

Christian, R. R., Naldi, M., and Viaroli, P., 1998. Construction and analysis of static, structured models of nitrogen cycling in coastal ecosystems. In: A. L. Koch, J. A. Robinson, and G. A. Milliken (eds.), *Mathematical Modelling in Microbial Ecology*. Chapman & Hall, New York, pp. 162–192.

Clark, C. W., 1990. *Mathematical Bioeconomics: The Optimal Management of Renewable Resources*, John Wiley & Sons, New York, 386pp.

Clark, C. W. and Mangel, M., 2000. *Dynamic State Variable Models in Ecology, Methods and Applications*, Oxford University Press, Oxford, 289pp.

Cloete, T. E. and Muyima, N. Y. O., 1997. *Microbial Community Analysis*, IWA Scientific and Technical Report no. 5, London, 98pp.

Clough, J. S., 2014. *AQUATOX (RELEASE 3.1 plus)*, US EPA Office of Water, Washington, DC, 102pp.

Cole, T. M. and Wells, S. A., 2015. *CE-QUAL-W2: A Two-Dimensional, Laterally Averaged, Hydrodynamic and Water Quality Model*, Version 3.72, Portland State University, Portland, OR, 797pp.

Cook, P. A., 1986. *Nonlinear Dynamical Systems*, Prentice Hall, Englewood Cliffs, NJ, 216pp.

Copp, J. B., 2002. *The COST Simulation Benchmark: Description and Simulator Manual*, Office for Official Publications of the European Community, Luxembourg, 154pp.

Corominas, L., Villez, K., Aguado, D., Rieger, L., Rosén, C., and Vanrolleghem, P. A., 2009. Evaluation of fault-detection strategies performance in wastewater treatment processes. In: *10th IWA Conference on Instrumentation, Control, and Automation*, Cairns, Australia, June 14–17.

Corominas, L., Villez, K., Aguado, D., Rieger, L., Rosén, C., and Vanrolleghem, P. A., 2011. Performance evaluation of fault detection methods for wastewater treatment processes. *Biotechnology and Bioengineering*, **108**, 333–344.

Crispi, G., Crise, A., and Solidoro, C., 1998. Three-dimensional oligotrophic ecosystem models driven by physical forcing: The Mediterranean Sea case. *Environmental Modelling and Software*, **13**, 483–490.

Dahlquist, G. and Bjorck, A., 2003. *Numerical Methods*, Dover Publication, London, 573pp.

De Biasi, A. M., Benedetti-Cecchi, L., Pacciardi, L., Maggi, E., Vaselli, S., and Bertocci, I., 2003. Spatial heterogeneity in the distribution of plants and benthic invertebrates in the lagoon of Orbetello (Italy). *Oceanologica Acta*, **26**, 39–46.

De Boor, C., 2001. *A Practical Guide to Splines*, Springer-Verlag, New York, 346pp.

De Pauw, D., 2005. *Optimal experimental design for calibration of bioprocess models: A validated software toolbox*. PhD Thesis, University of Gent (B), Gent, Belgium. Available on the web at http://biomath.ugent.be/publications/download/depauwdirk_phd.pdf.

Di Toro, D. M., 2001. *Sediment Flux Modeling*, Wiley-Interscience, New York, 624pp.

Di Toro, D. M., O'Connor, D. J., Thomann, R. V., and Mancini, J. L., 1975. Phytoplankton-Zooplankton-Nutrient interaction model for western Lake Erie. In: B. C. Patten (ed.), *Systems Analysis and Simulation in Ecology*, Vol. 3. Academic Press, New York, pp. 424–475.

Dillon, W. R. and Goldstein, M., 1984. *Multivariate Analysis Methods and Applications*, John Wiley & Sons, New York, 587pp.

Ding, F. and Chen, T., 2005. Identification of Hammerstein nonlinear ARMAX systems. *Automatica*, **41**, 1479–1489.

Dircks, K., Beun, J. J., Loosdrecht, M. C. M., Heijnen, J. J., and Henze, M., 2001. Glycogen metabolism in aerobic mixed cultures. *Biotechnology and Bioengineering*, **73**, 85–94.

Dochain, D. and Vanrolleghem, P. A., 2001. *Dynamic Modelling and Estimation in Wastewater Treatment Processes*, IWA Publishing, London, 342pp.

Dormand, J. R., 1996. *Numerical Methods for Differential Equations: A Computational Approach*, CRC Press, Boca Raton, FL, 384pp.

Drazin, P. G., 1992. *Nonlinear Systems*, Cambridge University Press, Cambridge, 317pp.

Droop, M. R., 1983. 25 years of algal growth kinetics: A personal view. *Botanica Marina*, **26**, 99–112.

Duhamel, P. and Vetterli, M., 1990. Fast Fourier transforms: A tutorial review and a state of the art. *Signal Processing*, **19**, 259–299.

Dunteman, G. H., 1989. *Principal Component Analysis*, SAGE Publication, Newbury Park, CA, 487pp.

Dürrenmatt, D. J. and Gujer, W., 2011. Data-driven modeling approaches to support wastewater treatment plant operation. *Environmental Modelling & Software*, **30**, 47–56.

Elaydi, S., 2005. *An Introduction to Difference Equations*, Springer Science & Business Media, New York, 540pp.

Elgavish, A., Elgavisd, G. A., and Berman, T., 1980. Intracelluar phosphorus pools in intact algal cells. *FEBS Letters*, **117**, 1–6.

Falkowski, P. G. and Raven, J. A., 2007. *Aquatic Photosynthesis*, Princeton University Press, Princeton, NJ, 484pp.

Fedorov, V. V., 1972. *Theory of Optimal Experiments*, Academic Press, New York, 292pp.

Feigenbaum, M. J., 1978. Quantitative universality for a class of nonlinear transformations. *Journal of Statistical Physics*, **19**, 25–52.

Fisher, B. H., 1967. The mechanics of dispersion in natural streams. *Journal of the Hydraulics Division, ASCE*, **93**, 187–216.

Fisher, H. B., List, E. J., Koh, R. C. J., Imberger, J., and Brooks, N. H., 1979. *Mixing in Inland and Coastal Waters*, Academic Press, San Diego, CA, 483pp.

France, J. and Thornley, J. H. M., 1984. *Mathematical Models in Agriculture*, Butterworths, London, 335pp.

Freedman, H. I., 1980. *Deterministic Mathematical Models in Population Ecology*, Marcel Dekker, New York, 254pp.

Freeman, M. M. R., 2008. Challenges of assessing cetacean population recovery and conservation status. *Endangered Species Research*, **6**, 173–184.

Friedlander, B. and Porat, B., 1984. The modified Yule-Walker method of ARMA spectral estimation. *IEEE Transactions on Aerospace Electronic Systems*, **20**, 158–173.

Frigo, M. and Johnson, S. G., 2005. The design and implementation of FFTW3. *Proceedings of the IEEE*, **93**, 216–231.

Fukuda, S., 2009. Consideration of fuzziness: Is it necessary in modelling fish habitat preference of Japanese medaka (*Oryzias latipes*)? *Ecological Modelling*, **220**, 2877–2884.

Fukuda, S., De Baets, B., Mouton, A. M., Waegeman, W., Nakajima, J., Mukai, T., Hiramatsu, K., and Onikura, N., 2011. Effect of model formulation on the optimization of a genetic Takagi–Sugeno fuzzy system for fish habitat suitability evaluation. *Ecological Modelling*, **222**, 1401–1413.

Fukuda, S. and Hiramatsu, K., 2008. Prediction ability and sensitivity of artificial intelligence-based habitat preference models for predicting spatial distribution of Japanese medaka (*Oryzias latipes*). *Ecological Modelling*, **215**, 301–313.

Fuller, R., 1999. *Introduction to Neuro-Fuzzy Systems*, Physica-Verlag, Berlin, Germany, 289pp.

Ganoulis, J., 2009. *Risk Analysis of Water Pollution*, Wiley-VCH, Weinheim, Germany, 311pp.

Ganoulis, J. G., 1994. *Engineering Risk Analysis of Water Pollution: Probabilities and Fuzzy Sets*, Wiley-VCH, Weinheim, Germany, 306pp.

Gantmacher, F. R., 2000. *Matrix Theory*, Vol. 2, American Mathematical Society, Providence, RI, 276pp.

Gause, G. F., 1934. *The Struggle for Existence*, Williams & Wilkins, Baltimore, MD.

Gernaey, K. V., van Loosdrecht, M. C. M., Henze, M., Lind, M., and Jørgensen, S. E., 2004. Activated sludge wastewater treatment plant modelling and simulation: State of the art. *Environmental Modelling & Software*, **19**, 763–783.

Gijzen, H. J., 2001. Anaerobes, aerobes and phototrophs: A winning team for wastewater management. *Water Science & Technology*, **44**, 123–132.

Gijzen, H. J. and Mulder, A., 2001. The nitrogen cycle out of balance. *Water* **21**(3), 38–40.

Gilat, A., 2005. *MATLAB: An Introduction with Applications*, John Wiley & Sons, New York, 343pp.

Ginzburg, L. R. and Golemberg, E. M., 1985. *Lectures in Theoretical Population Biology*, Prentice Hall, Englewood Cliffs, NJ, 246pp.

Giordani, G., Zaldívar, J. M., and Viaroli, P., 2009. Simple tools for assessing water quality and trophic status in transitional water ecosystems. *Ecological Indicators*, **9**, 982–991.

Giusti, E. and Marsili-Libelli, S., 2005. Modelling the interactions between nutrients and the submersed vegetation in the Orbetello lagoon. *Ecological Modelling*, **184**, 141–161.

Giusti, E. and Marsili-Libelli, S., 2006. An integrated model for the Orbetello lagoon ecosystem. *Ecological Modelling*, **196**, 379–394.

Giusti, E. and Marsili-Libelli, S., 2009. Spatio-temporal dissolved oxygen dynamics in the Orbetello lagoon by fuzzy pattern recognition. *Ecological Modelling*, **220**, 2415–2426.

Giusti, E., Marsili-Libelli, S., and Gualchieri, A., 2011a. Modelling a coastal lake for flood and quality management. In: *Watermatex 2011: 8th IWA Symposium on Systems Analysis and Integrated Management*, San Sebastian, Spain, pp. 396–404.

Giusti, E., Marsili-Libelli, S., and Mattioli, S., 2011b. A fuzzy quality index for the environmental assessment of a restored wetland. *Water Science & Technology*, **63**, 2061–2070.

Giusti, E., Marsili-Libelli, S., Renzi, M., and Focardi, S., 2010. Assessment of spatial distribution of submerged vegetation in the Orbetello lagoon by means of a mathematical model. *Ecological Modelling*, **221**, 1484–1493.

Giusti, E., Marsili-Libelli, S., and Spagni, A., 2011c. Modelling microbial population dynamics in nitritation processes. *Environmental Modelling & Software*, **26**, 938–949.

Goh, B. S., 1976. Nonvulnerability of ecosystems in unpredictable environments. *Theoretical Population Biology*, **10**, 83–95.

Goh, B. S., 1977. Global stability in many-species systems. *American Naturalist*, **111**, 135–143.

Goh, B. S., 1979. Robust stability concepts for ecosystem models. In: E. Halfon (ed.), *Theoretical Systems Ecology: Advances and Case Studies*. Academic Press, San Diego, CA, pp. 467–489.

Goh, B. S., 1980. *Management and Analysis of Biological Populations*, Elsevier Science B.V., Amsterdam, the Netherlands, 299pp.

Goldberg, D. E., 1989. *Genetic Algorithms in Search, Optimization, and Machine Learning*, Addison-Wesley, Reading, MA, 412pp.

Goldberg, S., 2010. *Introduction to Difference Equations*, Dover Books on Mathematics, New York, 288pp.

Gonzalez, A. and Descamp-Julien, B., 2004. Population and community variability in randomly fluctuating environments. *Oikos*, **106**, 105–116.

Goswami, J. C. and Chan, A. K., 2008. In: K. Chang (ed.), *Fundamentals of Wavelets: Theory, Algorithms and Applications*, Wiley, New York, pp. 359–361.

Gotelli, N. J., 2001. *A Primer of Ecology*, Sinauer Associates, Sunderland, MA, 265pp.

Gotelli, N. J. and Ellison, A. M., 2004. *A Primer of Ecological Statistics*, Sinauer Associates, Sunderland, MA, 510pp.

Graps, A., 1995. Introduction to wavelets. *IEEE Computational Science & Engineering*, **2**, 50–61.

Gu, K., Kharitonov, V. L., and Chen, J., 2003. *Stability of Time-Delay Systems*, Birkhauser, Boston, MA, 353pp.

Guerrero, J., Guisasola, A., and Baeza, J. A., 2011. The nature of the carbon source rules the competition between PAO and denitrifiers in systems for simultaneous biological nitrogen and phosphorus removal. *Water Research*, **45**, 4793–4802.

Gujer, W., 2006. Activated sludge modelling: Past, present and future. *Water Science & Technology*, **53**, 111–119.

Gujer, W., 2008. *System Analysis for Water Technology*, Springer-Verlag, Berlin, Germany, 462pp.

Gujer, W., Henze, M., Mino, T., and van Loosdrecht, M. C. M., 1999. Activated sludge model no. 3. *Water Science & Technology*, **39**, 183–193.

Gustafson, D. and Kessel, W., 1979. Fuzzy clustering with a fuzzy covariance matrix. In: *1978 IEEE Conference on Decision & Control*, San Diego, CA, pp. 761–766.

Haefner, J. W., 2005. *Modeling Biological Systems: Principles and Applications*, Springer Science & Business Media, New York, 475pp.

Haimi, H., Mulas, M., Corona, F., and Vahala, R., 2013. Data-derived soft-sensors for biological wastewater treatment plants: An overview. *Environmental Modelling & Software*, **47**, 88–107.

Hajek, P., 2010. Fuzzy logic. *Stanford Encyclopaedia of Philosophy*. Available on the web at http://plato.stanford.edu/archives/fall2010/entries/logic-fuzzy/.

Halfon, E., 1983. Is there a best model structure? II. Comparing the model structures of different fate models. *Ecological Modelling*, **20**, 153–163.

Hallam, T. G. and Levin, S. A., 1986. *Mathematical Ecology*, Springer-Verlag, Berlin, Germany, 457pp.

Han, C., Gu, X., Geng, J., Hong, Y., Zhang, R., Wang, X., and Gao, S., 2010. Production and emission of phosphine gas from wetland ecosystems. *Journal of Environmental Sciences*, **22**, 1309–1311.

Hawkes, F. R., Rozzi, A. G., Black, K., Guwy, A. J., and Hawkes, D. L., 1992. The stability of anaerobic digesters operating on a food-processing wastewaters. *Water Science & Technology*, **25**, 73–82.

Heathwaite, A. L., 2003. Making process-based knowledge useable at the operational level: A framework for modelling diffuse pollution from agricultural land. *Environmental Modelling & Software*, **18**, 753–760.

Heil, C., 1993. Ten lectures on wavelets (Ingrid Daubechies). *SIAM Review*, **35**, 666–669.

Hellinga, C., Schellen, A. A. J. C., Mulder, J. W., van Loosdrecht, M. C. M., and Heijnen, J. J., 1998. The SHARON process: An innovative method for nitrogen removal from ammonium rich waste water. *Water Science & Technology*, **37**, 135–142.

Hellinga, C., van Loosdrecht, M. C. M., and Heijnen, J. J., 1999. Model based design of a novel process for nitrogen removal from concentrated flow. *Mathematical and Computer Modelling of Dynamical Systems*, **5**, 351–371.

Hemond, H. F. and Fechner, E. J., 1999. *Chemical Fate and Transport in the Environment*, Academic Press, San Diego, CA, 433pp.

Henze, M., Gujer, W., Mino, T., and van Loosdrecht, M. C. M., 2000. *Activated Sludge Models: ASM1, ASM2, ASM2d and ASM3*, IWA Scientific and Technical Report no. 9, London, 121pp.

Hesselmann, R. P. X., Werlen, C., Hahn, D., van der Meer, J. R., and Zehnder, A. J. B., 1999. Enrichment, phylogenetic analysis and detection of a bacterium that performs enhanced biological phosphate removal in activated sludge. *Systematic and Applied Microbiology*, **22**, 454–465.

Hildebrand, F. B., 1974. *Introduction to Numerical Analysis*, McGraw-Hill, New York, 669pp.

Himmelblau, D. M., 1972. *Applied Nonlinear Programming*, McGraw-Hill, New York, 498pp.

Hirota, K., 1993. *Industrial Applications of Fuzzy Technology*, Springer-Verlag, Tokyo, Japan, 310pp.

Holland, J. H., 1992. *Adaptation in Natural and Artificial Systems*, MIT Press, Cambridge, MA, 211pp.

Holmberg, A., 1982. On the practical identifiability of microbial growth models incorporating Michaelis-Menten type non-linearities. *Mathematical Biosciences*, **62**, 23–43.

Hotelling, H., 1931. The generalization of student's ratio. *Annals of Mathematical Statistics*, **2**, 360–378.

Hotelling, H., 1947. Multivariate quality control illustrated by the testing of sample bombsights. In: O. Eisenhart (ed.), *Selected Techniques of Statistical Analysis*. McGraw-Hill, New York, pp. 113–184.

House, J. E., 2007. *Principles of Chemical Kinetics*, Elsevier Academic Press, Amsterdam, the Netherlands, 336pp.

Houston, P. L., 2006. *Chemical Kinetics and Reaction Dynamics*, Dover Publication, New York, 330pp.

Hu, L., Wang, J., Wen, X., and Qian, Y., 2005. Study on performance characteristics of SBR under limited dissolved oxygen. *Process Biochemistry*, **40**, 293–296.

Iacopozzi, I., Innocenti, V., Marsili-Libelli, S., and Giusti, E., 2007. A modified activated sludge model no. 3 (ASM3) with two-step nitrification-denitrification. *Environmental Modelling & Software*, **22**, 847–861.

Irvine, R. L. and Ketchum, L. H., 1989. Sequencing Batch Reactors for biological wastewater treatment. *CRC Critical Reviews Environmental Control*, **18**, 255–294.

Jamshidi, M., Titli, A., Zadeh, L. A., and Boverie, S., 1997. *Applications of Fuzzy Logic: Towards High Machine Intelligence Quotient Systems*, Prentice Hall, Upper Saddle River, NJ, 423pp.

Jamu, D. M. and Piedrahita, R. H., 2002. An organic matter and nitrogen dynamics model for the ecological analysis of integrated aquaculture/agriculture systems: I. Model development and calibration. *Environmental Modelling & Software*, **17**, 571–582.

Jang, J. R., 1993. ANFIS: Adaptive-network-based fuzzy inference system. *IEEE Transaction on Systems, Man, and Cybernetics*, **23**, 665–685.

Jang, J. S. R., Sun, C. T., and Mizutani, A. Y., 1996. *Neuro-Fuzzy and Soft Computing: A Computational Approach to Learning and Machine Intelligence*, Prentice Hall, San Diego, CA, 614pp.

Jassby, A. D. and Platt, T., 1976. Mathematical formulation of the relationship between photosynthesis and light for phytoplankton. *Limnology and Oceanography*, **21**, 540–547.

Jeppsson, U., 1996. *Modelling aspects of wastewater treatment processes.* PhD Thesis, University of Lund (S), Lund, Sweden.

Jeppsson, U., Pons, M.-N., Nopens, I., Alex, J., Copp, J. B., Gernaey, K. V., Rosen, C., Steyer, J.-P., and Vanrolleghem, P. A., 2007. Benchmark simulation model no. 2: General protocol and exploratory case studies. *Water Science & Technology*, **56**, 67–78.

John, E. H. and Flynn, K. J., 2000. Modelling phosphate transport and assimilation in microalgae: How much complexity is warranted? *Ecological Modelling*, **125**, 145–157.

Jolliffe, I. T., 2002. *Principal Component Analysis,* Second Edition, Springer Series in Statistics, Springer-Verlag Berlin, Germany, 487pp.

Jorgensen, S. E., 2008. Eco-exergy as an ecosystem health indicator. In: *Encyclopedia of Ecology.* Elsevier B.V., Amsterdam, the Netherlands, pp. 977–979.

Jorgensen, S. E. and Bendoricchio, G., 2001. *Fundamentals of Ecological Modelling*, Elsevier B.V., Amsterdam, the Netherlands, 530pp.

Jorgensen, S. E., Odum, H. T., and Brown, M. T., 2004. Emergy and exergy stored in genetic information. *Ecological Modelling*, **178**, 11–16.

Jorgensen, S. E. and Svirezhev, Y., 2004. *Towards a Thermodynamic Theory for Ecological Systems*, Elsevier B.V., Amsterdam, the Netherlands, 366pp.

Kadlec, R. H., 2003. Pond and wetland treatment. *Water Science & Technology*, **48**, 1–8.

Kadlec, R. H. and Wallace, S. D., 2009. *Treatment Wetlands*, CRC Press, Boca Raton, FL, 1016pp.

Kandel, A. and Langholz, G., 1994. *Fuzzy Control Systems*, CRC Press, Boca Raton, FL, 624pp.

Kaplan, D. and Glass, L., 1995. *Understanding Nonlinear Dynamics*, Springer, New York, 420pp.

Kaymak, U. and Setnes, M., 2002. Fuzzy clustering with volume prototypes and adaptive cluster merging. *IEEE Transactions on Fuzzy Systems*, **10**, 705–711.

Kecman, V., 2001. *Learning and Soft Computing*, MIT Press, Boston, MA, 541pp.

Kharitonov, V. L. and Zhabko, A. P., 2003. Lyapunov-Krasovskii approach to the robust stability analysis of time-delay systems. *Automatica*, **39**, 15–20.

Kiirikki, M. and Haapama, J., 1998. Linking the growth of filamentous algae to the 3D-ecohydrodynamic model of the Gulf of Finland. *Environmental Modelling & Software*, **13**, 503–509.

Kim, J. H., Chen, M., Kishida, N., and Sudo, R., 2004. Integrated real-time control strategy for nitrogen removal in swine wastewater treatment using sequencing batch reactors. *Water Research*, **38**, 3340–3348.

Kincanon, R. and McAnally, S. A., 2004. Enhancing commonly used model predictions for constructed wetland performance: As-built design considerations. *Ecological Modelling*, **174**, 309–322.

Kirk, J. T. O., 2011. *Light and Photosynthesis in Aquatic Ecosystems*, Cambridge University Press, Cambridge, 649pp.

Klir, G. J. and Folger, T. A., 1988. *Fuzzy Sets, Uncertainty, and Information*, Prentice Hall, New York, 355pp.

Knyazikhin, Y., Martonchik, J. V., Myneni, R. B., Diner, D. J., and Running, S. W., 1998. Synergistic algorithm for estimating vegetation canopy leaf area index and fraction of absorbed photosynthetically active radiation from MODIS and MISR data. *Journal of Geophysical Research*, **103**, 32257–32276.

Kolk, W. R. and Lerman, R. A., 1992. *Nonlinear System Dynamics*, Van Nostrand Reinhold, New York, 350pp.

Kong, Z., Vanrolleghem, P. A., Willems, P., and Verstraete, W., 1996. Simultaneous determination of inhibition kinetics of carbon oxidation and nitrification with a respirometer. *Water Research*, **30**, 825–836.

Kosko, B., 1992. *Neural Networks and Fuzzy Systems*, Prentice Hall, London, 449pp.

Krasovskii, N. N., 1956. On the application of the second method of Lyapunov for equations with time delays. *Prikladnaya Matematika i Mekhanika*, **20**, 315–327.

Krebs, C. J., 1998. *Ecological Methodology*, Benjamin Cummings, San Francisco, CA, 624pp.

Krebs, C. J., 2009. *Ecology: The Experimental Analysis of Distribution and Abundance*, Benjamin Cummings, San Francisco, CA, 655pp.

Krishnapuram, R. and Keller, J., 1993. A possibilistic approach to clustering. *IEEE Transactions on Fuzzy Systems*, **1**, 98–110.

Krishnapuram, R. and Kim, J., 1999. A note on the Gustafson-Kessel and adaptive fuzzy clustering algorithms. *IEEE Transactions on Fuzzy Systems*, **7**, 453–461.

Kuester, J. L. and Mize, J. H., 1974. *Optimization Techniques with FORTRAN*, McGraw-Hill, New York, 500pp.

Kuo, J. T., Lung, W. S., Yang, C. P., Liu, W. C., Yang, M. D., and Tang, T. S., 2006. Eutrophication modelling of reservoirs in Taiwan. *Environmental Modelling & Software*, **21**, 829–844.

Lakin, W. D. and Gross, C. E., 1992. A nonlinear haemodynamic model for the arterial pulsatile component of the intracranial pulse wave. *Neurological Research*, **14**, 219–225.

Lamouroux, N. and Capra, H., 2002. Simple predictions of instream habitat model outputs for target fish populations. *Freshwater Biology*, **47**, 1543–1556.

Lamouroux, N. and Souchon, Y., 2002. Simple predictions of instream habitat model outputs for fish habitat guilds in large streams. *Freshwater Biology*, **47**, 1531–1542.

Langergraber, G., 2003. Simulation of subsurface flow constructed wetlands-results and further research needs. *Water Science & Technology*, **48**, 157–166.

Langergraber, G., Giraldi, D., Mena, J., Meyer, D., Peña, M., Toscano, A., Brovelli, A., and Korkusuz, E. A., 2009a. Recent developments in numerical modelling of subsurface flow constructed wetlands. *Science of the Total Environment*, **407**, 3931–3943.

Langergraber, G., Rousseau, D. P. L., García, J., and Mena, J., 2009b. CWM1: A general model to describe biokinetic processes in subsurface flow constructed wetlands. *Water Science & Technology*, **59**, 1687–1697.

Lanham, A. B., Oehmen, A., Saunders, A. M., Carvalho, G., Nielsen, P. H., and Reis, M. A. M., 2014. Metabolic modelling of full-scale enhanced biological phosphorus removal sludge. *Water Research*, **66**, 283–295.

Lapointe, B. E., Tenore, K. R., and Dawes, C. J., 1984. Interactions between light and temperature on the physiological ecology of Gracilaria tikvahiae (Gigartinales: Rhodophyta). *Marine Biology*, **80**, 161–170.

Legendre, P. and Legendre, L., 2012. *Numerical Ecology*, Elsevier B. V., Amsterdam, the Netherlands, 990pp.

Lennox, J. and Rosen, C., 2002. Adaptive multiscale principal components analysis for online monitoring of wastewater treatment. *Water Science & Technology*, **45**, 227–235.

Lenzi, M., Palmieri, R., and Porrello, S., 2003. Restoration of the eutrophic Orbetello lagoon (Tyrrhenian Sea, Italy): Water quality management. *Marine Pollution Bulletin*, **46**, 1540–1548.

Leslie, P. H., 1945. The use of matrices in certain population mathematics. *Biometrika*, **33**, 183–212.

Leslie, P. H., 1957. An analysis of the data for some experiments carried out by Gause with populations of the protozoa, *Paramecium aurelia* and *Paramecium caudatum*. *Biometrika*, **44**, 314–327.

Leslie Grady, C. P., Daigger, G., Love, N. G., and Filipe, C. D. M., 2011. *Biological Wastewater Treatment*, IWA Publishing; CRC Press, Boca Raton, FL, 991pp.

Levenspiel, O., 1972. *Chemical Reaction Engineering*, John Wiley & Sons, New York, 578pp.

Levin, S. A., Hallam, T. G., and Gross, L. J., 1989. *Applied Mathematical Ecology*, Springer-Verlag, Berlin, Germany, 491pp.

Lilliefors, H. W., 1967. On the Kolmogorov-Smirnov test for normality with mean and variance unknown. *Journal of the American Statistical Association*, **62**, 399–402.

Lindenschmidt, K. E., 2006. The effect of complexity on parameter sensitivity and model uncertainty in river water quality modelling. *Ecological Modelling*, **190**, 72–86.

Lippmann, R. P., 1987. An introduction to computing with neural nets. *IEEE ASSP Magazine*, **4**, 4–22.

Ljung, L., 1999. *System Identification: Theory for the User*, Prentice Hall, Upper Saddle River, NJ, 609pp.

Ljung, L. and Glad, T., 1994. On global identifiability for arbitrary model parametrizations. *Automatica*, **30**, 265–276.

Lobry, J. R. and Flandrois, J.-P., 1991. Comparison of estimates of Monod's growth model from the same data set. *Binary*, **3**, 20–23.

Lopez-Vazquez, C. M., Oehmen, A., Hooijmans, C. M., Brdjanovic, D., Gijzen, H. J., Yuan, Z., and van Loosdrecht, M. C. M., 2009. Modeling the PAO-GAO competition: Effects of carbon source, pH and temperature. *Water Research*, **43**, 450–462.

Lu, Y. Z., 1996. *Industrial Intelligent Control: Fundamentals and Applications*, John Wiley & Sons, New York, 325pp.

Luccarini, L., Porrà, E., Spagni, A., Ratini, P., Grilli, S., Longhi, S., and Bortone, G., 2002. Soft sensors for control of nitrogen and phosphorus removal from wastewaters by neural networks. *Water Science & Technology*, **45**, 101–107.

Ludwig, D., Aronson, D. G., and Weinberger, H. F., 1979. Spatial patterning of the spruce budworm. *Journal of Mathematical Biology*, **8**, 217–258.

Ludwig, D., Jones, D. D., and Holling, C. S., 1978. Qualitative analysis of insect outbreak systems: Spruce budworm and forest. *Journal of Animal Ecology*, **47**, 315–332.

Lyons, R. G., 2011. *Understanding Digital Signal Processing*, Prentice Hall, Upper Saddle River, NJ, 954pp.

Macura, V., Škrinár, A., Kaluz, K., Jalčovíková, M., and Škrovinová, M., 2012. Influence of the morphological and hydraulic characteristics of mountain streams on fish habitat suitability curves. *River Research and Applications*, **28**, 1161–1178.

Madigan, M. T., Martinko, J. M., Dunlap, P. V., and Clark, D. P., 2009. *Brock Biology of Microorganisms*, Pearson, San Francisco, CA, 1056pp.

Magurran, A. E., 2004. *Measuring Biological Diversity*, Blackwell, Oxford, 256pp.

Marchini, A., Facchinetti, T., and Mistri, M., 2009. F-IND: A framework to design fuzzy indices of environmental conditions. *Ecological Indicators*, **9**, 485–496.

Mareels, I. M. Y., Bitmead, R. R., Gevers, M., Johnson, C. R., Kosut, R. L., and Poubelle, M. A., 1987. How exciting can a signal really be? *Systems and Control Letters*, **8**, 197–204.

Marin, G. B. and Yablonsky, G. S., 2011. *Kinetics of Chemical Reactions*, Wiley-VCH, Weinheim, Germany, 428pp.

Marsaglia, G., Tsang, W. W., and Wang, J., 2003. Evaluating Kolmogorov's distribution. *Journal of Statistical Software*, **8**, 1–4.

Marsili-Libelli, S., 1986. Modelling batch BOD exertion curves. *Environmental Technology Letters*, **7**, 341–350.

Marsili-Libelli, S., 1991. Modelling photosynthetic oxygen production in rivers. *Environmental Technology*, **12**, 59–67.

Marsili-Libelli, S., 1992. Parameter estimation of ecological models. *Ecological Modelling*, **62**, 233–258.

Marsili-Libelli, S., 1993. Dynamic modelling of sedimentation in the activated sludge process. *Civil Engineering Systems*, **10**, 207–224.

Marsili-Libelli, S., 1997. Simple model of a transport/diffusion system. *Proceedings of the IEEE, Control Theory & Applications, Part. D*, **144**, 459–465.

Marsili-Libelli, S., 1998. Adaptive fuzzy monitoring and fault detection. *International Journal of COMADEM*, **1**, 31–38.

Marsili-Libelli, S., 2004. Fuzzy pattern recognition of circadian cycles in ecosystems. *Ecological Modelling*, **174**, 67–84.

Marsili-Libelli, S., 2006. Control of SBR switching by fuzzy pattern recognition. *Water Research*, **40**, 1095–1107.

Marsili-Libelli, S. and Alba, P., 2000. Adaptive mutation in genetic algorithms. *Soft Computing*, **4**, 76–80.

Marsili-Libelli, S. and Arrigucci, S., 2004. Circadian patterns recognition in ecosystems by wavelet filtering and Fuzzy clustering. In: *Proceedings of the iEMSs 2004 International Congress: "Complexity and Integrated Resources Management,"* Osnabruck (D), Germany, pp. 63–68.

Marsili-Libelli, S. and Beni, S., 1996. Shock load modelling in the anaerobic digestion model. *Ecological Modelling*, **84**, 215–232.

Marsili-Libelli, S. and Castelli, M., 1987. An adaptive search algorithm for numerical optimization. *Applied Mathematics and Computation*, **23**, 341–357.

Marsili-Libelli, S. and Checchi, N., 2005. Identification of dynamic models for horizontal subsurface constructed wetlands. *Ecological Modelling*, **187**, 201–218.

Marsili-Libelli, S. and Cianchi, P., 1996. Fuzzy ecological models. In: W. Pedrycz (ed.), *FUZZY MODELING. Paradigms and Practices*. Kluver Academic Publications, Boston, MA, pp. 141–164.

Marsili-Libelli, S. and Colzi, A., 1998. Fuzzy control of a transport/diffusion system. *Control Engineering Practice*, **6**, 707–713.

Marsili-Libelli, S. and Giusti, E., 2008. Water quality modelling for small river basins. *Environmental Modelling & Software*, **23**, 451–463.

Marsili-Libelli, S., Giusti, E., and Nocita, A., 2013. A new instream flow assessment method based on fuzzy habitat suitability and large scale river modelling. *Environmental Modelling & Software*, **41**, 27–38.

Marsili-Libelli, S., Guerrizio, S., and Checchi, N., 2003. Confidence regions of estimated parameters for eco-logical systems. *Ecological Modelling*, **165**, 127–146.

Marsili-Libelli, S., Limberti, A., and Conti, A., 1997. Modelling algal growth rate for trophic testing. In: *Proceedings of the 11th Forum for Applied Biotechnology*. Universiteit Gent (B), Ghent, Belgium, pp. 1609–1615.

Marsili-Libelli, S. and Müller, A., 1996. Adaptive fuzzy pattern recognition in the anaerobic digestion process. *Pattern Recognition Letters*, **17**, 651–659.

Marsili-Libelli, S., Nocita, A., Giusti, E., and Saccà, M., 2008. A new definition of minimum sustainable flow based on water quality modelling and fuzzy processing. In: *iEMSs 2008 International Congress on Environmental Modelling and Software*. Barcelona (ES), Spain, pp. 288–294.

Marsili-Libelli, S., Ratini, P., Spagni, A., and Bortone, G., 2001. Implementation, study and calibration of a modified ASM2d for the simulation of SBR processes. *Water Science & Technology*, **43**, 69–76.

Marsili-Libelli, S., Spagni, A., and Susini, R., 2008. Intelligent monitoring system for long-term control of sequencing batch reactors. *Water Science & Technology*, **57**, 431–438.

Marsland, S., 2015. *Machine Learning: An Algorithmic Perspective*, CRC Press, Boca Raton, FL, 437pp.

Mashauri, D. A. and Kayombo, S., 2002. Application of the two coupled models for water quality management: Facultative pond cum constructed wetland models. *Physics and Chemistry of the Earth*, **27**, 773–781.

Mason, R. L., Young, J. C., and Charles, L., 2001. Applying Hotelling's T statistic to batch processes. *Journal of Quality Technology*, **33**, 466–479.

Massey, F. J., 1951. The Kolmogorov-Smirnov test for goodness of fit. *Journal of the American Statistical Association*, **46**, 68–78.

May, R. M., 1974. Biological populations with non-overlapping generations: Stable points, stable cycles, and chaos. *Science*, **186**, 645–647.

May, R. M., 1976a. Simple mathematical models with very complicated dynamics. *Nature*, **261**, 459–467.

May, R. M., 1976b. *Theoretical Ecology: Principles and Applications*, Blackwell, Oxford, 489pp.

May, R. M., 2001. *Stability and Complexity in Model Ecosystems*, Princeton University Press, Princeton, NJ, 304pp.

Maynard Smith, J., 1974. *Models in Ecology*, Cambridge University Press, Cambridge, 157pp.

McMahon, T. E., 1982. *Habitat Suitability Index Models: Creek Chub*, FWS/OBS-82/10.4, Biological Services Program, Fish and Wildlife Service, Department of Interior, Fort Collins, CO, 34pp.

McMahon, T. E., Gebbart, G., Maughan, O. E., and Nelson, P. C., 1984. *Habitat Suitability Index Models and Instream Flow Suitability Curves: Warmouth*, FWS/OBS-82/10.67, Division of Biological Service, U.S. Department of the Interior, Washington, DC, 33pp.

Melia, P., Nizzoli, D., Bartoli, M., Naldi, M., Gatto, M., and Viaroli, P. L., 2003. Assessing the potential impact of clam rearing in dystrophic lagoons: An integrated oxygen balance. *Chemistry and Ecology*, **00**, 1–18.

Metcalf & Eddy Inc., 2003. *Wastewater Engineering: Treatment and Reuse*, McGraw-Hill, New York, 1819pp.

Meyer, D., Chazarenc, F., Claveau-Mallet, D., Dittmer, U., Forquet, N., Molle, P., Morvannou, A. et al., 2015. Modelling constructed wetlands: Scopes and aims—A comparative review. *Ecological Engineering*, **80**, 205–213.

Michener, C. D. and Sokal, R. R., 1957. A quantitative approach to a problem of classification. *Evolution*, **11**, 130–162.

Mino, T., Van Loosdrecht, M. C. M., and Heijnen, J. J., 1998. Microbiology and biochemistry of the enhanced biological phosphate removal process. *Water Research*, **32**, 3193–3207.

Misiti, M., Misiti, Y., Oppenheim, G., and Poggi, J. M., 2010. *Wavelets and Their Applications*, Wiley-ISTE, New York, 352pp.

Mitchell, C. and McNevin, D., 2001. Alternative analysis of BOD removal in subsurface flow constructed wet-lands employing Monod kinetics. *Water Research*, **35**, 1295–1303.

Mittelbach, G. G., 2012. *Community Ecology*, Sinauer Associates, Sunderland, MA, 400pp.

Moir, H. J., Gibbins, C. N., Soulsby, C., and Youngson, A. F., 2005. PHABSIM modelling of Atlantic salmon spawning habitat in an upland stream: Testing the influence of habitat suitability indices on model output. *River Research and Applications*, **21**, 1021–1034.

Monod, J., 1949. The growth of bacterial cultures. *Annual Review of Microbiology*, **3**, 371–394.

Morales-Zárate, M. V., Arreguín-Sánchez, F., López-Martínez, J., and Lluch-Cota, S. E., 2004. Ecosystem trophic structure and energy flux in the Northern Gulf of California, México. *Ecological Modelling*, **174**, 331–345.

Morris, R. F. and Miller, C. A., 1954. The development of life tables for the spruce budworm. *Canadian Journal of Zoology*, **32**, 283–301.

Morris, W. F. and Doak, D. F., 2002. *Quantitative Conservation Biology*, Sinauer Associates, Sunderland, MA, 480pp.

Mouton, A. M., Alcaraz-Hernández, J. D., De Baets, B., Goethals, P. L. M., and Martínez-Capel, F., 2011. Data-driven fuzzy habitat suitability models for brown trout in Spanish Mediterranean rivers. *Environmental Modelling & Software*, **26**, 615–622.

Mouton, A. M., De Baets, B., and Goethals, P. L. M., 2009. Knowledge-based versus data-driven fuzzy habitat suitability models for river management. *Environmental Modelling & Software*, **24**, 982–993.

Mouton, A. M., Schneider, M., Depestele, J., Goethals, P. L. M., and De Pauw, N., 2007. Fish habitat modelling as a tool for river management. *Ecological Engineering*, **29**, 305–315.

Mouton, A. M., Schneider, M., Kopecki, I., and Goethals, P. L. M., 2006. Application of MesoCASiMiR: Assessment of Baetis rhodani habitat suitability. In: *3rd Biennial Meeting of the International Environmental Modelling and Software Society*. Burlington, VT. Available on the web at http://www.iemss.org/iemss2006/sessions/all.html.

Mouton, A. M., Schneider, M., Peter, A., Holzer, G., Müller, R., Goethals, P. L. M., and De Pauw, N., 2008. Optimisation of a fuzzy physical habitat model for spawning European grayling (*Thymallus thymallus* L.) in the Aare river (Thun, Switzerland). *Ecological Modelling*, **215**, 122–132.

Mueller, A., Marsili-Libelli, S., Aivasidis, A., Lloyd, T., Kroner, S., and Wandrey, C., 1997. Fuzzy control of disturbances in a wastewater treatment process. *Water Research*, **31**, 3157–3167.

Mueller, L. D. and Joshi, A., 2000. *Stability in Model Populations*, Princeton University Press, Princeton, NJ, 321pp.

Mulder, J. W., van Loosdrecht, M. C., Hellinga, C., and van Kempen, R., 2001. Full-scale application of the SHARON process for treatment of rejection water of digested sludge dewatering. *Water Science & Technology*, **43**, 127–134.

Müller, A., Marsili-Libelli, S., Aivasidis, A., Lloyd, T., Kroner, S., and Wandrey, C., 1997. Fuzzy control of disturbances in a wastewater treatment process. *Water Research*, **31**, 3157–3167.

Nagy-Kiss, A. M. and Schutz, G., 2013. Estimation and diagnosis using multi-models with application to a wastewater treatment plant. *Journal of Process Control*, **23**, 1528–1544.

Nelder, J. A. and Mead, R., 1965. A simplex method for function minimization. *Computer Journal*, **7**, 308–313.

Nelson, D. L. and Cox, M. M., 2008. *Lehninger Principles of Biochemistry*, W.H. Freeman, New York, 1100pp.

Nguyen, H. T., Sugeno, M. R. T., and Yager, R. R., 1995. *Theoretical Aspects of Fuzzy Control*, John Wiley & Sons, New York, 359pp.

Nicholson, A. J., 1954. An outline of the dynamics of animal populations. *Australian Journal of Zoology*, **2**, 9–65.

Nizzoli, D., Bartoli, M., and Viaroli, P. L., 2006. Nitrogen and phosphorous budgets during a farming cycle of the Manila clam *Ruditapes philippinarum*: An in situ experiment. *Aquaculture*, **261**, 98–108.

Nocita, A., Massolo, A., Vannini, M., and Gandolfi, G., 2009. The influence of calcium concentration on the distribution of the river bullhead *Cottus gobio* L. (Teleostes, Cottidae). *Italian Journal of Zoology*, **76**, 348–357.

Nopens, I., Batstone, D. J., Copp, J. B., Jeppsson, U., Volcke, E., Alex, J., and Vanrolleghem, P. A., 2009. An ASM/ADM model interface for dynamic plant-wide simulation. *Water Research*, **43**, 1913–1923.

Norton, J. P., 1986. *An Introduction to Identification*, Academic Press, London, 310pp.

Nowak, O., Svardal, K., and Schweighofer, P., 1995. The dynamic behaviour of nitrifying activated sludge system influenced by inhibiting wastewater compounds. *Water Science & Technology*, **31**, 115–124.

Nyholm, N., 1978. A simulation model for phytoplankton growth and nutrient cycling in eutrophic, shallow lakes. *Ecological Modelling*, **4**, 297–310.

Nyholm, N., 1985. Response variable in algal growth inhibition tests—Biomass or growth rate? *Water Research*, **19**, 273–279.

Nyholm, N. and Lingby, J. E., 1988. Algal bioassay in eutrophication research—A discussion in the framework of a mathematical analysis. *Water Research*, **22**, 1293–1300.

Obach, M., Wagner, R., Werner, H., and Schmidt, H. H., 2001. Modelling population dynamics of aquatic insects with artificial neural networks. *Ecological Modelling*, **146**, 207–217.

Odum, E. and Barrett, G. W., 2004. *Fundamentals of Ecology*, Cengage Learning, Andover, 624pp.

Odum, H. T., 1983. *Ecosystems Ecology: An Introduction*, Wiley-Interscience, New York, 644pp.

Odum, H. T., 1988. Self-organization, transformity, and information. *Science*, **242**, 1132–1139.

Odum, H. T., 2000a. *Emergy Accounting*, Center for Environmental Policy, University of Florida, Gainesville, FL, 20pp.

Odum, H. T., 2000b. *Handbook of Emergy Evaluation: A Compendium of Data for Emergy Computation Folio # 2: Emergy of Global Processes*, Center for Environmental Policy, University of Florida, Gainesville, FL, 28pp.

Odum, H. T., Brown, M. T., and Brandt-Williams, S., 2000. *Handbook of Emergy Evaluation Folio # 1: Introduction and Global Budget*, Center for Environmental Policy, University of Florida, Gainesville, FL, 17pp.

Odum, H. T. and Odum, E. C., 2000. *Modeling for All Scales*, Academic Press, San Diego, CA, 458pp.

Oehmen, A., Lemos, P. C., Carvalho, G., Yuan, Z., Keller, J., Blackall, L. L., and Reis, M. A. M., 2007. Advances in enhanced biological phosphorus removal: From micro to macro scale. *Water Research*, **41**, 2271–2300.

Olsson, G. and Newell, B., 1999. *Wastewater Treatment Systems*, IWA Publishing, London, 742pp.

Oppenheim, A. V. and Schafer, R. W., 2010. *Discrete-Time Signal Processing*, Prentice Hall, Upper Saddle River, NJ, 1120pp.

Orhon, D. and Artan, N., 1994. *Modelling of Activated Sludge Systems*, Technomics Publications, Basel, Switzerland, 589pp.

Ossenbruggen, P. J., Spanjers, H., and Klapwik, A., 1996. Assessment of a two-step nitrification model for activated sludge. *Water Research*, **30**, 939–953.

Padhi, N. P. and Simon, S. P., 2015. *Soft Computing with MATLAB Programming*, Oxford University Press, Oxford, 550pp.

Pal, N. R. and Bezdek, J. C., 1994. Measuring fuzzy uncertainty. *IEEE Transactions on Fuzzy Systems*, **2**, 107–118.

Pal, N. R., Pal, K., Keller, J. M., and Bezdek, J. C., 2005. A possibilistic fuzzy c-means clustering algorithm. *IEEE Transactions on Fuzzy Systems*, **13**, 517–530.

Pálfy, T. G. and Langergraber, G., 2014. The verification of the constructed wetland model no. 1 implementation in HYDRUS using column experiment data. *Ecological Engineering*, **68**, 105–115.

Palmeri, L., Barausse, A., and Jorgensen, S. E., 2014. *Ecological Processes Handbook*, CRC Press, Boca Raton, FL, 386pp.

Park, R. A. and Clough, J. S., 2014. *Aquatox (Release 3.1 plus) Technical Documentation*, US EPA, Office of Water, Washington, DC, 344pp.

Park, S. S. and Jaffé, P. R., 1996. Development of a sediment redox potential model for the assessment of post-depositonal metal mobility. *Ecological Modelling*, **91**, 169–181.

Passino, K. M. and Yurkovic, S., 1998. *Fuzzy Control*, Addison-Wesley, Reading, MA, 475pp.

Pastor, J., 2008. *Mathematical Ecology*, John Wiley & Sons, Oxford, 329pp.

Patyra, M. J. and Mlynek, D. M., 1996. *Fuzzy Logic: Implementations and Applications*, John Wiley & Sons, New York, 317pp.

Pavlostatis, S. G. and Giraldo-Gomez, E., 1991. Kinetics of anaerobic treatment: A critical review. *Critical Reviews in Environmental Control*, **21**, 411–490.

Pearl, J., 1978. On the connection between the complexity and credibility of inferred models. *International Journal of General Systems*, **4**, 255–264.

Pedrycz, W., 1993. *Fuzzy Control and Fuzzy Systems*, John Wiley & Sons, New York, 350pp.

Pedrycz, W., 1995. *Fuzzy Sets Engineering*, CRC Press, Boca Raton, FL, 332pp.

Pedrycz, W., 1996. *Fuzzy Modelling Paradigm and Practices*, Kluver Academic Publishers, Boston, MA, 394pp.

Pelletier, F. J., 2000. Mathematics of fuzzy logic. *The Bulletin of Symbolic Logic*, **6**, 342–346.

Pelletier, G. J., Chapra, S. C., and Tao, H., 2006. QUAL2Kw—A framework for modeling water quality in streams and rivers using a genetic algorithm for calibration. *Environmental Modelling & Software*, **21**, 419–425.

Pentz, M. J., 1972. *Science Foundation Courses: Unit 20 Species and Populations*, The Open University Press, Oxford, 86pp.

Persson, J. and Wittgren, H. B., 2003. How hydrological and hydraulic conditions affect performance of ponds. *Ecological Engineering*, **21**, 259–269.

Petersen, B., 2000. *Calibration, identifiability and optimal experiment design of activated sludge models*. PhD Thesis. University of Gent (B), Gent, Belgium. Available on the web at http://biomath.rug.ac.be/publications/download/petersenbritta_phd.pdf.

Petersen, B., Gernaey, K., Devisscher, M., Dochain, D., and Vanrolleghem, P. A., 2003. A simplified method to assess structurally identifiable parameters in Monod-based activated sludge models. *Water Research*, **37**, 2893–904.

Petersen, B., Gernaey, K., and Vanrolleghem, P. A., 2001. Practical identifiability of model parameters by combined respirometric-titrimetric measurements. *Water Science & Technology*, **43**, 347–355.

Pielou, E. C., 1977. *Mathematical Ecology*, John Wiley & Sons, New York, 385pp.

Pielou, E. C., 1984. *The Interpretation of Ecological Data: A Primer on Classification and Ordination*, John Wiley & Sons, New York, 263pp.

Pohjanpalo, H., 1978. System identifiability based on the power series expansion of the solution. *Mathematical Biosciences*, **41**, 21–33.

Poli, R., Langdon, W. B., and Mcphee, N. F., 2008. *A Field Guide to Genetic Programming*. Lulu, Raleigh, NC, 233pp.

Pratap, R., 2006. *Getting Started with MATLAB 7*, Oxford University Press, Oxford, 244pp.

Prato, T., 2007. Assessing ecosystem sustainability and management using fuzzy logic. *Ecological Economics*, **61**, 171–177.

Press, W. H., Flannery, B. P., Teukolsky, S. A., and Vetterling, W. T., 1986. *Numerical Recipes: The Art of Scientific Computing*, Cambridge University Press, Cambridge, 818pp.

Procyk, T. J. and Mamdani, E. H., 1979. A linguistic self-organizing controller. *Automatica*, **15**, 15–30.

Quinney, D., 1987. *An Introduction to the Numerical Solution of Differential Equations*, John Wiley & Sons, New York, 283pp.

Raleigh, R. F., Zuckermann, L. D., and Nelson, P. C., 1986. Habitat suitability index models and instream flow suitability curves: Brown trout. *U.S. Fish and Wildlife Service Biological Report*, **82**, 36–43.

Rashkovsky, I. and Margaliot, M., 2007. Nicholson's blowflies revisited: A fuzzy modeling approach. *Fuzzy Sets and Systems*, **158**, 1083–1096.

Reichert, P., Borchardt, D., Henze, M., Rauch, W., Shanahan, P., Somlyòdy, L., and Vanrolleghem, P. A., 2001. *River Water Quality Model No. 1*, IWA Scientific and Technical Report no. 12, London, 131pp.

Reichert, P. and Vanrolleghem, P. A., 2001. Identifiability and uncertainty analysis of the river water quality model no. 1 (RWQM1). *Water Science & Technology*, **43**, 329–338.

Renshaw, E., 1991. *Modelling Biological Populations in Space and Time*, Cambridge University Press, Cambridge, 403pp.

Ricklefs, R. E. and Miller, G. L., 1999. *Ecology*, W.H. Freeman, New York, 822pp.

Rieger, L., Alex, J., Gujer, W., and Siegrist, H., 2006. Modelling of aeration systems at wastewater treatment plants. *Water Science & Technology*, **53**, 439–447.

Rinaldi, S., Soncini-Sessa, R., Stehfest, H., and Tamura H., 1979. *Modeling and Control of River Quality*, McGraw-Hill, New York, 380pp.

Ringnér, M., 2008. What is principal component analysis? *Nature Biotechnology*, **26**, 303–304.

Rioul, O. and Vetterli, M., 1991. Wavelets and signal processing. *IEEE Signal Processing Magazine*, **8**, 14–38.

Rockwood, L. L., 2006. *Introduction to Population Ecology*, Blackwell, Oxford, 339pp.

Roels, J. A., 1983. *Energetics and Kinetics in Biotechnology*, Elsevier, Amsterdam, the Netherlands, 330pp.

Roger, J. S. and Sun, C. T., 1997. *Neuro-Fuzzy and Soft Computing: A Computational Approach to Learning and Machine Intelligence*, Prentice Hall, New York, 614pp.

Rosati, A., Metcalf, S. G., and Lampinen, B. D., 2004. A simple method to estimate photosynthetic radiation use efficiency of canopies. *Annals of Botany*, **93**, 567–574.

Rosen, C., Röttorp, J., and Jeppsson, U., 2003. Multivariate on-line monitoring: Challenges and solutions for modern wastewater treatment operation. *Water Science & Technology*, **47**, 171–179.

Ross, T. J., 1995. *Fuzzy Logic with Engineering Applications*, McGraw-Hill, New York, 600pp.

Rousseau, D. P. L., Vanrolleghem, P. A, and De Pauw, N., 2004. Model-based design of horizontal subsurface flow constructed treatment wetlands: A review. *Water Research*, **38**, 1484–1493.

Royama, T., 1984. Population dynamics of the spruce budworm *Choristoneura fumiferana*. *Ecological Monographs*, **54**, 429–462.

Royama, T., MacKinnon, W. E., Kettela, E. G., Carter, N. E., and Harting, L. K., 2005. Analysis of spruce budworm outbreak cycles in New Brunswick, Canada, since 1952. *Ecology*, **86**, 1212–1224.

Rozzi, A. G., 1984. Modelling and control of anaerobic digestion processes. *Transactions of the Institute of Measurement and Control*, **6**, 153–159.

Rozzi, A. G., Merlini, S., and Passino, R., 1985. Development of a four population model of the anaerobic degradation of carbohydrates. *Environmental Technology Letters*, **6**, 610–619.

Ruan, S., 2009. On nonlinear dynamics of predator-prey models with discrete delay. *Mathematical Modelling of Natural Phenomena*, **4**, 140–188.

Ruano, M. V., Ribes, J., De Pauw, D. J. W., and Sin, G., 2007. Parameter subset selection for the dynamic calibration of activated sludge models (ASMs): Experience versus systems analysis. *Water Science & Technology*, **56**, 107–115.

Rumelhart, D. E. and McClelland, J. L., 1986. *Parallel Distributed Processing*, MIT Press, Cambridge, MA, 567pp.

Russell, E. L., Chiang, L. H., and Braatz, R. D., 2000. Fault detection in industrial processes using canonical variate analysis and dynamic principal component analysis. *Chemometrics and Intelligent Laboratory Systems*, **51**, 81–93.

Sawyer, C. N., McCarty, P. L., and Parkin, G. F., 2003. *Chemistry for Environmental Engineering and Science*, McGraw-Hill, New York, 752pp.

Schneider, M., Jorde, K., Zoellner, F., and Kerle, F., 2001. Development of a user-friendly software for ecological investigations on river systems, integration of a fuzzy rule-based approach. In: *15th International Symposium Informatics for Environmental Protection*, Zürich, Switzerland, pp. 354–360.

Schraa, O., Tole, B., and Copp, J. B., 2006. Fault detection for control of wastewater treatment plants. *Water Science & Technology*, **53**, 375.

Seber, G. A. F. and Wild, C. J., 1989. *Nonlinear Regression*, Wiley-Interscience, New York, 792pp.

Seborg, D. E., Edgar, T. F., and Mellichamp, D. A., 1989. *Process Dynamics and Control*, John Wiley & Sons, New York, 717pp.

Seviour, R. J., Mino, T., and Onuki, M., 2003. The microbiology of biological phosphorus removal in activated sludge systems. *FEMS Microbiology Reviews*, **27**, 99–127.

Shang, Y., Johnson, B. R., and Sieger, R., 2005. Application of the IWA Anaerobic Digestion Model (ADM1) for simulating full-scale anaerobic sewage sludge digestion. *Water Science & Technology*, **52**, 487–492.

Shanno, D. F., 1970. Conditioning of quasi-Newton methods for function minimization. *Mathematics of Computing*, **24**, 647–656.

Shannon, C. E., 1948. A mathematical theory of communication. *Bell System Technical Journal*, **27**, 379–423.

Shimkin, N. and Feuer, A., 1987. Persistency of excitation in continuous-time systems. *Systems and Control Letters*, **9**, 225–233.

Sin, G., van Hulle, S. W. H., De Pauw, D. J. W., van Griensven, A., and Vanrolleghem, P. A., 2005. A critical comparison of systematic calibration protocols for activated sludge models: A SWOT analysis. *Water Research*, **39**, 2459–2474.

Snell, T. W. and Serra, M., 1998. Dynamics of natural rotifer populations. *Hydrobologia*, **368**, 29–35.

Sokal, R. R., 1961. Distance as a measure of taxonomic similarity. *Systematic Zoology*, **10**, 70–79.

Sokal, R. R. and Michener, C. D., 1958. A statistical method for evaluating systematic relationships. *Kansas University Science Bulletin*, **38**, 1409–1438.

Sokal, R. R. and Rohlf, F. J., 1962. The comparison of dendrograms by objective methods. *Taxon*, **11**, 33–40.

Solé, R. V. and Bascompte, J., 2006. *Self-Organization in Complex Ecosystems*, Princeton University Press, Princeton, NJ, 371pp.

Spagni, A., Buday, J., Ratini, P., and Bortone, G., 2001. Experimental considerations on monitoring ORP, pH, conductivity and dissolved oxygen in nitrogen and phosphorus biological removal processes. *Water Science & Technology*, **43**, 197–204.

Spagni, A. and Marsili-Libelli, S., 2009. Nitrogen removal via nitrite in a sequencing batch reactor treating sanitary landfill leachate. *Bioresource Technology*, **100**, 609–614.

Spagni, A. and Marsili-Libelli, S., 2010. Artificial intelligence control of a sequencing batch reactor for nitrogen removal via nitrite from landfill leachate. *Journal of Environmental Science and Health. Part A*, **45**, 1085–1091.

Spagni, A., Marsili-Libelli, S., and Lavagnolo, M. C., 2008. Optimisation of sanitary landfill leachate treatment in a sequencing batch reactor. *Water Science & Technology*, **58**, 337–343.

Spindler, A., 2014. Structural redundancy of data from wastewater treatment systems: Determination of individual balance equations. *Water Research*, **57**, 193–201.

Starfeld, A. M. and Bleloch, A. L., 1986. *Building Models for Conservation and Wildlife Management*, Macmillan, New York, 304pp.

Steele, J. H., 1962. Environmental control of photosynthesis in the sea. *Limnology and Oceanography*, **7**, 115–271.

Stephanopoulos, G., 1984. *Chemical Process Control: An Introduction to Theory and Practice*, Prentice Hall, New York, 696pp.

Sterner, R. W. and Grover, J. P., 1998. Algal growth in warm temperate reservoirs: Kinetic examination of nitrogen, temperature, light, and other nutrients. *Water Research*, **32**, 3539–3548.

Stevens, S. A. and Lakin, W. D., 2005. A mathematical model of the systemic circulatory system with logistically defined nervous system regulatory mechanisms. *Mathematical and Computer Modelling of Dynamical Systems*, **12**, 555–576.

Stojanovic, S. B., Debeljkovic, D. L., and Mladenovic, I., 2007. A Lyapunov-Krasovskii methodology for asymptotic stability of discrete time delay system. *Serbian Journal of Electrical Engineering*, **4**, 109–117.

Strang, G., 1994. Wavelets. *American Scientist*, **82**, 250–255.

Strogatz, S. H., 1994. *Nonlinear Dynamics and Chaos*, Perseus Books Publications, Cambridge, MA, 498pp.

Sudhakar, K., Srivastava, T., Satpathy, G., and Premalatha, M., 2013. Modelling and estimation of photosynthetically active incident radiation based on global irradiance in Indian latitudes. *International Journal of Energy and Environmental Engineering*, **4**, 1–8.

Sugeno, M., 1985. *Industrial Applications of Fuzzy Control*, North Holland, Amsterdam, the Netherlands, 269pp.

Svobodova, Z., Lloyd, R., Machova, J., and Vykusova, B., 1993. *Water Quality and Fish Health*, EIFAC Technical Paper. No. 54, FAO, Rome, Italy, 59pp.

Takács, I., Patry, G. G., and Nolasco, D., 1991. A dynamic model of the clarification thickening process. *Water Research*, **25**, 1263–1271.

Takagi, T. and Sugeno, M., 1985. Fuzzy identification of systems and its applications to modeling and control. *IEEE Transaction on Systems, Man, and Cibernetics*, **15**, 116–132.

Tay, J. H. and Zhang, X., 1999. Neural fuzzy modeling of anaerobic biological wastewater treatment systems. *ASCE Journal of Environmental Engineering*, **125**, 1149–1159.

Tay, J. H. and Zhang, X., 2000. A fast predicting neural fuzzy model for high-rate anaerobic wastewater treatment system. *Water Research*, **34**, 2849–2860.

Tchobanoglous, G. and Schroeder, E. D., 1985. *Water Quality*, Addison-Wesley, Reading, MA, 768pp.

Theodoridis, S., Pikrakis, A., Koutroumbas, K., and Cavouras, D., 2010. *Introduction to Pattern Recognition: A MATLAB Approach*, Academic Press, Burlington, VT, 219pp.

Thomann, R. V. and Mueller, J. A., 1987. *Principles of Surface Water Quality Modeling and Control*, Harper & Row Publications, Cambridge, MA, 644pp.

Tryon, R. C., 1939. *Cluster Analysis*, Edwards Brothers, Ann Arbor, MI.

Tsoukalas, L. H., Uhrig, R. E., and Zadeh, L. A., 1997a. *Fuzzy and Neural Approaches in Engineering*, Wiley-Interscience, New York, 600pp.

Tsoukalas, L. H., Uhrig, R. E., and Zadeh, L. A., 1997b. *Fuzzy and Neural Approaches in Engineering, MATLAB Supplement*, Wiley-Interscience, New York.

Turchin, P., 2003. *Complex Population Dynamics: A Theoretical/Empirical Synthesis*, Princeton University Press, Princeton, NJ, 450pp.

Tych, W. and Young, P. C., 2012. A MATLAB software framework for dynamic model emulation. *Environmental Modelling & Software*, **34**, 19–29.

van Broekhoven, E., Adriaenssens, V., De Baets, B., and Verdonschot, P. F. M., 2006. Fuzzy rule-based macroinvertebrate habitat suitability models for running waters. *Ecological Modelling*, **198**, 71–84.

van Dongen, U., Jetten, M. S. M., and Van Loosdrecht, M. C. M., 2001. The SHARON-Anammox process for treatment of ammonium rich wastewater. *Water Science & Technology*, **44**, 153–160.

van Loosdrecht, M. C. and Henze, M., 1999. Maintenance, endogenous respiration, lysis, decay and predation. *Water Science & Technology*, **39**, 107–117.

van Tongeren, O. F. R., 1995. Data analysis or simulation model: A critical evaluation of some methods. *Ecological Modelling*, **78**, 51–60.

Vandermeer, J. H. and Goldberg, D. E., 2013. *Population Ecology: First Principles*, Princeton University Press, Princeton, NJ, 263pp.

Vanrolleghem, P. A. and Keesman, K. J., 1996. Identification of biodegradation models under model and data uncertainty. *Water Science & Technology*, **33**, 91–105.

Versyck, K. J., Claes, J. E., and Van Impe, J. F., 1998. Optimal experimental design for practical identification of unstructured growth models. *Mathematics and Computers in Simulation*, **46**, 621–629.

Vidakovic, B. and Mueller, P., 1991. Wavelets for kids: A tutorial introduction. *Victoria*, 1–35.

Villegas, T., Fuente, M., and Rodríguez, M., 2010. Principal component analysis for fault detection and diagnosis: Experience with a pilot plant. In: *Proceedings of the 9th WSEAS, Advances in Computational Intelligence, Man-Machine Systems and Cybernetics*, Merida, Venezuela, pp. 147–152.

Von Bertalanffy, 1969. *General System Theory: Foundations, Development, Applications*, George Brazziler, New York, 269pp.

Vrecko, D., Gernaey, K. V., Rosen, C., and Jeppsson, U., 2006. Benchmark simulation model no. 2 in MATLAB-Simulink: Towards plant-wide WWTP control strategy evaluation. *Water Science & Technology*, **54**, 65–72.

Walker, J. S., 2008. *A Primer on Wavelets and Their Scientific Applications*, Chapman & Hall/CRC Press, New York, 320pp.

Walsby, A. E., 1997. Numerical integration of phytoplankton photosynthesis through time and depth in a water column. *New Phytologist*, **136**, 189–209.

Weijers, S. R. and Vanrolleghem, P. A., 1997. A procedure for selecting best identifiable parameters in calibrating activated sludge model no.1 to full scale plant data. *Water Science & Technology*, **36**, 69–79.

Werner, T. M. and Kadlec, R. H., 1996. Application of residence time distributions to stormwater treatment systems. *Ecological Engineering*, **7**, 213–234.

Werner, T. M. and Kadlec, R. H., 2000. Wetland residence time distribution modeling. *Ecological Engineering*, **15**, 77–90.

Wieland, R., Mirschel, W., Groth, K., Pechenick, A., and Fukuda, K., 2011. A new method for semi-automatic fuzzy training and its application in environmental modeling. *Environmental Modelling & Software*, **26**, 1568–1573.

Wilderer, P. A., Irvine, R. L., and Goronszy, M. C., 2001. *Sequencing Batch Reactor Technology*, IWA Scientific and Technical Report no. 10, London, 76pp.

Willems, J. C., Rapisarda, P., Markovsky, I., and Moor, B. L. M. De, 2005. A note on persistency of excitation. *Systems and Control Letters*, **54**, 325–329.

Wilson, H. and Recknagel, F., 2001. Towards a generic artificial neural network model for dynamic predictions of algal abundance in freshwater lakes. *Ecological Modelling*, **146**, 69–84.

Witten, I. H. and Frank, E., 2005. *Data Mining*, Elsevier, Amsterdam, the Netherlands, 524pp.

Wu, W., Dandy, G. C., and Maier, H. R., 2014. Protocol for developing ANN models and its application to the assessment of the quality of the ANN model development process in drinking water quality modelling. *Environmental Modelling and Software*, **54**, 108–127.

Wynn, T. M. and Liehr, S. K., 2001. Development of a constructed subsurface-flow wetland simulation model. *Ecological Engineering*, **16**, 519–536.

Yager, R. R. and Filev, D. P., 1994. *Essentials of Fuzzy Modeling and Control*, John Wiley & Sons, New York, 388pp.

Yager, R. R., Ovchinnikov, S., Tong, R., and Nguyen, H. T., 1987. *Fuzzy Sets and Applications: Selected Papers by L.A. Zadeh*, John Wiley & Sons, New York, 684pp.

Yager, R. R. and Zadeh, L. A., 1994. *Fuzzy Sets, Neural Networks, and Soft Computing*, Van Nostrand Reinhold, New York, 440pp.

Yajnik, A., 2012. *Wavelet Analysis and Its Applications: An Introduction*, Alpha Science Intern, London, 150pp.

Yoo, C. K., Choi, S. W., and Lee, I., 2002. Disturbance detection and isolation in the activated sludge process. *Water Science & Technology*, **45**, 217–226.

Yoo, C. K., Vanrolleghem, P. A., and Lee, I., 2006. Fault detection, monitoring and diagnosis of a sequencing batch reactor for integrated wastewater treatment management system. *Environmental Engineering Research*, **11**, 1–14.

Young, O. R., Freeman, M. M. R., Osherenko, G., Andersen, R. R., Caulfield, R. A., Friedheim, R. A., Langdon, S. J., Ris, M., and Usher, P. J., 1994. Subsistence, sustainability, and sea mammals: Reconstructing the international whaling regime. *Ocean & Coastal Management*, **23**, 117–127.

Young, P. C., 1970. An instrumental variable method for real-time identification of a noisy process. *Automatica*, **6**, 271–287.

Young, P. C., 2006. The CAPTAIN toolbox for MATLAB. In: *14th IFAC Symposium on System Identification*, Newcastle, Australia. pp. 410–415.

Young, W. J., Lam, D., Ressel, V., and Wong, I., 2000. Development of an environmental flows decision support system. *Environmental Modelling and Software*, **15**, 257–265.

Zadeh, L. A., 1965. Fuzzy sets. *Information and Control*, **8**, 338–353.

Zeng, X., Rasmussen, T. C., Beck, M. B., Parker, A. K., and Lin, Z., 2006. A biogeochemical model for metabolism and nutrient cycling in a Southeastern Piedmont impoundment. *Environmental Modelling and Software*, **21**, 1073–1095.

Zhang, J. and Jørgensen, S. E., 2005. Modelling of point and non-point nutrient loadings from a watershed. *Environmental Modelling & Software*, **20**, 561–574.

Index

Note: Locators followed by '*f*' and '*t*' denote figure and table in the text.